Concepts and Strategies in Plant Sciences

Series Editor
Chittaranjan Kole, Raja Ramanna Fellow, Government of India, ICAR-National Institute for Plant Biotechnology, Pusa, Delhi, India

This book series highlights the spectacular advances in the concepts, techniques and tools in various areas of plant science. Individual volumes may cover topics like genome editing, phenotyping, molecular pharming, bioremediation, miRNA, fast-track breeding, crop evolution, IPR and farmers' rights, to name just a few. The books will demonstrate how advanced strategies in plant science can be utilized to develop and improve agriculture, ecology and the environment. The series will be of interest to students, scientists and professionals working in the fields of plant genetics, genomics, breeding, biotechnology, and in the related disciplines of plant production, improvement and protection.

Interested in editing a volume? Please contact Prof. Chittaranjan Kole, Series Editor, at ckoleorg@gmail.com

Chittaranjan Kole • Anurag Chaurasia
Kathleen L. Hefferon • Jogeswar Panigrahi
Editors

Tools & Techniques of Plant Molecular Farming

Springer

Editors
Chittaranjan Kole
Prof. Chittaranjan Kole Foundation for Science and Society
Kolkata, India

Anurag Chaurasia
ICAR-Indian Institute of Vegetable Research
Varanasi, India

Kathleen L. Hefferon
Department of Microbiology
Cornell University
Ithaca, NY, USA

Jogeswar Panigrahi
Berhampur University
Berhampur, India

ISSN 2662-3188 ISSN 2662-3196 (electronic)
Concepts and Strategies in Plant Sciences
ISBN 978-981-99-4858-1 ISBN 978-981-99-4859-8 (eBook)
https://doi.org/10.1007/978-981-99-4859-8

This Springer imprint is published by the registered company Springer Nature Singapore Pte Ltd.
The registered company address is: 152 Beach Road, #21-01/04 Gateway East, Singapore 189721, Singapore

Foreword

As president of the International Society for Plant Molecular Farming (ISPMF), I am very pleased to write a foreword of the two volumes of a book on plant molecular farming, entitled "Tools & Techniques of Plant Molecular Farming" and "Applications of Plant Molecular Farming in Agriculture and Food Industry." The production of valuable proteins, peptides, and small molecules through the use of plant biotechnology is an important issue of today but still has a way to go to increase its impact. In 1989, molecular farming had burst onto the cover of *Nature* with Andy Hiatt's "monoclonal antibody expression in plants" paper demonstrating a revolutionary new use of plants. This book shows impressively how the field has developed since then. Apart from the great progress in the scientific exploration of the field, several companies based on molecular farming have emerged and established themselves on the international market. Protalix Therapeutics in Israel with one marketed product and several in development or BioApp in Korea with an approved vaccine against the classical swine fever are only two examples, with others in the United States, Canada, Europe, South Africa, and Thailand. Most of them focus on vaccines and research agents. Many of the scientists involved in the field are members of the ISPMF. Since its foundation in 2014, the membership of our Society has increased to close to 200 members from 21 countries. As an academic society, the ISPMF's main objective is to support excellence in research, scholarship, and practice in plant molecular farming. One of our objectives is to provide a meeting place for all the players involved in plant molecular farming online to exchange up-to-date news and ideas as well as face-to-face in the form of biennial scientific conferences. The other objective is to facilitate the exchange of students and scientists to support their scientific carrier. The ISPMF is pleased to see how the increasing importance of plant molecular farming becomes reflected in these two excellent books.

International Society for Plant Molecular Farming (ISPMF) Inge Broer
London, UK

Preface

Plants have played a substantial role in the field of medicine since the beginning of mankind. Many of our modern drugs can trace their origins from plants. It is no surprise, then, that even in contemporary medical treatments, plants continue to contribute in ways that are both concrete and substantial. Plant molecular farming could be defined as the use of plants or plant tissues as production platforms for the expression of biologics, such as vaccines, therapeutic proteins, and monoclonal antibodies. Plant molecular farming now encompasses a growing discipline of scientific research, with subdomains converging on nanomedicine, immunotherapy, process engineering, and innovations in plant breeding and growth technologies. Plants as bioreactors are easy to scale up, inexpensive to grow, and environmentally sustainable. The benefits of molecular farming could be enormous, as more robust plant production platforms are brought forward. Vaccines and other biologics produced in plants have been demonstrated to be efficacious and exhibit stability at room temperatures. This makes them an appealing alternative for the development of therapeutics for personalized medicine, as well as for the stockpiling of vaccines and other biologics for potential pandemics. The opportunities of plant molecular farming as a viable option for drug development in low- to middle-income countries are vast. This book series exhaustively discusses plant molecular farming in its many facets, from its unique history to its position within the current regulatory landscape, from the use of plant cell culture to the production within vertical farms, and from downstream process manufacturing to challenges associated with clinical trial development. This two-volume set will provide reference material for the coming years, for an exciting field that is only beginning to approach its full potential.

The first volume entitled "Tools & Techniques of Plant Molecular Farming" deliberates on various concepts and strategies currently being used for plant molecular farming along with the existing challenges, opportunities, and future perspectives. Recent tools and techniques of plant molecular farming including genome editing-assisted PMF, transplastomics, seed-based production of recombinant proteins, and plant-based antibody manufacturing, and use of turnip mosaic virus as a versatile tool for nanoparticle production, have been elaborated. The book has a special coverage on topical relevance of PMF, especially for the developing countries with an example on cost-effective, rapid, bulk production of COVID-19 vaccine. Finally, biosafety, risk, and regulatory issues of PMF, its social acceptability, and constraints coming in the way of bulk industrial production and commercialization of PMF products have been elaborated over 15 chapters contributed by 61 international experts from 14 countries.

The second volume on "Applications of Plant Molecular Farming in Agriculture and Food Industry" explores the different crop systems currently being used for PMF with special coverage on rice, tobacco, duckweed, microalgae, cannabis, and legumes. Plant viral vectors, microbes, plant cell suspension cultures, bioreactor engineering, plant expression system, plant virus nanoparticles, and genetically modified seeds have also been explored for various PMF purposes, like molecular farming of industrial enzymes, antimicrobial peptides for plant protection and stress tolerance, viruslike particles, recombinant pharmaceutical proteins, drugs, and, for humans, specifically antigen-specific immunotherapy for allergic and autoimmune diseases and to overcome the global impact of neglected tropical diseases and veterinary vaccines. The 24 chapters of this volume are contributed by 81 global experts from 13 countries.

We express our sincere thanks and gratitude to all the contributing authors of these two volumes for their exquisite contributions, magnificent deliberations, and excellent cooperation since inception to publication of these two books. We sincerely thank Prof. Inge Broer, President, International Society for Plant Molecular Farming (ISPMF), for contributing a Foreword to these volumes, in which she underlines the importance of both the volumes in covering the developments, which have taken place in the area of plant molecular farming research during the last four decades, and also predicts a very bright future for PMF industries particularly for the developing countries under the present pandemic situation. We are also thankful to the staff of Springer Nature, specifically Dr. Aakanksha Tyagi, Senior Editor, Books on Medicine and Life Sciences; Ms. Yogitha Subramani, Project Coordinator (Books); Mr. Ashok Kumar; and Mr. Chihiro Haraguchi for their timely cooperation and useful professional assistance.

Kolkata, India — Chittaranjan Kole
Varanasi, India — Anurag Chaurasia
Ithaca, NY, USA — Kathleen L. Hefferon
Berhampur, India — Jogeswar Panigrahi

Contents

Editors and Contributors

About the Editors

Chittaranjan Kole is an internationally reputed scientist with an illustrious professional career spanning over 38 years and original contributions in the fields of plant genomics, biotechnology, and molecular breeding.

Anurag Chaurasia is a biotechnologist working with the Indian Council of Agricultural Research (ICAR), Government of India.

Kathleen L. Hefferon received her PhD from the University of Toronto and is currently a Faculty in the Department of Microbiology, Cornell University.

Jogeswar Panigrahi completed his PhD degree from Utkal University and is presently working as Professor and Head of PG Department of Biotechnology, Berhampur University.

Contributors

Neelam Batool Department of Biochemistry, Faculty of Biological Sciences, Quaid-i-Azam University, Islamabad, Pakistan

Barry Bratcher KBio, Inc., Owensboro, KY, USA

Max Brennan ZabBio, Inc., San Diego, CA, USA

Nicolau Brito da Cunha Universidade de Brasília, Brasília, Brazil

Universidade Católica de Brasília, Brasília, Brazil

Sara Mesquite De Albuquerque Applied Diagnostics Research and Evaluations Unit, Institute for Infection and Immunity, St George's University of London, London, UK

Amanda Lopes Ferreira Embrapa Arroz e Feijão, Santo Antônio de Goiás, Brazil

Sebastian Fuller Health Systems Collaborative, Nuffield Department of Medicine, University of Oxford, Oxford, UK

Josh Fuqua KBio, Inc., Owensboro, KY, USA

Wen Cong Gan Division of Applied Biomedical Sciences and Biotechnology, International Medical University, Kuala Lumpur, Malaysia

Silvia Grimaldi School of Design and Service Futures Lab, University of the Arts London, London, UK

Krystal Hamorsky KBio, Inc., Owensboro, KY, USA

Kathleen L. Hefferon Department of Microbiology, Cornell University, Ithaca, NY, USA

Rosie Hornbuckle School of Design and Service Futures Lab, University of the Arts London, London, UK

Steve Hume KBio, Inc., Owensboro, KY, USA

Fatima Ijaz Department of Biochemistry, Faculty of Biological Sciences, Quaid-i-Azam University, Islamabad, Pakistan

Kaya İşleyen Department of Genetics and Bioengineering, Faculty of Engineering, Yeditepe University, Istanbul, Turkey

Sandra Jordaan Cape Biologix Technologies Pty Ltd., Unit 3, The Powder Mill, Ndabeni, Cape Town, South Africa

Kirthikah Kadiresen Division of Applied Biomedical Sciences and Biotechnology, International Medical University, Kuala Lumpur, Malaysia

Beyza Kocaoğlu Department of Genetics and Bioengineering, Faculty of Engineering, Yeditepe University, Istanbul, Turkey

Chalinee Kongsawat National Center for Genetic Engineering and Biotechnology, National Science and Technology Development Agency, Pathum Thani, Thailand

Lilya Kopertekh Julius Kühn Institute (JKI), Federal Research Centre for Cultivated Plants, Institute for Biosafety in Plant Biotechnology, Quedlinburg, Germany

Rakchanok Koto Faculty of Science, Srinakharinwirot University, Bangkok, Thailand

Asqwin Uthaya Kumar Division of Applied Biomedical Sciences and Biotechnology, International Medical University, Kuala Lumpur, Malaysia

Cristiano Lacorte Embrapa Recursos Genéticos e Biotecnologia, Brasília, Brazil

Mariba Lebeko Cape Biologix Technologies Pty Ltd., Unit 3, Cape Town, Western Cape, South Africa

Maribanyana Lebeko Cape Biologix Technologies Pty Ltd., Unit 3, The Powder Mill, Ndabeni, Cape Town, South Africa

Anna Pick Kiong Ling Division of Applied Biomedical Sciences and Biotechnology, International Medical University, Kuala Lumpur, Malaysia

Andreas Gunter Lössl Department of Applied Plant Science and Plant Biotechnology, University of Natural Resources and Life Sciences (BOKU), Vienna, Austria

Muhammad Suleman Malik Department of Biochemistry, Faculty of Biological Sciences, Quaid-i-Azam University, Islamabad, Pakistan

Henry I. Miller American Council on Science and Health, New York, NY, USA

Josh Morton KBio, Inc., Owensboro, KY, USA

Aline Melro Murad Universidade Federal de Santa Catarina, Florianópolis, Brazil

Benita Ortega-Berlanga Department of Plant Biotechnology, Institute of Plant Genetics of the Polish Academy of Sciences, Poznań, Poland

División de Biología Molecular, Instituto Potosino de Investigación Científica y Tecnológica, San Luis Potosí, Mexico

Bahar Soğutmaz Özdemir Department of Genetics and Bioengineering, Faculty of Engineering, Yeditepe University, Istanbul, Turkey

Michael H. Pauly ZabBio, Inc., San Diego, CA, USA

Francisco Pera Cape Bio Pharms Pty Ltd., Unit 3, The Powder Mill, Ndabeni, Cape Town, South Africa

Cape Biologix Technologies Pty Ltd., Unit 3, The Powder Mill, Ndabeni, Cape Town, South Africa

Patricia Valle Pinheiro Embrapa Arroz e Feijão, Santo Antônio de Goiás, Brazil

Tomasz Pniewski Department of Plant Biotechnology, Institute of Plant Genetics of the Polish Academy of Sciences, Poznań, Poland

Gregory P. Pogue KBio, Inc., Owensboro, KY, USA

IC2 Institute, The University of Texas at Austin, Austin, TX, USA

Fernando Ponz Centro de Biotecnología y Genómica de Plantas (CBGP), Instituto Nacional de Investigación y Tecnología Agraria y Alimentaria (INIA-CSIC), Universidad Politécnica de Madrid (UPM), Madrid, Spain

Alison Prendiville School of Design and Service Futures Lab, University of the Arts London, London, UK

Tarin Ramsaroop Cape Biologix Technologies Pty Ltd., Unit 3, The Powder Mill, Ndabeni, Cape Town, South Africa

Cape Bio Pharms Pty Ltd, Unit 3, Cape Town, Western Cape, South Africa

Sara Rincón Centro de Biotecnología y Genómica de Plantas (CBGP), Instituto Nacional de Investigación y Tecnología Agraria y Alimentaria (INIA-CSIC), Universidad Politécnica de Madrid (UPM), Madrid, Spain

Joseph Rininger Latham BioPharm Group, Inc., Elkridge, MD, USA

Joshua M. Royal KBio, Inc., Owensboro, KY, USA

Kiran Saba Department of Biochemistry, Shaheed Benazir Bhutto Women University, Peshawar, Pakistan

Department of Biochemistry, Faculty of Biological Sciences, Quaid-i-Azam University, Islamabad, Pakistan

Muhammad Sameeullah Center for Innovative Food Technologies Development, Application and Research, Bolu Abant Izzet Baysal University, Bolu, Turkey

Lauren Schoukroun-Barnes Latham BioPharm Group, Inc.,, Elkridge, MD, USA

Ke Sin Seow Division of Applied Biomedical Sciences and Biotechnology, International Medical University, Kuala Lumpur, Malaysia

Tamlyn Shaw Cape Bio Pharms Pty Ltd., Unit 3, The Powder Mill, Ndabeni, Cape Town, South Africa

Cape Biologix Technologies Pty Ltd., Unit 3, The Powder Mill, Ndabeni, Cape Town, South Africa

Sean Stevens KBio, Inc., Owensboro, KY, USA

Kelsi Swope KBio, Inc., Owensboro, KY, USA

Fumio Takaiwa Soul Signal Institute, Shiraoi, Hokkaido, Japan

Daniel A. Truchado Centro de Biotecnología y Genómica de Plantas (CBGP), Instituto Nacional de Investigación y Tecnología Agraria y Alimentaria (INIA-CSIC), Universidad Politécnica de Madrid (UPM), Madrid, Spain

Deniz Uras Department of Genetics and Bioengineering, Faculty of Engineering, Yeditepe University, Istanbul, Turkey

Mohammad Tahir Waheed Department of Biochemistry, Faculty of Biological Sciences, Quaid-i-Azam University, Islamabad, Pakistan

Kevin Whaley ZabBio, Inc., San Diego, CA, USA

Ralf Wilhelm Julius Kühn Institute (JKI), Federal Research Centre for Cultivated Plants, Institute for Biosafety in Plant Biotechnology, Quedlinburg, Germany

Ying Pei Wong Division of Applied Biomedical Sciences and Biotechnology, International Medical University, Kuala Lumpur, Malaysia

Larry Zeitlin ZabBio, Inc., San Diego, CA, USA

Lucía Zurita Centro de Biotecnología y Genómica de Plantas (CBGP), Instituto Nacional de Investigación y Tecnología Agraria y Alimentaria (INIA-CSIC), Universidad Politécnica de Madrid (UPM), Madrid, Spain

Chapter 1
Plant Molecular Farming: Concept and Strategies

Kirthikah Kadiresen, Ke Sin Seow, Asqwin Uthaya Kumar, Wen Cong Gan, Ying Pei Wong, and Anna Pick Kiong Ling

Abstract Plant molecular farming refers to a plant-based approach to produce recombinant proteins and secondary metabolites. The utilization of plants as a bioreactor for the production of valuable recombinant proteins has shown some unique advantages in comparison to other expression systems. With the high cost and increase in demand for therapeutics and industrial proteins, plant molecular farming has gained interests over the years for its promising low-cost productions at large scales. This chapter starts off with an introduction to plant molecular farming and discussions on its recent updates. It is then followed by describing the principle of the approaches, which explains the processes involved in plant molecular farming, different plant species that are utilized as expression hosts for the production of recombinant proteins and secondary metabolites, and types of recombinant proteins produced. This chapter also discusses various transformation strategies that are utilized as well as the advantages and challenges of plant molecular farming.

Keywords Edible vaccine · Genetic engineering · Nuclear transformation · Plastid transformation · Recombinant proteins

Abbreviations

2,4,5-T	2,4,5-Trichlorophenoxyacetic acid
2,4-D	2,4-Dichlorophenoxyacetic acid
2iP(IPA)	*N*^{6}-(2-isopentenyl)adenine
BAP	6-Benzylaminopurine
CAMV-35S	Cauliflower mosaic virus 35S

K. Kadiresen · K. S. Seow · A. U. Kumar · W. C. Gan · Y. P. Wong · A. P. K. Ling (✉)
Division of Applied Biomedical Sciences and Biotechnology, International Medical University, Kuala Lumpur, Malaysia
e-mail: kirthikah.kadiresen@student.imu.edu.my; seow.kesin@student.imu.edu.my; gan.wencong@student.imu.edu.my; YingPei_Wong@imu.edu.my; anna_ling@imu.edu.my

C. Kole et al. (eds.), *Tools & Techniques of Plant Molecular Farming*, Concepts and Strategies in Plant Sciences, https://doi.org/10.1007/978-981-99-4859-8_1

COVID-19	Coronavirus disease-19
ELP	Elastin-like polypeptide
FDA	Food and Drug Administration
GMP	Good manufacturing practice
GR2	Second-generation Golden Rice
HFBI	Hydrophobin I
hGH	Human growth hormone
HPV	Human papillomavirus
IAA	Indole-3-acetic acid
mAb	Monoclonal antibody
NAA	1-Naphthaleneacetic acid
NHL	Non-Hodgkin's lymphoma
PEG	Polyethylene glycol
PHAs	Polyhydroxyalkanoates
PMF	Plant molecular farming
PVX	Potato virus X
scFv	Single-chain variable fragment
T-DNA	Transfer DNA
TMV	Tobacco mosaic viruses
USDA	United States Department of Agriculture
VLP	Virus like particles

1.1 Introduction

The concept of "farming" is no longer restricted to the conventional perception of just the process of rearing animals or planting crops for food. Advancements in biotechnological fields such as molecular biology and genetic engineering have paved the way for molecular farming, which allows for a whole new spectrum of resources that can be harnessed for a variety of potential applications. Efforts in developing targeted gene editing tools in plants have laid the groundwork for plant molecular farming (PMF), to efficiently and precisely produce organisms that express specific recombinant proteins and secondary metabolites as desired (Mao et al. 2019). The concept of PMF differs slightly in other approaches of plant genetic engineering. Plants can be genetically engineered to confer specific desired qualities or characteristics to the plant, typically seen in agriculture, wherein to increase the yield or durability of crops, the resultant recombinant protein is induced into the plant to increase the production of a desired metabolite (Fischer and Schillberg 2016). Rather than adding value to the plant, PMF aims to specifically produce the desired products via a commercialized, large-scale approach (Fischer and Schillberg 2016). Achieving this feat would allow for inexpensive access to much-needed medicines and pharmaceuticals to virtually everyone globally.

The use of plants in molecular farming has now seen almost 40 years of development since the first plant-produced pharmaceutical protein (human growth hormone) from tobacco and sunflower calluses in 1986 (Barta et al. 1986) and the first plant-produced antibody in transgenic plants in 1989 (Hiatt et al. 1989). It was only 6 years later that the first oral immunization was observed on mice after consuming transgenic potatoes, making it the first edible plant-based vaccine to be produced (Haq et al. 1995). However, it took more than 20 years for a plant-based vaccine to gain the world's first regulatory approval when a Newcastle's disease vaccine for poultry produced in transgenic tobacco cell lines was approved by the United States Department of Agriculture (USDA) in 2006 (Sparrow et al. 2007). Unfortunately, the vaccine was never commercialized and brought to the market by the developing company (Thomas et al. 2011). It was also in the same year that the production of a monoclonal antibody (CB-Hep.1) used to produce a vaccine for hepatitis B was approved by Cuba's Medication Quality Control Agency (Sparrow et al. 2007), marking the first approval granted for the commercial use of plants in molecular farming. Further success in human clinical trials was observed as a plant-based vaccine produced in tobacco was effective in eliciting an immune response against follicular lymphoma (McCormick et al. 2008; Thomas et al. 2011).

Despite the considerably long developmental history of plant-based biopharmaceutics, the commercialization of such products in the market is still at its infancy. So far, there has only been one such product that was authorized to enter the market after its completion of phase I–III clinical trials (Schillberg and Finnern 2021; Zimran et al. 2018). Another notably historic use of a PMF product was during the not-so-recent Ebola outbreak, where ZMapp, a cocktail of monoclonal antibodies produced in transgenic tobacco (*N. benthamiana*), was authorized for emergency use in humans despite lacking the necessary human clinical trials at the time (Fanunza et al. 2018; Hiatt et al. 2015; Qiu et al. 2014). While ZMapp showed great promise during the initial use, further clinical trials conducted after the outbreak remained inconclusive (The PREVAIL II Writing Group for the Multi-National PREVAIL II Study Team 2016). Fortunately, the efforts in PMF development were not discouraged, as the call for action for mass-produced biopharmaceuticals arrived with the onset of the coronavirus disease-19 (COVID-19) pandemic. The urgency for preventative measures and treatments for COVID-19 were the focus of the efforts of researchers globally, including biopharmaceutical manufacturers such as Medicago, Kentucky Bioprocessing, iBio, and Newcotiana (Kumar et al. 2021). This highlights the potential reliance on a stable, efficient, and high-throughput method for the production of pharmaceuticals and recombinant proteins alike in the future, a niche that is perfect for the expansion of PMF (see also Chap. 12).

In comparison to other expression systems such as bacteria, yeast, mammalian cells, or insect cells, plants have been the popular platform used in molecular farming as it offers significant benefits over the other expression systems (Diamos et al. 2020; Schillberg and Finnern 2021; Shanmugaraj et al. 2020). The robustness, scalability cost, and relative safety of plant recombinant proteins (which shall be discussed in further sections) certainly boost the potential of plants in molecular farming and are the reasons why plants are the general preferred option in molecular farming. This chapter serves as an introduction to the various concepts within PMF as well as the challenges faced in its implementation. Further, the diverse approaches and optimization strategies undertaken to further develop PMF's viability shall also be discussed.

1.2 Concept

1.2.1 Plant as Expression System for Recombinant Protein Production

Utilization of transgenic plants for the production of various recombinant proteins and secondary metabolites offers an appealing alternative to other commonly used expression systems. In general, the process of PMF involves the development of transgenic plants as bioreactors through recombinant DNA technology. This entire process begins with the selection of a gene of interest. This candidate gene encodes for proteins that are able to express therapeutic properties such as vaccines and antibodies as well as industrial proteins such as enzymes or even diagnostic reagents. Following the selection of the gene of interest, the gene is inserted into a vector. The key to successful PMF is increased expression of recombinant proteins and secondary metabolites. Hence, expression cassettes play an essential role in ensuring an improved level of recombinant protein production (Makhzoum et al. 2014). Generally, these plant expression vectors consist of several components including the origin of replication, multiple cloning sites, and plant selectable markers (Low et al. 2018). The origin of replication site is an A + T-rich region where the replication process is initiated by unwinding the DNA (Low et al. 2018; Rajewska et al. 2012). The multiple cloning site has multiple restriction sites that can be cut by specific restriction enzymes, which permits the insertion of gene of interest allowing easy cloning of the gene (Crook et al. 2011). The plant selectable markers on the other hand play an essential role in providing proof of successful insertion of gene into vector. In terms of plant transformations, commonly employed vectors include the *Agrobacterium tumefaciens*-based Ti plasmid vector and plant RNA virus vectors to produce transient transformations allowing the transgene DNA to be expressed in the host, but it is not integrated into the host's germline DNA (Clark and Maselko 2020).

The vector containing the gene of interest is then introduced into plant cells. Prior to the introduction, a suitable plant species and its tissue need to be selected. This step is crucial as the physiological, morphological, biochemical, and genetic properties of each plant species may influence its success in terms of biomass and recombinant protein yield, capability in posttranslational modifications, and determination of polypeptide structural stabilities (Leite et al. 2019). Utilizing plant species with watery tissues such as tomato would also lead to a lower cost of downstream processing due to an easier process of extracting recombinant proteins compared to plants with dry tissues (Obembe et al. 2011). Once a suitable plant species has been selected, the transgene will be integrated into the plant cells by either stable or transient transformation process. A stable transformation refers to the integration of the gene of interest into the host cell genome, which results in the production of transgenic lines (Stepanenko and Heng 2017). This stable transformation can be done by nuclear transformation or plastid transformation. In transient transformation on the other hand, the gene of interest is present in an extrachromosomal manner, and it does not affect the host genome stability (Honda et al. 2019; Ji et al. 2014). This transient transformation can be done by agroinfiltration, viral infection, or magnification system, which will be discussed further in the following sections. Successful transformants are identified and selected to be regenerated into transgenic plants.

The regeneration of transgenic plants is carried out by culturing successful transformants in plant growth regulators. Two of the major plant growth regulators utilized in plant tissue culture are cytokinins and auxins (Abiri et al. 2016). Auxins play an essential role in the formation of an unorganized and undifferentiated mass of cells known as callus (Rahayu et al. 2016). Commonly employed auxins include 2,4-dichlorophenoxyacetic acid (2,4-D), 1-naphthaleneacetic acid (NAA), indole-3-acetic acid (IAA), picloram, dicamba, and 2,4,5-trichlorophenoxyacetic acid (2,4,5-T) (Abiri et al. 2016). Cytokinins on the other hand are essential to promote cell division and cell expansion in the development of leaf cells (Wu et al. 2021). Commonly used cytokinins include zeatin, kinetin, 6-benzylaminopurine (BAP), thidiazuron, and N^6-(2-isopentenyl)adenine (2iP(IPA)) (Bhatia 2015). An increased auxin-to-cytokinin ratio promotes the regeneration of roots, while an increased cytokinin-to-auxin ratio promotes the regeneration of shoots (Ikeuchi et al. 2013). This method of inserting a gene of interest into plant cells or tissue from explants and regenerating transgenic plants is utilized in stable transformation strategies. Alternatively, in transient transformation strategies, the gene of interest in vectors such as *Agrobacterium tumefaciens* and plant viruses are directly inserted into the intracellular space of the plant leaves by vacuum infiltration, syringe infiltration, or viral injection (Floss and Conrad 2013; Shanmugaraj et al. 2021). As a result, the plants will produce the desired recombinant proteins. Subsequently, the harvest will undergo downstream processing to extract and purify the desired recombinant proteins. The processes involved in plant molecular farming are summarized in Fig. 1.1.

1.2.2 Plant as a Bioreactor for Production of Recombinant Proteins and Secondary Metabolites

For a long time, breakthroughs in plant biotechnology were directed at the significant enhancement and transformation of specific plant components such as carbohydrates, proteins, lipids, and vitamins. This enhancement is referred to as "transgenic plants" where the plants can produce a broad range of foreign genes and enable massive protein and metabolite synthesis in pharmaceutical and agricultural environments (Franken et al. 1997; Sharma and Sharma 2009). The need for transgenic plants as bioreactors, on the other hand, is a comparatively modern approach that is rapidly developing. It comprises genetic altering of the host plant by introducing and expressing the gene of interest. Industrial enzymes, pharmaceutical proteins, secondary metabolites, bioactive peptides, vaccine antigens, and antibodies are some of the often-produced proteins and metabolites in plant bioreactors (Lau and Sun 2009; Sharma and Sharma 2009). For instance, during this SARS-CoV-2

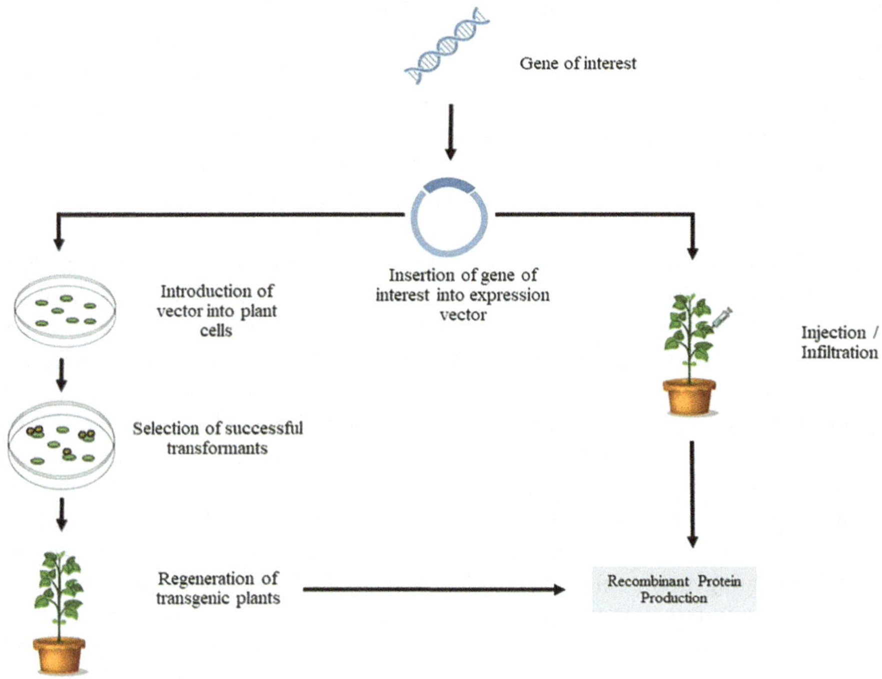

Fig. 1.1 Plant molecular farming process

pandemic era, the plant-based vaccine has obtained more attention for its cost-effective and easy manufacturing process. Researchers were racing to produce an efficient vaccine against the virus using plants as bioreactors (Kumar et al. 2021). One such company, Medicago, worked on using a close relative of the tobacco plant. They utilized the viruslike particle (VLP) technology and grew within the plant species to produce the vaccine candidate against the pathogen. Medicago also has successfully developed a VLP-based influenza vaccine using the Proficia plant as a bioreactor (Kumar et al. 2021). This indicates the success of using plants as bioreactors to produce vaccine proteins or candidates. Pereira et al. have successfully demonstrated that plant bioreactors are exhibiting promising results for enzyme production. They worked with the *lipA* gene originating from *Acremonium alcalophilum* that encodes for a lipase, which has multiple industrial uses such as in the field of biofuels and food processing industries (Pereira et al. 2013). The *lipA* gene has successfully been expressed in the tobacco plant, and the plant-produced recombinant LipA enzyme is highly active. This indicates that the plant bioreactor also exhibits promising outcomes in producing industrial enzymes, and the tobacco plant is a suitable bioreactor for fungal enzyme production (Pereira et al. 2013; see also chapter "Plant Molecular Farming for the Bulk Production of Industrial Enzymes").

However, the expression hosts of the recombinant proteins and secondary metabolites are not limited to tobacco plant. Researchers are discovering and studying multiple potential plant expression hosts that can be utilized as a bioreactor. They have identified and established some other plant bioreactors as the expression host, which will be discussed in the following sections (see also chapter "Molecular Farming of Pharmaceutical Proteins in Different Crop Systems: A Way Forward").

1.2.2.1 Tobacco

Tobacco is a widely used plant as a bioreactor in the field of plant biotechnology and is known as a "white mouse." It is the most preferred plant system for genetic modification and has been the major system for recombinant protein and secondary metabolites for the past 30 years. Tobacco is identified as one of the best plant bioreactors for its large biomass production and massive soluble protein production compared to any other plant bioreactors (Tremblay et al. 2010). The advantages of using tobacco as bioreactors include high yield; ability to produce a wide range of therapeutic, industrial, and agricultural proteins; and a short time for large-scale production (Gutiérrez et al. 2013; Tremblay et al. 2010). One considerable downside of using tobacco as a bioreactor is the shelf life of the plant. The transgenic tobacco plant should have shorter half time after reaching maturity and needs urgent processing of the protein (Sharma and Sharma 2009). Despite the advantages of producing a high yield, there are limitations, whereby certain recombinant protein yields will be significantly low. To overcome this, the transgenic tobacco plant could be designed together with other proteins or fusion tags to increase the yield. To demonstrate this, a study conducted by Gutierrez et al. has fused elastin-like polypeptide (ELP), hydrophobin I (HFBI), and Zera protein to increase the desired recombinant protein level. In their study, they tested the protein body formation in *N. tabacum* plant (Gutiérrez et al. 2013). These tags have shown promising results where the protein body level is significantly high and suggest that these two tags could be a potential fusion partner in tobacco plants to increase the yield of the recombinant protein (Gutiérrez et al. 2013). As mentioned earlier, tobacco transgenic plants are used widely in multiple fields; for instance, they can also be used to produce recombinant silk protein. Spider silk is one of the powerful natural fibers and has multiple uses in the field of medicine including facilitating healing and connecting the wounded or cut skin (Kirsh 2017). However, the availability of the protein is limited and requires an alternative way to produce it in large amounts. Scheller et al. have demonstrated that the silk protein could be produced using the tobacco plant as a bioreactor. They have transformed many recombinant proteins from *Nephila clavipes*, which are responsible for silk production. They have shown promising data, where the silk protein observed in the plant tissues was approximately 100 kDa, and when produced in the plant, the protein showed strong heat stability (Scheller et al. 2001). These data show that tobacco is a good candidate for silk protein production and can be deduced that a foreign gene can be expressed in the plant easily without many complications (see also chapter "Tobacco Plants as a Versatile Host for the Expression of Glycoproteins").

1.2.2.2 Rice

Rice is another available plant bioreactor option to deliver desired protein or metabolites to humans in the form of edible transgenic rice. It is the first-ever food crop to ever completely sequence and the genome available for the public view (McCouch and Herdt 2004). One of the well-known achievements of plant biotechnology and usage of rice bioreactors is the development of "Golden Rice." Vitamin A deficiency is a major nutritional condition that can be observed in children around the globe, especially in underdeveloped countries. To overcome this, a food-based vitamin A supplement was a sustainable option for the deficiency (Tang et al. 2009). Therefore, researchers worked on developing the Golden Rice, which is rich in beta-carotene, which in turn will be converted into vitamin A by the human body upon consumption of the rice. In 2004, the researchers Ingo Potrykus and Peter Beyer who developed Golden Rice performed their first field trial, which has shown promising results with the beta-carotene level being higher compared to the other grains. The first-generation Golden Rice had approximately 6 μg beta-carotene, which was lower than the recommended intake, but this was overcome in the same year with the second-generation Golden Rice (GR2), with approximately 35 μg beta-carotene (Tang et al. 2009; The Golden Rice Story n.d.). Tang et al. have confirmed that beta-carotene obtained from the Golden Rice is successfully converted into vitamin A upon consumption. They have tested the blood samples of healthy volunteers after consumption of Golden Rice serving and observed the retinol (vitamin A) level in the blood. The results showed a promising and significant result with 0.24–0.94 mg retinol after consumption indicating that the conversion of the beta-carotene was successful (Tang et al. 2009). Some of the benefits of using rice as a bioreactor include the following: (a) rice seed offers higher biosafety than other plant species, (b) rice endosperm has higher potential to efficiently express the recombinant protein of interest, and (c) it has the ability to express complex and functional proteins (Ou et al. 2014; Tang et al. 2009). Some of the limitation involved in using transgenic rice as a bioreactor is the public concern, the time required, and the attention provided to grow the crops. The crops require proper care and time to be fully developed and are sensitive to environmental factors such as pests and predator issues if grown in the field (McCouch and Herdt 2004). Besides the development of Golden Rice, rice could be used for other purposes as well. For instance, Yang et al. have used transgenic rice as a bioreactor to produce lipase B originated from *Candida antarctica*. They have successfully demonstrated the expression of lipase B in the rice seeds. The rice was transformed with the gene of interest, and a significant amount of lipase B has been expressed in the seeds. The transformed rice recorded the highest lipolytic activity with 85 units/g of dry seeds (Yang et al. 2014). The functions and effectiveness of the enzyme are identical to those used in industrial applications, which suggests that rice seed could be also used to produce industrial enzymes besides using as a food-based supplement (see also chapter "The Use of Rice Seed as Bioreactor").

1.2.2.3 Maize/Corn

Maize or also known as corn is a highly produced crop across the globe and is majorly used for multiple purposes such as for human food, animal feed, and raw material for industrial use. Over the years, maize has been used in multiple transformation studies to encounter the high demand for maize grains for various purposes (Yadava et al. 2017). Some of the benefits of using maize as a bioreactor compared to other plants are the size of the seed, which allows the desired recombinant protein to accumulate in high concentration, which is beneficial in downstream processing. There will be low interference of other substances in the expression of the recombinant protein. The limitation of maize plants would be low yield despite the high concentration compared to other plant systems such as tobacco (Ramessar et al. 2008). Various studies have been performed on maize to achieve various goals; for instance, Barros et al. have succeeded in producing human monoclonal antibody 2G12 which is proposed to use against HIV infection. In this study, transgenic maize containing the 2G12 was bred together with South African maize lines, which resulted in a hybrid. The outcome was promising, where the hybrid transgenic maize expressed a higher level of antibody compared to the non-hybrid transgenic line (Barros and Nelson 2010). This shows another alternate option of increasing the recombinant protein expression in transgenic maize. This discovery could lead to large-scale production of neutralizing antibodies against HIV and could revolutionize the pharmaceutics industry (Barros and Nelson 2010). This finding also indicates the effectiveness and yield of the recombinant protein and secondary metabolite production. Table 1.1 lists multiple plant expression hosts that are used as plant bioreactors together with their advantages and disadvantages.

Table 1.1 Plant expression host with their advantages and disadvantages

Plant expression host	Advantages	Disadvantages	References
Tobacco	High-yield biomass, cost effective, time efficient	Half shelf life, extraction time point, low protein stability	Gutiérrez et al. (2013), Ramessar et al. (2008), Sharma and Sharma (2009), Tremblay et al. (2010)
Rice	Higher biosafety, efficient expression, expression of complex protein	Time consuming, public opinion	McCouch and Herdt (2004), Ou et al. (2014), Tang et al. (2009)
Maize	High concentration, low interference	Lower yield	Ramessar et al. (2008)

1.2.3 *Types of Recombinant Proteins Produced by PMF*

With the diverse options of plant systems available and being studied to produce recombinant proteins, there are a lot of successful plant systems that exist for a wide range of products, with a large influx of commercially viable products (Xu et al. 2018). The aim of developing a plant system as bioreactors is to produce a broad range of beneficial recombinant proteins. The type of proteins that are studied or successfully produced using a plant system includes edible vaccines, industrial proteins, diagnostic reagents, biopolymers, and therapeutic proteins, which will be discussed in the following sections.

1.2.3.1 Edible Vaccines

Vaccines are essential biological proteins that have been developed to provide protection and enhance the immune system's function against pathogens and disease. The vaccines could provide long-term or short-term protection based on the system used to develop them (Kurup and Thomas 2020). Conventional vaccines involve various tedious downstream processes, such as propagation, isolation, purification, and isolation of vaccine proteins. Another limitation includes that the conventional vaccine requires adjuvant as it needs to be administered through parenteral route and requires specific storage conditions such as the vaccine needs to be placed in cold temperature (Kurup and Thomas 2020). To overcome these issues, the concept of an edible vaccine has been established. The idea of edible vaccine refers to the usage of genetically modified plant or transgenic plants to carry and produce vaccine subunits against specific pathogens or diseases. The brief concept is that the vaccine proteins are introduced to the body via the intake of transgenic plants orally that are modified to produce vaccine proteins (Saxena and Rawat 2014). The advantages of edible vaccines include avoiding the need of using the invasive approach to administer vaccine where the intake is performed orally and avoiding the use of inactivated virus or toxins in the production of the vaccine protein (Saxena and Rawat 2014). Up to now, multiple plants have been studied and used as potential candidates to produce vaccine proteins, which include tobacco, potato, rice, maize, tomato, lettuce, spinach, and banana (Gunasekaran and Gothandam 2020; Saxena and Rawat 2014). For instance, an edible vaccine against hepatitis B has been produced using potatoes. Arntzen and his team from Arizona State University in Tempe have developed an edible vaccine that produces hepatitis B virus antigen, which upon intake will initiate the human immune system response that behaves as the booster shot against the virus. The development could replace the need for the conventional vaccine as the booster shot for hepatitis B (Khamsi 2005). Another discovery, wherein scientists from the University of Tokyo and Chiba University have entered the first human trial of edible cholera vaccine (MucoRice-CTB) using powdered rice, has shown promising results. The vaccine is developed by using Japanese short-grain rice plants, where it is genetically modified with an insertion of a gene that produces a region of cholera toxin B, which is nontoxic to humans. The portion was selected to ensure that the protein can be identified by the human immune system and initiate a response. The vaccine is proved to be stable at room temperature, and the participants with high doses have exhibited greater immune response against the virus (Edible Cholera Vaccine Made of Powdered Rice | EurekAlert! n.d.; see also chapter "Delivery of Drugs and Vaccines Through Plant Molecular Farming").

1.2.3.2 Industrial Proteins

Industrial proteins are proteins that are often used for industrial application in huge quantities and are required to be cost effective for production. The proteins include enzymes and diagnostic proteins (Hood and Woodard 2002; Xu et al. 2018). The plant expression system is preferred due to the cost-efficient process in the production of the industrial enzyme, the high protein expression level of protein of interest at a specific region of the plant, and the easy harvesting process (Hood and Woodard 2002). Among various available transgenic plants, field plants are the most preferred type to produce industrial proteins. This is because of the numerous benefits it possesses, including being cost effective, ability to store the protein at the desired organ to achieve high stability, and fast production (Xu et al. 2018). One of the first-ever proteins produced for industrial purposes was avidin developed and sold in 1998. One factor that bothered the production of avidin was that the stability of protein produced after ten generations became lower and insufficient. Over time, the researchers overcame this by the selection process and modification in germplasm to maintain the stability of avidin for multiple generations without any depletion (Hood and Woodard 2002; Xu et al. 2018). Some of the studied and available commercialized proteins and enzymes include hydrolases, cellulase, α-amylase, β-glucuronidase, and trypsin (Xu et al. 2018). One of the famous and widely used enzymes in industries is papain derived from the papaya fruit. It is widely applied in the food industry, where it is a frequently used type of protease to tenderize the meat (Shouket et al. 2020; see also chapter "Molecular Farming of Industrial Enzymes: Products and Applications").

1.2.3.3 Therapeutic Proteins

Proteins are one of the important biological molecules which can be categorized based on their function and pharmacological activity. Therapeutic proteins are proteins that can be found naturally in the human body and can be uniquely designed for use in pharmaceutics and treatment purposes. Therapeutic proteins can be classified according to their molecular types, and some of the examples include monoclonal antibody, blood factors, growth factors, enzymes, hormones, interferons, and interleukins (Awwad et al. 2018; Dimitrov 2012). The requirement and need for these therapeutic proteins are increasing annually, and the conventional production of these proteins is time consuming and very expensive. The plant's expression system is used to overcome these issues, where plants can produce these proteins in large amounts, at the same time being low-cost, safer, and stable protein compared to those produced in mammalian cells or other expression systems (Xu et al. 2018). Monoclonal antibody (mAb) is one of the most known therapeutic proteins, where it is produced from B cells-lymphocytes and is known to replace antibodies in the host. mAb can improve, modify, and mimic the host immune system to attack pathogens or cancerous cells (Monoclonal Antibody Drugs for Cancer: How They Work—Mayo Clinic n.d.; Tabll et al. 2021). One of the studies performed by Lai et al. has proven that mAb developed from the plant system cures West Nile virus infection. The transgenic tobacco plant has been designed to produce humanized mAb that has a high affinity towards West Nile virus envelope protein, which results in viral fusion inhibition (Lai et al. 2010). The purity of the mAb produced via the transgenic tobacco plant is more than 95%, and the neutralization activity of the mAb was noticeably high (Lai et al. 2010). Another type of therapeutic protein produced using a plant system is the growth factor, where it is a released bioactive compound that can influence cell development and growth (Stone et al. 2021). One of the studies conducted by Musiychuk et al. has produced human erythropoietin growth factors from the plant system to promote hematopoietic stem cell differentiation to red blood cells. They also analyzed another three human growth factors, which are stem cell factor, interleukin, and insulin-like growth factor-1, using transgenic tobacco plant. The researchers have used the tobacco mosaic virus vector approach to introduce the desired gene to produce these growth factors. The EC50 values of these four human growth factors were substantially high and showed promising results (Musiychuk et al. 2013).

1.3 Transformation Strategies

1.3.1 Nuclear Transformation

Stable nuclear transformation is the most common and widely used method to date for genetic manipulation in plants to produce recombinant protein (Horn et al. 2004; Shanmugaraj et al. 2020). Nuclear transformation is the introduction of the genes of interest into the nuclear genome of plants to cause the alteration of genetic structures and ultimately express the transgene within the host genome (Alireza and Nader 2015). The foreign gene can be incorporated into the in vitro plantlets in a stable manner as a transgene in the plant expression vector by *Agrobacterium tumefaciens*-mediated transformation or particle bombardment, biolistic method (Horn et al. 2004). *Nicotiana* genus is widely used as the expression system to produce most of the recombinant proteins due to its rapid growth rate and easy manipulation of the gene. Furthermore, crops such as rice, maize, tomato, and potato also showed their potential as expression systems in PMF (Burnett and Burnett 2020). Upon the development of the pool of transgenic lines, the best transgenic line will be screened for protein production. The succession of the integration of transgene into the plant genome would result in the recombinant protein production in successive generations (Shanmugaraj et al. 2020).

This method has been used to produce the largest number of recombinant proteins in PMF. The recombinant proteins such as erythropoietin, collagen, and human-secreted alkaline phosphatase were produced via *Agrobacterium*-mediated method with tobacco as the expression system (Komarnytsky et al. 2000; Matsumoto et al. 1995; Ruggiero et al. 2000). Human growth hormone (hGH) and α1-antitrypsin were produced via biolistic method with rice as the expression system (Kim et al. 2008; Terashima et al. 1999). Exploitation of this method has shown the advantage to perform in crop species such as grains, rice, cereals, and corn. As the protein products accumulate in the seed, it can be harvested in a dry condition and stored for a long term until the accomplishment of processing (Delaney 2002). However, the potential of some grains to cross with the native species or food products has affected the social acceptance towards the use of this method (Commandeur et al. 2003).

1.3.2 Plastid Transformation

Plant cells contain three DNA-harboring organelles, which are the nucleus, plastid, and mitochondria (Yu et al. 2020). The success in engineering the genome of chloroplast for disease, herbicides, and insect resistance as well as for the production of biopharmaceuticals had led to a new era of plant biotechnology to a more environmental-friendly direction (Daniell et al. 1998). This is because the genome of chloroplast defied the laws of Mendelian inheritance that resulted in avoiding the outcrossing of transgenes with native species and reducing the toxicity of transgenic pollen to the non-targeted insects (Cosa et al. 2001; Svab et al. 1990). The genome of plastids is a circular double-stranded DNA that is inherited maternally and presents in many copies in the organelle (Adem et al. 2017). Thus, it is protected from gene silencing, which would result in lowering the transgene expression and thereby allowing the accumulation of high level of foreign protein, 5–40% of total soluble protein (Roudsari et al. 2009).

Svab et al. were the first to describe the system for plastid transformation using higher plant, tobacco, after the success of plastid transformation on unicellular algae, *Chlamydomonas reinhardtii*. Chloroplast genetic engineering started to develop in 1980s where isolated intact chloroplasts were able to introduce into protoplasts (Svab et al. 1990). The crop species that can be used for plastid transformations include tobacco, tomato, potato, eggplant, and soybean (Ding et al. 2006; Ma et al. 2015; Shanmugaraj et al. 2020; Zhou et al. 2004). There are three main steps in plastid transformation as illustrated in Fig. 1.2. Firstly, the transgene is delivered to cells of an explant. Secondly, the transgene is integrated into the chloroplast genome by going through homologous recombination at a specific site (Yu et al. 2020). Lastly, the ideal transformants are screened repeatedly and thoroughly on

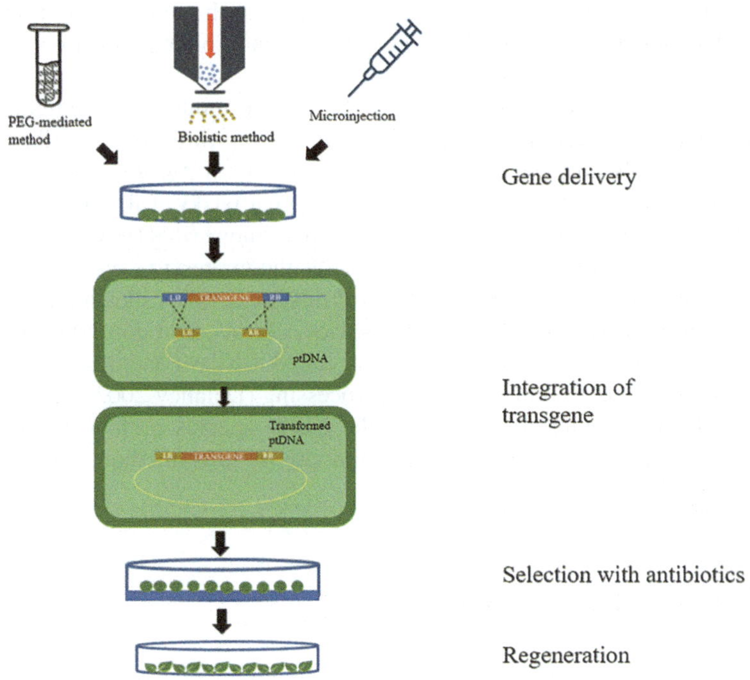

Fig. 1.2 Plastid transformation process

selection medium until a state of homoplasmy (fully eliminate the wild-type genome in the pool) (Lu et al. 2006). The selection process can be conducted on the medium containing spectinomycin (Alireza and Nader 2015). The positive transformants will then be regenerated into a stable transgenic line.

Biolistic transformation and polyethylene glycol (PEG)-mediated transfection are the two most common methods for introducing the foreign DNA into chloroplasts (Lu et al. 2006; Yu et al. 2020). PEG-mediated method only works on protoplasts by culturing the protoplasts with the presence of PEG vesicles containing foreign DNA to allow the DNA uptake by protoplasts (Lu et al. 2006). However, the development of gene gun by John Sanford to be used as a transformation device allows direct transformation to the plant cells instead of isolating protoplasts (Daniell 1993). Recently, nanoparticles were introduced as a new strategy for plastid transformation to allow the introduction of DNA without the need of other instrument or isolation of protoplasts (Kwak et al. 2019).

Antigen vaccines, protein-based drugs, and industrial enzymes are able to be produced from the chloroplast of the plant via plastid transformation (Yu et al. 2020). Protein-based drug insulin is able to be produced in transplastomic tobacco plants with the expression of 14.3% of total soluble protein (Kwon et al. 2013). This might help to solve the issues of high cost of insulin produced by yeast due to the additional cost needed to maintain the yeast suspension (Daniell et al. 2016).

Besides, plant chloroplast is able to perform a proper posttranslational modification of insulin to promote the proper folding like phosphorylation and disulfide bond formation (Řepková 2010).

The culprit of cervical cancer is human papillomavirus (HPV), which kills more than 250,000 of women annually. The protein E7 antigen from HPV type 16 (HPV-16 E7) has been selected as a candidate to produce therapeutic vaccine (Morgenfeld et al. 2009, 2014). It has been reported that HPV-16 E7 can be expressed in the tobacco plant via plastid transformation (Morgenfeld et al. 2009). Studies also showed that mice and rabbits developed specific immune responses after being injected with chloroplast-derived vaccines (Molina et al. 2005). However, the accumulation of E7 protein in the chloroplast is relatively low due to the instability of natively unfolded protein. Thus, the strategy of fusing E7 coding sequence with potato virus X coat protein was done to improve the stability and immunogenicity of the antigen (Morgenfeld et al. 2014; see also chapter " Development of Oral Prophylactic and Therapeutic Vaccines Against HPV on the Basis of Plant Expression System").

In addition, genetically engineered chloroplast genome can be used to produce biomaterials. The biodegradable polyester biopolymers such as polyhydroxyalkanoates (PHAs) are a perfect alternative to petroleum-based plastics that are naturally synthesized by microorganisms (Svab et al. 1990). However, transgenic crop via plastid transformation like tobacco has the potential to be used as a source of PHAs. Transgenic crops can achieve an accumulation level of PHA up to 40% dry cell weight of the leaf and 20% dry cell weight of the seed (Dobrogojski et al. 2018). Industrial enzymes such as β-glucosidase and xylanase can be obtained via chloroplast production in tobacco plants as well (Jin et al. 2011; Kolotilin et al. 2013).

Despite the great potential of plastid transformation, only tobacco is practically possible for this method. However, due to the toxicity of the plant with full of poisonous alkaloid, the use of tobacco in the production of edible recombinant protein or vaccine is not feasible (Horn et al. 2004; see also Chap. 9).

1.3.3 Agroinfiltration

Agroinfiltration is based on the *Agrobacterium*-based approach for transient transformation that is commonly used to transfer the foreign gene into somatic cells of targeted plants such as leaves (Debler et al. 2021; Kaur et al. 2021). Early discovery of *Agrobacterium tumefaciens* in the late eighteenth century has played a prominent role in plant biotechnology. Initially, it was developed to use as a tool for the investigation of plant-virus interaction (Chen et al. 2013; Kaur et al. 2021). However, the first transgenic plant produced via *Agrobacterium*-mediated transformation process in the year 1983 opened a door for plant genetic transformation for plant farming interventions (Ramkumar et al. 2020). In addition, *Agrobacterium*-mediated transformation is a relatively effective, simple, and less time-consuming transient expression method compared to biolistic method and PEG-mediated transfection of protoplast as it does not involve the isolation of protoplast (Belhaj et al. 2013; Norkunas et al. 2018). Agroinfiltration works by the principle of delivering the gene of interest carried by recombinant *Agrobacterium* in the extracellular spaces of the intact leaves via physical or vacuum infiltration (Donini and Marusic 2019). It is preferable because this method has high level of expression, which yields up to 50% of TSP compared to stable transformation, which only yields less than 1% of total TSP. Human growth hormone (hGH), aprotinin, and human fibroblast growth factor 8b are the examples of transient expression application (Sheludko 2008).

It is an indirect gene delivery method, in which *Agrobacteria* species has the ability to transfer DNA into the plant cells. *Agrobacterium tumefaciens* is a pathogen of the plant. It has a tumor-inducing (Ti) plasmid containing transfer DNA (T-DNA) that is able to transfer into plant host cells by infecting the plant cells (Chen et al. 2013). It would result in causing tumors and crown gall in the plant as Ti plasmid is the factor of tumorigenesis and part of the plasmids, and T-DNA was incorporated into the plant genome (Gelvin 2003). On the other hand, *Agrobacterium rhizogenes* harbors root-inducing (Ri) plasmid that induces abnormal root growth (Hwang et al. 2017). Therefore, it was then proposed that Ti or Ri plasmids can be used as a vector to introduce transgene into plant cells by developing the *Agrobacterium* strains that do not cause tumor formation but have the capability to transfer foreign gene (Chen et al. 2013; Gelvin 2003). T-DNA and virulence region are the two genetic components that are responsible for the transformation (Chen et al. 2013).

There are few plant hosts that could be used including *Lactuca sativa* (Lai et al. 2012), *Nicotiana tabacum*, *Arabidopsis thaliana*, *Glycine max* (King et al. 2015), and *Pisum sativum* (Guy et al. 2016). However, *Nicotiana benthamiana* is the most preferred experimental host due to the high susceptibility to a diverse range of viruses and other plant-pathogenic agents including bacteria, oomycetes, and fungi (Goodin et al. 2008). In addition, *Nicotiana benthamiana* has gained a great attention to serve as "biofactories" in producing recombinant proteins due to its ability to perform posttranslational modifications for appropriate protein folding to result in functional biological activity (Bally et al. 2018). The expression of the transgene could be controlled by a suitable promoter such as cauliflower mosaic virus 35S (CAMV-35S) (Debler et al. 2021). However, it is not easy to clone a foreign gene into T-DNA due to large size of Ti plasmids and lack of unique restriction endonuclease cleavage sites on T-DNA and elsewhere (Gelvin 2003). Thus, two approaches were developed to facilitate the introduction of foreign gene into T-DNA including co-integrating system and binary vector system. But binary vector system is the most performed method of molecular genetic modification (Mardanova et al. 2017).

Co-integrating system generally refers to cloning the gene by an indirect mean into Ti plasmid in a way that the new gene was in *cis* with the virulence genes resulted on the same plasmid (Gelvin 2003). A region of DNA containing unique restriction endonuclease sites targeted for disruption is cloned into a broad host range plasmid, which is able to replicate in *E. coli* and in *Agrobacterium*. It is known as an intermediate plasmid. The foreign gene with antibiotic resistance marker is cloned into a unique restriction endonuclease site. A disarmed Ti plasmid contains the vir function maintained in *Agrobacterium*. The intermediate plasmid is then conjugated with *Agrobacterium* containing disarmed Ti plasmid and undergoes homologous recombination. *Agrobacterium* with co-integrated plasmid is selected and introduced into the plant (Garfinkel et al. 1981; Gelvin 2003). However, the protocols are sophisticated and therefore binary system is generally a preferred method in agroinfiltration.

Binary vector system revolutionized the *Agrobacterium*-mediated transformation method (Gelvin 2003). pGreenII is one of the most widely used binary vectors (Hellens et al. 2000). It generally refers to the system in which *vir* and T-DNA regions of the Ti plasmids can be split into two separate replicons but located in the same *Agrobacterium* cell (Simpson et al. 1986). Therefore, this system consisted of two plasmids: T-binary vector and the vir helper plasmid. T-DNA region containing foreign gene is located on the binary vector, and non-T-DNA region contains the origins of replication that could perform replication in both *E. coli* and in *Agrobacterium* as well as antibiotic resistance genes for selection process (Komori et al. 2007). It is also known as a small artificial T-DNA since this vector does not contain tumor-inducing gene and vir genes, and thus it has smaller size compared to Ti plasmids. The vir gene containing replicon is served as a helper plasmid to synthesize vir protein (Hwang et al. 2017). The vir region serves a function in participating in the events in the host cell that involve cytoplasmic trafficking of T-DNA, nuclear targeting, and integration into host genome as well as improvement of the virulence (Howard et al. 1992). As a result, the foreign gene could be easily introduced to small T-DNA regions within binary vectors. Then, the characterization and verification of the constructs are done in *E. coli*. T-binary vector is able to mobilize into *Agrobacterium* strain containing vir helper plasmids. This system has simplified the protocol in generating transgenic plants compared to co-integrating system (Lee and Gelvin 2008). The schematic diagram of co-integrating vector and binary vector is illustrated in Fig. 1.3.

Syringe infiltration and vacuum infiltration are the two most common agroinfiltration methods (Kaur et al. 2021). Syringe infiltration involves the use of needleless syringe to introduce *Agrobacterium* into the leaves of the plant, and Tween-20 could be used to improve the efficacy (Zhao et al. 2017). *Agrobacterium* together with infiltration medium is injected into the nick-created leaves. It can either transfer a target gene into the entire leaf entities or introduce multiple genes in different areas of a leaf for multiple assay purposes (Vaghchhipawala et al. 2011). On the other hand, vacuum infiltration is done by submerging the plant leaves into the infiltration media containing *Agrobacterium* in a vacuum chamber under a negative atmospheric pressure (Simmons et al. 2009). The air of interstitial spaces of the leaves is drawn out by vacuum, and air filled by *Agrobacterium*-containing media occupies the spaces when the vacuum is released to achieve agroinfiltration (Chen et al. 2013). Vacuum filtration is more robust compared to syringe filtration as large number of plants can be infiltrated in a shorter time and it can be used for those plants that cannot be infiltrated by syringe (Rivera et al. 2012). According to the studies, vacuum infiltration requires a shorter timeframe for infiltration, which only takes 3 min for the process, compared to syringe infiltration, which needs 15 min

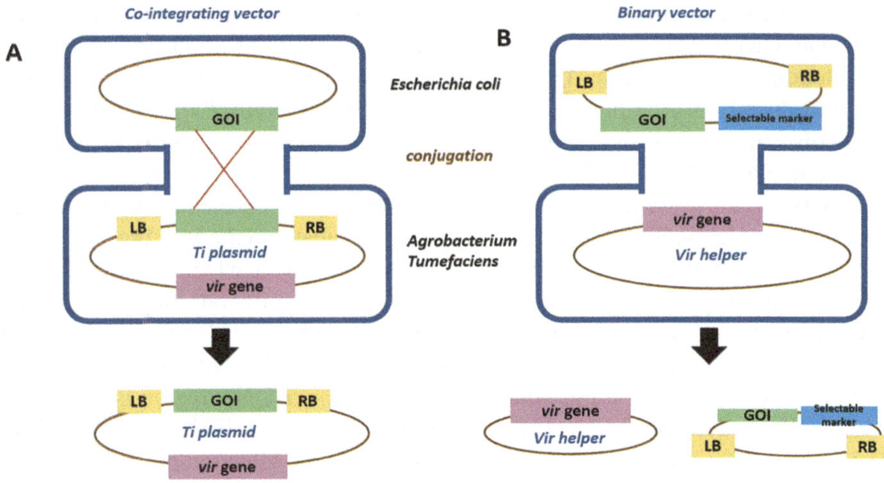

Fig. 1.3 Agroinfiltration process. *GOI* gene of interest, *LB* left border, *RB* right border

(Leuzinger et al. 2013). Hence, this method has facilitated the development of a protein production platform that is fast, safe, and economical.

To date, *Agrobacterium*-mediated transient expression has considerably drawn the attention of scientists as an approach for large-scale production of recombinant protein using plant systems due to its effectiveness and safety (Sheludko 2008). It has been practically utilized for the transient expression of recombinant protein. Furthermore, agroinfiltration is also applicable for a virulence gene discovery and various studies in plant: gene silencing, resistance mechanism, promoter characterization, vaccine production, as well as phyto-sensing studies (Chen et al. 2013; Sheludko 2008).

1.3.4 Viral Infection

Transient expression based on virus-based vectors is another alternative to complement transgenic plants, which offers high expression level of protein in the time of few days (Shanmugaraj et al. 2020). Plant viruses are not just used as pathogens but had evolved to be used as a tool for recombinant protein expression that is of industrial importance such as vaccine antigens and antibodies (Yuri Gleba et al. 2007; Ibrahim et al. 2019). It has also been used for chimeric viral vaccine production and as an agent of nanoparticles for drug delivery (Ibrahim et al. 2019). Tobacco mosaic viruses (TMVs) and X potato viruses are the most prominent viruses to be used as a basic tool for protein production in PMF (Alireza and Nader 2015; Bamogo et al. 2019). The general approach is that the viruses are engineered to contain a gene of interest upon introduction into the host and that will replicate in the host and produce the recombinant protein in a significant amount in infected plants (Bamogo et al. 2019; Gleba et al. 2007). The plant viruses can be engineered to result in expressing subunit vaccines or to serve as epitope presentation systems (Bamogo et al. 2019). The viral infection processes are illustrated in Fig. 1.4.

Ever since the emergence of virology, TMV was the first exploited virus to be developed as a viral vector to produce various and diverse types of recombinant proteins and became the most prominent choice to develop natural vectors (Ibrahim et al. 2019; Kagale et al. 2012). It began with a Russian scientist, Dmitri Ivanovsky, who described that infected sap of leaves with tobacco mosaic diseases retained the infectious properties (Lustig and Levine 1992) with the ability to replicate within the plant cells even after extracted sap had passed through bacteria-retaining filters since the filtrate is infectious to infect tobacco plants (Van Regenmortel 2008). It was then followed by the success of precipitating the pathogen of tobacco mosaic disease in 1927, which came to reveal that the pathogen contained RNA as well as protein DNA, which is the TMV (Rifkind and Freeman 2005). TMV belongs to *Tobamovirus* genus with a single-stranded RNA and a positive sense (Dunigan and Zaitlin 1990). Viruses with RNA genome are preferably to be exploited due to the ability of the virus to build the capsid around the genome, which results in less constraint on the size of transgene insertion (Yusibov et al. 2006). Within the untranslated region of the genome, it harbored transfer RNA like structure that encoded for protein needed for RNA replication, movement protein as well as coat protein (Ibrahim et al. 2019). The promoter of viral RNA can be manipulated to synthesize the recombinant messenger RNAs in the whole plant. Immunogenic molecules could be expressed using TMV vectors such as anti-idiotypic single-chain variable fragment (scFv) antibodies, and the antibodies were shown to be protective against

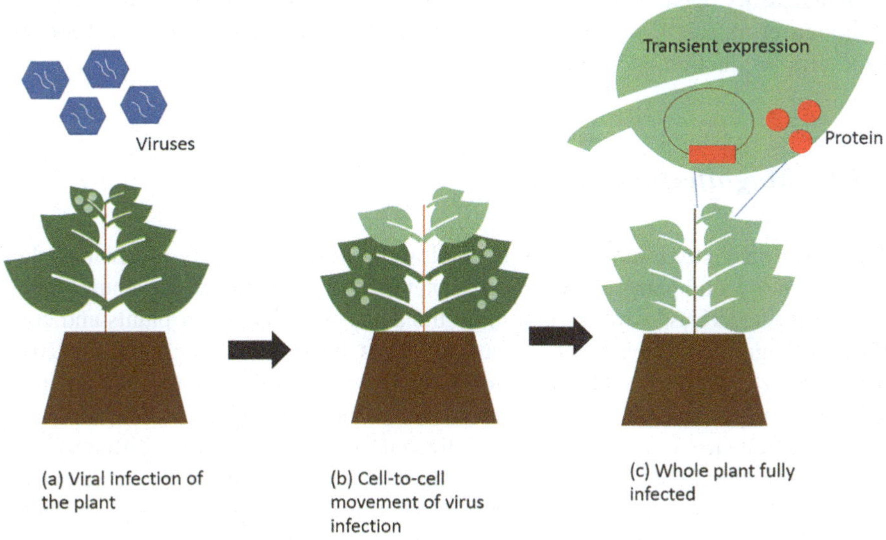

Fig. 1.4 Viral infection process

cancer by recognizing the immunoglobulin on human tumor cells (McCormick et al. 2003).

Potato virus X (PVX) vectors were developed for transient expression of foreign genes in plants in 1992 (Chapman et al. 1992). It is a positive-sense RNA plant virus with flexuous and rod shape. The molecular event of PVX infection had paved the way in developing expression vectors, and it is ultimately applied in functional genomics studies as well as biosynthesis of heterologous recombinant protein (Lico et al. 2015). The strategy is that the foreign gene is inserted between a duplication of the subgenomic promoter sequence directed at the production of mRNA, in which the coat protein is translated (Lacomme and Chapman 2008). The capsid protein gene and triple-gene block are engineered to deconstruct vectors. PVX can be used to express full-length proteins, fusion proteins, as well as epitope for the use in medical purpose (Bamogo et al. 2019).

These plant viral vectors can be categorized into different categories based on the manner in which they are designed. The first-generation vector, also the first strategy, is the "full virus strategy" which uses the unmodified native virus to maintain its properties completely. Hence, this strategy is to design the viruses that are fully functional and retain infectivity and virulence despite carrying and expressing heterologous protein sequences (Awram et al. 2002; Gleba et al. 2004). However, due to the limitation of the first-generation vector and safety issues that can spread in the environment, another strategy called "deconstructed virus strategy" has been established in which the deconstructed vectors are composed solely of the genome that is required for virus replication by removing unnecessary and retaining only the essential and required part of the vector to function (Gleba et al. 2014; Pogue et al. 2002). This would also allow large size of the gene to be incorporated into the genome of virus and with high level of protein expression since the protein coded for unnecessary open reading frame was deleted and replaced by the gene of interest (Bamogo et al. 2019; see also chapter "Plant Viral Vectors: Important Tools for Biologics Production").

1.3.5 Magnifection System

Magnifection system, magnICON technology, was developed by a German plant biotechnology company named Icon Genetics that utilized the advantages of a "deconstructed virus" strategy that lacks the ability to infect other plants and combine with agroinfiltration for the gene delivery (Gleba et al. 2005; Hefferon 2017; Marillonnet et al. 2005). Magnifection system is a scalable protocol for recombinant protein production in plant that relies on the transient amplification of the deconstructed viral vectors, which are delivered by *Agrobacterium* to different areas of a plant body (Gleba et al. 2005). It is a new generation of transfection that effectively addresses most of the drawbacks of other available transformation technologies in PMF. It could be used to produce recombinant antigen and antibodies in *Nicotiana benthamiana* (Klimyuk et al. 2012). This technology has combined the advantages of three biological systems: vector efficiency, ability of *Agrobacterium* in its systemic delivery, and rate and expression level of a virus as well as the post-translational capabilities (Gleba et al. 2005; Hiatt and Pauly 2006).

The general approach of this technology is to supply a plant cell with an RNA or linear DNA molecule by *Agrobacterium* that assembles inside the cells assisted by the site-specific recombinase instead of supplying a complete viral vector mature viral particle (Giritch et al. 2006). Thus, in vitro transcription in generating RNA-based vectors can be avoided as engineered in vitro RNA synthesis is the time and effort bottleneck of functional genomics using viral vectors for the actual expression (Fitzmaurice et al. 2002). *Agrobacterium* carrying T-DNAs encoding for RNA replicons is prepared in a diluted suspension in order to infiltrate into the plant. The bacteria generally function as a primary infection to the plant for systemic movement. On the other hand, the viral vectors are responsible for cell-to-cell spread, amplification, as well as high-level expression of recombinant protein (Marillonnet et al. 2004). However, the coat proteins of the virus strains are removed to prevent its systemic delivery to infect the whole plant (Alireza and Nader 2015). This approach is preferable because it provides very high yield of protein with up to 80% of total soluble protein with a very short period of expression, which takes up to 3–4 days, and most importantly, it is inherently scalable (Marillonnet et al. 2004).

This new technology can be applied for the plant vaccine production. A deconstructed TMV vector is employed to generate E7 oncoprotein of human papillomavirus (HPV) (Plchova et al. 2011) as well as M2e epitope of influenza (Denis et al. 2008). Then, the deconstructed TMV vector that generates antigen is housed within an *Agrobacterium tumefaciens* binary vector that would result in generating vaccines against influenza virus and cholera. Furthermore, magnICON deconstructed vector is one of the vectors that are used to develop personalized medicine against non-Hodgkin's lymphoma (NHL) (McCormick et al. 2008). NHL occurs due to the overproliferation of B cells that present a unique cell surface idiotype that is only specific to that individual (Singh et al. 2020). Hence, NHL can be vaccinated with their own idiotype. Thus, personalized medicine can be generated using the magnifection method. Deconstructed magnICON vectors express the TMV constructs composed of scFv subunit and full-length idiotype Ig molecules as heavy and light chains that will be assembled into full immunoglobulins in the plant. The vaccine constructs have successfully passed phase I clinical trial in which they have been demonstrated to be safe and elicit few adverse effects (Hefferon 2017; see also chapter "Antigen-Specific Immunotherapy for Allergic and Autoimmune Diseases Using Plant-Made Antigens").

As the summary, Table 1.2 highlights the advantages and disadvantages of the five main transformation strategies used in plant molecular farming.

Table 1.2 Advantages and disadvantages of transformation strategies

Transformation strategy	Advantages	Disadvantages	References
Nuclear transformation	• Simple and widely used method in many plant species • Reduced cost of storage • Increased capacity of scale-up • Has unique glycosylation pattern	• Low levels of expression • Risk of gene silencing • Possibility of transgene contamination	Burnett and Burnett (2020), Moon et al. (2020)
Plastid transformation	• Transgene containment • High levels of expression • High stability and reduced risk of gene silencing • Ease in ecological controls	• Lack of glycosylation in plastids • Lack of field release of transgenic plant	Gleba and Giritch (2012), Maliga (2017)
Agroinfiltration	• Shorter timeline of process • Increased expression of recombinant protein • Potential of introducing more than one T-DNA into plant cells	• Challenges in scalability • Some plant species are not compatible with *Agrobacterium* • Some plant species are not amenable to agroinfiltration • Challenges in RNA silencing due to the host recognizing T-DNA as foreign	Canto (2016), Menassa et al. (2012)
Viral infection	• Increased expression of recombinant protein • Allows the screening of multiple construct variants in various plant species • Capability of expressing targeted genes at a specific stage of plant growth by differing inoculation timing • Small size of viral genome and ease in manipulation • Shorter timeline of process than stable transformation strategies	• Gene of interest introduced could be lost due to mutation or deletion over time • Chances of adverse effects on plant host or interaction of plant viral vector with other viruses • Chances of transmission to the environment	Abrahamian et al. (2020), Cañizares et al. (2005)

(continued)

Table 1.2 (continued)

Transformation strategy	Advantages	Disadvantages	References
Magnifection system	• Shorter timeline of process • Increased expression of recombinant protein • High yield of protein reduces downstream processing cost • Versatility in genes expressed • Permits the co-expression of several proteins that are required for hetero-oligomeric protein assembly • Vector efficiency	• Challenges in expressing an economically acceptable level of gene as high level of expression of certain proteins may be toxic to host plant • Challenges in expressing oligomultimeric proteins as the manipulation of viral vector requires polypeptides to be expressed in equimolar	Burnett and Burnett (2020), Gleba et al. (2005)

1.4 Advantages of PMF

Given the diversity in protein expression systems that are currently available, there is no surprise that each system's characteristics are compared head-to-head to determine their functions and suitability of use. While other expression systems, such as mammalian cell lines, bacteria, yeast, and insects, certainly have their niche advantages, plants often take the upper hand when it comes to molecular farming.

Often, the first factor to be brought up for discussion is the commerciality of plants in molecular farming. The cost of operations is essential for sustainable production of recombinant proteins. Molecular farming on an industrial scale using cell lines, inclusive of mammalian, bacterial, or fungal (yeast), relies on using large fermenters, bioreactors, and other equipment, which contributes to its high cost (Alireza and Nader 2015; Fischer and Schillberg 2016). In contrast, the cost of PMF is relatively cheaper compared to other expression systems as transgenic plants can be grown in greenhouses or vertical farms, allowing PMF to reach agricultural scalability and space efficiency due to the ease of growing additional plants (Buyel 2019; Fischer and Schillberg 2016). Hence, this lets PMF to have a relatively higher production ceiling and ease of manufacturing. Running costs such as the expensive media typically used in cell cultures, especially in mammalian cell expression systems, are also much higher compared to the defined fertilizer solutions for cultivating plants (Buyel and Fischer 2012; Xu et al. 2017).

The ease of storage and transportation of PMF products drives its costs lower. The robustness and durability of plants result in lower storage requirements, which also consequently reduces costs for transportation. Additionally, the utilization of plants in molecular farming allows the manufacture process of recombinant proteins to be conducted on-site (through the use of greenhouses and farms, as mentioned previously), skipping the requirements of existing or constructing new laboratory facilities (Chung et al. 2021). This not only reduces transport requirements, but would also provide accessibility to rural regions with pressing needs for biopharmaceuticals. Utilizing PMF for rapid deployment of emergency medicines or vaccines was shown to be possible during the 2014 Ebola outbreak in West Africa where ZMapp, an experimental treatment for Ebola produced via transient expression in *N. benthamiana*, received authorization for emergency deployment before clinical trials were conducted (Schillberg and Finnern 2021; Zhang et al. 2014). As such, many look to PMF as a potential pandemic response that may be able to help combat the ongoing COVID-19 pandemic (Kumar et al. 2021; Leblanc et al. 2021; Tusé et al. 2020).

The safety of recombinant proteins and molecular farmed products is a highly controversial issue that concerns consumers, researchers, and investors alike. Plants, which are innately immune to most human pathogens and endotoxins, present little to no risk of contamination during the cultivation of transgenic plants for PMF (Alireza and Nader 2015; Shanmugaraj et al. 2020; Wang and Ma 2012). This lends credibility of PMF products' relative safety compared to other expression systems such as bacteria or yeast. The manufacturing process is also simplified using plants when PMF produces edible products in the forms of fruits and vegetables. In these cases, post-processing and purification steps can be skipped almost entirely during manufacturing (Wang and Ma 2012). Not only would this be beneficial by reducing production costs, but it also allows convenient oral administration to consumers.

Additionally, plant recombinant proteins are able to undergo relatively complex posttranslational modifications, such as glycosylation, phosphorylation, and disulfide bridging, enabling further complex proteins to be produced in PMF. This gives plants a further edge compared to bacterial, fungal, or insect expression systems, where only simple modifications can be performed, or none at all. That said, plants still fall short compared to mammalian cell lines in terms of the similarity of modifications to humans. The differences in posttranslational modifications, such as differing glycosylation patterns or chemically different sugars involved, may lead to immunogenicity or loss of function of the protein (Marintcheva 2018). Solutions to overcome this problem involve expressing relevant mammalian posttranslational modification enzymes in the plant system itself and inversely silencing the relevant plant systems that are still in development (Marintcheva 2018).

1.5 Challenges of PMF

Although PMF approach is advantageous in many aspects, there exists a bottleneck for the utilization of PMF in the production of recombinant proteins commercially (Schillberg and Finnern 2021; Xu et al. 2018). One of the major limitations of PMF is its low productivity, which is perceived to limit its successful commercialization. The levels of recombinant proteins produced are reported to hardly exceed 100 μg/kg of the plant tissue fresh weight or 100 μg/L of plant cell suspension cultures (Permyakova et al. 2021; Schillberg and Finnern 2021). This level of recombinant proteins produced in plant expression systems are relatively considered to be much lower than other expression systems. However, plants are able to yield a higher biomass when grown in the field compared to conventional expression platforms such as mammalian cells, bacteria, and yeast that involve fermentation (Twyman et al. 2013). The yield produced by the plants can be affected by a number of factors including environmental factors such as temperature, humidity, and light; protein targeting strategies; expression strategies; and protease inhibitor co-expressions (Schillberg et al. 2019).

Although challenges in the expression of transcript have been identified, there are ways of optimizing the level of transcriptions. A commonly utilized approach is the utilization of cauliflower mosaic virus (CaMV) 35S promoter, which is suitable for dicotyledonous plants, and maize-1 ubiquitin promoter, which is suitable for monocotyledonous plants (Alireza and Nader 2015). Apart from the promoter, the polyadenylation site also plays an essential role in increasing the transcription levels, and some of the commonly used polyadenylation sites are from CaMV 35S transcript, pea SSU gene, and *Agrobacterium tumefaciens* nopaline synthase gene (Singh et al. 2021). The silencing or degradation of RNA may also affect the stability of gene transcripts, which leads to a lower recombinant protein yield (Alireza and Nader 2015). In terms of maintaining stable activity of recombinant proteins in transgenic plants, targeting transgenes to be expressed in its storage organs such as tubers or seeds is beneficial (Moustafa et al. 2016). In fact, activity of human coagulation factor IX protein, which was expressed in the seeds of soybean plants, was reported to be functionally stable for a period of 6 years (Moustafa et al. 2016; see also Chaps. 5 and 6).

Glycosylation refers to covalent binding of various glycans such as saccharides, sugar, or carbohydrates to a protein (Li et al. 2019). This posttranslational modification is essential for the function, stability, and activity of proteins. Although plants are advantageous compared to bacteria and yeast in this aspect, the difference in the carbohydrate pattern of glycosylation in plants when compared to mammalian cells is a challenge. The glycosylation pattern that occurs in plants may give rise to glycan-specific antibodies and subsequent undesirable clearance of the plant-made therapeutics when administered (Gomord et al. 2010; Schoberer and Strasser 2018). The N-glycan glycoprotein epitopes including Le^a, α-1,3-fucose-containing, and β-1,3-xylose-containing epitopes are typically immunogenic (Gomord et al. 2010). Apart from these, the selection of a suitable plant species as the host is also critical. Although there are various options for a host, each plant has its advantages and disadvantages (which was discussed in previous sections). The phenol and toxic alkaloid contents, amount of biomass, and tight regulatory requirements for certain plant species should be taken into consideration when selecting a host (Burnett and Burnett 2020).

Another major hurdle in PMF is challenges with regulatory approval and social acceptance. PMF products are managed by different regulations in different parts of the world based on its risk assessments. Concerns regarding PMF that have been raised include the risk of contamination of feed or food crops that may pose a threat to food industries and farmers, the risk of possible horizontal gene transfer which may result in the development of antibiotic-resistant microbes, and the risk of transgene spread or unwanted exposures to wild-type plants (Obembe et al. 2011). Hence, biosafety strategies and regulations are essential. The need to follow good manufacturing practice (GMP) guidelines that have been set based on mammalian and bacterial cell expression systems instead of developing specific guidelines for plants is known to be challenging (Menary et al. 2020). In the USA, the United States Department of Agriculture's (USDA) Animal and Plant Health Inspection Service and the Food and Drug Administration (FDA) are key regulatory bodies that regulate the cultivation of plants and the safety of pharmacological products, respectively (Breyer et al. 2009; see also Chaps. 14 and 15).

1.6 Concluding Remarks

PMF has shown to possess great potential to be utilized for a variety of applications. Plants including tobacco, cereals, fruits, and vegetables have been utilized as bioreactors successfully. With an increase in the number of studies investigating the potential of PMF, various stable and transient transformation strategies have been developed. Utilizing plants as expression hosts is suggested to greatly reduce the production and transportation cost of recombinant proteins. With these advantages comes challenges in terms of productivity. However, the development of optimization strategies has enabled successful PMF utilization, and it would promise various potential applications in the future.

Acknowledgements The authors would like to thank Prof. Hoe I. Ling of Columbia University (New York, USA) for his editorial input.

References

Abiri R, Valdiani A, Maziah M, Shaharuddin NA, Mahbod S, Yusof ZNB, Atabaki N, Talei D (2016) A critical review of the concept of transgenic plants: insights into pharmaceutical biotechnology and molecular farming. Curr Issues Mol Biol 18:21–42

Abrahamian P, Hammond RW, Hammond J (2020) Plant virus–derived vectors: applications in agricultural and medical biotechnology. Annu Rev Virol 7:513–535

Adem M, Beyene D, Feyissa T (2017) Recent achievements obtained by chloroplast transformation. Plant Methods 13:1–11

Alireza T, Nader RE (2015) Molecular farming in plants. In: El-Shemy H (ed) Plants for the future. IntechOpen

Awram P, Gardner RC, Forster RL, Bellamy AR (2002) The potential of plant viral vectors and transgenic plants for subunit vaccine production. Adv Virus Res 58:81–124

Awwad S, Ginn C, Brocchini S (2018) The case for protein PEGylation. In: Parambath A (ed) Engineering of biomaterials for drug delivery systems. Elsevier, pp 27–49

Bally J, Jung H, Mortimer C, Naim F, Philips JG, Hellens R, Bombarely A, Goodin MM, Waterhouse PM (2018) The rise and rise of Nicotiana benthamiana: a plant for all reasons. Annu Rev Phytopathol 56:405–426

Bamogo PKA, Brugidou C, Sérémé D, Tiendrébéogo F, Djigma FW, Simpore J, Lacombe S (2019) Virus-based pharmaceutical production in plants: an opportunity to reduce health problems in Africa. Virol J 16:1–16

Barros E, Nelson SW (2010) Creation of a high-yielding recombinant maize hybrid for the production of a microbicide to prevent HIV-1 transmission. S Afr J Sci 106:5–6

Barta A, Sommergruber K, Thompson D, Hartmuth K, Matzke MA, Matzke AJM (1986) The expression of a nopaline synthase—human growth hormone chimaeric gene in transformed tobacco and sunflower callus tissue. Plant Mol Biol 6:347–357

Belhaj K, Chaparro-Garcia A, Kamoun S, Nekrasov V (2013) Plant genome editing made easy: targeted mutagenesis in model and crop plants using the CRISPR/Cas system. Plant Methods 9:1–10

Bhatia S (2015) Plant tissue culture. In: Bhatia S, Sharma K, Dahiya R, Bera T (eds) 2015. Academic Press, Modern applications of plant biotechnology in pharmaceutical sciences, pp 31–107

Breyer D, Goossens M, Herman P, Sneyers M (2009) Biosafety considerations associated with molecular farming in genetically modified plants. J Med Plant Res 3:825–838

Burnett MJB, Burnett AC (2020) Therapeutic recombinant protein production in plants: challenges and opportunities. Plants People Planet 2:121–132

Buyel JF (2019) Plant molecular farming – integration and exploitation of side streams to achieve sustainable biomanufacturing. Front Plant Sci 9:1893

Buyel JF, Fischer R (2012) Predictive models for transient protein expression in tobacco (Nicotiana tabacum l.) can optimize process time, yield, and downstream costs. Biotechnol Bioeng 109:2575–2588

Cañizares MC, Nicholson L, Lomonossoff GP (2005) Use of viral vectors for vaccine production in plants. Immunol Cell Biol 83:263–270

Canto T (2016) Transient expression systems in plants: potentialities and constraints. In: Vega M (ed) Advanced technologies for protein complex production and characterization. Springer, Cham, pp 287–301

Chapman S, Kavanagh T, Baulcombe D (1992) Potato virus X as a vector for gene expression in plants. Plant J 2:549–557

Chen Q, Lai H, Stahnke J, Leuzinger K, Dent M (2013) Agroinfiltration as an effective and scalable strategy of gene delivery for production of pharmaceutical proteins. Adv Exp Med Biol 1:103

Chung YH, Church D, Koellhoffer EC, Osota E, Shukla S, Rybicki EP, Pokorski JK, Steinmetz NF (2021) Integrating plant molecular farming and materials research for next-generation vaccines. Nat Rev Mater 7:372

Clark M, Maselko M (2020) Transgene biocontainment strategies for molecular farming. Front Plant Sci 11:210

Commandeur U, Twyman R, Fischer R (2003) The biosafety of molecular farming in plants. CAB International

Cosa BD, Moar W, Lee S, Miller M, Daniell H (2001) Overexpression of the Bt cry2Aa2 operon in chloroplasts leads to formation of insecticidal crystals. Nat Biotechnol 19:71–74

Crook NC, Freeman ES, Alper HS (2011) Re-engineering multicloning sites for function and convenience. Nucleic Acids Res 39:e92

Daniell H (1993) Foreign gene expression in chloroplasts of higher plants mediated by tungsten particle bombardment. Methods Enzymol 217:536–548

Daniell H, Datta R, Varma S, Gray S, Lee S (1998) Containment of herbicide resistance through genetic engineering of the chloroplast genome. Nat Biotechnol 16:345–348

Daniell H, Lin CS, Yu M, Chang WJ (2016) Chloroplast genomes: diversity, evolution, and applications in genetic engineering. Genome Biol 17:1–29

Debler JW, Henares BM, Lee RC (2021) Agroinfiltration for transient gene expression and characterisation of fungal pathogen effectors in cool-season grain legume hosts. Plant Cell Rep 40:805–818

Delaney DE (2002) Choice of crop species and development of transgenic product lines. In: Hood EE, Howard JA (eds) Plants as factories for protein production. Kluwer Academic, pp 139–158

Denis J, Acosta-Ramirez E, Zhao Y, Hamelin ME, Koukavica I, Baz M, Abed Y, Savard C, Pare C, Lopez Macias C, Boivin G, Leclerc D (2008) Development of a universal influenza A vaccine based on the M2e peptide fused to the papaya mosaic virus (PapMV) vaccine platform. Vaccine 26:3395–3403

Diamos AG, Hunter JGL, Pardhe MD, Rosenthal SH, Sun H, Foster BC, DiPalma MP, Chen Q, Mason HS (2020) High level production of monoclonal antibodies using an optimized plant expression system. Front Bioeng Biotechnol 7:472

Dimitrov DS (2012) Therapeutic proteins. In: Voynov V, Caravella J (eds) Methods in molecular biology. Humana Press, Totowa, NJ, pp 1–26

Ding SH, Huang LY, Wang YD, Sun HC, Xiang ZH (2006) High-level expression of basic fibroblast growth factor in transgenic soybean seeds and characterization of its biological activity. Biotechnol Lett 28:869–875

Dobrogojski J, Spychalski M, Luciński R, Borek S (2018) Transgenic plants as a source of polyhydroxyalkanoates. Acta Physiol Plant 40:1–17

Donini M, Marusic C (2019) Current state-of-the-art in plant-based antibody production systems. Biotechnol Lett 41:335–346

Dunigan DD, Zaitlin M (1990) Capping of tobacco mosaic virus RNA: analysis of viral-coded guanylyltransferase-like activity. J Biol Chem 265:7779–7786

Edible Cholera Vaccine Made of Powdered Rice | EurekAlert! (n.d.). https://www.eurekalert.org/news-releases/842189

Fanunza E, Frau A, Corona A, Tramontano E (2018) Antiviral agents against Ebola virus infection: repositioning old drugs and finding novel small molecules. Annu Rep Med Chem 51:135–173

Fischer R, Schillberg S (2016) Molecular farming. In: Encyclopedia of applied plant sciences, vol 2, 2nd edn. Elsevier

Fitzmaurice WP, Holzberg S, Lindbo JA, Padgett HS, Palmer KE, Wolfe GM, Pogue GP (2002) Epigenetic modification of plants with systemic RNA viruses. OMICS 6:137–151

Floss DM, Conrad U (2013) Plant molecular pharming, veterinary applications plant molecular pharming veterinary applications. In: Christou P, Savin R, Costa-Pierce BA, Misztal I, Whitelaw CBA (eds) Sustainable food production. Springer, New York, p 135801365

Franken E, Teuschel U, Hain R (1997) Recombinant proteins from transgenic plants. Curr Opin Biotechnol 8:411–416

Garfinkel DJ, Simpson RB, Ream LW, White FF, Gordon MP, Nester EW (1981) Genetic analysis of crown gall: fine structure map of the T-DNA by site-directed mutagenesis. Cell 27:143–153

Gelvin SB (2003) Agrobacterium-mediated plant transformation: the biology behind the "gene-jockeying" tool. Microbiol Mol Biol Rev 67:16

Giritch A, Marillonnet S, Engler C, Van Eldik G, Botterman J, Klimyuk V, Gleba Y (2006) Rapid high-yield expression of full-size IgG antibodies in plants coinfected with noncompeting viral vectors. Proc Natl Acad Sci U S A 103:14701–14706

Gleba YY, Giritch A (2012) Vaccines, antibodies, and pharmaceutical proteins. In: Altman A, Hasegwa PM (eds) Plant biotechnology and agriculture. Academic, pp 465–479

Gleba Y, Marillonnet S, Klimyuk V (2004) Engineering viral expression vectors for plants: the "full virus" and the "deconstructed virus" strategies. Curr Opin Plant Biol 7:182–188

Gleba Y, Klimyuk V, Marillonnet S (2005) Magnifection - a new platform for expressing recombinant vaccines in plants. Vaccine 23:2042–2048

Gleba Y, Klimyuk V, Marillonnet S (2007) Viral vectors for the expression of proteins in plants. Curr Opin Biotechnol 18:134–141

Gleba YY, Tusé D, Giritch A (2014) Plant viral vectors for delivery by Agrobacterium. Curr Top Microbiol Immunol 375:155–192

Gomord V, Fitchette A-C, Menu-Bouaouiche L, Saint-Jore-Dupas C, Plasson C, Michaud D, Faye L (2010) Plant-specific glycosylation patterns in the context of therapeutic protein production. Plant Biotechnol J 8:564–587

Goodin MM, Zaitlin D, Naidu RA, Lommel SA (2008) Nicotiana benthamiana: its history and future as a model for plant–pathogen interactions. Mol Plant Microbe Interact 21:1015–1026

Gunasekaran B, Gothandam KM (2020) A review on edible vaccines and their prospects. Braz J Med Biol Res 53:e8749

Gutiérrez SP, Saberianfar R, Kohalmi SE, Menassa R (2013) Protein body formation in stable transgenic tobacco expressing elastin-like polypeptide and hydrophobin fusion proteins. BMC Biotechnol 13:40

Guy E, Boulain H, Aigu Y, Le Pennec C, Chawki K, Morlière S, Schädel K, Kunert G, Simon JC, Sugio A (2016) Optimization of agroinfiltration in Pisum sativum provides a new tool for studying the salivary protein functions in the pea aphid complex. Front Plant Sci 7:1171

Haq TA, Mason HS, Clements JD, Arntzen CJ (1995) Oral immunization with a recombinant bacterial antigen produced in transgenic plants. Science 268:714–716

Hefferon K (2017) Plant virus expression vectors: a powerhouse for global health. Biomedicine 5:44

Hellens RP, Anne Edwards E, Leyland NR, Bean S, Mullineaux PM (2000) pGreen: a versatile and flexible binary Ti vector for Agrobacterium-mediated plant transformation. Plant Mol Biol 42:819–832

Hiatt A, Pauly M (2006) Monoclonal antibodies from plants: a new speed record. Proc Natl Acad Sci U S A 103:14645–14646

Hiatt A, Cafferkey R, Bowdish K (1989) Production of antibodies in transgenic plants. Nature 342:76–78

Hiatt A, Pauly M, Whaley K, Qiu X, Kobinger G, Zeitlin L (2015) The emergence of antibody therapies for Ebola. Hum Antibodies 23:49–56

Honda Y, Tanigawa E, Tsukihara T, Nguyen DX, Kawabe H, Sakatoku N, Watari J, Sato H, Yano S, Tachiki T, Irie T, Watanabe T, Watanabe T (2019) Stable and transient transformation, and a promoter assay in the selective lignin-degrading fungus, Ceriporiopsis subvermispora. AMB Express 9:92

Hood EE, Woodard SL (2002) Industrial proteins produced from transgenic plants. In: Hood EE, Howard JA (eds) Plants as factories for protein production. Springer, Netherlands, pp 119–135

Horn ME, Woodard SL, Howard JA (2004) Plant molecular farming: systems and products. Plant Cell Rep 22:711–720

Howard EA, Zupan JR, Citovsky V, Zambryski PC (1992) The VirD2 protein of A. tumefaciens contains a C-terminal bipartite nuclear localization signal: implications for nuclear uptake of DNA in plant cells. Cell 68:109–118

Hwang HH, Yu M, Lai EM (2017) *Agrobacterium*-mediated plant transformation: biology and applications. Arabidopsis Book 15:e0186

Ibrahim A, Odon V, Kormelink R (2019) Plant viruses in plant molecular pharming: toward the use of enveloped viruses. Front Plant Sci 10:803

Ikeuchi M, Sugimoto K, Iwase A (2013) Plant callus: mechanisms of induction and repression. Plant Cell 25:3159–3173

Ji X, Zheng L, Liu Y, Nie X, Liu S, Wang Y (2014) A transient transformation system for the functional characterization of genes involved in stress response. Plant Mol Biol Rep 32:732–739

Jin S, Kanagaraj A, Verma D, Lange T, Daniell H (2011) Release of hormones from conjugates: chloroplast expression of β-glucosidase results in elevated phytohormone levels associated with significant increase in biomass and protection from aphids or whiteflies conferred by sucrose esters. Plant Physiol 155:222

Kagale S, Uzuhashi S, Wigness M, Bender T, Yang W, Hossein Borhan M, Rozwadowski K (2012) TMV-Gate vectors: gateway compatible tobacco mosaic virus based expression vectors for functional analysis of proteins. Sci Rep 2:1–10

Kaur M, Manchanda P, Kalia A, Ahmed FK, Nepovimova E, Kuca K, Abd-Elsalam KA (2021) Agroinfiltration mediated scalable transient gene expression in genome edited crop plants. Int J Mol Sci 22:10882

Khamsi R (2005) Potatoes pack a punch against hepatitis B. Nature. https://doi.org/10.1038/news050214-2

Kim TG, Baek MY, Lee EK, Kwon TH, Yang MS (2008) Expression of human growth hormone in transgenic rice cell suspension culture. Plant Cell Rep 27:885–891

King JL, Finer JJ, McHale LK (2015) Development and optimization of agroinfiltration for soybean. Plant Cell Rep 34:133–140

Kirsh D (2017) Artificial spider silk: why you should care about it. Medical Design & Outsourcing. https://www.medicaldesignandoutsourcing.com/artificial-spider-silk-used-medical-applications/

Klimyuk V, Pogue G, Herz S, Butler J, Haydon H (2012) Production of recombinant antigens and antibodies in Nicotiana benthamiana using 'magnifection' technology: GMP-compliant facilities for small- and large-scale manufacturing. Curr Top Microbiol Immunol 375:127–154

Kolotilin I, Kaldis A, Pereira EO, Laberge S, Menassa R (2013) Optimization of transplastomic production of hemicellulases in tobacco: effects of expression cassette configuration and tobacco cultivar used as production platform on recombinant protein yields. Biotechnol Biofuels 6:1–15

Komarnytsky S, Borisjuk NV, Borisjuk LG, Alam MZ, Raskin I (2000) Production of recombinant proteins in tobacco guttation fluid. Plant Physiol 124:927–934

Komori T, Imayama T, Kato N, Ishida Y, Ueki J, Komari T (2007) Current status of binary vectors and superbinary vectors. Plant Physiol 145:1155

Kumar AU, Kadiresen K, Gan WC, Ling APK (2021) Current updates and research on plant-based vaccines for coronavirus disease 2019. Clin Exp Vaccine Res 10:13–23

Kurup VM, Thomas J (2020) Edible vaccines: promises and challenges. Mol Biotechnol 62:79–90

Kwak S, Lew T, Sweeney C, Koman V, Wong M, Bohmert-Tatarev K, Snell K, Seo J, Chua N, Strano M (2019) Chloroplast-selective gene delivery and expression in planta using chitosan-complexed single-walled carbon nanotube carriers. Nat Nanotechnol 14:447–455

Kwon K, Nityanandam R, New J, Daniell H (2013) Oral delivery of bioencapsulated exendin-4 expressed in chloroplasts lowers blood glucose level in mice and stimulates insulin secretion in beta-TC6 cells. Plant Biotechnol J 11:77–86

Lacomme C, Chapman S (2008) Use of potato virus x (PVX)–based vectors for gene expression and virus-induced gene silencing (VIGS). Curr Protoc Microbiol 8:161.1–161.3

Lai H, Engle M, Fuchs A, Keller T, Johnson S, Gorlatov S, Diamond MS, Chen Q, Arntzen CJ (2010) Monoclonal antibody produced in plants efficiently treats West Nile virus infection in mice. Proc Natl Acad Sci U S A 107:2419–2424

Lai H, He J, Engle M, Diamond MS, Chen Q (2012) Robust production of virus-like particles and monoclonal antibodies with geminiviral replicon vectors in lettuce. Plant Biotechnol J 10:95–104

Lau OS, Sun SSM (2009) Plant seeds as bioreactors for recombinant protein production. Biotechnol Adv 27:1015–1022

Leblanc Z, Waterhouse P, Bally J (2021) Plant-based vaccines: the way ahead? Viruses 13:1–13

Lee LY, Gelvin SB (2008) T-DNA binary vectors and systems. Plant Physiol 146(2):325

Leite ML, Sampaio KB, Costa FF, Franco OL, Dias SC, Cunha NB (2019) Molecular farming of antimicrobial peptides: available platforms and strategies for improving protein biosynthesis using modified virus vectors. An Acad Bras Ciênc 91:e20180124

Leuzinger K, Dent M, Hurtado J, Stahnke J, Lai H, Zhou X, Chen Q (2013) Efficient agroinfiltration of plants for high-level transient expression of recombinant proteins. J Vis Exp 77:50521

Li Y, Tran AH, Danishefsky SJ, Tan Z (2019) Chemical biology of glycoproteins: from chemical synthesis to biological impact. In: Shukla AK (ed) . Academic, Methods in enzymology, pp 213–229

Lico C, Benvenuto E, Baschieri S (2015) The two-faced potato virus X: from plant pathogen to smart nanoparticle. Front Plant Sci 6:1009

Low LY, Yang SK, Kok DXA, Ong-Abdullah J, Tan NP, Lai KS (2018) Transgenic plants: gene constructs, vector and transformation method. In: Çelik Ö (ed) New visions in plant science. InTech

Lu X, Yin W, Hu Z (2006) Chloroplast transformation. Methods Mol Biol 318:285–303

Lustig A, Levine AJ (1992) One hundred years of virology. J Virol 66:4629–4631

Ma JKC, Drossard J, Lewis D, Altmann F, Boyle J, Christou P, Cole T, Dale P, van Dolleweerd CJ, Isitt V, Katinger D, Lobedan M, Mertens H, Paul MJ, Rademacher T, Sack M, Hundleby PAC, Stiegler G, Stoger E, Twyman RM, Vclear B, Fischer R (2015) Regulatory approval and a first-in-human phase I clinical trial of a monoclonal antibody produced in transgenic tobacco plants. Plant Biotechnol J 13:1106–1120

Makhzoum A, Benyammi R, Moustafa K, Trémouillaux-Guiller J (2014) Recent advances on host plants and expression cassettes' structure and function in plant molecular pharming. BioDrugs 28:145–159

Maliga P (2017) Transformation in plastids. In: Thomas B, Murray BG, Murphy DJ (eds) Encyclopedia of applied plant sciences. Academic, pp 255–261

Mao Y, Botella JR, Liu Y, Zhu JK (2019) Gene editing in plants: progress and challenges. Natl Sci Rev 6:421–437

Mardanova ES, Blokhina EA, Tsybalova LM, Peyret H, Lomonossoff GP, Ravin NV (2017) Efficient transient expression of recombinant proteins in plants by the novel pEff vector based on the genome of potato virus X. Front Plant Sci 8:247

Marillonnet S, Giritch A, Gils M, Kandzia R, Klimyuk V, Gleba Y (2004) In planta engineering of viral RNA replicons: efficient assembly by recombination of DNA modules delivered by Agrobacterium. Proc Natl Acad Sci U S A 101:6852–6857

Marillonnet S, Thoeringer C, Kandzia R, Klimyuk V, Gleba Y (2005) Systemic *Agrobacterium tumefaciens*-mediated transfection of viral replicons for efficient transient expression in plants. Nat Biotechnol 23:718–723

Marintcheva B (2018) Viral tools for protein expression and purification. In: Marintcheva B (ed) Harnessing the power of viruses. Academic, pp 103–131

Matsumoto S, Ikura K, Ueda M, Sasaki R (1995) Characterization of a human glycoprotein (erythropoietin) produced in cultured tobacco cells. Plant Mol Biol 27:1163–1172

McCormick AA, Reinl SJ, Cameron TI, Vojdani F, Fronefield M, Levy R, Tusé D (2003) Individualized human scFv vaccines produced in plants: humoral anti-idiotype responses in vaccinated mice confirm relevance to the tumor Ig. J Immunol Methods 278:95–104

McCormick AA, Reddy S, Reinl SJ, Cameron TI, Czerwinkski DK, Vojdani F, Hanley KM, Garger SJ, White EL, Novak J, Barrett J, Holtz RB, Tusé D, Levy R (2008) Plant-produced idiotype vaccines for the treatment of non-Hodgkin's lymphoma: safety and immunogenicity in a phase I clinical study. Proc Natl Acad Sci U S A 105:10131–10136

McCouch SR, Herdt RW (2004) Potentials and limitations of biotechnology in rice background and introduction. FAO Rice Conference

Menary J, Hobbs M, Mesquita de Albuquerque S, Pacho A, Drake PMW, Prendiville A, Ma JKC, Fuller SS (2020) Shotguns vs Lasers: identifying barriers and facilitators to scaling-up plant molecular farming for high-value health products. PLoS One 15:e0229952

Menassa R, Ahmad A, Joensuu JJ (2012) Transient expression using agroinfiltration and its applications in molecular farming. In: Wang A, Ma S (eds) Molecular farming in plants: recent advances and future prospects. Springer, Netherlands, pp 183–198

Molina A, Veramendi J, Hervas-Stubbs S (2005) Induction of neutralizing antibodies by a tobacco chloroplast-derived vaccine based on a B cell epitope from canine parvovirus. Virology 342:266–275

Monoclonal antibody drugs for cancer: how they work—Mayo Clinic. (n.d.). https://www.mayoclinic.org/diseases-conditions/cancer/in-depth/monoclonal-antibody/art-20047808

Moon KB, Park JS, Park YI, Song IJ, Lee HJ, Cho HS, Jeon JH, Kim HS (2020) Development of systems for the production of plant-derived biopharmaceuticals. Plants 9:30

Morgenfeld M, Segretin ME, Wirth S, Lentz E, Zelada A, Mentaberry A, Gissmann L, Bravo-Almonacid F (2009) Potato virus X coat protein fusion to human papillomavirus 16 E7 oncoprotein enhance antigen stability and accumulation in tobacco chloroplast. Mol Biotechnol 43:243–249

Morgenfeld M, Lentz E, Segretin ME, Alfano EF, Bravo-Almonacid F (2014) Translational fusion and redirection to thylakoid lumen as strategies to enhance accumulation of human papillomavirus E7 antigen in tobacco chloroplasts. Mol Biotechnol 56:1021–1031

Moustafa K, Makhzoum A, Trémouillaux-Guiller J (2016) Molecular farming on rescue of pharma industry for next generations. Crit Rev Biotechnol 36:840–850

Musiychuk K, Sivalenka R, Jaje J, Bi H, Flores R, Shaw B, Jones RM, Golovina T, Schnipper J, Khandker L, Sun R, Li C, Kang L, Voskinarian-Berse V, Zhang X, Streatfield S, Hambor J, Abbot S, Yusibov V (2013) Plant-produced human recombinant erythropoietic growth factors support erythroid differentiation in vitro. Stem Cells Dev 22:2326–2340

Norkunas K, Harding R, Dale J, Dugdale B (2018) Improving agroinfiltration-based transient gene expression in Nicotiana benthamiana. Plant Methods 14:1–14

Obembe OO, Popoola JO, Leelavathi S, Reddy SV (2011) Advances in plant molecular farming. Biotechnol Adv 29:210–222

Ou J, Guo Z, Shi J, Wang X, Liu J, Shi B, Guo F, Zhang C, Yang D (2014) Transgenic rice endosperm as a bioreactor for molecular pharming. Plant Cell Rep 33:585–594

Pereira EO, Tsang A, McAllister TA, Menassa R (2013) The production and characterization of a new active lipase from Acremonium alcalophilum using a plant bioreactor. Biotechnol Biofuels 6:111

Permyakova NV, Marenkova TV, Belavin PA, Zagorskaya AA, Sidorchuk YV, Uvarova EA, Kuznetsov VV, Rozov SM, Deineko EV (2021) Assessment of the level of accumulation of the dIFN protein integrated by the knock-in method into the region of the histone H3.3 gene of Arabidopsis thaliana. Cell 10:2137

Plchova H, Moravec T, Hoffmeisterova H, Folwarczna J, Cerovska N (2011) Expression of Human papillomavirus 16 E7ggg oncoprotein on N- and C-terminus of Potato virus X coat protein in bacterial and plant cells. Protein Expr Purif 77:146–152

Pogue GP, Lindbo JA, Garger SJ, Fitzmaurice WP (2002) Making an ally from an enemy: plant virology and the new agriculture. Annu Rev Phytopathol 40:45–74

Qiu X, Wong G, Audet J, Bello A, Fernando L, Alimonti JB, Fausther-Bovendo H, Wei H, Aviles J, Hiatt E, Johnson A, Morton J, Swope K, Bohorov O, Bohorova N, Goodman C, Kim D, Pauly MH, Velasco J, Pettitt J, Olinger GG, Whaley K, Xu B, Strong JE, Zeitlin L, Kobinger GP (2014) Reversion of advanced Ebola virus disease in nonhuman primates with ZMapp. Nature 514:47–53

Rahayu S, Roostika I, Bermawie N (2016) The effect of types and concentrations of auxins on callus induction of Centella asiatica. Nusantara Biosci 8:283–287

Rajewska M, Wegrzyn K, Konieczny I (2012) AT-rich region and repeated sequences – the essential elements of replication origins of bacterial replicons. FEMS Microbiol Rev 36:408–434

Ramessar K, Sabalza M, Capell T, Christou P (2008) Maize plants: an ideal production platform for effective and safe molecular pharming. Plant Sci J 174:409–419

Ramkumar TR, Lenka SK, Arya SS, Bansal KC (2020) A short history and perspectives on plant genetic transformation. Methods Mol Biol 2124:39–68

Řepková J (2010) Potential of chloroplast genome in plant breeding. Czech J Genet Plant Breed 46:103–113

Rifkind D, Freeman GL (2005) Tobacco mosaic virus. The Nobel Prize Winning Discoveries in Infectious Diseases. pp 81–84

Rivera AL, Gómez-Lim M, Fernández F, Loske AM (2012) Physical methods for genetic plant transformation. Phys Life Rev 9:308–345

Roudsari FM, Salmanian AH, Mousavi A, Sohi HH, Jafari M (2009) Regeneration of glyphosate-tolerant Nicotiana tabacum after plastid transformation with a mutated variant of bacterial aroA gene. Iran J Biotechnol 7:247–253

Ruggiero F, Exposito JY, Bournat P, Gruber V, Perret S, Comte J, Olagnier B, Garrone R, Theisen M (2000) Triple helix assembly and processing of human collagen produced in transgenic tobacco plants. FEBS Lett 469:132–136

Saxena J, Rawat S (2014) Edible vaccines. In: Ravi I, Baunthiyal Saxena MJ (eds) Advances in biotechnology. Springer, New Dehli, pp 207–226

Scheller J, Gührs KH, Grosse F, Conrad U (2001) Production of spider silk proteins in tobacco and potato. Nat Biotechnol 19:573–577

Schillberg S, Finnern R (2021) Plant molecular farming for the production of valuable proteins – critical evaluation of achievements and future challenges. J Plant Physiol 258–259:153359

Schillberg S, Raven N, Spiegel H, Rasche S, Buntru M (2019) Critical analysis of the commercial potential of plants for the production of recombinant proteins. Front Plant Sci 10:720

Schoberer J, Strasser R (2018) Plant glyco-biotechnology. Semin Cell Dev Biol 80:133–141

Shanmugaraj B, Bulaon CJI, Phoolcharoen W (2020) Plant molecular farming: a viable platform for recombinant biopharmaceutical production. Plants 9:842

Shanmugaraj B, Bulaon CJI, Malla A, Phoolcharoen W (2021) Biotechnological insights on the expression and production of antimicrobial peptides in plants. Molecules 26:4032

Sharma AK, Sharma MK (2009) Plants as bioreactors: recent developments and emerging opportunities. Biotechnol Adv 27:811–832

Sheludko Y (2008) *Agrobacterium*-mediated transient expression as an approach to production of recombinant proteins in plants. Recent Pat Biotechnol 2:198–208

Shouket HA, Ameen I, Tursunov O, Kholikova K, Pirimov O, Kurbonov N, Ibragimov I, Mukimov B (2020) Study on industrial applications of papain: a succinct review. Earth Environ Sci 614:012171

Simmons CW, Vandergheynst JS, Upadhyaya SK (2009) A model of Agrobacterium tumefaciens vacuum infiltration into harvested leaf tissue and subsequent in planta transgene transient expression. Biotechnol Bioeng 102:965–970

Simpson RB, Spielmann A, Margossian L, McKnight TD (1986) A disarmed binary vector from Agrobacterium tumefaciens functions in Agrobacterium rhizogenes. Plant Mol Biol 6:403–415

Singh R, Shaik S, Negi BS, Rajguru JP, Patil PB, Parihar AS, Sharma U (2020) Non-Hodgkin's lymphoma: a review. J Family Med Prim Care 9:1834

Singh AA, Pillay P, Tsekoa TL (2021) Engineering approaches in plant molecular farming for global health. Vaccine 9:1270

Sparrow PAC, Irwin JA, Dale PJ, Twyman RM, Ma JKC (2007) Pharma-Planta: road testing the developing regulatory guidelines for plant-made pharmaceuticals. Transgenic Res 16:147–161

Stepanenko AA, Heng HH (2017) Transient and stable vector transfection: pitfalls, off-target effects, artifacts. Mutat Res Rev Mutat Res 773:91–103

Stone WL, Leavitt L, Varacallo M (2021) Physiology, growth factor. In: StatPearls. StatPearls Publishing

Svab Z, Hajdukiewicz P, Maliga P (1990) Stable transformation of plastids in higher plants. Proc Natl Acad Sci U S A 87:8526–8530

Tabll AA, Shahein YE, Omran MM, Elnakib MM, Ragheb AA, Amer KE (2021) A review of monoclonal antibodies in COVID-19: role in immunotherapy, vaccine development and viral detection. Hum Antibodies 29:179–191

Tang G, Qin J, Dolnikowski GG, Russell RM, Grusak MA (2009) Golden rice is an effective source of vitamin A. Am J Clin Nutr 89:1776–1783

Terashima M, Murai Y, Kawamura M, Nakanishi S, Stoltz T, Chen L, Drohan W, Rodriguez RL, Katoh S (1999) Production of functional human α1-antitrypsin by plant cell culture. Appl Microbiol Biotechnol 52:516–523

The Golden Rice Story (n.d.) Golden Rice Humanitarian Board. https://www.goldenrice.org/Content1-Who/who2_history.php

The PREVAIL II Writing Group for the Multi-National PREVAIL II Study Team (2016) A randomized, controlled trial of ZMapp for Ebola virus infection. N Engl J Med 375:1448

Thomas DR, Penney CA, Majumder A, Walmsley AM (2011) Evolution of plant-made pharmaceuticals. Int J Mol Sci 12:3220

Tremblay R, Wang D, Jevnikar AM, Ma S (2010) Tobacco, a highly efficient green bioreactor for production of therapeutic proteins. Biotechnol Adv 28:214–221

Tusé D, Nandi S, McDonald KA, Buyel JF (2020) The emergency response capacity of plant-based biopharmaceutical manufacturing-what it is and what it could be. Front Plant Sci 11:1573

Twyman R, Schillberg S, Fischer R (2013) Optimizing the yield of recombinant pharmaceutical proteins in plants. Curr Pharm Des 19:5486–5494

Vaghchhipawala Z, Rojas CM, Senthil-Kumar M, Mysore KS (2011) Agroinoculation and agroinfiltration: simple tools for complex gene function analyses. Methods Mol Biol 678:65–76

Van Regenmortel MHV (2008) Tobacco mosaic virus. In: Mahy BWJ, Van Regenmortel MHV (eds) Encyclopedia of virology, 3rd edn. Academic, pp 54–60

Wang A, Ma S (2012) Molecular farming in plants: recent advances and future prospects. Springer. https://doi.org/10.1007/978-94-007-2217-0

Wu W, Du K, Kang X, Wei H (2021) The diverse roles of cytokinins in regulating leaf development. Hortic Res 8:118

Xu S, Gavin J, Jiang R, Chen H (2017) Bioreactor productivity and media cost comparison for different intensified cell culture processes. Biotechnol Prog 33:867–878

Xu J, Towler M, Weathers PJ (2018) Platforms for plant-based protein production. In: Pavlov A, Bley T (eds) Bioprocessing of plant in vitro systems. Springer International Publishing AG, pp 509–548

Yadava P, Abhishek A, Singh R, Singh I, Kaul T, Pattanayak A, Agrawal PK (2017) Advances in maize transformation technologies and development of transgenic maize. Front Plant Sci 7:1949

Yang Y, Wang D, Zhang X, Fang J, Shen Z, Lin C (2014) Transgenic rice as bioreactor for production of the Candida antarctica lipase B. Plant Biotechnol J 12:963–970

Yu Y, Yu PC, Chang WJ, Yu K, Lin CS (2020) Plastid transformation: how does it work? Can it be applied to crops? What can it offer? Int J Mol Sci 21:1–21

Yusibov V, Rabindran S, Commandeur U, Twyman RM, Fischer R (2006) The potential of plant virus vectors for vaccine production. Drugs R D 7:203–217

Zhang YF, Li DP, Jin X, Huang Z (2014) Fighting Ebola with ZMapp: spotlight on plant-made antibody. Sci China Life Sci 57:987–988

Zhao H, Tan Z, Wen X, Wang Y (2017) An improved syringe agroinfiltration protocol to enhance transformation efficiency by combinative use of 5-Azacytidine, Ascorbate Acid and Tween-20. Plants 6:444

Zhou JY, Cheng LQ, Zheng XJ, Wu JX, Shang SB, Wang JY, Chen JG (2004) Generation of the transgenic potato expressing full-length spike protein of infectious bronchitis virus. J Biotechnol 111:121–130

Zimran A, Gonzalez-Rodriguez DE, Abrahamov A, Cooper PA, Varughese S, Giraldo P, Petakov M, Tan ES, Chertkoff R (2018) Long-term safety and efficacy of taliglucerase alfa in pediatric Gaucher disease patients who were treatment-naïve or previously treated with imiglucerase. Blood Cells Mol Dis 68:163–172

Chapter 2
Plant Molecular Pharming: Opportunities, Challenges, and Future Perspectives

Benita Ortega-Berlanga and Tomasz Pniewski

Abstract Over the last decades, plant-based expression systems have emerged as a novel platform for the production of recombinant proteins due to a number of advantages compared to the conventional established expression systems based on bacteria, yeast, or mammalian cell cultures. These advantages include low cost, high scalability, considerable productivity, rapid production, safety, capacity to produce multimeric or glycosylated proteins, and for certain biopharmaceuticals the option of distribution at ambient temperature and needle-free oral administration. Several molecular pharming products have reached the market-ready stage, but the number of success stories has been limited by industrial inertia driven by regulatory hurdles that create barriers to translation. This chapter discusses the advantages and opportunities offered by the use of plant-based expression systems for biopharmaceutical production. The plant-based systems appear as a meaningful alternative during global economic and ecological crisis, especially important in developing countries. The high cost of therapeutics produced by existing methods promotes consideration of the challenges and potential future directions to enable the broader application of production platforms based on plants.

Keywords Plant-based biopharmaceuticals · Plant expression systems · Plant molecular pharming · Stable transformation · Transient expression

B. Ortega-Berlanga
Plant Biotechnology Team, Institute of Plant Genetics of the Polish Academy of Sciences, Poznań, Poland

División de Biología Molecular, Instituto Potosino de Investigación Científica y Tecnológica, San Luis Potosí, México
e-mail: benita.ortega@ipicyt.edu.mx

T. Pniewski (✉)
Plant Biotechnology Team, Institute of Plant Genetics of the Polish Academy of Sciences, Poznań, Poland
e-mail: tpni@igr.poznan.pl

C. Kole et al. (eds.), *Tools & Techniques of Plant Molecular Farming*, Concepts and Strategies in Plant Sciences, https://doi.org/10.1007/978-981-99-4859-8_2

Abbreviations

ADH	Alcohol dehydrogenase
AEC	Anion exchange chromatography
AFC	Affinity chromatography
AIDS	Acquired immunodeficiency syndrome
AMV	Alfalfa mosaic virus
ApoA1	Apolipoprotein A-I
CaMV35S	Cauliflower mosaic virus 35S RNA promoter
CEC	Cation exchange chromatography
CHO	Chinese hamster ovary cell line
CMV	Cucumber mosaic virus
COVID-19	Coronavirus disease 2019
CP	Coat protein
CPMV	Cowpea mosaic virus
CRISPR/Cas9	Clustered regularly interspaced short palindromic repeats, the process carried out by Cas9 complex
Ct	Calcitonin
DNA	Deoxyribonucleic acid
EGF	Epidermal growth factor
ELPs	Elastin-like polypeptides
EPO	Erythropoietin
ER	Endoplasmic reticulum
ETEC	Enterotoxigenic *Escherichia coli*
EU	European Union
EVD	Ebola virus disease
Fab	Fragment antigen-binding region
Fc	Fragment crystallizable
FDA	Food and Drug Administration
FW	Fresh weight
GCP	Good clinical practice
Glb-1	Globulin protein
GLP	Good laboratory practice
GluB-1,-4	Glutelin proteins
GM	Genetically modified
GMP	Good manufacturing practice
GST	Glutathione-*S*-transferase
HBcAg	Hepatitis B core antigen
HBV	Hepatitis B virus
HEK293	Human embryonic kidney cell line
hGH	Human growth hormone
HIC	Hydrophobic interaction chromatography
HIV	Human immunodeficiency virus
HPV	Human papillomavirus
HSP	Heat-shock protein

IEC	Ion-exchange chromatography
Ig	Immunoglobulin
ILs	Interleukins
kDa	Kilo daltons
kg	Kilogram
LicKM	Lichenase, 1,3-1,4-glucanase
mAbs	Monoclonal antibodies
MBP	Maltose-binding protein
MMC	Mixed-mode chromatography
MP	Movement protein
NDV	Newcastle disease virus
NOS	Nopaline synthase
OPRX-106	Tumor necrosis factor—Fc fusion, the form of TNF produced by Protalix Biotherapeutics
OST	Oligosaccharyl transferase
PBs	Protein bodies
PTGS	Posttranscriptional gene silencing
PVX	Potato virus X
QVLP	Quadrivalent influenza vaccine
RNA	Ribonucleic acid
SARS-CoV-2	Severe acute respiratory syndrome coronavirus 2
scFvs	Single-chain antibody variable-region fragments
SEC	Size-exclusion chromatography
SUMO	Small ubiquitin-related modifier
TGS	Transcriptional gene silencing
TMV	Tobacco mosaic virus
TRX	Thioredoxin
TSP	Total soluble protein
USA	*United States* of America
USDA	United States Department of Agriculture
UTRs	Untranslated regions
VHHs	Heavy chain-only antibodies
VLPs	Viruslike particles
WHO	World Health Organization

2.1 Introduction

Genetic engineering has opened up new opportunities for using plants as production factories for recombinant proteins ranging from pharmaceutical therapeutics to non-pharmaceutical products. The first plant-produced protein of pharmaceutical interest was human growth hormone, produced in transgenic tobacco in 1986 (Barta et al. 1986). In 1989, a functional murine full-size IgG1 antibody was the first antibody produced in transgenic tobacco plants (Hiatt et al. 1989). Later, the avidin and β-glucuronidase were produced in transgenic maize plants (Hood et al. 1997, 1999).

Both proteins are used as research reagents, and avidin is still sold by Merck via its subsidiary MilliporeSigma. Since then, over 100 recombinant proteins such as human serum proteins, growth regulators, vaccines, cytokines, antibodies, and enzymes have been produced in different plant species, and several of them have reached the late stages of commercial development (Lienard 2007). Importantly, in 2012, "Elelyso," a recombinant enzyme produced in carrot cells and commercialized by Protalix Biotherapeutics (Karmiel, Israel), was approved by the Food and Drug Administration (FDA) for treatment of Gaucher's disease (Tekoah et al. 2015). Other two products which have been licensed are (1) the plant-made scFV mAb used in the production of a recombinant HBV vaccine in Cuba and (2) the Newcastle disease virus (NDV) vaccine for poultry approved by the US Department of Agriculture (USDA) (Naderi and Baratali 2015). Other plant-produced biopharmaceuticals have been tested in preclinical and clinical trials, such as OPRX-106, tumor necrosis factor, produced by Protalix Biotherapeutics; vaccines against lymphomas developed by ICON Genetics (Halle, Germany); Covifenz®, VLP-based vaccine against SARS-CoV-2 produced by Medicago (Quebec City, Canada); or ZMapp antibodies against Ebola virus, provided by Kentucky BioProcessing (Owensboro, USA), subsidiary of Mapp Biopharmaceutical (San Diego, USA) (Tusé et al. 2015; Almon et al. 2021). Many more plant-based biopharmaceuticals are subjects of R&D programs and scientific research. Plant systems can be used for the production of various monoclonal antibody forms (full size, scFv, Fab, etc.), vaccines, adjuvants, immunomodulators, hormones, and other proteins of biomedical importance, including their complexes such as virus like particles (VLPs) for further processing, and then uses as versatile scaffolds, vehicles, and transporters (Moon et al. 2019; Shanmugaraj et al. 2020).

A unique feature of plant-based protein expression is the diversity of plant species and systems used for production. Whereas conventional biopharmaceutical manufacturing involves a small number of well-established platforms, such as the bacterium *Escherichia coli*, the yeast *Saccharomyces cerevisiae*, Chinese hamster ovary (CHO) cells, murine NSO and Sp2/0 cells, and human embryonic kidney (HEK293) cells (Schillberg and Finnern 2021), many different plant species have been proposed as production platforms, including many cereal crops, fruits, and vegetables such as tobacco, rice, maize, soybean, lettuce, tomato, carrot, potato, alfalfa, suspension cultures, microalgae, and others. The choice of plant species for production of a given biopharmaceutical depends usually on its purpose. Biopharmaceuticals considered for oral administration can be produced in various transgenic or—if possible—transplastomic edible plants, but also in tobacco or carrot suspension cultures, and then delivered after minimal processing, such as lyophilization. However, most biopharmaceuticals are required to be purified before parenteral or transmucosal administration; hence, they are usually produced in high-yield systems based on tobacco species or suspension cultures.

Nicotiana tabacum and its close relative *N. benthamiana* are the two species most commonly used for the expression of recombinant proteins due to their easy genetic transformation and rapid development as well as well-established systems of viral infection or *Agrobacterium* infiltration, which are the foundation of nowadays commonly exploited technologies of transient expression (Spiegel et al. 2018; Sainsbury 2020). Importantly, tobacco is a nonfood, which minimizes regulatory barriers by eliminating the risk of plant-made recombinant proteins entering the food supply. Tobacco is usually used for production of biopharmaceuticals which are extracted and purified, due to the presence of toxic alkaloids in tobacco vegetative tissues. However, there are also low-alkaloid varieties that can be considered for the production of orally administered pharmaceutical proteins. Plant systems may be utilized as whole plants or in organ cultures—such as hairy roots, and cell suspension cultures, especially commonly used BY-2 tobacco cell line or established carrot suspension culture, e.g., exploited in ProCellEx™ platform (Protalix Biotherapeutics, Karmiel, Israel).

The expression strategies used in molecular pharming are of the stable or transient type (Burnett and Burnett 2020). The first method can be further subdivided into techniques involving *Agrobacterium*-mediated transformation of the nuclear genome or the biolistic (particle bombardment) method enabling stable modification of nuclear or chloroplast (plastid) genomes, whereas transient expression may be achieved using viral vectors or by agroinfiltration, i.e., infiltration with suspension of *Agrobacterium* carrying dedicated vectors (Moon et al. 2019).

Stable methods comprise the transgene insertion into nuclear or chloroplast genome. A majority of the recombinant proteins to date have been produced by nuclear *Agrobacterium*-mediated transformation, which takes advantage of the bacteria's ability to transfer DNA segments into the host genome, thereby conferring heritable traits to the progeny (Ortega-Berlanga and Pniewski 2022). The main advantages of the production system based on transgenic plants are high scalability, low cost, and flexibility regarding the expression of recombinant proteins in various hosts or plant organs depending on a specific purpose, comprised by type and function of a protein, its processing, distribution, storage, and administration route. For example, several proteins can be produced in dry seeds of cereals, which prolongs their half-life since no cold chain is required for their conservation (Moon et al. 2019). Moreover, a target protein can be produced in an edible plant, which could be considered as a feedstock for oral vaccine, useful especially in developing countries thanks to the lower production costs and no specialist equipment or facilities required for their storage and application. However, as the result of *Agrobacterium*-mediated transformation, the insertion of a transgene into genomic DNA via heterologous recombination is random, which can lead to positional and "gene dosage" effects, both inducing silencing. Accompanying effects such as disruption of essential genes or epigenetic alterations, although rarer, can also occur. Hence, an unpredictable level of a biopharmaceutical expression in obtained transgenic plants is the main problem, which requires labor- and time-consuming selection of the most efficient producer lines (Ortega-Berlanga and Pniewski 2022). This hindrance can be partially omitted by appropriate construction of a transgene encoding a protein, which comprises selection of promoters and other regulatory sequences, codon usage, and signaling/targeting peptides resulting in protein deposition in specific compartments or secretion. Nevertheless, the recombinant protein production in transgenic plants remains relatively low, usually few to several tens of micrograms per gram of fresh weight (μg/g FW). Higher content of a biopharmaceutical can be achieved using long-time breeding techniques like selection, backcrossing, introgression, and hybridization of selected lines (Hayden et al. 2015). Another concern is the potential risk of genetically modified plants crossing with native species or food crops, although in case of biopharmaceutical production, transgenic plants are usually cultivated in controlled contained facilities.

Alternatively, although much less frequently, plants can be transformed using the biolistic method to obtain transplastomic plants with modified plastid DNA. In contrast to *Agrobacterium*-mediated nuclear transformation, transgenes are inserted into spacers—specific sites in plastid genome (plastome) between functional genes—via homologous recombination between defined plastome sequences and identical flanking sequences carried by a vector and encompassing a factual gene of interest (Jin and Daniell 2015). This method represents additional advantages over nuclear transformation, which includes a high recombinant protein yield because of the high number of DNA copies per plastid and then plastids per a vegetal cell, expression of polycistronic genes, and natural transgene containment since plastid genes are not usually transmitted through pollen and lack of position effect/gene silencing (Jin and Daniell 2015; Burnett and Burnett 2020). Bombardment devices deliver to cells the nucleic acids adsorbed on tungsten or more often on gold spherical particles, usually between 0.6 and 1 μm in diameter by high-pressure shots of helium (Canto 2016). This method, however, also causes a degree of mechanical damage to the targeted tissue, and only some of the cells where the metal particles are introduced survive the mechanical stress and express the exogenous genes (Canto 2016). The extent of tissue damage and the number of cells that express these genes will depend on parameters such as the type of bombardment gun used, how tender-leaved the plant species is, the distance of the device to the leaf surface, the type of particle used, or the pressure used for shooting. Some devices currently under use are available commercially, such as the Bio-Rad Helios® Gene Gun system, while others are manufactured from researchers themselves. The main advantage of biolistic bombardment over other *Agrobacterium*-mediated transient or nuclear expression systems is that it delivers nucleic acids into live plant cells through a mechanical process that does not require interaction between the plant species and a compatible biological agent, such as bacteria or viruses. The technique requires adapting bombardment conditions to the specific plant host, to achieve its maximum efficiency. However, the use of this technique to produce transplastomic plants is still limited to very few species like tobacco, tomato, carrot, or lettuce. Progress and expanding of species spectrum can be expected though, since protein production can reach even several hundred μg/g FW.

Transient expression systems possess multiple advantages compared with stable methods. These offer rapid and high protein expression within a few days and hence are considered as a suitable convenient platform, especially for the rapid production of vaccine antigens or antibodies during a pandemic situation. The first approach of this system had been based on the use of vectors derived from plant viruses to deliver foreign genes without integration into the plant genome. The main viruses used here are those of the RNA type, such as tobacco mosaic virus (TMV), potato virus X (PVX), alfalfa mosaic virus (AlMV), cucumber mosaic virus (CMV), and cowpea mosaic virus (CPMV) (Laere et al. 2016). Generally, the limit imposed by "full virus" vectors on the size of the insert is about 1 kb (Gleba et al. 2005). Another, at present more common approach is agroinfiltration—delivery of hybrid vectors, e.g., magnICON system or pEAQ vector series composed of elements coming from plant viruses and *Agrobacterium* vectors (Sainsbury et al. 2009; Peyret and Lomonossoff 2015). Thanks to that, this system has the capacity to express longer genes, up to 2.3 kb inserts or up to 80 kDa proteins (Gleba et al. 2005). The agroinfiltration strategy combines the advantages of three biological systems: the speed and expression level/yield of the virus, the transfection efficiency of *Agrobacterium*, and the posttranslational capabilities and low production cost of plants (Gleba et al. 2005). Depending on the vector used, the host organism, and the initial density of *Agrobacterium*, the process takes from 4 to 10 days and the expression levels, depending on the nature of the protein encoded by the gene of interest, can reach a yield of up to 5 g recombinant protein per kg of fresh leaf biomass or over 50% of total soluble protein. This process is in essence an infiltration of leaves or whole mature plants with a diluted suspension of *Agrobacterium* carrying vectors with T-DNAs encoding RNA replicons (Gleba et al. 2005). The infiltration of plants is usually achieved by vacuum infiltration for 10–30 s or by syringe infiltration (Fujiki et al. 2008). The main advantage of agroinfiltration is that most plant cells inside the area infiltrated with the bacterial culture will receive the T-DNAs and express the desired genes. Additionally, in some type of vectors, e.g., magnICON or TRBO, contained MP coding sequence (movement protein) allows to spread a vector from cell to cell and consequently systemic expression (Peyret and Lomonossoff 2015). One important factor that constrains transient expression from T-DNAs is their being recognized as foreign by the plant, which elicits an RNA-based silencing response that depresses both the steady-state levels of the transcript messenger RNA encoded by the T-DNA and those of the protein product it may encode (Johansen and Carrington 2001). To neutralize this silencing response and enhance the transient steady-state levels of T-DNA-encoded genes, co-expression of proteins functioning as RNA silencing "suppressors" is used routinely. Most if not all plant viruses express at least one suppressor factor (Canto 2016). All these characteristics allow the simple scale-up of the procedure by increasing the infiltrated surfaces to produce large amounts of the T-DNA-derived product/s, thus opening the possibility of large-scale applications (Canto 2016).

While agroinfiltration is the main method of transient expression which makes possible production of biopharmaceutical full-size proteins and their complexes, the use of plant viruses in molecular pharming is being intensively developed and their importance is constantly growing. Virus production is routine and easy; hence, virus-derived particles—viroparticles—can be available in virtually unlimited quantities and, after processing and/or functionalization, versatilely used. Although the capacity of proteins genetically fused to viral coat protein is limited (see above), exploitation of viroparticles as vaccine scaffolds is still valid, since in many cases even truncated epitopes are sufficient for effective immunization (Balke and Zeltins 2019). In turn, decoration of coat protein with small linkers enables attaching

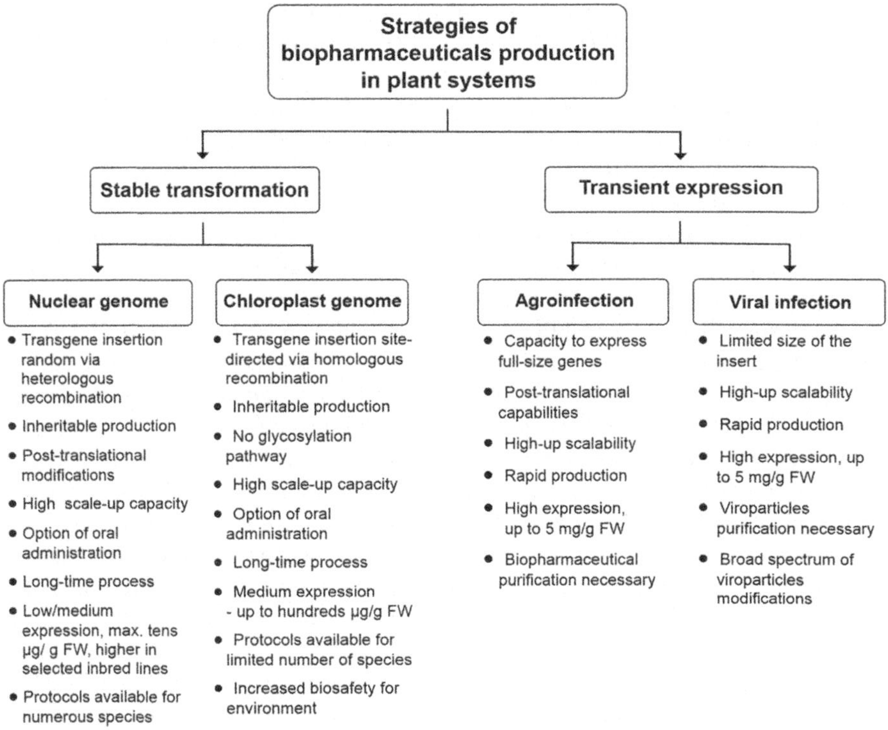

Fig. 2.1 Strategies used for recombinant protein production in plants

in vitro covalently or non-covalently larger epitopes or antibodies for active or passive immunization, respectively. Viroparticles bearing linkers can also be used as carriers of a plethora of various ligands for auxiliary or indirect biomedical purposes—as fluorescence or imaging agents, for antibody purification, biocatalysis, and other applications (Ibrahim et al. 2019; Venkataraman and Hefferon 2021). Finally, plant viruses as module nanoparticles can be dis- and reassembled and then derived VLPs are used as adjuvants or loaded with drugs, contrasting agents, or other nanoparticles for targeting, infection and cancer therapy, and imaging or biosensing (Eiben et al. 2019; Evtushenko et al. 2020). It should be added that also VLPs originated from non-plant viruses but produced in plants, e.g., formed by hepatitis B core antigen (HBcAg), can be used for above-described purposes, beside typical use as vaccines or adjuvants (Rybka et al. 2019; Pyrski et al. 2019). Figure 2.1 summarizes the plant expression approaches.

2.2 Opportunities of Plant Molecular Pharming

Expression of recombinant proteins in plant-based systems has a number of advantages over the traditional expression systems (Burnett and Burnett 2020). The main advantages are the following:

Low Cost. Substantially lower production costs than with microbial bioreactors and especially when compared with mammalian cell lines or transgenic animals (Daniell et al. 2001; Horn et al. 2004). Both capital and running costs are significantly lower than those of cell-based production systems because there is no need for fermenters or the skilled personnel to run them. It is estimated that recombinant proteins can be produced in plants at 2–10% of the cost of microbial fermentation systems and at 0.1% of the cost of mammalian cell cultures, although this depends on the product yield (Giddings 2001). Chloroplast transformation and transient expression offer better yields compared with nuclear expression. However, even yields of 0.1–1.0% total soluble protein (TSP), the typical levels observed for the production of pharmaceutical proteins using nuclear methods, are sufficiently competitive with other expression systems to make plants economically viable (Hood et al. 2002).

High Scalability. Cultivation of transgenic plants in particular can, in theory, be expanded to the agricultural scale. Moreover, infrastructure and expertise already exist for the planting, harvesting, and processing of plant material (Buyel 2019). The scale of plant-based production can be modulated rapidly in response to market demand simply by using more or less space (greenhouse facilities) as required, whereas fermentation systems and transgenic animals have limited potential in this respect. The speed of scale-up is also important. It can take several years to achieve tenfold scale-up in a herd of transgenic sheep using natural breeding cycles, but a field of transgenic plants can be scaled up more than 1000-fold in a single generation owing to the prolific seed output (Schillberg et al. 2002). Moreover transgenic plant lines can be stored indefinitely and inexpensively as seeds (see also Chaps. 5 and 6).

Optional Oral Administration of Plant-Produced Biopharmaceuticals. Nutraceuticals can be directly consumed, as Golden Rice enriched in β-carotene (Beyer et al. 2002), currently provided by Syngenta (Basel, Switzerland), or rice seeds containing lactoferrin, lysozyme, and other human milk proteins for infant feeding (Nandi et al. 2002; Lönnerdal 2002), currently provided by Ventria Bioscience (USA). However, effective immunization regime using plant tissue bearing bioencapsulated antigens, especially in case of those originated from blood-transmitted pathogens, still requires meticulous studies (Chan and Daniell 2015; Pniewski et al. 2018) as harsh environment of the gut, low permeability of intestinal membrane, and immune tolerance mechanisms appeared together as a significant barrier for oral vaccination (Rezende and Weiner 2017; Pantazica et al. 2021; Zhang et al. 2021). Nevertheless, orally administered formulation in the form of lyophilized tissue, which can be converted into tablets or capsules, has been established (Pniewski et al. 2011; Lakshmi et al. 2013). This formulation can be efficaciously applied for certain biopharmaceuticals of simpler structure and action mechanisms, targeted for the therapy of various diseases, for instance OPRX-106, tumor necrosis factor fusion protein produced in BY-2 tobacco suspension culture by Protalix Biotherapeutics (Karmiel, Israel) for therapy of ulcerative colitis (Almon et al. 2021). Lyophilized formulations of microalgae are currently commercially optimized for various biopharmaceuticals by TransAlgae (Rehovot, Israel). Lyophilization as minimal plant processing allows to further decrease the costs of plant-produced biopharmaceuticals, as well as storage of formulas stable at room temperature and needle-free oral delivery enable to reduce costs of their distribution and administration thanks to elimination of cold chain and additional equipment while medical personnel could be moved to more necessary and demanding tasks (see also chapter "Plant Molecular Farming for Vaccine Development").

Bioequivalence of Biopharmaceuticals. Higher plants generally synthesize proteins from eukaryotes with correct folding and similar glycosylation pattern, thus with preserved full activity. Moreover, several engineered plants and plant cell cultures have been established, e.g., using CRISPR/Cas9 gene edition technology, that allow the recombinant protein to receive an authentic human glycosylation pattern (Yusibov et al. 2016; Hanania et al. 2017; Mercx et al. 2017; see also Chap. 3). Plants can also be used to produce intrinsically disordered proteins which are naturally abundant in plants (Buyel 2019) or to facilitate the manufacturing of products like viscumin, a lectin with anticancer activity (Gengenbach et al. 2019). These cannot be synthesized efficiently in mammalian cells or prokaryotes due to their toxicity and complex structure, respectively (Buyel 2019).

Stability of Produced Biopharmaceuticals. Plant cells can direct proteins to compartments that reduce degradation and therefore increase stability (Horn et al. 2004). The use of signal peptides and targeting motifs can direct and keep the recombinant proteins to different subcellular compartments (endoplasmic reticulum, chloroplast, apoplast); in this way, degradation is avoided due to the less proteolytic activities at these locations. Another approach is to exploit the ability of vegetative tissues (e.g., leaves or roots) or expression systems (e.g., suspension cultures) to secrete recombinant proteins in their exudates, enabling proteins to be collected continuously (see also chapter "Scaling Up the Plant Molecular Farming via Bioprocessing of Plant Cell Suspension Culture").

Rapid production. Transient expression systems allow the recombinant protein production to be achieved ~8 weeks after receiving the corresponding DNA sequence (Horn et al. 2004). This can be especially useful to deal with pandemics, where the production of a vaccine or adequate treatment in a short time could save many lives (see also Chap. 12).

Expression of Multiple Genes. Multiple gene engineering can be achieved in different ways: transgenic lines expressing one gene can be crossed together to combine all the genes responsible for the trait under study in a single plant, co-transformation of multiple gene expression can be done in one single plant, or multiple genes can be expressed also in the form of multiple expression cassettes linked together, each expression cassette with its own promoter and terminator

(Ahmad 2014). Regarding transient expression, co-infiltration of *Agrobacterium* cultures carrying particular vectors and subsequent co-expression of encoded proteins enable formation of multicomponent structures, including mosaic or multilayer VLPs (Thuenemann et al. 2013). Alternative approach to produce mosaic VLPs, which is possible also to apply in stable transformation strategy, is the construction of heterodimers. An appropriate coding sequence of a tandem protein comprises linearly arranged sequences of a monomer fused like an epitope, linker (usually composed of glycine and serine residues), and unmodified monomer. Then, heterodimers assembly into VLPs displaying attached epitopes without steric hindrance (Peyret et al. 2015). Co-expression in plants can also be obtained using the strategy based on 2A peptides, which autocleavage a precursor polyprotein into functional proteins (Zhang et al. 2017; see also chapters "Plant Molecular Farming-Production of Virus-Like Particles in Plants" and 8).

Biosafety. Low risk of both plant production and process-related contamination of biopharmaceuticals comes from the fact that no human or animal pathogens replicate in plants (Commandeur et al. 2003).

Use of Secondary Metabolites. Plants produce a diverse array of secondary metabolites, which could be exploited not only as pharmaceuticals directly, but also as their respective chemical precursors. For example, residual biomass, mostly lipids and lignocellulose, which is usually used as biofuel feedstock could then be processed to yield further chemical building blocks, as has been the focus of extensive research over the last decade (Dewick 2009; Espro et al. 2021; Karagoz et al. 2023).

Plant molecular pharming has a remarkable potential for saving time and labor requirements and improving productivity and scalability. Moreover, this technology has a large variety of production systems and the existing of different genetic transformation methods available for this purpose. The choice of the expression system and plant host depends on the nature and purpose of the recombinant protein to be expressed (see also chapter "Molecular Farming of Pharmaceutical Proteins in Different Crop Systems: A Way Forward").

2.3 Challenges and Solutions Faced by Plant Expression Systems

The Yield Challenge: The maximum yield of each system is a major consideration. At the beginning of the plant molecular pharming, the first major technical hurdle was the low yields that were initially achieved in transgenic plants compared to established platforms. However, there are many strategies now available to increase the yields of recombinant proteins in plants. Moreover, protein yield can also be enhanced by increasing the production volume or by the use of highly productive plant systems.

Some strategies used to improve the yield of the recombinant protein are the following:

- *Highly Productive Plant Systems*: Attempts to increase biosynthesis of recombinant proteins have dated back to the very beginning of biopharming. The efforts led to the development of systems alternate to commonly used transgenic plants as above-described transplastomic plants or transient expression, which intrinsic properties ensure tens to hundred times and even larger production scale than in transgenic plants. On the other hand, continuous improvement of transgene construction described below, adaptation of plant cell suspension cultures together with novel hosts, and optimized facilities caused that the use of transgenic plants is still a considerable approach for biopharmaceutical production. Alternative hosts as fast-growing plants as duckweed (*Lemna minor*), moss (Fischer and Schillberg 2004; Cox et al. 2006; Tan et al. 2022), or microalgae (Rosales-Mendoza et al. 2021; Grahl and Reumann 2021), thanks to innate high protein biosynthesis and biomass production, can be alternative for the most efficient transient expression systems, despite requirements specific for aquacultures or bioreactors. It can be stated that present technological progress, although far from finishing, achieved level when practically any biopharmaceutical can be efficiently produced in a plant system. Developed tool kits can be selected or stacked on purpose for a specific biopharmaceutical (see also chapters "Duckweed, an Efficient Green Bio-factory for the Production of Recombinant Proteins" and "Microalgae as a Bioreactor for Molecular Farming of Oral Edible Vaccines Against Infectious Diseases of Humans and Animals").
- *Codon Usage Optimization*: Codon optimization of recombinant protein sequences in favor of the codon usage of the expression hosts has been shown in numerous studies as an effective means of increasing translation efficiency and protein yield (Wang et al. 2021).
- *Promoters, Terminators, and Untranslated Regions (UTRs)*: Strong and robust constitutive promoters are usually preferred in plant-based expression systems for the maximum transcript accumulation and protein yield. The constitutive cauliflower mosaic virus 35S RNA (CaMV35S) promoter and the nopaline synthase (NOS) terminator are among the most universally adopted pairing of regulatory sequences found in plant expression constructs (Liu and Timko 2022). For the production of pharmaceutically relevant proteins in crops, especially cereals, tissue-specific promoters were preferred more than the constitutive ones for targeted high-level and stable protein yield while not affecting the normal growth and development (Rybicki 2010). The gene promoters of the native seed storage proteins, including glutelin (*GluB-1* and *GluB-4*), prolamin, and globulin (*Glb-1*), are strong and specific promoters usually used for protein expression in the endosperm tissue of rice seeds (Wakasa and Takaiwa 2013; see also chapter "The Use of Rice Seed as Bioreactor"). Both 5′- and 3′-UTRs also play an important role in determining mRNA stability and translation (Diamos et al. 2016; Mayr 2017). For example, the use of alcohol dehydrogenase (*ADH*) 5′-UTR together with the heat-shock protein (*HSP*) terminator from *Arabidopsis* gave the

best yield and highest activity of the recombinant human glucocerebrosidase among the tested expression constructs in *N. benthamiana* (Limkul et al. 2015).

- *Subcellular and Apoplast Targeting*: Previous studies have reported that the subcellular and apoplast localization of a recombinant protein can have crucial impact on its accumulation level due to lower proteolytic activities at these locations, in addition to the facilitated folding and glycosylation along the secretory pathway. Compared to cytosol localization, recombinant pharmaceutical proteins generally have a much higher yield when targeted to the apoplast, the endoplasmic reticulum (ER), or the chloroplast (Palaniswamy et al. 2016; Liu and Timko 2022). For some recombinant proteins, highest accumulation is achieved by retention in the ER. For example, carboxy-terminal fusion of the KDEL signal peptide to single-chain antibody variable-region fragments (scFvs) and subsequent ER retention has been found to increase antibody levels by a factor of up to 10–100 compared with either extracellular secretion or expression in the cytosol (Doran 2000).
- *Fusion of Protein Partners/Carrier Molecules*: The glutathione-*S*-transferase (GST), maltose-binding protein (MBP), thioredoxin (TRX), and small ubiquitin-related modifier (SUMO) are well-known protein partners widely used for enhancing protein solubility and expression (Liu and Timko 2022). In addition, the elastin-like polypeptides (ELPs) have been demonstrated to significantly enhance the accumulation of a number of heterologous recombinant proteins, such as monoclonal antibodies (mAbs), interleukins (ILs), and human erythropoietin (EPO) in both transiently and stably transformed tobacco (Conley et al. 2009; Liu and Timko 2022). Some carrier molecules like lichenase (LicKM) (1,3-1,4-glucanase) of *Clostridium thermocellum* have a thermostable feature of maintaining activity at high temperatures, thus allowing for fast and cost-effective purification of its fused proteins by simple heat treatment and increased recovery of recombinant proteins (Musiychuk et al. 2013; Ortega-Berlanga et al. 2015). Coat protein (CP) of alfalfa mosaic virus (AMV) is another example of carrier molecule that permits increase in the yield of recombinant proteins due to its multiple union sites (Ortega-Berlanga et al. 2015). In this sense, plant-produced VLPs have emerged as a promising alternatives for the production of many therapeutic proteins, notably candidate vaccines against infectious disease agents, such as influenza viruses, hepatitis B virus (HBV), human immunodeficiency virus (HIV), human papillomavirus (HPV), and severe acute respiratory syndrome coronavirus 2 (SARS-CoV-2) (Liu and Timko 2022). VLP's formation results in a multivalent antigen-presenting structure that mimics the morphological and immunological properties of the original virus but without any viral genetic material. Therefore, plant-derived VLPs are considered safe and efficient in the immune response stimulation and, more importantly, exhibit great potential for the presentation of a wide range of candidate antigens through chimeric fusion. Zera®, the N-terminal proline-rich region from zein, the storage protein which forms protein bodies (PBs) in maize seeds, which primarily contains eight repeats of the PPPVHL hexapeptide unit, also has been shown to effectively stabilize several pharmaceutical proteins, such as calcitonin (Ct), epidermal

growth factor (EGF), and human growth hormone (hGH), by the intracellular encapsulation into the ER-derived PB-like organelles in plant and other eukaryotic cells (Torrent et al. 2009). In fact, the recombinant vaccine made of the E7 protein of HPV fused with Zera® was successfully produced in *N. benthamiana* with high yield and specific immunogenicity (Whitehead et al. 2014; see also chapter "Medical Applications of Plant Virus Nanoparticles").
- *Deactivation of Proteases*: Directly blocking protease activity is one of the most effective strategies to protect target proteins. There are some main approaches to achieve this: (1) co-expression of protease inhibitors, (2) protease gene knockdown/knockout, and (3) use of broad-spectrum protease inhibitors during protein purification process (Liu and Timko 2022).
- *Silencing Suppressors*: Plants have a sophisticated natural defense system to prevent pathogen invasion and remove foreign (pathogen) genetic materials, and this involves transcriptional gene silencing (TGS) and posttranscriptional gene silencing (PTGS) mechanisms (Liu and Timko 2022). Disrupting the innate RNA silencing pathway may lead to an increased stability of the transgenic RNA and protein yield. Perhaps one of the most widely used viral silencing suppressors is P19 protein, first identified from Tombusvirus (Silhavy et al. 2002). Numerous studies have reported boosted expression of recombinant proteins by co-expression of the P19 silencing suppressor (Boivin et al. 2010; Garabagi et al. 2012; Chiong et al. 2021). Another effective strategy for increasing the level of recombinant protein production is through the full or partial repression of the RNA silencing pathway in the host plant. However, to create PTGS deficiency requires various genetic engineering processes. Butaye et al. (2004) first reported the use of *Arabidopsis* RDR6 mutants (sgs2 and sgs3) with silenced PTGS as a platform to achieve high-level transgene expression, and this genetic background was subsequently used to structurally characterize the oligosaccharyl transferase (OST) subunits in *Arabidopsis*.

Glycosylation Challenge: Glycosylation is the most important posttranslational modification that ensures the integrity and functionality of the glycoproteins among eukaryotes. Hence, this is another important technical aspect influencing the quality of the product because glycan structures can affect the stability, targeting, immunogenicity, pharmacokinetic properties, and biological activity of a protein. Glycosylation pattern differs among organisms; however, plants have an advantage over microorganisms as biopharmaceutical producers since glycosylation does not occur in bacteria, while yeast high-mannosidic glycans are extremely distinct from those characteristics of mammalian proteins. Plant glycosylation pattern can be considered even more similar to the mammalian one than that in insect cells and derived baculovirus-based expression system. Nonetheless, there are several differences in the biosynthesis and structure of protein glycans between plants and mammals. Addition in the ER of high-mannose glycans at specific asparagine (N) sites on proteins is identical in mammalian and plant cells; however, subsequent trimming of the sugar residues in the ER and Golgi generates complex N-glycans of different structures and properties (Doran 2000). The xylosyl and fucosyl residues on plant N-linked complex glycans have been demonstrated to be the key epitopes responsible for the allergenicity of plant glycoproteins in humans (Doran 2000). Strategies to avoid the formation of immunogenic plant glycans may be necessary for plant proteins administered systemically. Two possible approaches include the retaining of recombinant glycoproteins in the ER—usually via addition of KDEL retention signal to prevent plant-specific modifications in the Golgi, and the modification of Golgi activity by removing or adding specific enzymes, either via RNAi strategy or recently via gene editing using CRISPR/Cas9 technology (Cox et al. 2006; Bosch and Schots 2010; Karki et al. 2021; see also Chap. 4). Consequently, the humanization of glycan chains in plants by knocking out plant glycosyltransferases and introducing their human counterparts may result in avoidance of any adverse reactions when plant-derived biopharmaceuticals are injected into patients (Montero-Morales and Steinkellner 2018). In contrast, vaccines and certain biopharmaceuticals for cancer immunotherapy may benefit from plant-specific glycans acting as adjuvants because their immunogenicity stimulates the activity of antigen-presenting cells, particularly via lectins or mannose/fucose receptors on the surface of dendritic cells (Rosales-Mendoza et al. 2016). In turn, glycoengineering enables also creation of new glycan types of novel immunogenic properties and further—rational design of biopharmaceuticals (Bosch and Schots 2010).

The Purification Challenge: Recombinant proteins accumulated inside plant cells have to be extracted from the plant material, requiring the elimination of large quantities of insoluble debris and soluble plant host cell proteins during downstream processing. Even when recombinant proteins are secreted as in plant cell suspension cultures, the medium also contains several secreted host cell proteins that complicate product purification. To overcome this problem, there are some supporting technologies such as flocculation, filter aids, and pre-coat filtration techniques, which have recently been adapted for plant-based systems, reducing the associated costs by more than 75% (Buyel and Fischer 2014). Treatment at moderate temperatures (~65 °C), at low pH (~5.5), or by ultrafiltration/diafiltration (100–300 kDa) can reduce impurities' content in extracts by more than 90%, facilitating product purification and reducing production costs (Hassan et al. 2008; Lightfoot et al. 2008). VLPs being large structures can also be reasonably pre-purified using density gradient centrifugation followed by dialysis. Chromatography is a usual technique for product purification; there are many different conventional chromatography resins using different GMP-ready base matrices (e.g., Sepharose, cellulose, and polymethacrylate). These matrices are paired with ligands suitable for hydrophobic interaction chromatography (HIC), mixed-mode chromatography (MMC), size-exclusion chromatography (SEC, also called gel filtration), affinity chromatography (AFC), and ion-exchange chromatography (IEC), the latter including anion exchange chromatography (AEC) and cation exchange chromatography (CEC) (Buyel 2015, 2019). Differences in the selectivity of these ligands for target proteins are used to purify the target, typically using multiple operations with orthogonal separation mechanisms. Another strategy developed by SemBioSys Genetics to reduce protein purification costs in plants is the oleosin platform that allows recombinant proteins to be isolated from the lipid fraction of seeds followed by endoprotease cleavage (Boothe et al. 2010). This approach has been used successfully to purify a wide range of pharmaceutical proteins, including hirudin, growth hormone, apolipoprotein A-I, and insulin, which was found to be bioequivalent to commercially available human insulin produced in bacteria (Boothe et al. 2010). Another approach is the use of protein-based polymer encoded by synthetic gene and composed of 251 repeats of elastomeric pentapeptide GVGVP as a fusion protein to facilitate single-step purification without the use of chromatography. (GVGVP)251 at low temperatures exists as an extended molecule, but, upon raising the temperature above the transition range, the polymer hydrophobically folds into dynamic structures called β-spirals that further aggregate by hydrophobic association to form twisted filaments (Daniell et al. 2001).

Environmental Contamination Challenge: Perhaps the biggest challenge facing protein expression in plants is the concerns around genetically modified (GM) crops. Major concerns include the spread of recombinant genes through seed dispersal, pollen dispersal, viral transfer, or horizontal transfer; therapeutic proteins getting into the food supply of humans or animals; and adverse effects on organisms in the environment (Ma et al. 2003; Obembe et al. 2011). Given an appropriate choice of host species, the only way to fully avoid transgene spread from field plants to compatible crops and wild species is by contained systems. The containment may be physical and based on habitat barriers; for example, growing crops inside appropriately managed greenhouses, hydroponic growth rooms, using cell suspension cultures, or chloroplast expression system can provide an effective and economical means of containing GM plant material (Ma et al. 2003; Obembe et al. 2011). In some cases, natural genetic barriers have been exploited. For example, pharmaceutical production in self-pollinating species (e.g., rice, wheat, pea) or crops with no sexually compatible wild relatives near the site of production provides a first level of defense gene flow. Similarly, crops with asynchronous flowering times or atypical growing seasons are useful (Fischer and Schillberg 2004). The use of male sterility in GM plant lines, self-pollinating species, producing of non-germinating seeds, and producing of inactive fusion proteins that are activated by post-purification processing are also other strategies used to handle this challenge (Daniell et al. 2001; Ma et al. 2003; Obembe et al. 2011). Figure 2.2 shows opportunities and challenges of plant molecular pharming.

2.4 Recent Examples of Recombinant Proteins Produced in Plants

As previously described, an important breakthrough was achieved in 2012 when the first molecular pharming product was approved for use in humans: Elelyso®, the enzyme taliglucerase alfa, a recombinant form of human glucocerebrosidase developed by Protalix Biotherapeutics for the treatment of the lysosomal storage disorder Gaucher's disease (Grabowski et al. 2014). Then, two clinical trial applications for

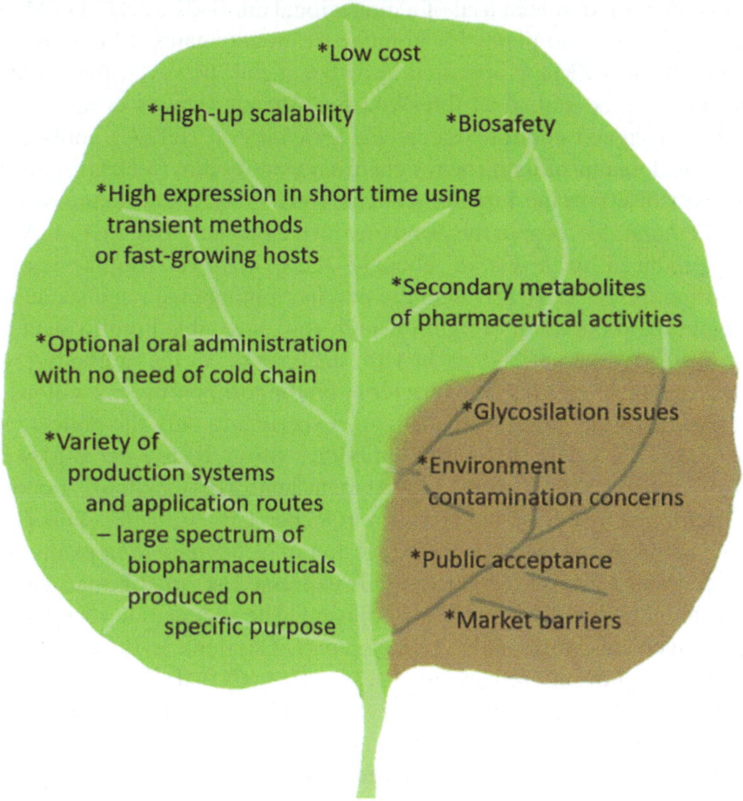

Fig. 2.2 Opportunities (green color) and challenges (ocher color) of the plant-based systems

plant-derived pharmaceuticals were also approved in the European Union (EU), one for insulin produced in safflower (developed by SemBioSys Genetics) and another for an HIV-neutralizing monoclonal antibody produced in tobacco; it was developed by a publicly funded consortium (Pharma-Planta; see also Chap. 14).

Moore et al. (2021) expressed a broadly neutralizing antibody (N6) against HIV in glycoengineered line of *N. benthamiana* plants (pN6) and compared to the mammalian cell-expressed equivalent (mN6). pN6 yield was 49 mg/kg (fresh leaf tissue). The binding kinetics of pN6 and mN6, measured by surface plasmon resonance, were similar for HIV gp120. pN6 had a tenfold higher affinity for FcγRIIIa, which was reflected in an antibody-dependent cellular cytotoxicity assay, where pN6 induced a more potent response from effector cells than that of mN6. pN6 demonstrated the same potency and breadth of neutralization as mN6, against a panel of HIV strains (Moore et al. 2021). In another study, one plant-derived influenza vaccine (QVLP) was tested in a randomized phase 3 trial; the results have shown that QVLP vaccine provided substantial protection against respiratory illness and influenza-like illness caused by influenza viruses in adults. Moreover, QVLP vaccine was well tolerated, and no major safety signal arose in participants (Ward et al.

2020). In another randomized trial of a monoclonal antibody cocktail (ZMapp) produced in *N. benthamiana* plants for immune treatment against Ebola virus disease (EVD), the effect of ZMapp appeared to be beneficial; however, the result did not meet the prespecified statistical threshold for efficacy (Davey et al. 2016). Virdi et al. (2013) designed enterotoxigenic *Escherichia coli* (ETEC) antibodies; they fused variable domains of llama heavy chain-only antibodies (VHHs) against ETEC to the Fc part of a porcine immunoglobulin (IgG or IgA) and expressed them in *Arabidopsis thaliana* seeds. In this way, four VHH-IgG and four VHH-IgA antibodies were produced to levels of about 3–0.2% of seed weight, respectively. Co-transformation of VHH-IgA with the porcine joining chain and secretory component led to the production of light-chain devoid, assembled multivalent dimeric, and secretory IgA-like antibodies. In vitro analysis of all the antibody-producing seed extracts showed inhibition of bacterial binding to porcine gut villous enterocytes. However, in the piglet feed challenge experiment, only the piglets receiving feed containing the VHH-IgA-based antibodies (dose 20 mg per pig) were protected. In February 2022, Canada Government approved Covifenz, a coronavirus vaccine developed by Medicago company (Quebec City, Canada). This is a plant-based vaccine produced by transient expression in *N. benthamiana* plants and composed of recombinant spike (S) glycoprotein expressed as viruslike particles (VLPs) co-administered with GSK's pandemic adjuvant. The vaccination regimen calls for two doses given intramuscularly 21 days apart. Covifenz is indicated for active immunization to prevent coronavirus disease 2019 (COVID-19) caused by severe acute respiratory syndrome coronavirus 2 (SARS-CoV-2) in individuals 18–64 years of age and currently is only authorized in Canada; however, it has been submitted for approval by the WHO. All these studies demonstrate to date the potential for a plant-based platform to produce recombinant proteins with pharmacology activity that can be safe, immunogenic, and effective. More than 100 field trials for the large-scale production of plant-derived molecules are currently awaiting approval by regulatory agencies (Joshi and Lopez 2005). Table 2.1 shows some examples of biopharming companies and their major(s) products. Figure 2.3 shows a schematic representation of biopharmaceutical production in plants.

2.5 Future Perspectives

Along with growing human population, the demands of biopharmaceuticals will increase considerably. In this sense, plant molecular pharming emerges as an attractive alternative to answer growing human needs for the treatment of various pathologies and new threats. Therefore, the production of recombinant proteins in plant-based systems for pharmaceutical, veterinary, and industrial purposes is one of the most promising challenges of molecular pharming. Compared with the traditional protein production systems, plants offer crucial advantages including rapid scalability, low cost, speed, and reduced risk for human contamination. So far, many important biopharmaceuticals have been produced in plant-based pharming

Table 2.1 Molecular pharming companies

Company name	Location	Production platform	Major(s) product
Medicago	Quebec City, Quebec, Canada	*N. benthamiana*, transient expression	Seasonal flu vaccine, coronavirus vaccine
PlantForm	Toronto, Ontario, Canada	*N. benthamiana*, transient expression	Trastuzumab, antibodies for HIV/AIDS, antibodies for Ebola, butyrylcholinesterase
SemBioSys	Calgary, Alberta, Canada	Transgenic safflower seeds	Insulin, ApoA1
iBio Inc.	Newark, Delaware, USA	*N. benthamiana*, transient expression	Antibodies, growth factors, receptors, plasma products
Kentucky BioProcessing	Owensboro, Kentucky, USA	*N. benthamiana*, transient expression	Aprotinin, vaccines, and antibodies
Mapp Biopharmaceutical	San Diego, California, USA	*N. benthamiana*, transient expression	ZMapp
Monsanto	St. Louis, Missouri, USA	Transgenic maize seeds	Avidin
Ventria Bioscience	Fort Collins, Colorado, USA	Transgenic rice seeds	Lactoferrin
Protalix Biotherapeutics	Karmiel, Israel	Transgenic tobacco and carrot suspension cell cultures	Elelyso®—taliglucerase alfa
TransAlgae	Rehovot, Israel	Transgenic microalgae	High-value products for human pharma, animal health, and crop protection
Icon Genetics	Halle (Saale), Germany	*N. benthamiana*, transient expression	Vaccines against infectious diseases and neoplasms
Pharma-Planta	Europe	Transgenic tobacco, maize, and other plants	Vaccines and antibodies
Samabriva	Amiens, France	Rhizosecretion and root cultures	Recombinant proteins and secondary metabolites

systems. Major shortcomings, however, still need to be addressed such as low production levels of proteins—at least in systems based on stable transformation, high rate of proteolysis, and biosafety and public acceptability issues related to the potential dispersal of transgenes to nontarget organisms. Each of these concerns needs to be treated on a case-by-case basis involving the host plant and the target protein. Due to the lack of information, the public is concerned that drugs obtained in plant systems may pose a health risk by triggering allergic reactions. However, it should be emphasized that proteins manufactured in plants are subjected to the same quality control standards as pharmaceuticals produced in bacterial, animal, or yeast systems. Biopharmaceuticals produced in the world, in particular, those approved

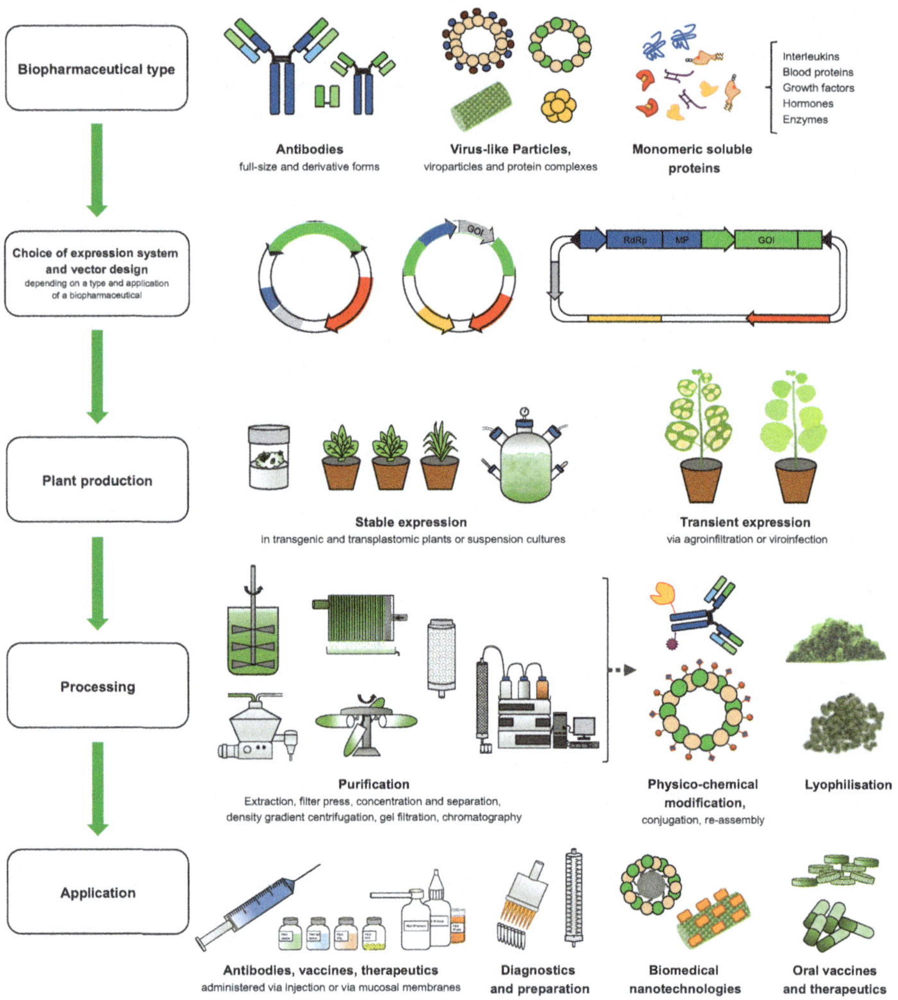

Fig. 2.3 Overview of plant-based biopharmaceutical production in plants

by the WHO, the EU, or the USA, must meet certain requirements referred to as good manufacturing practice (GMP), good laboratory practice (GLP), and finally good clinical practice (GCP). The FDA also has domain over human drug and biological products produced in plant systems. The biopharmaceutical production process is strictly controlled at every stage, and what is more, the finished product is tested for toxicity or presence of contaminants. Another important aspect that slows down the wider use of molecular pharming in the production of biopharmaceuticals is the fact that the industry prefers to rely on known and well-established technologies. To address this constraint, the Pharma Factory project (https://pharmafactory.org/) was created from EU funds for large-scale commercial use of molecular pharming. The main goal of this project is to support innovation in the field of molecular pharming and, above all, to remove technical regulations that hinder public acceptance and exit from the laboratory research phase to the market. Considering the enormous potential of plants as producers of therapeutic proteins, it seems reasonable that, apart from raising public awareness of this topic, there is a great need to support research groups and the pharmaceutical industry in their pursuit of the commercialization of as many necessary plant-derived drugs as possible.

2.6 Conclusions

In emergency situations like the recent COVID-19 pandemic, protein production capacities become scarce very quickly, because the manufacturing of other drugs and diagnostics cannot be stopped or delayed in the face of a new disease. Transient expression in plants provides a strategy to close production gaps quickly: the plants can be grown while the pathogen's genome sequence is investigated and are then ready for protein production as soon as antigen sequences are available. The extent to which molecular pharming can assert itself over other protein production systems remains to be seen, but the unique benefits of transient expression should allow plants to carve an important niche market for recombinant proteins as diagnostic and therapeutic agents against infectious diseases. Overall, transgenic or transplastomic plants will likely be the favorite expression system with proteins that do not express well in traditional systems and are given in large doses, for which production costs make them too expensive to bring to market, or if option of oral administration of a vaccine or other biopharmaceutical is approved. A combination of strong and adaptable regulatory oversight with technological solutions is required if the twin goals of realizing the full potential of biopharming and safeguarding the food system and the environment are to be met.

Acknowledgments The research leading to these results has received funding from the Norwegian Financial Mechanism 2014-2021 and the POLS project, No. 2020/37/K/NZ7/02387.

References

Ahmad K (2014) Molecular farming: strategies, expression systems and bio-safety considerations. Czech J Genet Plant Breed 50:1–10. https://doi.org/10.17221/187/2013-CJGPB

Almon E, Shaaltiel Y, Sbeit W, Fich A, Schwartz D, Waterman M, Szlaifer M, Reuveni H, Amit-Cohen B, Alon S, Chertkoff R, Paz A, Ilan Y (2021) Novel orally administered recombinant anti-TNF alpha fusion protein for the treatment of ulcerative colitis. J Clin Gastroenterol 55:134–140. https://doi.org/10.1097/MCG.0000000000001314

Balke I, Zeltins A (2019) Use of plant viruses and virus-like particles for the creation of novel vaccines. Adv Drug Deliv Rev 145:119–129. https://doi.org/10.1016/j.addr.2018.08.007

Barta A, Sommergruber K, Thompson D, Hartmuth K, Matzke MA, Matzke AJ (1986) The expression of a nopaline synthase—human growth hormone chimaeric gene in transformed tobacco and sunflower callus tissue. Plant Mol Biol 6:347–357. https://doi.org/10.1007/BF00034942

Beyer P, Al-Babili S, Ye X, Lucca P, Schaub P, Welsch R, Potrykus I (2002) Golden Rice: introducing the beta-carotene biosynthesis pathway into rice endosperm by genetic engineering to defeat vitamin A deficiency. J Nutr 132:506S–510S. https://doi.org/10.1093/jn/132.3.506S

Boivin EB, Lepage E, Matton DP (2010) Transient expression of antibodies in suspension plant cell suspension cultures is enhanced when co-transformed with the tomato bushy stunt virus p19 viral suppressor of gene silencing. Biotechnol Prog 26:1534–1543. https://doi.org/10.1002/btpr.485

Boothe J, Nykiforuk C, Shen Y, Zaplachinski S, Szarka S, Kuhlman P, Murray E, Morck D, Moloney MM (2010) Seed-based expression systems for plant molecular pharming. Plant Biotechnol J 8:588–606. https://doi.org/10.1111/j.1467-7652.2010.00511.x

Bosch D, Schots A (2010) Plant glycans: friend or foe in vaccine development? Expert Rev Vaccine 9:835–842. https://doi.org/10.1586/ERV.10.83

Burnett MJ, Burnett AC (2020) Therapeutic recombinant protein production in plants: challenges and opportunities. Plants People Planet 2:121–132. https://doi.org/10.1002/ppp3.10073

Butaye KM, Goderis IJ, Wouters PF, Pues JM, Delauré SL, Broekaert WF, Depicker A, Cammue BP, De Bolle MF (2004) Stable high-level transgene expression in Arabidopsis thaliana using gene silencing mutants and matrix attachment regions. Plant J 39:440–449. https://doi.org/10.1111/j.1365-313X.2004.02144.x

Buyel JF (2015) Process development strategies in plant molecular farming. Curr Pharm Biotechnol 16:966–982. https://doi.org/10.2174/138920101611150902115413

Buyel JF (2019) Plant molecular farming - integration and exploitation of side streams to achieve sustainable biomanufacturing. Front Plant Sci 18:1893. https://doi.org/10.3389/fpls.2018.01893

Buyel JF, Fischer R (2014) Downstream processing of biopharmaceutical proteins produced in plants: the pros and cons of flocculants. Bioengineered 5:138–142. https://doi.org/10.4161/bioe.28061

Canto T (2016) Transient expression systems in plants: potentialities and constraints. In: Vega MC (ed) Advanced technologies for protein complex production and characterization, Advances in experimental medicine and biology, vol 896. Springer, Cham, pp 287–301. https://doi.org/10.1007/978-3-319-27216-0_18

Chan H-T, Daniell H (2015) Plant-made oral vaccines against human infectious diseases—are we there yet? Plant Biotechnol J 13:1056–1070. https://doi.org/10.1111/pbi.12471

Chiong KT, Cody WB, Scholthof HB (2021) RNA silencing suppressor-influenced performance of a virus vector delivering both guide RNA and Cas9 for CRISPR gene editing. Sci Rep 11:6769. https://doi.org/10.1038/s41598-021-85366-4

Commandeur U, Twyman RM, Fischer R (2003) The biosafety of molecular farming in plants. CABI Rev 5:1–9

Conley AJ, Joensuu JJ, Jevnikar AM, Menassa R, Brandle JE (2009) Optimization of elastin-like polypeptide fusions for expression and purification of recombinant proteins in plants. Biotechnol Bioeng 103:562–573. https://doi.org/10.1002/bit.22278

Cox KM, Sterling JD, Regan JT, Gasdaska JR, Frantz KK, Peele CG, Black A, Passmore D, Moldovan-Loomis C, Srinivasan M, Cuison S, Cardarelli PM, Dickey LF (2006) Glycan optimization of a human monoclonal antibody in the aquatic plant *Lemna minor*. Nat Biotechnol 24:1591–1597. https://doi.org/10.1038/nbt1260

Daniell H, Streatfield SJ, Wycoff K (2001) Medical molecular farming: production of antibodies, biopharmaceuticals and edible vaccines in plants. Trends Plant Sci 6:219–226. https://doi.org/10.1016/S1360-1385(01)01922-7

Davey RT, Dodd L, Proschan MA, Neaton J, Neuhaus Nordwall J, Koopmeiners JS, Beigel J, Tierney J, Lane HC, Fauci AS, Massaquoi MBF, Sahr F, Malvy D (2016) A randomized, controlled trial of ZMapp for Ebola virus infection. N Engl J Med 375:1448–1456. https://doi.org/10.1056/NEJMoa1604330

Dewick PM (2009) Secondary metabolism: the building blocks and construction mechanisms. In: Dewick PM (ed) Medicinal natural products: a biosynthetic approach, 3rd edn. Wiley, pp 7–38. https://doi.org/10.1002/9780470742761.ch2

Diamos AG, Rosenthal SH, Mason HS (2016) 5′ and 3′ untranslated regions strongly enhance performance of geminiviral replicons in *Nicotiana benthamiana* leaves. Front Plant Sci 7:200. https://doi.org/10.3389/fpls.2016.00200

Doran PM (2000) Foreign protein production in plant tissue cultures. Curr Opin Biotechnol 11:199–204. https://doi.org/10.1016/s0958-1669(00)00086-0

Eiben S, Koch C, Altintoprak K, Southan A, Tovar G, Laschat S, Weiss IM, Wege C (2019) Plant virus-based materials for biomedical applications: trends and prospects. Adv Drug Deliv Rev 145:96–118. https://doi.org/10.1016/j.addr.2018.08.011

Espro C, Paone E, Mauriello F, Gotti R, Uliassi E, Bolognesi ML, Rodríguez-Padrónae D, Luque R (2021) Sustainable production of pharmaceutical, nutraceutical and bioactive compounds from biomass and waste. Chem Soc Rev 50:11191–11207. https://doi.org/10.1039/D1CS00524C

Evtushenko EA, Ryabchevskaya EM, Nikitin NA, Atabekov JG, Karpova OV (2020) Plant virus particles with various shapes as potential adjuvants. Sci Rep 10:10365. https://doi.org/10.1038/s41598-020-67023-4

Fischer R, Schillberg S (eds) (2004) Molecular farming: plant-made pharmaceuticals and technical proteins. Wiley-VCH Verlag GmbH & Co. KgaA, Weinheim

Fujiki M, Kaczmarczyk JF, Yusibov V, Rabindran S (2008) Development of a new cucumber mosaic virus-based plant expression vector with truncated 3a movement protein. Virology 381:136–142. https://doi.org/10.1016/j.virol.2008.08.022

Garabagi F, Gilbert E, Loos A, McLean MD, Hall JC (2012) Utility of the P19 suppressor of gene-silencing protein for production of therapeutic antibodies in *Nicotiana* expression hosts. Plant Biotechnol J 10:1118–1128. https://doi.org/10.1111/j.1467-7652.2012.00742.x

Gengenbach BB, Keil LL, Opdensteinen P, Müschen CR, Melmer G, Lentzen H, Bührmann J, Buyel JF (2019) Comparison of microbial and transient expression (tobacco plants and plant-cell packs) for the production and purification of the anticancer mistletoe lectin viscumin. Biotechnol Bioeng 116:2236–2249. https://doi.org/10.1002/bit.27076

Giddings G (2001) Transgenic plants as protein factories. Curr Opin Biotechnol 12:450–454. https://doi.org/10.1016/S0958-1669(00)00244-5

Gleba Y, Klimyuk V, Marillonnet S (2005) Magnifection—a new platform for expressing recombinant vaccines in plants. Vaccine 23:2042–2048. https://doi.org/10.1016/j.vaccine.2005.01.006

Grabowski GA, Golembo M, Shaaltiel Y (2014) Taliglucerase alfa: an enzyme replacement therapy using plant cell expression technology. Mol Genet Metab 112:1–8. https://doi.org/10.1016/j.ymgme.2014.02.011

de Grahl I, Reumann, S (2021) Stramenopile microalgae as "green biofactories" for recombinant protein production. World J Microbiol Biotechnol 37: 163. https://doi.org/10.1007/s11274-021-03126-y

Hanania U, Ariel T, Tekoah Y, Fux L, Sheva M, Gubbay Y, Weiss M, Oz D, Azulay Y, Turbovski A, Forster Y, Shaaltiel Y (2017) Establishment of a tobacco BY2 cell line devoid of plant-specific xylose and fucose as a platform for the production of biotherapeutic proteins. Plant Biotechnol J 15:1120–1129. https://doi.org/10.1111/pbi.12702

Hassan S, van Dolleweerd CJ, Ioakeimidis F, Keshavarz-Moore E, Ma JK-C (2008) Considerations for extraction of monoclonal antibodies targeted to different subcellular compartments in transgenic tobacco plants. Plant Biotechnol J 6:733–748. https://doi.org/10.1111/j.1467-7652.2008.00354.x

Hayden CA, Fischer ME, Andrews BL, Chilton HC, Turner DD, Walker JH, Tizard IR, Howard JA (2015) Oral delivery of wafers made from HBsAg-expressing maize germ induces long-term immunological systemic and mucosal responses. Vaccine 33(2015):2881–2886. https://doi.org/10.1016/j.vaccine.2015.04.080

Hiatt A, Caffferkey R, Bowdish K (1989) Production of antibodies in transgenic plants. Nature 342:76–78. https://doi.org/10.1038/342076a0

Hood EE, Witcher DR, Maddock S, Meyer T, Baszczynski C, Bailey M, Flynn P, Register J, Marshall L, Bond D, Kulisek E, Kusnadi A, Evangelista R, Nikolov Z, Wooge C, Mehigh RJ, Hernan R, Kappel WK, Ritland D, Li CP, Howard JA (1997) Commercial production of avidin from transgenic maize: characterization of transformant, production, processing, extraction and purification. Mol Plant Breed 3:291–306. https://doi.org/10.1023/A:1009676322162

Hood EE, Kusnadi A, Nikolov Z, Howard JA (1999) Molecular farming of industrial proteins from transgenic maize. Adv Exp Med Biol 464:127–147. https://doi.org/10.1007/978-1-4615-4729-7_11

Hood EE, Woodard SL, Horn ME (2002) Monoclonal antibody manufacturing in transgenic plants – myths and realities. Curr Opin Biotechnol 13:630–635. https://doi.org/10.1016/s0958-1669(02)00351-8

Horn ME, Woodard SL, Howard JA (2004) Plant molecular farming: systems and products. Plant Cell Rep 22:711–720. https://doi.org/10.1007/s00299-004-0767-1

Ibrahim A, Odon V, Kormelink R (2019) Plant viruses in plant molecular pharming: toward the use of enveloped viruses. Front Plant Sci 10:803. https://doi.org/10.3389/fpls.2019.00803

Jin S, Daniell H (2015) The engineered chloroplast genome just got smarter. Trends Plant Sci 20:622–640. https://doi.org/10.1016/j.tplants.2015.07.004

Johansen LK, Carrington JC (2001) Silencing on the spot. Induction and suppression of RNA silencing in the *Agrobacterium*-mediated transient expression system. Plant Physiol 126:930–938. https://doi.org/10.1104/pp.126.3.930

Joshi L, Lopez LC (2005) Bioprospecting in plants for engineered proteins. Curr Opin Plant Biol 8:223–226. https://doi.org/10.1016/j.pbi.2005.01.003

Karagoz P, Khiawjan S, Marques MPC, Santzouk S, Bugg TDH, Lye GJ (2023) Pharmaceutical applications of lignin-derived chemicals and lignin-based materials: linking lignin source and processing with clinical indication. Biomass Conv Bioref. https://doi.org/10.1007/s13399-023-03745-5.pdf?pdf=core

Karki U, Fang H, Guo W, Unnold-Cofre C, Xu J (2021) Cellular engineering of plant cells for improved therapeutic protein production. Plant Cell Rep 40:1087–1099. https://doi.org/10.1007/s00299-021-02693-6

Laere E, Ling APK, Wong YP, Koh RY, Lila MAM, Hussein S (2016) Plant-based vaccines: production and challenges. J Bot 2016:4928637. https://doi.org/10.1155/2016/4928637

Lakshmi PS, Verma D, Yang X, Lloyd B, Daniell H (2013) Low cost tuberculosis vaccine antigens in capsules: expression in chloroplasts, bio-encapsulation, stability and functional evaluation *in vitro*. PLoS One 8:e54708. https://doi.org/10.1371/journal.pone.0054708

Lienard D (2007) Pharming and transgenic plants. Biotechnol Annu Rev 13:115–147. https://doi.org/10.1016/S1387-2656(07)13006-4

Lightfoot EN, Root TW, O'Dell JL (2008) Emergence of ideal membrane cascades for downstream processing. Biotechnol Prog 24:599–605. https://doi.org/10.1021/bp0703351

Limkul J, Misaki R, Kato K, Fujiyama K (2015) The combination of plant translational enhancers and terminator increase the expression of human glucocerebrosidase in *Nicotiana benthamiana* plants. Plant Sci 240:41–49. https://doi.org/10.1016/j.plantsci.2015.08.018

Liu H, Timko MP (2022) Improving protein quantity and quality—the next level of plant molecular farming. Int J Mol Sci 23:1326. https://doi.org/10.3390/ijms23031326

Lönnerdal B (2002) Expression of human milk proteins in plants. J Am Coll Nutr 3(Suppl):218S–221S. https://doi.org/10.1080/07315724.2002.10719269

Ma JK-C, Drake PMW, Christou P (2003) The production of recombinant pharmaceutical proteins in plants. Nat Rev Genet 4:794–805. https://doi.org/10.1038/nrg1177

Mayr C (2017) Regulation by 3′-untranslated regions. Annu Rev Genet 51:171–194. https://doi.org/10.1146/annurev-genet-120116-024704

Mercx S, Smargiasso N, Chaumont F, De Pauw E, Boutry M, Navarre C (2017) Inactivation of the beta(1,2)-xylosyltransferase and the alpha(1,3)-fucosyltransferase genes in *Nicotiana tabacum* BY-2 cells by a multiplex CRISPR/Cas9 strategy results in glycoproteins without plant-specific glycans. Front Plant Sci 8:403. https://doi.org/10.3389/fpls.2017.00403

Montero-Morales L, Steinkellner H (2018) Advanced plant-based glycan engineering. Front Bioeng Biotechnol 6:81. https://doi.org/10.3389/fbioe.2018.00081

Moon KB, Park JS, Park YI, Song IJ, Lee HJ, Cho HS, Jeon JH, Kim HS (2019) Development of systems for the production of plant-derived biopharmaceuticals. Plants (Basel) 9:30. https://doi.org/10.3390/plants9010030

Moore CM, Grandits M, Grünwald-Gruber C, Altmann F, Kotouckova M, The AY-H, Ma JK-C (2021) Characterisation of a highly potent and near pan-neutralising anti-HIV monoclonal antibody expressed in tobacco plants. Retrovirology 18:17. https://doi.org/10.1186/s12977-021-00560-6

Musiychuk K, Sivalenka R, Jaje J, Bi H, Flores R, Shaw B, Jones RM, Golovina T, Schnipper J, Khandker L, Sun R, Li C, Kang L, Voskinarian-Berse V, Zhang X, Streatfield S, Hambor J, Abbot S, Yusibov V (2013) Plant-produced human recombinant erythropoietic growth factors support erythroid differentiation in vitro. Stem Cells Dev 22:2326–2340. https://doi.org/10.1089/scd.2012.0489

Naderi S, Baratali F (2015) Overview of plant-based vaccines. Res J Fish Hydrobiol 10:275–289

Nandi S, Suzuki YA, Huang J, Yalda D, Pham P, Wu L, Bartley G, Huang N, Lönnerdal B (2002) Expression of human lactoferrin in transgenic rice grains for the application in infant formula. Plant Sci 163:713–722. https://doi.org/10.1016/S0168-9452(02)00165-6

Obembe OO, Popoola JO, Leelavathi S, Reddy SV (2011) Advances in plant molecular farming. Biotechnol Adv 29:210–222. https://doi.org/10.1016/j.biotechadv.2010.11.004

Ortega-Berlanga B, Pniewski T (2022) Plant-based vaccines in combat against coronavirus diseases. Vaccines (Basel) 10:138. https://doi.org/10.3390/vaccines10020138

Ortega-Berlanga B, Musiychuk K, Shoji Y, Chichester JA, Yusibov V, Patiño-Rodríguez O, Noyola DE, Alpuche-Solís ÁG (2015) Engineering and expression of a RhoA peptide against respiratory syncytial virus infection in plants. Planta 243:451–458. https://doi.org/10.1007/s00425-015-2416-z

Palaniswamy H, Syamaladevi DP, Mohan C, Philip A, Petchiyappan A, Narayanan S (2016) Vacuolar targeting of r-proteins in sugarcane leads to higher levels of purifiable commercially equivalent recombinant proteins in cane juice. Plant Biotechnol J 14:791–807. https://doi.org/10.1111/pbi.12430

Pantazica A-MM, Cucos L-M, Stavaru C, Clarke J-L, Branza-Nichita N (2021) Challenges and prospects of plant-derived oral vaccines against hepatitis B and C viruses. Plants 10:2037. https://doi.org/10.3390/plants10102037

Peyret H, Lomonossoff GP (2015) When plant virology met *Agrobacterium*: the rise of the deconstructed clones. Plant Biotechnol J 13:1121–1135. https://doi.org/10.1111/pbi.12412

Peyret H, Gehin A, Thuenemann EC, Blond D, El Turabi A, Beales L, Clarke D, Gilbert RJC, Fry EE, Stuart DI, Holmes K, Stonehouse NJ, Whelan M, Rosenberg W, Lomonossoff GP, Rowlands DJ (2015) Tandem fusion of hepatitis B core antigen allows assembly of virus-like particles in bacteria and plants with enhanced capacity to accommodate foreign proteins. PLoS One 10:e0120751. https://doi.org/10.1371/journal.pone.0120751

Pharma Factory (n.d.). https://pharmafactory.org/

Pniewski T, Kapusta J, Bociąg P, Wojciechowicz J, Kostrzak A, Gdula M, Fedorowicz-Strońska O, Wójcik P, Otta H, Samardakiewicz S, Wolko B, Płucienniczak A (2011) Low-dose oral immunization with lyophilized tissue of herbicide-resistant lettuce expressing hepatitis B surface antigen for prototype plant-derived vaccine tablet formulation. J Appl Genet 52:125–136. https://doi.org/10.1007/s13353-010-0001-5

Pniewski T, Milczarek M, Wojas-Turek J, Pajtasz-Piasecka E, Wietrzyk J, Czyż M (2018) Plant lyophilisate carrying S-HBsAg as an oral booster vaccine against HBV. Vaccine 36:6070–6076. https://doi.org/10.1016/j.vaccine.2018.09.006

Pyrski M, Mieloch AA, Plewiński A, Basińska-Barczak A, Gryciuk A, Bociąg P, Murias M, Rybka JD, Pniewski T (2019) Parenteral–oral immunization with plant-derived HBcAg as a potential therapeutic vaccine against chronic hepatitis B. Vaccine 7:211. https://doi.org/10.3390/vaccines7040211

Rezende RM, Weiner HL (2017) History and mechanisms of oral tolerance. Semin Immunol 30:3–11. https://doi.org/10.1016/j.smim.2017.07.004

Rosales-Mendoza S, Salazar-Gonzalez JA, Decker EL, Reski R (2016) Implications of plant glycans in the development of innovative vaccines. Expert Rev Vaccines 15:915–925. https://doi.org/10.1586/14760584.2016.1155987

Rosales-Mendoza S, Solís-Andrade KI, Márquez-Escobar VA, González-Ortega O, Bañuelos-Hernandez B (2021) Current advances in the algae-made biopharmaceuticals field. Expert Opin Biol Ther 20:751–766. https://doi.org/10.1080/14712598.2020.1739643

Rybicki EP (2010) Plant-made vaccines for humans and animals. Plant Biotechnol J 8:620–637. https://doi.org/10.1111/j.1467-7652.2010.00507.x

Rybka JD, Mieloch AA, Plis A, Pyrski M, Pniewski T, Giersig M (2019) Assembly and characterization of HBc derived virus-like particles with magnetic core. Nanomaterials (Basel) 9:155. https://doi.org/10.3390/nano9020155

Sainsbury F (2020) Innovation in plant-based transient protein expression for infectious disease prevention and preparedness. Curr Opin Biotechnol 61:110–115. https://doi.org/10.1016/j.copbio.2019.11.002

Sainsbury F, Thuenemann EC, Lomonossoff GP (2009) pEAQ: versatile expression vectors for easy and quick transient expression of heterologous proteins in plants. Plant Biotechnol J 7:682–693. https://doi.org/10.1111/j.1467-7652.2009.00434.x

Schillberg S, Finnern R (2021) Plant molecular farming for the production of valuable proteins—critical evaluation of achievements and future challenges. J Plant Physiol 258:153359. https://doi.org/10.1016/j.jplph.2020.153359

Schillberg S, Emans N, Fischer R (2002) Antibody molecular farming in plants and plant cells. Phytochem Rev 1:45–54. https://doi.org/10.1023/A:1015880218651

Shanmugaraj B, Bulaon CJI, Phoolcharoen W (2020) Plant molecular farming: a viable platform for recombinant biopharmaceutical production. Plants (Basel) 9:842. https://doi.org/10.3390/plants9070842

Silhavy D, Molnar A, Lucioli A, Szittya G, Hornyik C, Tavazza M, Burgyán J (2002) A viral protein suppresses RNA silencing and binds silencing-generated, 21- to 25-nucleotide double-stranded RNAs. EMBO J 21:3070–3080. https://doi.org/10.1093/emboj/cdf312

Spiegel H, Stoger E, Twyman RM, Buyel JF (2018) Ch1: Current status and perspectives of the molecular farming landscape. In: Kermode AR, Jiang L (eds) Molecular farming: applications, challenges, and emerging areas. Wiley, pp 3–23. https://doi.org/10.1002/9781118801512

Tan X, Chen S, Fang Y, Liu P, Hu Z, Jin Y, Yi Z, He K, Li X, Zhao L, Wang H, Zhao H (2022) Rapid and highly efficient genetic transformation and application of interleukin-17B expressed in duckweed as mucosal vaccine adjuvant. Biomolecules 12:1881. https://doi.org/10.3390/biom12121881

Tekoah Y, Shulman A, Kizhner T, Ruderfer I, Fux L, Nataf Y, Bartfeld D, Ariel T, Gingis-Velitski S, Hanania U, Shaaltiel Y (2015) Large-scale production of pharmaceutical proteins in plant cell culture-the Protalix experience. Plant Biotechnol J 13:1199–1208. https://doi.org/10.1111/pbi.12428

Thuenemann EC, Meyers AE, Verwey J, Rybicki EP, Lomonossoff GP (2013) A method for rapid production of heteromultimeric protein complexes in plants: assembly of protective bluetongue virus-like particles. Plant Biotechnol J 11:839–846. https://doi.org/10.1111/pbi.12076

Torrent M, Llompart B, Lasserre-Ramassamy S, Llop-Tous I, Bastida M, Marzabal P, Westerholm-Parvinen A, Saloheimo M, Heifetz PB, Ludevid MD (2009) Eukaryotic protein production in designed storage organelles. BMC Biol 7:5. https://doi.org/10.1186/1741-7007-7-5

Tusé D, Ku N, Bendandi M, Becerra C, Collins R Jr, Langford N, Inogés SS, López-Díaz de Cerio A, Pastor F, Kandzia R, Thieme F, Jarczowski F, Krause D, Ma JK-C, Pandya S, Klimyuk V, Gleba Y, Butler-Ransohoff JE (2015) Clinical safety and immunogenicity of tumor-targeted, plant-made Id-KLH conjugate vaccines for follicular lymphoma. Biomed Res Int 2015:648143. https://doi.org/10.1155/2015/648143

Venkataraman S, Hefferon K (2021) Application of plant viruses in biotechnology, medicine, and human health. Viruses 13:1697. https://doi.org/10.3390/v13091697

Virdi V, Coddens A, De Buck S, DEpicker A (2013) Orally fed seeds producing designer IgAs protect weaned piglets against enterotoxigenic *Escherichia coli* infection. Proc Natl Acad Sci U S A 110:11809–11814. https://doi.org/10.1073/pnas.1301975110

Wakasa Y, Takaiwa F (2013) The use of rice seeds to produce human pharmaceuticals for oral therapy. Biotechnol J 8:1133–1143. https://doi.org/10.1002/biot.201300065

Wang X, Karki U, Abeygunaratne H, UnnoldCofre C, Xu J (2021) Plant cell-secreted stem cell factor stimulates expansion and differentiation of hematopoietic stem cells. Process Biochem 100:39–48. https://doi.org/10.1016/j.procbio.2020.09.029

Ward BJ, Makarkov A, Séguin A, Pillet S, Trépanier S, Dhaliwall J, Libman MD, Vesikari T, Landry N (2020) Efficacy, immunogenicity, and safety of a plant-derived, quadrivalent, virus-like particle influenza vaccine in adults (18–64 years) and older adults (≥65 years): two multicentre, randomised phase 3 trials. Lancet 396:1491–1503. https://doi.org/10.1016/S0140-6736(20)32014-6

Whitehead M, Ohlschlager P, Almajhdi FN (2014) Human papillomavirus (HPV) type 16 E7 protein bodies cause tumour regression in mice. BMC Cancer 14:367. https://doi.org/10.1186/1471-2407-14-367

Yusibov V, Kushnir N, Streatfield SJ (2016) Antibody production in plants and green algae. Annu Rev Plant Biol 67:669–701. https://doi.org/10.1146/annurev-arplant-043015-111812

Zhang B, Rapolu M, Kumar S, Gupta M, Liang Z, Han Z, Williams P, Su WW (2017) Coordinated protein co-expression in plants by harnessing the synergy between an intein and a viral 2A peptide. Plant Biotechnol J 15:718–728. https://doi.org/10.1111/pbi.12670

Zhang Y, Li M, Du G, Chen X, Sun X (2021) Advanced oral vaccine delivery strategies for improving the immunity. Adv Drug Deliv Rev 177:113928. https://doi.org/10.1016/j.addr.2021.113928

Chapter 3
Improving Plant Molecular Farming via Genome Editing

Rakchanok Koto and Chalinee Kongsawat

Abstract Once plant genetic modification began in the 1980s, plants have provided humans with useful agriculture products, but in the last 30 years, they have also been developed as production platforms for economically valuable recombinant proteins, including chemical building blocks, polymers, and renewable energy. All these applications can be described as "plant molecular farming" (PMF). PMF provides a low-cost and simple system for the high-value recombinant protein production on a large scale. They have numerous advantages in terms of simplicity because of no requirement for sterilization, economy for cultivation and processing, safety from any human pathogen and bacterial toxins, and applicability to the agricultural scale easily. Additionally, plants can perform the posttranslational modifications similar to mammalian systems that are required for the bioactivity and pharmacokinetics of recombinant therapeutic proteins, and also there is possibility of using breeding methods and sexual crosses to obtain active recombinant multichain proteins. In the last decade, the use of genome editing (GE) technologies with site-specific nucleases (SSNs) has successfully demonstrated precise gene editing in both animal and plant systems, in contrast to the transgenic approach, which leads to random insertions and very often random phenotypes. GE technology is equipped with a powerful toolbox of molecular scissors to cut DNA at a predetermined site with different efficiencies for designing an approach that best suits the objectives of each plant breeding strategy. This technology not only revolutionizes plant biology, but also provides the means to solve challenges related to plant architecture, food security, nutrient content, adaptation to the environment, resistance to diseases, and production of plant-based materials. This chapter illustrates how these technologies could make plant molecular farming improved, safe, and sustainable.

R. Koto (✉)
Faculty of Science, Srinakharinwirot University, Bangkok, Thailand

C. Kongsawat
National Center for Genetic Engineering and Biotechnology, National Science and Technology Development Agency, Pathum Thani, Thailand
e-mail: chalinee@biotec.or.th

C. Kole et al. (eds.), *Tools & Techniques of Plant Molecular Farming*, Concepts and Strategies in Plant Sciences, https://doi.org/10.1007/978-981-99-4859-8_3

Keywords Genome editing · Plant molecular farming · Meganucleases · CRISPR/Cas9 · Plastid genome

Abbreviations

CRISPR/Cas	Clustered regularly interspaced short palindromic repeats/CRISPR-associated protein
DdCBE	DddA-derived cytosine base editor
ER	Endoplasmic reticulum
ERAD	Endoplasmic reticulum-associated degradation
FucT	(1,3)-Fucosyltransferase
HDR	Homology-directed repair
IRE1	Inositol-requiring enzyme 1
NHEJ	Nonhomologous end joining
PPOs	Polyphenol oxidases
RIDD	Regulated IRE1-dependent decay
SSNs	Sequence-specific nucleases
TALENs	Transcription activator-like effector nucleases
TALEs	Transcription activator-like effectors
tracrRNA	*Trans*-activating crRNA
UPR	Unfolded protein response
XylT	β(1,2)-Xylosyltransferase
ZFNs	Zinc finger nucleases
ZIP	Iron-regulated transporter-like

3.1 Structure and Mechanism of Genome Editing Tools

Genome editing also called gene editing is a method for making specific changes to the DNA of organism, such as the insertion, deletion, or replacement of DNA through sequence-targeted recombination that has increased the speed, ease, and reproducibility of DNA changes (Miladinovic et al. 2021). Based on engineering, genome editing technologies have been developed at a rapid pace over the past 10 years and have begun to show extraordinary utility in various fields (Cornu et al. 2017).

There are four major types of programmed nuclease-based technologies: meganucleases, ZFNs, TALENs, and CRISPR/Cas9. Nucleases generate targeted nicks in the form of double-stranded breakages (DSB) in nuclear DNA, which in turn triggers a repair by one of the two major mechanisms that occur in almost all cell types and organisms: nonhomologous end-joining (NHEJ) and homology-directed repair (HDR).

One of the first tools in the GE toolbox was the zinc finger nucleases (ZFNs). ZFNs, chimeric proteins designed to cut at specific DNA sequences, contain motifs capable of binding to specific DNA sequences. This class of nucleases were first discovered as a fragment of the transcription factor IIIa in clawed frog oocytes (Miller et al. 1985) and share 30-amino acid-long ZF motifs that form one alpha-helix and two antiparallel beta sheets (Pabo et al. 2001). Zinc finger domains can be assembled in large modules to recognize longer DNA fragments (multiples of three bases) and fused to the nonspecific DNA cleavage domain from the FokI nuclease, generating what is known as a zinc finger nuclease (ZFN), to produce DSBs in the genome with extremely high specificity (Kim et al. 1996).

The second tool to edit genes, the transcription activator-like effector nucleases (TALENs), is composed of transcription activator-like effectors (TALEs) and FokI endonuclease. TALE proteins are DNA-binding domains derived from various plant bacterial pathogens of the genus *Xanthomonas* to manipulate host gene transcription and promote successful infection. Like ZFNs, TALENs function as pairs in a similar way to create a break at a specific DNA sequence recognized by the TALE domain. TALENs for DNA recognition use a tandem array of 16 (or more) nearly identical protein modules, each of which targets one nucleotide at the DNA target site, thus making TALENs highly specific (Christian et al. 2010).

Meganucleases or homing endonucleases are sequence-specific endonucleases recognizing cleavage long sequences (typically 18–30 base pairs) that occur only once in any given genome and for this reason are rare-cutting enzymes. They generate DSBs, and the site-specific I-SceI is the prototypical meganuclease that has been used as a tool for genome engineering. For genome editing purposes, thousands of meganucleases have been redesigned and mutants created with new specificities (Silva et al. 2011).

Recently, a novel method for site-directed mutagenesis using an adaptive bacterial and archaeal immune systems has emerged to respond and eliminate invading viruses and plasmids, namely CRISPR/Cas (clustered regularly interspaced short palindromic repeats/CRISPR-associated protein). In bacteria, CRISPR system provides protection against DNA from invading viruses and plasmids via RNA-guided DNA cleavage by Cas proteins (Sorek et al. 2013). Short segments of foreign DNA are integrated within the CRISPR locus and transcribed into CRISPR RNA (crRNA), which then anneal to trans-activating crRNA (tracrRNA) to direct sequence-specific degradation of pathogenic DNA by the Cas9 protein. CRISPR/Cas system is an easier and more efficient genome editing tool than the engineered ZFNs and TALENs (Jinek et al. 2012).

3.2 Using Genome Editing in Plants

Genome editing techniques have been developed to introduce precise and predictable genomic change into plants to obtain desired traits, and they are giving rise to precision breeding techniques that are defining the next generation of plant breeding (Chen et al. 2019). This technique is carried out using programmable sequence-specific nucleases (SSNs). SSNs include engineered meganucleases, ZFNs, TALENs, and the CRISPR/Cas system. All nuclease proteins generate targeted nicks as double-stranded breakages (DSBs) in genomic DNA, in turn triggering a repair mechanism called nonhomologous end joining (NHEJ) in the absence of a repair template. The repairing mechanism NHEJ directly rejoins the 5′ end resection of nicked DNA by either inserting or deleting nucleotides regardless of the nucleotide sequence, resulting in gene expression knockout and permanent loss of function; it is more prevalent in the G1 phase. Moreover, introduction of two targeted DSBs at the same time may lead to sequence deletions or other chromosomal aberrations involving the nuclease recognition and cleavage loci (Iliakis et al. 2004). Therefore, to avoid these genetic abnormalities, DSBs can be repaired by using a repair DNA template via a process called homologous directed repair (HDR). HDR needs the 5′ ends of the DSB to undergo partial 5′–3′ nucleolytic degradation to generate 3′ single-stranded overhangs, a process known as 5′ resection (Pâques and Haber 1999), in order to start the pair process, and is most frequent during the S and G2 phases of the cell cycle after DNA replication since a sister chromatid can serve as template for repair (Osakabe and Osakabe 2015).

Meganuclease-, ZFN-, and TALEN-based technologies cleave DNA at specific sites through DNA-protein interaction. These modified proteins are required for each target sequence; therefore, multiple modified proteins will be required to edit multiple target sequences, which is generally expensive in terms of both time and cost (Li et al. 2021; Ng et al. 2020). Furthermore, CRISPR/Cas system relies on RNA-DNA pairing between specific guide RNAs (gRNAs) and targeted genome sites, which offers a simple and efficient method for genome editing (Li et al. 2021). Differences between gene editing technologies that rely on meganucleases, ZFNs, TALENs, and CRISPR/Cas, along with some advantages and limitations, have been summarized in Table 3.1.

Development of genome editing methodologies began ~20 years ago; four types of engineered nucleases are used for genome editing: engineered meganucleases, ZFNs, TALENs, and CRISPR/Cas. In particular, TALEN and CRISPR/Cas are now used widely in various organisms especially in plants.

3.2.1 Using Meganucleases in Plant

Meganucleases, also known as homing endonucleases, were the first sequence-specific nucleases deployed for targeted double-stranded breaks and are found in prokaryotes, archaea, and unicellular eukaryotes. They selectively cleave DNA at genomic targets with specific nucleotide sequences of 14–40 base pairs (bp) (Carroll 2017).

Overall, plant engineering via meganucleases is very accessible and can be successfully deployed for targeted genome modification in plants. Moreover, they are smaller in size of 40 kDa, making them compatible in some viral vectors with shorter coding sequences. But their application is not as widespread as others because certain limitations cause the DNA binding and cleavage domain to overlap

Table 3.1 Comparison of meganuclease, ZFN, TALEN, and CRISPR/Cas systems (modified from Iqbal et al. 2020)

Characteristic features	Meganucleases	ZFNs	TALEN	CRISPR/Cas
Origin	Microbial genetic element	Eukaryotic gene regulator	*Xanthomonas* bacteria	Adaptive bacterial and archaeal immune systems
Core components	DNA-binding domain and DNA cleavage domain	Zinc finger domain with *Fok*I nuclease	TALE-DNA-binding domain with *Fok*I nuclease	crRNA, Cas protein
Dimerization requirement	Dimeric	Dimeric	Dimeric	Monomeric
Targeted sequence	12–45 bp	18–36 bp	24–59 bp	20–22 bp
Mode of action	Single/chimeric work mode per target	Paired work mode per target	Paired work mode per target	Single work mode per target
Catalytic domain	DNA-binding site contains the catalytic domain	*Fok*I catalytic domain	*Fok*I catalytic domain	RUVC and HNH catalytic domain
Protein engineering criteria	Required	Required	Required	Not difficult to test gRNA
Cloning criteria	Essential	Essential	Essential	Not essential
gRNA requirement	Not required	Not required	Not required	gRNA required and is easy to produce
Genome altering	Generates DBS in target DNA	Generates DBS in target DNA	Generates DBS in target DNA	Generates single-strand nicks or DBS in target DNA
Delivery ease	Easy in vivo delivery; small size allows use in a variety of viral vectors	Easy in vivo delivery; small size allows use in a variety of viral vectors	Difficult in vivo delivery; large size and repetitive nature cause unwanted recombination events	Moderately easy in vivo delivery
Target recognition efficiency	Low to moderate	Moderate	Moderate	High
Mutation rate	High	Medium	Medium	Low
Multiplexing	Not possible	Difficult	Difficult	Easier
Methylation sensitivity	High	High	High	Low
Cytotoxic effect	Variable to high	Variable to high	Low	Low

(continued)

Table 3.1 (continued)

Characteristic features	Meganucleases	ZFNs	TALEN	CRISPR/Cas
Vector packaging	Easy	Difficult	Difficult	Moderate
Cost-effectiveness	Not cost effective	Not cost effective	Moderately cost effective	Highly cost effective

with each other and sometimes meganucleases are prone to sequence degeneracy, which increases the probability of off-target binding (Argast et al. 1998).

3.2.2 Using ZFNs in Plant

ZFNs, chimeric DNA, belong to the foremost generation of genome editing tools and are basically zinc finger-based DNA recognition elements coupled with DNA cleavage domain of the *Fok*I endonuclease. ZFNs rely upon chimerically engineered nucleases that are generated on the basis of functional principles of Cys2-His2 zinc finger domain. ZFN monomers have Cys2-His2 domain at the N-terminal and *Fok*I DNA cleavage domain at the C-terminal (Kim et al. 1996). By combining various zinc finger repeats, the DNA-binding domain can be programmed to recognize a specific nucleotide sequence of 9–18 bases. The *Fok*I enzyme can only cut DNA when dimerized; a pair of ZFNs that recognize sites in close proximity are used to cut DNA at the intended genomic target (Segal et al. 2003). Compared with meganucleases, ZFNs are more flexible site-directed nuclease because they can be designed to target any genomic site.

ZFNs had been successfully deployed to modify *Arabidopsis thaliana*, soybean, *Nicotiana*, petunia, rapeseed, rice, apple, maize, and fig (Ran et al. 2017). The discovery and implementation of ZFNs to plant genomes have transformed the field of plant genome editing by manipulating sites of interest. ZFNs are advantageous over other genome editing techniques in relation to high efficiency and specificity. Moreover, availability of restricted number of target sites makes ZFN application sometimes off-target due to nonspecific DNA binding (DeFrancesco 2011).

3.2.3 Using TALENs in Plant

TALENs are chimeric DNA-cutting enzymes formed by coupling of 13–28 transcriptional activator-like effector (TALE) repeats with *Fok*I endonuclease (Miller et al. 2011; Zhang et al. 2011). The DNA-binding domain of a TALEN originates from the virulence transcription activator-like (TAL) effectors from the bacterial plant pathogen *Xanthomonas*, and its virulence in cotton and rice, and consists of up to about 30 near-identical repeats (Bogdanove et al. 2010). Each TALE repeat targets only one nucleotide, which allows flexible target design, which increases the number of potential target sites (Moscou and Bogdanove 2009).

TALENs have been implicated in *Nicotiana*, *Arabidopsis*, rapeseed, potato, flax, soybean, sugarcane, barley, rice, maize, tomato, and wheat (Ran et al. 2017). This technology is still valued for plant genome engineering because of its programmability, efficiency, and target specificity (Gaj et al. 2013). However, there are certain disadvantages as the size of the cDNA encoding TALEN is approximately 3 kb making it problematic to deliver and express a pair of TALENs in a plant cell.

3.2.4 Using CRISPR/CAS Systems in Plant

Recently, a novel method for site-directed mutagenesis using an adaptive bacterial and archaeal immune system has emerged: the CRISPR/Cas system against invasive viruses by cutting the viral DNA (Barrangou and Marraffini 2014). When infected with viruses, bacteria or archaea capture small part of the viral DNA and particularly insert them into their own DNA in a pattern segment known as CRISPR arrays. The CRISPR sequence can accumulate to several hundred small unique sequences (~30–40 bp), called protospacers. When the viruses attack again, the CRISPR locus is transcribed and processed to generate mature CRISPR RNAs (crRNAs). Each crRNA associates with Cas effector proteins that use crRNAs as guides to invader to produce DSBs on infected viral DNA, causing to silence foreign genetic elements (Jinek et al. 2012).

CRISPR/Cas systems are highly diverse and have been classified into two main classes, based on the structural variation of the Cas genes and their organization style. Class 1 members all use a multisubunit crRNA-effector complex for CRISPR/Cas immunity. The members of this class are type I, type III, and type IV systems, which are characterized by the presence of Cas3, Cas10, and Csf1, respectively (Makarova et al. 2020).

Specifically, class 1 CRISPR/Cas systems consist of multiprotein effector complexes, whereas class 2 systems comprise only a single effector protein. The single protein and a tracRNA perform the biological functions of the multisubunit crRNA-effector complex for immunity. This class is composed of type II, type V, and type VI CRISPR/Cas systems, whose signature proteins are Cas9, Cpf1 (Cas12), and Cas13, respectively (Makarova et al. 2020).

3.2.4.1 CRISPR/Cas9 System

The most frequently used is the type II CRISPR/Cas9 system, which depends on a single Cas protein from *Streptococcus pyogenes* (SpCas9) targeting particular DNA sequences and is therefore an attractive gene editing tool (Doudna and Charpentier 2014). These CRISPR arrays along with the spacers are transcribed upon combating foreign DNA and are processed to form 40 nt long small interfering crRNAs. They then associate with trans-activating CRISPR RNA (tracrRNA) to stimulate and guide the Cas9 nuclease. The different crRNA/tracrRNA duplexes remain together after processing and associate with the Cas9 protein, containing PI, REC1, RuvC, and NUC domains. The 20 nucleotides at the 5′ end of the crRNA will guide the Cas9 to the complementary protospacer target as long as the target contains a PAM; in the case of *S. pyogenes*, Cas9 consists of a 3 nt sequence, 5′-NGG-3′ (N = A, T, C, or G) (Gasiunas et al. 2012). The Cas9 protein contains an HNH nuclease domain and a RuvC-like nuclease domain that cut the DNA strands into complementary and noncomplementary sequence, respectively. The cut is very precise and happens between the third and fourth nucleotides upstream from the PAM to create blunt ends (Horvath and Barrangou 2010). Following the DSB, DNA-DSB repair mechanisms initiate genome repair. With the CRISPR/Cas9 system, through pathways of NHEJ or high-fidelity HDR, targeted genomic modifications, including the introduction of small insertions and deletions (indels), can be made (Ran et al. 2017).

3.2.4.1.1 Using Cas9 as a Nuclease

Engineered CRISPR/Cas9 systems use the same system as the bacteria, but the tracrRNA and crRNA molecules are replaced by a single hybrid RNA molecule known as single guide RNA (sgRNA) (Cong et al. 2013). In order to perform genome editing, it is necessary to deliver both components to the plant nucleus, which is achieved by genetic transformation to integrate a cassette containing the Cas9 cDNA and the sgRNA, under the control of strong promoters, such as the cauliflower mosaic virus (CaMV) 35S promoter, for dicots and some monocot species, or the maize ubiquitin promoter for monocots, into the plant genome. For a given target, genetic constructs are extremely simple to assemble by introducing the 20 bp targeting sequence at the 5′ end of the sgRNA in the CRISPR cassette. After the recovery of transgenic lines and verification of the existence of mutations in the intended target, the CRISPR transgene can be easily removed by Mendelian segregation in those species with sexual reproduction (Mao et al. 2017). The recognition specificity can be easily changed by modifying the variable region of the guide RNA, which makes CRISPR/Cas a highly programmable tool (Zhu et al. 2020).

The most direct application of CRISPR/Cas9 system in plants involves the generation of gene knockouts. These mutants are achieved by incorporating indels resulting in frameshift mutations to form premature stop codons. Till date, CRISPR/Cas9 editing tool has been implemented in rice, maize, grapes, wheat, barley, sorghum, tomato, flax, Camelina, cotton, cucumber, rapeseed, lettuce, grapefruit, apple, soybean, oranges, potato, and watermelon (Ricroch et al. 2017).

3.2.4.1.2 Using Cas9 as a Base Editor

Most important agronomic traits are determined by single nucleotide polymorphisms (SNPs) (Zhao et al. 2011), or in some cases, introduction of specific SNPs into elite varieties from wild relatives is a long and arduous process, but genome editing technologies could greatly accelerate this process. Beyond DSB-mediated genome editing, CRISPR/Cas-derived base editors have emerged as powerful tools for generating programmable single-DNA base changes. There are two main classes of base editors—both of which are based on deamination activities: cytidine base editors (CBEs) that convert cytosine (C) to thymine (T) and adenine base editors (ABEs) that convert adenine (A) to guanine (G) (Komor et al. 2016).

Base editors are fusions of catalytically impaired Cas9 (nCas9 D10A) nucleases with single-stranded DNA (ssDNA)-specific deaminases, such as rAPOBEC1 and PmCDA1 cytidine deaminases in CBEs or laboratory-evolved TadA deoxyadenosine deaminase in ABEs, which catalyze C·G to T·A or A·T to G·C transitions in the ssDNA strand of the R-loop induced by CRISPR/Cas at target sites, respectively (Gaudelli et al. 2017; Komor et al. 2016). The base editor systems have been successfully applied in several plant species, including *Arabidopsis*, rice, wheat, maize, tomato, and cotton (Mao et al. 2019). Moreover, dual base editors that combine the functional domains of CBEs and ABEs can induce simultaneous C·G to T·A and A·T to G·C changes at the same target site (Li et al. 2020), further broadening the scope of base editing in plants.

3.2.4.1.3 Using Cas9 as Prime Editors

Current base editors are limited to base transitions (C·G to T·A and A·T to G·C), but eight DNA base transversion mutations, that is, C → A, C → G, G → C, G → T, A → C, A → T, T → A, and T → G, are not possible to perform accurately until the advent of prime editors (PEs). The idea behind PEs is to combine a nickase with a reverse transcriptase (RT) and an extended gRNA, the prime editing gRNA (pegRNA), containing the genetic information to correct the target sequence. A standard prime editor contains an engineered reverse transcriptase enzyme along with a Cas9 nickase and a prime editing guide RNA (pegRNA) (Anzalone et al. 2019). Prime editing has been demonstrated in different plants such as rice, wheat, maize, tomato, and potato (Butt et al. 2020; Lin et al. 2020; Jiang et al. 2020; Lu et al. 2021).

3.2.4.1.4 Using Cas9 as a DNA-Binding Protein

The CRISPR/Cas9 system efficiently enables targeting of DNA and thus provides a tool to address almost any desired site. Through the mutation of both the HNH and RuvC catalytic domains of the Cas9 enzyme, it can be converted into a DNA-binding protein, called "deadCas9" (dCas9). The ability to direct Cas effectors to precise chromosomal loci using sgRNA molecules can be exploited in a variety of ways to alter gene expression patterns (Gao et al. 2014). The dCas9-based transcriptional activation systems have shown a high activation rate are synergistic activator mediator (SAM) and Supernova tagging (SunTag) (Chavez et al. 2015).

Plant-based activators such as the EDLL motif found on some ethylene response factor/apetala 2 family members (Tiwari et al. 2012) seem to provide stronger gene activation when fused to dCas9 (Tiwari et al. 2012; Piatek et al. 2015). Based on the SAM system, expression of a dCas9-VP64-T2A-MS2-EDLL polypeptide that would ultimately yield two different proteins, a dCas9-VP64 fusion and a MS2-EDLL fusion, produced spectacular results with RNA transcript levels reaching up to 30- and 34-fold increase in two different targeted genes in both dicots and monocots (Lowder et al. 2018). The SunTag system has been used in *Arabidopsis* to induce robust and specific activation of several genes in diverse chromatin contexts (Papikian et al. 2019). Co-expression of three sgRNAs targeting WRKY30, RLP23, and CDG1 successfully achieved activation of all targeted genes in *Arabidopsis* using the CRISPR-TV system. Similarly, transcriptional activation of three rice genes (Os03g01240, Os04g39780, and Os11g35410) was achieved using the CRISPR-Act2.0 system (Lowder et al. 2018).

3.2.4.1.5 Using Cas9 for Mediated Epigenome Editing

There are many layers for epigenetics to regulate the gene expression, which majorly contain DNA methylation, histone modification, and noncoding RNAs. DNA methylation is a common gene regulation mechanism, where a specific part of DNA sequence is methylated or demethylated. Generally, a gene with methylation is usually silenced. In plants, CG (or CpG), CHG, and CHH (H is A, T, or C) are three common sequence contexts with high frequency of methylation (Adli 2018). Thus, the CRISPR/Cas system has been approved to be an efficient genome editing tool, which attempts to fuse dCas9 with a methyltransferase or a demethylase. Despite dCas9 losing catalyzed function, it still tightly binds to the target site, and hence, these enzymes can methylate or demethylate the nucleotides. Papikian and colleagues (2019) developed a robust CRISPR/Cas9-based methylation targeting system for plants by utilizing the CRISPR/dCas-SunTag system with the catalytic domain of the Nicotiana tabacum DRM methyltransferase and this system can efficiently target DNA methylation to specific loci in plants.

Histone posttranslational modification is another major factor epigenetically controlling gene expression. Thus, employing CRISPR/Cas system to change the status of histone proteins will change the epigenetic and genetic features of chromatin and affect gene functions. Even dCas9 has to be fused with p300, to which a H3K27 histone acetyltransferase is associated with gene activation, resulting in the expression of both histone-modified constructs in *Arabidopsis* seedlings (Lee et al. 2019).

3.2.4.1.6 Type V CRISPR/Cas12 (Cpf1) System

Cas12 nucleases are the second most widely used Cas proteins in plants; particularly, Cas12a (formerly Cpf1) has been applied in many plant species. Even though Cas12- and Cas9-based CRISPR systems have similar principles and action, they also exhibit differences that give Cas12 an edge for some applications. Cas12 effectors tend to favor T-rich PAMs, providing additional targeting possibilities (Kim et al. 2017). Then, Cas12 systems do not need the intervention of a tracrRNA, using a single RNA molecule that can be engineered to a final length of ~42 nt, making it more economical to synthesize and easier to use for multiplex editing, instead of the ~100 nt in Cas9 sgRNAs (Zetsche et al. 2015).

The first demonstration to apply Cas12 systems into plant genome editing was performed in tobacco and rice (Endo et al. 2016). A codon-optimized FnCpf1 under the control of the parsley ubiquitin promoter was used in combination with six crRNAs to target two different tobacco genes, encoding the phytoene desaturase *NtPDS* and the STENOFOLIA ortholog *NtSTF*1. Furthermore, the FnCpf1 under the control of the maize ubiquitin promoter has been used to target the rice drooping leaf (*OsDL*) and acetolactone synthase (*OsALS*) genes (Endo et al. 2016).

3.2.4.1.7 Type VI CRISPR/Cas13 Systems

The immense amount of sequence information available today has allowed the use of data mining to discover putative new CRISPR systems. A very recent and highly useful addition to the CRISPR toolbox is Cas13, formerly known as C2c2 and C2c6 in case of Cas13a and Cas13b, respectively, which was first identified in 2015 (Shmakov et al. 2015). It is the first class II effector to have some kind of RNase activity, which is catalyzed by its two higher eukaryotes and prokaryotes nucleotide-binding (HEPN) domains. Confirmation of such RNase activity came in 2016 when it was confirmed that Cas13a can cleave ssRNA using an sgRNA as targeting device in a way similar to types II and V CRISPR systems. Even Cas13a can process pre-crRNA molecules into individual mature gRNAs by itself as Cas12a (Abudayyeh et al. 2016).

Otherwise, obvious application of Cas13 effectors is their potential to be used to confer virus resistance. Although CRISPR/Cas9-based strategies can be used to control DNA viruses, most plant viruses have RNA genomes. Using a plant codon-optimized Cas13a from *L. shahii*, under the control of the CaMV 35S promoter, transient expression experiments in *Nicotiana benthamiana* used agroinfiltration to target a recombinant turnip mosaic virus (TuMV) expressing GFP (Aman et al. 2018; see also Chaps. 8 and "Tobacco Plants as a Versatile Host for the Expression of Glycoproteins").

3.3 Strategies to Achieve Plants via Genome Editing for Plant Molecular Farming Applications

In the last 30 years, plants have also been developed as production platforms for small molecules and recombinant proteins (Fischer and Buyel 2020). This niche area has expanded with the global bioeconomy starting around 2010 to include chemical building blocks, polymers, and renewable energy (Buyel 2019). All these applications can be described as "plant molecular farming." Despite its potential to increase the sustainability of biologics manufacturing, PMF has yet to be embraced broadly by industry (see also Chap. 13). However, an additional important drawback is the limited adaptation of plants and plant cells to the requirements of industry-scale manufacturing. Limitations to use wild plant and plant cell are the following: (1) time consuming to produce stable transgenic plants (Sack et al. 2015) and the bottlenecks along the path to biosafety regulatory approval (Ma et al. 2015; see also Chap. 14); (2) low-yield protein production; (3) lack of human glycosylation profiles (Fischer et al. 2018); (4) unwanted product modifications or degradation that may occur during downstream processing (DSP) due to oxidation or proteolysis in the crude extract (Buyel et al. 2015a); (5) secondary metabolite and toxic contamination in the extracts, such as nicotine from tobacco (Buyel et al. 2015a); and finally (6) codon bias that causes a gene from other organisms to be poorly expressed in plant. However, some limitations can be overcome by genome editing, which provides complementary toolsets (see also Chap. 4).

3.3.1 DNA-Free Genome Editing/Cas9 Protein RNP Complexes In Vitro

The direct use of CRISPR/Cas ribonucleoproteins (RNPs) is the most obvious approach to achieve transgene-free genome editing. RNPs can be easily delivered of preassembled ribonucleoprotein complexes of Cas9 and in vitro-transcribed sgRNA into host cells using chemical or physical delivery methods. Then, efficient single-cell regenerated method is a major achievement after delivery (Liang et al. 2017).

3.3.1.1 Protoplast Transformation

Recently, efforts have been made to deliver CRISPR/Cas9 in RNP form into the protoplasts of *Arabidopsis*, lettuce, and tobacco, which revealed subsequent genome editing and regeneration from single protoplast cells (Woo et al. 2015). Even a CRISPR/Cpf1 RNP editing method has been recently developed for rice, soybeans, and wild tobacco protoplasts (Kim et al. 2017). In most cases, polyethylene glycol-calcium (PEG-Ca^{2+})-mediated cell transfection was the method used to deliver the RNPs into plant protoplasts (Andersson et al. 2018). Other useful strategies for the delivery of genes or proteins, electro-transfection of CRISPR/Cas9 RNPs into cabbage protoplasts, provided 1.6% increase in efficiency compared to PEG-mediated transfection (Lee et al. 2020), and lipofection was demonstrated to transport RNPs into negatively charged tobacco BY2 protoplasts by mixing the CRISPR/Cas9 RNPs with positively charged cationic lipids, resulting in a 6% editing efficiency (Liu et al. 2020). However, regeneration from protoplasts is tedious and quite inefficient in most plant species.

3.3.1.2 Particle Bombardment

Particle bombardment can be used to deliver CRISPR/Cas RNPs into various tissue types such as leaf discs, immature embryos, cell clump, and calli and is not limited by plant-host range (Altpeter et al. 2005). Major cereal crops, such as rice (Banakar et al. 2020), maize (Svitashev et al. 2016), and wheat (Liang et al. 2017), have been successfully transformed by bombardment with gold particles coated with CRISPR/Cas RNPs using a helium gene gun. Mutated plants were successfully generated from bombarded tissue in 6–8 weeks' selection (Liang et al. 2019).

3.3.1.3 Zygotes as Delivery Targets

Zygotes and pollen have the potential to avoid protoplast regeneration. Rice zygotes are created by uniting isolated egg and sperm cells and then induced gamete fusion. Cell walls are immature during the early stages of gamete fusion, allowing Toda et al. (2019) to perform PEG-mediated transfection of preassembled CRISPR/Cas9 RNPs. After 30–40 days of culture, 14–64% of the generated plants from the zygotes contained CRISPR-induced mutations (Toda et al. 2019).

3.3.1.4 Nanoparticles for Cargo Delivery

Nanoparticles have been successfully used to deliver DNA, RNA, and proteins into plant cells (Demirer et al. 2021). Polyethyleneimine (PEI)-coated Fe_3O_4 magnetic nanoparticles were used to carry exogenous DNA plasmids into the pollen grains of several dicot plants, including cotton, pepper, pumpkin, and cocozelle (Zhao et al. 2017). In mammalian cells, nanoparticle delivery of CRISPR/Cas9 RNPs has been accomplished; however, using the conjugated nanomaterials/RNPs as the delivery method is an attractive possibility for future research (Ahmar et al. 2021; Demirer et al. 2021).

3.3.2 *Targeted Gene Integration Platforms*

Conventional plant transformation (usually mediated by *Agrobacterium tumefaciens* or particle bombardment) generates random transgene insertion events. Position effects include transgene integration in genomic regions with different chromatin structures (euchromatin or heterochromatin) and in areas that lack the proximity of native regulatory elements. Furthermore, epigenetic modifications such as methylation at the integration site can reduce the long-term stability of transgene expression (Kumar et al. 2015). This limitation is overcome using SSNs by introducing a double-strand break (DSB) at a predetermined sequence; SSNs enable controlled transgene integration. SSN-mediated DNA insertion has been described in a handful of studies in different agronomic species including maize (Svitashev et al. 2015), barley (Watanabe et al. 2016), tobacco, rice (Li et al. 2016), soybean (Bonawitz et al. 2019), and potato (Forsyth et al. 2016).

Ideally, at least one safe-harbor locus should be identified as the site for targeted integration for each plant species and variety used for PMF, which allows sustained, high-level expression. Transgene integration would not cause any obvious deleterious phenotypic effects. Because targeted integration and replacement in plants are technically challenging and rather inefficient (Kumar et al. 2016), more endeavors have to be made in understanding how to improve the molecular processes before they can be routinely exploited, also for PMF purposes.

3.3.3 *Improving the Yields of Recombinant Protein*

Numerous plant species repurposed as recombinant protein production platforms include tobacco, potato, tomato, alfalfa, safflower, carrot, lettuce, strawberry, moss, duckweed, maize, wheat, and rice. The ability to tailor the plant system through cellular engineering techniques or process modification for yield improvement has to be described.

3.3.3.1 Codon Preferences and tRNA Pools

The phenomenon of codon bias can cause a gene from one (source) organism to be poorly expressed in another (host) due to the prevalence of disfavored codons (Gustafsson et al. 2004). Therefore, the yield of recombinant protein can be improved by maximizing codon preference, with replacing each codon in the mRNA with the preferred codon in the host, or harmonizing codon preference via replacing each codon in the mRNA with an equivalent codon in terms of usage frequency in the host (Tuller et al. 2010).

3.3.3.2 Suppression of Gene Silencing

Sequence-dependent RNA degradation or silencing can be directed against RNA transcribed from transgenes, thereby reducing the yield of recombinant proteins. However, this mechanism can be prevented by the co-expression of viral silencing suppressors (Brodersen and Voinnet 2006).

The CRISPR/Cas9 system has been used to knock out RNA-dependent RNA polymerase 6 (*RDR6*) in benthamiana tobacco. This enzyme is accelerating the synthesis of dsRNAs that are subsequently changed into siRNAs (Matsuo and Atsumi 2019). Interestingly, the *N. benthamiana* LAB strain from *Nicotiana benthamiana* Genome and Transcriptome Consortium (http://benthgenome.qut.edu.au/) carries a natural frameshift insertion in the *RDR1* gene that affects its response to viral infection and makes it an ideal host for viral expression vectors (Yang et al. 2004; see also chapter "Plant Viral Vectors: Important Tools for Biologics Production"). Genome editing may facilitate the transfer of these useful features of the *N. benthamiana* LAB strain to other production hosts, including other *Nicotiana* species (Bally et al. 2018).

3.3.3.3 Stress Resilience and Modified Degradation Pathways

The expression of recombinant proteins in plants often causes an imbalance between the amount of unfolded protein and the protein folding machinery of the endoplasmic reticulum (ER), which induces the ER stress and an unfolded protein response (UPR) (Arcalis et al. 2019).

Inositol-requiring enzyme 1 (IRE1) acts as a major signaling hub and comprises an endoribonuclease domain and a kinase domain. Moreover, the ribonuclease activity of IRE induces regulated IRE1-dependent decay (RIDD) of mRNAs encoding secretory proteins (Chen and Brandizzi 2013) and would be a target for gene disruption; the complete knockout of IRE1 is detrimental for plant development because the kinase activity of IRE1 plays a key role independent of the ribonuclease activity (Wakasa et al. 2012). Today's genome editing technologies will allow even more efficient targeted modifications, making it feasible to generate plant expression hosts lacking the RNase domain of IRE1 while maintaining its kinase activity (see also chapter "Plant Molecular Farming of Antimicrobial Peptides for Plant Protection and Stress Tolerance").

3.3.3.4 Modulation of Chaperone Expression

Chaperones mediate the folding of proteins within the endoplasmic reticulum (ER). These chaperones provide stringent quality control and ensure that misfolded protein is targeted for endoplasmic reticulum-associated degradation (ERAD) (Strasser 2018). Recombinant protein expression stresses the protein folding machinery and also induces UPR, as described above. The modulations of selected chaperones can be used as a strategy to improve recombinant protein production. The subtle adjustment of the ER-associated folding pathway was shown to increase recombinant protein yields. For example, in the presence of human CRT, several human viral glycoproteins accumulated to much higher levels in *N. benthamiana* compared to hosts with the plant chaperone machinery alone (Margolin et al. 2020).

3.3.3.5 Modulating Endogenous Protease Activity

One of the major problems of PMF is the relatively low yield of target recombinant protein compared to established production platforms, with presence of endogenous proteases. This enzyme is not only degrading the recombinant protein product *in planta*, in the supernatant of cell suspension cultures, or following the disruption of the plant tissue for product extraction, but also interfering with DSP and affecting product quality because the degradation products are difficult to remove (Puchol Tarazona et al. 2020).

The knockout of protease genes by genome editing is a more effective approach. Different proteases can contribute to the degradation of a target protein by attacking different regions, so multiplex editing would be required to remove all relevant proteases (Schiermeyer 2020).

Interestingly, due to the high throughput of the CRISPR/Cas9 system, it is now possible to conduct systematic screens of large numbers of genes, for example to identify protease inhibitors that enhance the accumulation of pharmaceutical proteins in *N. benthamiana* (Grosse-Holz et al. 2018), thereby facilitating their knockout and the establishment of a general-purpose expression host plant line.

3.3.3.6 Modulation of Endogenous Oxidase Activity

The extraction of recombinant protein product in plants requires homogenization of green leaf material; this process also releases the phenolic compounds. Polyphenol oxidases (PPOs) catalyze the formation complexes between the recombinant protein and the phenol compounds. It can result in the aggregation and precipitation of recombinant protein (Twyman et al. 2003). The adoptions of CRISPR/Cas-mediated knockout can solve this problem. For example, the application of the CRISPR/Cas9 system to induce mutations in the *StPPO2* gene in potato showed a lower PPO activity in the tuber with the consequent reduction of the enzymatic browning. Mutations induced in the four alleles of StPPO2 gene led to lines with a reduction of up to 69% in tuber PPO activity and a reduction of 73% in enzymatic browning (González et al. 2020).

3.3.3.7 Advanced Genome Editing of Plastid Genomes

Plastids have produced high yield of many proteins (from 5% to 70% of total soluble protein) and hence are an attractive structure for metabolic engineering and synthesis of biopharmaceuticals, biofuels, and biomaterials. Moreover, they are maternally inherited in most plants, providing natural transgene to environment (Daniell et al. 2009). Thus, plastid genome editing represents an important opportunity for plant molecular farming. Methods for editing these genes in organelles are in well demand for improving many traits. Regardless of this fact, there are no targeting and PAM site approaches for plastome editing, and homologous recombination-based plastid transformation may introduce point mutations (Bock 2015). Fortunately, Kang et al. (2021) developed a Golden Gate cloning system, namely DddA-derived cytosine base editor (DdCBE) plasmids, that is used to promote point mutagenesis in chloroplast DNA. This DdCBE triggers base editing at rates of up to 38% in lettuce or rapeseed chloroplasts (see also Chap. 9).

To improve consumer-specific traits like medicinal/industrial value in crops, editing of plastome harbors various biosynthetic pathways, including the shikimate, de novo fatty acid synthesis, and methylerythritol 4-phosphate pathways, which serves as a precursor for a wide range of commercially important secondary metabolites, including tocopherols, pigments, and many phytohormones (Li et al. 2021).

3.3.4 Modifying Posttranslational Modifications and Product Quality

3.3.4.1 Specific N-Linked and O-Linked Glycosylation Profiles

Glycosylation is one of the most important posttranslational modifications in the recombinant produces of PMF. The structure of sugar residues influences protein homogeneity, assembly, immunogenicity, and functionality, such as the ability of mAbs to trigger antibody-dependent cellular cytotoxicity (Chenoweth et al. 2020).

Animals and plants also have a glycosylation system but not the same profile; however, there are three main differences between the complex glycans of plants and humans: (1) plant glycans typically carry core α(1,3)fucose and β(1,2)xylose, which are not present in humans; (2) some recombinant proteins produced in plants, including human erythropoietin, are modified by adding β(1,3)galactose and α(1,4) fucose to the terminal GlcNAc residues, forming a structure known as the Lewis a (Le^a) trisaccharide (Castilho et al. 2013), which occurs only rarely on the glycoproteins of healthy adult humans (Parsons et al. 2013); and (3) paucimannosidic-type glycans lack the two terminal GlcNAc residues, which are trimmed off by specific β-*N*-acetylhexosaminidases (HEXO). These differences prevent the addition of homogeneous humanlike galactosylated N-linked and O-linked glycans on recombinant glycoproteins produced in plants (Kriechbaum et al. 2020).

The modification of the plant metabolism to avoid glycan synthesis is an important application of multiplex genome editing (MGE) in molecular farming (Ma et al. 2003). Knocking out the targeted gene that encoded the enzymes, that is, β(1,2)-xylosyltransferase (*XylT*) and (1,3)-fucosyltransferase (*FucT*), may produce a desirable recombinant protein without synthesis of the plant glycan. It was demonstrated that the mutant of *N. benthamiana* also manufactured a recombinant antibody without the synthesis of plant glycans by knockout of the two XylT and four FucT enzyme-encoding genes (Jansing et al. 2019). Similarly, CRISPR/Cas9-based knockouts of the XylT and FucT genes in tobacco L.cv Bright Yellow 2 (BY2) cell suspensions resulted in removal of plant glycans. It is irrefutable that the MGE technique offers a promising platform for manufacturing potent biopharmaceutical products (Hanania et al. 2017). This may offer an adequate chance to extend potential targets for genome editing.

Le^a epitopes were eliminated by HR-mediated knockout of the β(1,3)galactosyl transferase gene achieved using SSNs in higher plants. This improved β(1,4)galactosylation of proteins other than monoclonal antibodies (Strasser et al. 2009).

O-glycans are the second major type of glycosylation that is catalyzed within the secretory pathway (Gomord et al. 2010). O-glycans present on viral envelope glycoproteins, which are potential vaccine candidates (Wohlschlager et al. 2018). This modification is catalyzed by *N*-acetylgalactosaminyltransferases, which plants lack, along with the glycotransferases that are responsible for the elongation and branching of these O-glycans. The first step towards humanizing plant O-linked glycans is knockout of the P_4H genes. This is achieved by HR, allowing the production of human erythropoietin devoid of nonhuman prolyl-hydroxylation and without obvious phenotypic modifications. The multiplexing capability of the CRISPR/Cas9 system makes it the most suitable tool to inactivate multiple P_4H paralogs (Schoberer and Strasser 2018).

3.3.4.2 Modifying to Avoid Toxic Metabolites or Other Disadvantageous Molecules

Nonfood or plants like tobacco may produce toxic compounds, such as the alkaloid nicotine in the case of tobacco. The purification steps required for biopharmaceutical products ensure that small molecules and protein-based impurities are removed below the limit of detection (Ma et al. 2015). Developing a chassis for PMF that is devoid of such potentially toxic compounds is therefore appealing. In tobacco, this goal has been achieved by knocking out both alleles of all six genes coding for berberine bridge enzyme-like (BBL) proteins, which are responsible for the final oxidation step in the synthesis of nicotine (Schachtsiek and Stehle 2019). CRISPR/Cas9 was used for this approach, resulting in a >99.6% reduction of nicotine levels.

Accordingly, the modulation of secondary metabolism may cause unwanted side effects. For example, when a homospermidine synthase was overexpressed in tobacco to reduce spermidine levels, the transgenic plants showed a stunted phenotype (Kaiser et al. 2002). Instead of manipulating enzyme expression directly by gene knockout or overexpression, corresponding transcription factors can be targeted to control metabolite concentrations in a spatiotemporally regulated manner (Hayashi et al. 2020).

The unattractive odor of residual plant biomass can prevent subsequent building materials (Buyel et al. 2021). Thus far, many researchers have focused on the introduction of enzymes that enhance the production of aromatics, for example by overexpressing a monoterpene synthase in tobacco to increase limonene levels (Lucker et al. 2004) and thus alter the smell of the plants (El Tamer et al. 2003). PMF applications could be facilitated by introducing enzymes that degrade odorous volatile organic compounds (VOCs) (Hammerbacher et al. 2019), for example terpenes released during harvest and biomass decomposition (Schiavon et al. 2017), eliminating the odor of residual biomass and increasing consumer acceptance if the bagasse is used as a byproduct. Genome editing could also be used to terminate metabolic pathways at a point where interference with other relevant metabolites is limited (Buyel et al. 2015b).

3.3.5 Modification of Plant Habits to Increase Space–Time Yield

The shape and stature of plants not only affect biomass accumulation but can also be adapted to facilitate bioprocessing. For example, stunted growth can increase the volumetric productivity of vertical farms, and a high leaf-to-stem mass ratio can limit the processing of biomass with low product content (Buyel and Fischer 2012). Genetic modifications can also be used to control plant shape and stature; for example, the *Rht*1 and *Rht*2 genes controlled the wheat dwarfing phenotype responsible for ~60% of the increased grain yield during the Green Revolution in the 1960s (Khush 2001).

Genes controlling flowering and senescence can also improve the properties of PMF hosts by influencing stem elongation and biomass quality. CRISPR/Cas9 was used to inactivate the tobacco *FT*5 gene, which encodes a floral activator; the resulting plant would remain in the vegetative state and continue to accumulate biomass, providing twin advantages for PMF applications: high biomass production and an enhanced biosafety profile due to the absence of pollen and seed dispersal (Schmidt et al. 2020). Similar results were achieved when three other *FT* genes (*FT*1, *FT*2, and *FT*3) were overexpressed in tobacco because the corresponding proteins are floral repressors, causing the plants to remain in the vegetative growth phase (Harig et al. 2012).

3.4 Conclusions

During the last 30 years, great progress has been made in demonstrating the utility of plant production systems for PMF. Genome editing has the potential to alleviate many of the shortcomings of earlier genetic manipulation methods because it potentially facilitates the precise rather than random modification of genomes. Furthermore, targeted transgene integration at a safe-harbor locus in plants could also represent a groundbreaking advance from the regulatory perspective. Together, this chapter covers multiple aspects of the genome editing landscape, including scientific and technical characteristics and applications in different PMF contexts that must be navigated to use genome editing technologies for research and/or commercial purposes.

References

Abudayyeh OO, Gootenberg JS, Konermann S, Joung J, Slaymaker IM, Cox DB, Shmakov S, Makarova KS, Semenova E, Minakhin L, Severinov K, Regev A, Lander ES, Koonin EV, Zhang F (2016) C2c2 is a single-component programmable RNA-guided RNA-targeting CRISPR effector. Science 353:aaf5573

Adli M (2018) The CRISPR tool kit for genome editing and beyond. Nat Commun 9:1911

Ahmar S, Mahmood T, Fiaz S, Mora-Poblete F, Shafique MS, Chattha MS (2021) Advantage of nanotechnology-based genome editing system and its application in crop improvement. Front Plant Sci 12:663849

Altpeter F, Baisakh N, Beachy R, Bock R, Capell T, Christou P (2005) Particle bombardment and the genetic enhancement of crops: myths and realities. Mol Breed 15(3):305–327

Aman R, Ali Z, Butt H, Mahas A, Aljedaani F, Khan MZ, Ding SW, Mahfouz M (2018) RNA virus interference via CRISPR/Cas13a system in plants. Genome Biol 19:1

Andersson M, Turesson H, Olsson N, Fält A-S, Ohlsson P, Gonzalez MN (2018) Genome editing in potato via CRISPR-Cas9 ribonucleoprotein delivery. Physiol Plant 164(4):378–384

Anzalone AV, Randolph PB, Davis JR, Sousa AA, Koblan LW, Levy JM, Chen PJ, Wilson C, Newby GA, Raguram A, Liu DR (2019) Search-and-replace genome editing without double-strand breaks or donor DNA. Nature 576(7785):149–157

Arcalis E, Ibl V, Hilscher J, Rademacher T, Avesani L, Morandini F, Bortesi L, Pezzotti M, Vitale A, Pum D, De Meyer T (2019) Russell-like bodies in plant seeds share common features with prolamin bodies and occur upon recombinant protein production. Front Plant Sci 10:777

Argast GM, Stephens KM, Emond MJ, Monnat RJ Jr (1998) I-PpoI and I-CreI homing site sequence degeneracy determined by random mutagenesis and sequential in vitro enrichment. J Mol Biol 280(3):345–353

Bally J, Jung H, Mortimer C, Naim F, Philips JG, Hellens R, Bombarely A, Goodin MM, Waterhouse PM (2018) The rise and rise of Nicotiana benthamiana: a plant for all reasons. Annu Rev Phytopathol 56:405–426

Banakar R, Schubert M, Collingwood M, Vakulskas C, Eggenberger AL, Wang K (2020) Comparison of CRISPR-Cas9/Cas12a ribonucleoprotein complexes for genome editing efficiency in the rice phytoene desaturase (*OsPDS*) gene. Rice 13(1):4

Barrangou R, Marraffini LA (2014) CRISPR-Cas systems: prokaryotes upgrade to adaptive immunity. Mol Cell 54(2):234–244

Bock R (2015) Engineering plastid genomes: methods, tools, and applications in basic research and biotechnology. Annu Rev Plant Biol 66:211–241

Bogdanove AJ, Schornack S, Lahaye T (2010) TAL effectors: finding plant genes for disease and defense. Curr Opin Plant Biol 13(4):394–401

Bonawitz ND, Ainley WM, Itaya A, Chennareddy SR, Cicak T, Effinger K, Jiang K, Mall TK, Marri PR, Samuel JP, Sardesai N (2019) Zinc finger nuclease-mediated targeting of multiple transgenes to an endogenous soybean genomic locus via non-homologous end joining. Plant Biotechnol J 17(4):750–761

Brodersen P, Voinnet O (2006) The diversity of RNA silencing pathways in plants. Trends Genet 22(5):268–280

Butt H, Rao GS, Sedeek K, Aman R, Kamel R, Mahfouz M (2020) Engineering herbicide resistance via prime editing in rice. Plant Biotechnol J 18(12):2370

Buyel JF (2019) Plant molecular farming – integration and exploitation of side streams to achieve sustainable biomanufacturing. Front Plant Sci 9:1893

Buyel JF, Fischer R (2012) Predictive models for transient protein expression in tobacco (*Nicotiana tabacum* L.) can optimize process time, yield, and downstream costs. Biotechnol Bioeng 109(10):2575–2588

Buyel JF, Twyman RM, Fischer R (2015a) Extraction and downstream processing of plant-derived recombinant proteins. Biotechnol Adv 33(6):902–913

Buyel JF, Buyel JJ, Haase C, Fischer R (2015b) The impact of *Pseudomonas syringae* type III effectors on transient protein expression in tobacco. Plant Biol 17(2):484–492

Buyel JF, Stöger E, Bortesi L (2021) Targeted genome editing of plants and plant cells for biomanufacturing. Transgenic Res 30(4):401–426

Carroll D (2017) Focus: genome editing: genome editing: past, present, and future. Yale J Biol Med 90(4):653

Castilho A, Neumann L, Gattinger P, Strasser R, Vorauer-Uhl K, Sterovsky T, Altmann F, Steinkellner H (2013) Generation of biologically active multi-sialylated recombinant human EPOFc in plants. PLoS One 8(1):e54836

Chavez A, Scheiman J, Vora S, Pruitt BW, Tuttle M, Iyer EPR, Lin S, Kiani S, Guzman CD, Wiegand DJ, Ter-Ovanesyan D, Braff JL, Davidsohn N, Housden BE, Perrimon N, Weiss R, Aach J, Collins JJ, Church GM (2015) Highly efficient Cas9-mediated transcriptional programming. Nat Methods 12:326–328

Chen Y, Brandizzi F (2013) IRE1: ER stress sensor and cell fate executor. Trends Cell Biol 23(11):547–555

Chen K, Wang Y, Zhang R, Zhang H, Gao C (2019) CRISPR/Cas genome editing and precision plant breeding in agriculture. Annu Rev Plant Biol 70:667–697

Chenoweth AM, Wines BD, Anania JC, Mark Hogarth P (2020) Harnessing the immune system via FcγR function in immune therapy: a pathway to next-gen mAbs. Immunol Cell Biol 98(4):287–304

Christian M, Cermak T, Doyle EL, Schmidt C, Zhang F, Hummel A, Bogdanove AJ, Voytas DF (2010) Targeting DNA double-strand breaks with TAL effector nucleases. Genetics 186:757–761

Cong L, Ran FA, Cox D, Lin S, Barretto R, Habib N, Hsu PD, Wu X, Jiang W, Marraffini LA, Zhang F (2013) Multiplex genome engineering using CRISPR/Cas systems. Science 339(6121):819–823

Cornu TI, Mussolino C, Cathomen T (2017) Refining strategies to translate genome editing to the clinic. Nat Med 23:415–423

Daniell H, Singh ND, Mason H, Streatfield SJ (2009) Plant-made vaccine antigens and biopharmaceuticals. Trends Plant Sci 14(12):669–679

DeFrancesco L (2011) Move over ZFNs: a new technology for genome editing may put the zinc finger nuclease franchise out of business, some believe. Not so fast, say the finger people. Nat Biotechnol 29(8):681–685

Demirer GS, Silva TN, Jackson CT, Thomas JB, Ehrhardt DW, Mortimer JC (2021) Nanotechnology to advance CRISPR-Cas Genetic Engineering of Plants. Nat Nanotechnol 16(3):243–250

Doudna JA, Charpentier E (2014) The new frontier of genome engineering with CRISPR-Cas9. Science 346(6213):1258096

El Tamer MK, Smeets M, Holthuysen N, Lücker J, Tang A, Roozen J, Bouwmeester HJ, Voragen AG (2003) The influence of monoterpene synthase transformation on the odour of tobacco. J Biotechnol 106(1):15–21

Endo A, Masafumi M, Kaya H, Toki S (2016) Efficient targeted mutagenesis of rice and tobacco genomes using Cpf1 from *Francisella novicida*. Sci Rep 6:38169

Fischer R, Buyel JF (2020) Molecular farming–the slope of enlightenment. Biotechnol Adv 40:107519

Fischer R, Holland T, Sack M, Schillberg S, Stoger E, Twyman RM, Buyel JF (2018) Glyco-engineering of plant-based expression systems. In: Advances in glycobiotechnology. Springer, Cham, pp 137–166

Forsyth A, Weeks T, Richael C, Duan H (2016) Transcription activator-like effector nucleases (TALEN)-mediated targeted DNA insertion in potato plants. Front Plant Sci 7:1572

Gaj T, Gersbach CA, Barbas CF III (2013) ZFN, TALEN, and CRISPR/Cas-based methods for genome engineering. Trends Biotechnol 31(7):397–405

Gao X, Tsang JC, Gaba F, Wu D, Lu L, Liu P (2014) Comparison of TALE designer transcription factors and the CRISPR/dCas9 in regulation of gene expression by targeting enhancers. Nucleic Acids Res 42:e155

Gasiunas G, Barrangou R, Horvath P, Siksnys V (2012) Cas9–crRNA ribonucleoprotein complex mediates specific DNA cleavage for adaptive immunity in bacteria. Proc Natl Acad Sci U S A 109(39):E2579–E2586

Gaudelli NM, Komor AC, Rees HA, Packer MS, Badran AH, Bryson DI, Liu DR (2017) Programmable base editing of A·T to G·C in genomic DNA without DNA cleavage. Nature 551(7681):464–471

Gomord V, Fitchette AC, Menu-Bouaouiche L, Saint-Jore-Dupas C, Plasson C, Michaud D, Faye L (2010) Plant-specific glycosylation patterns in the context of therapeutic protein production. Plant Biotechnol J 8(5):564–587

González MN, Massa GA, Andersson M, Turesson H, Olsson N, Fält AS, Storani L, Décima Oneto CA, Hofvander P, Feingold SE (2020) Reduced enzymatic browning in potato tubers by specific editing of a polyphenol oxidase gene via ribonucleoprotein complexes delivery of the CRISPR/Cas9 system. Front Plant Sci 10:1649

Grosse-Holz F, Madeira L, Zahid MA, Songer M, Kourelis J, Fesenko M, Ninck S, Kaschani F, Kaiser M, van der Hoorn RA (2018) Three unrelated protease inhibitors enhance accumulation of pharmaceutical recombinant proteins in *Nicotiana benthamiana*. Plant Biotechnol J 16(10):1797–1810

Gustafsson C, Govindarajan S, Minshull J (2004) Codon bias and heterologous protein expression. Trends Biotechnol 22(7):346–353

Hammerbacher A, Coutinho TA, Gershenzon J (2019) Roles of plant volatiles in defence against microbial pathogens and microbial exploitation of volatiles. Plant Cell Environ 42(10):2827–2843

Hanania U, Ariel T, Tekoah Y, Fux L, Sheva M, Gubbay Y, Weiss M, Oz D, Azulay Y, Turbovski A, Forster Y (2017) Establishment of a tobacco BY2 cell line devoid of plant-specific xylose and fucose as a platform for the production of biotherapeutic proteins. Plant Biotechnol J 15(9):1120–1129

Harig L, Beinecke FA, Oltmanns J, Muth J, Müller O, Rüping B, Twyman RM, Fischer R, Prüfer D, Noll GA (2012) Proteins from the FLOWERING LOCUS T-like subclade of the PEBP family act antagonistically to regulate floral initiation in tobacco. Plant J 72(6):908–921

Hayashi S, Watanabe M, Kobayashi M, Tohge T, Hashimoto T, Shoji T (2020) Genetic manipulation of transcriptional regulators alters nicotine biosynthesis in tobacco. Plant Cell Physiol 61(6):1041–1053

Horvath P, Barrangou R (2010) CRISPR/Cas, the immune system of bacteria and archaea. Science 327(5962):167–170

Iliakis G, Wang H, Perrault AR, Boecker W, Rosidi B, Windhofer F, Pantelias G (2004) Mechanisms of DNA double strand break repair and chromosome aberration formation. Cytogenet Genome Res 104(1–4):14–20

Iqbal Z, Iqbal MS, Ahmad A, Memon AG, Ansari MI (2020) New prospects on the horizon: genome editing to engineer plants for desirable traits. Curr Plant Biol 24:100171

Jansing J, Sack M, Augustine SM, Fischer R, Bortesi L (2019) CRISPR/Cas9-mediated knockout of six glycosyltransferase genes in *Nicotiana benthamiana* for the production of recombinant proteins lacking β-1,2-xylose and core α-1,3-fucose. Plant Biotechnol J 17(2):350–361

Jiang YY, Chai YP, Lu MH, Han XL, Lin Q, Zhang Y, Zhang Q, Zhou Y, Wang XC, Gao C, Chen QJ (2020) Prime editing efficiently generates W542L and S621I double mutations in two ALS genes in maize. Genome Biol 21(1):1–10

Jinek M, Chylinski K, Fonfara I, Hauer M, Doudna JA, Charpentier E (2012) A programmable dual-RNA–guided DNA endonuclease in adaptive bacterial immunity. Science 337(6096):816–821

Kaiser A, Sell S, Hehl R (2002) Heterologous expression of a bacterial homospermidine synthase gene in transgenic tobacco: effects on the polyamine pathway. Arch Pharm 335(4):143–151

Kang BC, Bae SJ, Lee S, Lee JS, Kim A, Lee H, Kim JS (2021) Chloroplast and mitochondrial DNA editing in plants. Nat Plants 7(7):899–905

Khush GS (2001) Green revolution: the way forward. Nat Rev Genet 2(10):815–822

Kim YG, Cha J, Chandrasegaran S (1996) Hybrid restriction enzymes: zinc finger fusions to Fok I cleavage domain. Proc Natl Acad Sci U S A 93:1156–1160

Kim H, Kim S-T, Ryu J, Kang B-C, Kim J-S, Kim S-G (2017) CRISPR/Cpf1-mediated DNA-free Plant Genome Editing. Nat Commun 8(1):14406

Komor AC, Kim YB, Packer MS, Zuris JA, Liu DR (2016) Programmable editing of a target base in genomic DNA without double-stranded DNA cleavage. Nature 533:420–424

Kriechbaum R, Ziaee E, Grünwald-Gruber C, Buscaill P, van der Hoorn RA, Castilho A (2020) BGAL1 depletion boosts the level of β-galactosylation of N- and O-glycans in N. benthamiana. Plant Biotechnol J 18(7):1537–1549

Kumar S, Anunanthini P, Ramalingam S (2015) Epigenetic silencing in transgenic plants. Front Plant Sci 6:693

Kumar S, Worden A, Novak S, Lee R, Petolino JF (2016) A trait stacking system via intra-genomic homologous recombination. Planta 244(5):1157–1166

Lee JE, Neumann M, Duro DI, Schmid M (2019) CRISPR-based tools for targeted transcriptional and epigenetic regulation in plants. PLoS One 14(9):e0222778

Lee MH, Lee J, Choi SA, Kim Y-S, Koo O, Choi SH (2020) Efficient genome editing using CRISPR-Cas9 RNP delivery into cabbage protoplasts via electro-transfection. Plant Biotechnol Rep 14(6):695–702

Li J, Meng X, Zong Y, Chen K, Zhang H, Liu J, Li J, Gao C (2016) Gene replacements and insertions in rice by intron targeting using CRISPR–Cas9. Nat Plants 2(10):1–6

Li C, Zhang R, Meng X, Chen S, Zong Y, Lu C, Gao C (2020) Targeted, random mutagenesis of plant genes with dual cytosine and adenine base editors. Nat Biotechnol 38(7):875–882

Li S, Chang L, Zhang J (2021) Advancing organelle genome transformation and editing for crop improvement. Plant Commun 2(2):100141

Liang Z, Chen K, Li T, Zhang Y, Wang Y, Zhao Q, Liu J, Zhang H, Liu C, Ran Y, Gao C (2017) Efficient DNA-free genome editing of bread wheat using CRISPR/Cas9 ribonucleoprotein complexes. Nat Commun 8(1):1–5

Liang Z, Chen K, Gao C (2019) Biolistic delivery of CRISPR/Cas9 with ribonucleoprotein complex in wheat. Methods Mol Biol 1917:327–335

Lin Q, Zong Y, Xue C, Wang S, Jin S, Zhu Z, Wang Y, Anzalone AV, Raguram A, Doman JL, Gao C (2020) Prime genome editing in rice and wheat. Nat Biotechnol 38(5):582–585

Liu W, Rudis MR, Cheplick MH, Millwood RJ, Yang J-P, Ondzighi-Assoume CA (2020) Lipofection-mediated genome editing using DNA-free delivery of the Cas9/gRNA ribonucleoprotein into plant cells. Plant Cell Rep 39(2):245–257

Lowder LG, Zhou J, Zhang Y, Malzahn A, Zhong Z, Hsieh TF, Voytas DF, Zhang Y, Qi Y (2018) Robust transcriptional activation in plants using multiplexed CRISPR-Act2.0 and mTALE-Act systems. Mol Plant 11:245–256

Lu Y, Tian Y, Shen R, Yao Q, Zhong D, Zhang X, Zhu JK (2021) Precise genome modification in tomato using an improved prime editing system. Plant Biotechnol J 19(3):415

Lucker J, Schwab W, van Hautum B, Blaas J, van der Plas LH, Bouwmeester HJ, Verhoeven HA (2004) Increased and altered fragrance of tobacco plants after metabolic engineering using three monoterpene synthases from lemon. Plant Physiol 134(1):510–519

Ma JK, Drake PM, Christou P (2003) The production of recombinant pharmaceutical proteins in plants. Nat Rev Genet 4(10):794–805

Ma JKC, Drossard J, Lewis D, Altmann F, Boyle J, Christou P, Cole T, Dale P, van Dolleweerd CJ, Isitt V, Fischer R (2015) Regulatory approval and a first-in-human phase I clinical trial of a monoclonal antibody produced in transgenic tobacco plants. Plant Biotechnol J 13(8):1106–1120

Makarova KS, Wolf YI, Iranzo J, Shmakov SA, Alkhnbashi OS, Brouns SJ, Charpentier E, Cheng D, Haft DH, Horvath P, Koonin EV (2020) Evolutionary classification of CRISPR–Cas systems: a burst of class 2 and derived variants. Nat Rev Microbiol 18(2):67–83

Mao Y, Botella JR, Zhu JK (2017) Heritability of targeted gene modifications induced by plant-optimized CRISPR systems. Cell Mol Life Sci 74:1075–1093

Mao Y, Botella JR, Liu Y, Zhu JK (2019) Gene editing in plants: progress and challenges. Natl Sci Rev 6(3):421–437

Margolin E, Oh YJ, Verbeek M, Naude J, Ponndorf D, Meshcheriakova YA, Peyret H, van Diepen MT, Chapman R, Meyers AE, Lomonossoff GP (2020) Co-expression of human calreticulin significantly improves the production of HIV gp140 and other viral glycoproteins in plants. Plant Biotechnol J 18(10):2109–2117

Matsuo K, Atsumi G (2019) CRISPR/Cas9-mediated knockout of the RDR6 gene in Nicotiana benthamiana for efficient transient expression of recombinant proteins. Planta 250(2):463–473

Miladinovic D, Antunes D, Yildirim K, Bakhsh A, Cvejić S, Kondić-Špika A, Marjanovic Jeromela A, Opsahl-Sorteberg HG, Zambounis A, Hilioti Z (2021) Targeted plant improvement through genome editing: from laboratory to field. Plant Cell Rep 40(6):935–951

Miller J, McLachlan AD, Klug A (1985) Repetitive zinc-binding domains in the protein transcription factor IIIA from Xenopus oocytes. EMBO J 4(6):1609–1614

Miller JC, Tan S, Qiao G, Barlow KA, Wang J, Xia DF, Meng X, Paschon DE, Leung E, Hinkley SJ, Dulay GP (2011) A TALE nuclease architecture for efficient genome editing. Nat Biotechnol 29(2):143–148

Moscou MJ, Bogdanove AJ (2009) A simple cipher governs DNA recognition by TAL effectors. Science 326(5959):1501–1501

Ng IS, Keskin BB, Tan SI (2020) A critical review of genome editing and synthetic biology applications in metabolic engineering of microalgae and cyanobacteria. Biotechnol J 15(8):1900228

Osakabe Y, Osakabe K (2015) Genome editing with engineered nucleases in plants. Plant Cell Physiol 56(3):389–400

Pabo CO, Peisach E, Grant RA (2001) Design and selection of novel Cys_2His_2 zinc finger proteins. Annu Rev Biochem 70(1):313–340

Papikian A, Liu W, Gallego-Bartolome J, Jacobsen SE (2019) Site-specific manipulation of Arabidopsis loci using CRISPR-Cas9 SunTag systems. Nat Commun 10:729

Pâques F, Haber JE (1999) Multiple pathways of recombination induced by double-strand breaks in Saccharomyces cerevisiae. Microbiol Mol Biol Rev 63(2):349–404

Parsons J, Altmann F, Graf M, Stadlmann J, Reski R, Decker EL (2013) A gene responsible for prolyl-hydroxylation of moss-produced recombinant human erythropoietin. Sci Rep 3(1):1–8

Piatek A, Ali Z, Baazim H, Li L, Abulfaraj A, Al-Shareef S, Aouida M, Mahfouz MM (2015) RNA-guided transcriptional regulation in planta via synthetic dCas9-based transcription factors. Plant Biotechnol J 13:578–589

Puchol Tarazona AA, Lobner E, Taubenschmid Y, Paireder M, Torres Acosta JA, Göritzer K, Steinkellner H, Mach L (2020) Steric accessibility of the cleavage sites dictates the proteolytic vulnerability of the anti-HIV-1 antibodies 2F5, 2G12, and PG9 in plants. Biotechnol J 15(3):1900308

Ran Y, Liang Z, Gao C (2017) Current and future editing reagent delivery systems for plant genome editing. Sci China Life Sci 60(5):490–505

Ricroch A, Clairand P, Harwood W (2017) Use of CRISPR systems in plant genome editing: toward new opportunities in agriculture. Emerg Top Life Sci 1(2):169–182

Sack M, Rademacher T, Spiegel H, Boes A, Hellwig S, Drossard J, Stoger E, Fischer R (2015) From gene to harvest: insights into upstream process development for the GMP production of a monoclonal antibody in transgenic tobacco plants. Plant Biotechnol J 13(8):1094–1105

Schachtsiek J, Stehle F (2019) Nicotine-free, nontransgenic tobacco (*Nicotiana tabacum* l.) edited by CRISPR-Cas9. Plant Biotechnol J 17(12):2228

Schiavon M, Martini LM, Corrà C, Scapinello M, Coller G, Tosi P, Ragazzi M (2017) Characterisation of volatile organic compounds (VOCs) released by the composting of different waste matrices. Environ Pollut 231:845–853

Schiermeyer A (2020) Optimizing product quality in molecular farming. Curr Opin Biotechnol 61:15–20

Schmidt FJ, Zimmermann MM, Wiedmann DR, Lichtenauer S, Grundmann L, Muth J, Twyman RM, Prüfer D, Noll GA (2020) The major floral promoter NtFT5 in tobacco (*Nicotiana tabacum*) is a promising target for crop improvement. Front Plant Sci 10:1666

Schoberer J, Strasser R (2018) Plant glyco-biotechnology. In: Seminars in cell & developmental biology, vol 80. Academic, pp 133–141

Segal DJ, Beerli RR, Blancafort P, Dreier B, Effertz K, Huber A, Barbas CF (2003) Evaluation of a modular strategy for the construction of novel polydactyl zinc finger DNA-binding proteins. Biochemistry 42(7):2137–2148

Shmakov S, Abudayyeh OO, Makarova KS, Wolf YI, Gootenberg JS, Semenova E, Minakhin L, Joung J, Konermann S, Severinov K, Zhang F, Koonin EV (2015) Discovery and functional characterization of diverse class 2 CRISPR-Cas systems. Mol Cell 60:385–397

Silva G, Poirot L, Galetto R, Smith J, Montoya G, Duchateau P, Paques F (2011) Meganucleases and other tools for targeted genome engineering: perspectives and challenges for gene therapy. Curr Gene Ther 11(1):11–27

Sorek R, Lawrence CM, Wiedenheft B (2013) CRISPR-mediated adaptive immune systems in bacteria and archaea. Annu Rev Biochem 82:237–266

Strasser R (2018) Protein quality control in the endoplasmic reticulum of plants. Annu Rev Plant Biol 69:147–172

Strasser R, Castilho A, Stadlmann J, Kunert R, Quendler H, Gattinger P, Jez J, Rademacher T, Altmann F, Mach L, Steinkellner H (2009) Improved virus neutralization by plant-produced anti-HIV antibodies with a homogeneous β1,4-galactosylated N-glycan profile. J Biol Chem 284(31):20479–20485

Svitashev S, Young JK, Schwartz C, Gao H, Falco SC, Cigan AM (2015) Targeted mutagenesis, precise gene editing, and site-specific gene insertion in maize using Cas9 and guide RNA. Plant Physiol 169(2):931–945

Svitashev S, Schwartz C, Lenderts B, Young JK, Mark Cigan A (2016) Genome editing in maize directed by CRISPR–Cas9 ribonucleoprotein complexes. Nat Commun 7(1):1–7

Tiwari SB, Belachew A, Ma SF, Young M, Ade J, Shen Y, Marion CM, Holtan HE, Bailey A, Stone JK, Edwards L, Wallace AD, Canales RD, Adam L, Ratcliffe OJ, Repetti PP (2012) The EDLL motif: a potent plant transcriptional activation domain from AP2/ERF transcription factors. Plant J 70:855–865

Toda E, Koiso N, Takebayashi A, Ichikawa M, Kiba T, Osakabe K, Okamoto T (2019) An efficient DNA- and selectable-marker-free genome-editing system using zygotes in rice. Nat Plants 5(4):363–368

Tuller T, Carmi A, Vestsigian K, Navon S, Dorfan Y, Zaborske J, Pan T, Dahan O, Furman I, Pilpel Y (2010) An evolutionarily conserved mechanism for controlling the efficiency of protein translation. Cell 141(2):344–354

Twyman RM, Stoger E, Schillberg S, Christou P, Fischer R (2003) Molecular farming in plants: host systems and expression technology. Trends Biotechnol 21(12):570–578

Wakasa Y, Hayashi S, Ozawa K, Takaiwa F (2012) Multiple roles of the ER stress sensor IRE1 demonstrated by gene targeting in rice. Sci Rep 2(1):1–6

Watanabe K, Breier U, Hensel G, Kumlehn J, Schubert I, Reiss B (2016) Stable gene replacement in barley by targeted double-strand break induction. J Exp Bot 67(5):1433–1445

Wohlschlager T, Scheffler K, Forstenlehner IC, Skala W, Senn S, Damoc E, Holzmann J, Huber CG (2018) Native mass spectrometry combined with enzymatic dissection unravels glycoform heterogeneity of biopharmaceuticals. Nat Commun 9(1):1–9

Woo JW, Kim J, Kwon SI, Corvalán C, Cho SW, Kim H, Kim SG, Kim ST, Choe S, Kim JS (2015) DNA-free genome editing in plants with preassembled CRISPR-Cas9 ribonucleoproteins. Nat Biotechnol 33(11):1162–1164

Yang SJ, Carter SA, Cole AB, Cheng NH, Nelson RS (2004) A natural variant of a host RNA-dependent RNA polymerase is associated with increased susceptibility to viruses by *Nicotiana benthamiana*. Proc Natl Acad Sci U S A 101(16):6297–6302

Zetsche B, Volz S, Zhang F (2015) A split-Cas9 architecture for inducible genome editing and transcription modulation. Nat Biotechnol 33:139–142. https://doi.org/10.1038/nbt.3149

Zhang F, Cong L, Lodato S, Kosuri S, Church GM, Arlotta P (2011) Efficient construction of sequence-specific TAL effectors for modulating mammalian transcription. Nat Biotechnol 29(2):149–153

Zhao K, Tung CW, Eizenga GC, Wright MH, Ali ML, Price AH (2011) Genome-wide association mapping reveals a rich genetic architecture of complex traits in *Oryza sativa*. Nat Commun 2:467

Zhao X, Meng Z, Wang Y, Chen W, Sun C, Cui B (2017) Pollen magnetofection for genetic modification with magnetic nanoparticles as gene carriers. Nat Plants 3(12):956–964

Zhu H, Li C, Gao C (2020) Applications of CRISPR–Cas in agriculture and plant biotechnology. Nat Rev Mol Cell Biol 21(11):661–677

Chapter 4
Recent Genome Editing Tool-Assisted Plant Molecular Farming

Kaya İşleyen, Deniz Uras, Beyza Kocaoğlu, and Bahar Soğutmaz Özdemir

Abstract Since ancient times, plant systems have been used to provide useful products. In recent years, with the development of modern biotechnological tools, plant systems have become factories for small molecules and recombinant protein production. For the production of these valuable products, plant systems have unique advantages in comparison with mammalian cell systems. Being not able to replicate human pathogens makes plants safer than mammalian systems. Large-scale production is another advantage as a result of the potential for open-field cultivation by using transgenics, and besides, gram quantities of product might be obtained in less than 4 weeks by transient expression. With all these benefits, plant molecular farming uses engineered plants to produce large quantity of recombinant industrial and pharmaceutical proteins. Although traditional transgenic technologies have been used for molecular farming so far, a new biotechnological tool, genome editing, has been an alternative by providing the ability of targeted genome manipulation. Using site-specific nucleases like CRISPR/Cas (clustered regularly interspaced short palindromic repeat) systems, plants can be modified for safe production of pharmaceutically and economically valuable molecules by overcoming the bottlenecks. In this review, this new aspect of science, genome editing applications, will be discussed in the view of plant molecular farming.

Keywords Genome editing · CRISPR/Cas systems · Plant molecular farming/pharming · Transgenic technology · Plant-derived valuable products · Plant tissue culture · Cell suspension culture

---

K. İşleyen · D. Uras · B. Kocaoğlu · B. Soğutmaz Özdemir (✉)
Department of Genetics and Bioengineering, Faculty of Engineering, Yeditepe University, Istanbul, Turkey
e-mail: bahar.sogutmaz@yeditepe.edu.tr

C. Kole et al. (eds.), *Tools & Techniques of Plant Molecular Farming*, Concepts and Strategies in Plant Sciences, https://doi.org/10.1007/978-981-99-4859-8_4

Abbreviations

AGO2	Argonaute RISC catalytic component 2
AT	Adenine-thymine
BR55-2	Anti-Lewis Y recombinant antibody
CHO	Chinese hamster ovary
CMV	Cucumber mosaic virus
CRISPR/Cas	Clustered regularly interspaced short palindromic repeat/Cas
CRISPRa	CRISPR activation
CRISPRi	CRISPR interference
crRNA	CRISPR RNA
CTB	Cholera toxin B subunit
dCas9	Deactivated Cas9
DCL2	Dicer-like 2
DCL3	Dicer-like 3
DNA	Deoxyribonucleic acid
DSB	Double-strand break
dsDNA	Double-stranded DNA
FT5 gene	Folate transporter 5 gene
G2	Growth 2 phase
GBSS	Granule-bound starch synthase
GM	Genetically modified
GMO	Genetically modified organism
gRNA	Guide RNA
HDR	Homology-directed repair
hG-CSF	Human granulocyte-colony-stimulating factor
HIV	Human immunodeficiency virus
IFN-α2b	Interferon-alpha 2b
LIG4	DNA ligase 4
LTB	Heat-labile enterotoxin
MMEJ	Microhomology-mediated end joining
mRNA	microRNA
NHEJ	Nonhomologous end joining
NUC	Nuclear
OsSWEET14	*Oryza sativa* SWEET14
PAM	Protospacer adjacent motif region
PHA	Polyhydroxyalkanoate
PHB	Polyhydroxybutyrate
PMF	Plant molecular farming
PPO	Polyphenol oxidase
PVX	Potato virus X
REC	Recognition
RNA	Ribonucleic acid
RNAi	RNA interference
RNPs	Ribonucleoproteins

S	Synthesis phase
SARS	Severe acute respiratory syndrome
scFvT84.66	Single-chain FV antibody
sgRNA	Single-guide RNA
siRNAs	Small interfering RNAs
SSN	Site-specific nucleases
TALENs	Transcription activator-like effector nucleases
TALEs	Transcription activator-like effectors
TMV	Tobacco mosaic virus
tracrRNA	*Trans*-activating crRNA
ZFNs	Zinc finger nucleases

4.1 Plant Transgenic Technology: Milestones

A milestone study introduced the DNA recombination technology, which provides the transfer of DNA molecules among different species (Cohen et al. 1973). The recombinant DNA technology development accelerated genetic engineering studies in various areas including agriculture. Modern plant biotechnology applications provided many solutions for the drawbacks encountered in conventional plant breeding. Adaptation of recombinant DNA technology to plants was followed by development of the first transgenic plants in the early 1980s that became a milestone for plant biotechnology (Zerbini et al. 2014). In 1983, three independent research groups developed the first genetically modified plants, which were antibiotic-resistant tobacco and petunias (Bevan and Chilton 1982; Herrera-Estrella et al. 1983; Fraley 1983). Another important example of transgenic plant biotechnology is Golden Rice, which is a variety of rice (*Oryza sativa*) engineered to biosynthesize beta-carotene, a precursor of vitamin A in the edible part (Ye et al. 2000). Big populations of the underdeveloped countries who depend on rice as their main food supply suffer from life-threatening diseases related to nutritional obstacles such as vitamin A deficiency.

For plant transgenic technology, two transformation strategies became prominent as *Agrobacterium*-mediated transformation and particle bombardment (biolistics) (Barampuram and Zhang 2011; Klein et al. 1987). For both strategies, callus tissue is widely used in transformation since it can regenerate genetically identical plants from callus to the whole plant. As a result of this, optimization of plant tissue culture protocols for callus initiation, regeneration, micropropagation, rooting, and adaptation of plantlets to greenhouse and field conditions are extremely crucial to finalize an accomplished transgenic plant technology (Azegami et al. 2020).

In the last 25 years, the success of transgenic plant technology has been supported by over a 100-fold increase in production of GM crops with the aforementioned strategies (Mathur et al. 2017). Plant biotechnology enables increased yield, improved nutritional quality, as well as development of plants with abiotic or biotic stress-resistant plants. Additionally, transgenic plants can also be developed for the production of valuable compounds, which are commonly used in food supplements, vaccines, antibodies, industrial enzymes, and therapeutic proteins (Fischer et al.

2013). With the expansion in the global biobased economy starting around 2010, this niche area is recognized as plant molecular farming (PMF) (Buyel 2019).

4.2 Genome Editing Technology

In classical transgenic technology, transferred genes are randomly integrated into the genome which can cause frameshifts in important genes. Also, the transgene copy number cannot be controlled which limits the expression. The main difference of genome editing applications is that the integration is targeted and controlled. A double-strand break (DSB) created at a targeted region of the genome by utilizing site-specific nucleases (SSNs) is one of the key elements of genome editing technology. Figure 4.1 shows that upon the formation of a DSB, the cell's repair mechanisms step in, and either nonhomologous end joining (NHEJ) or homology-directed repair (HDR) is used to repair the break. NHEJ is the dominant repair mechanism in eukaryotes, which is error prone and often leads to insertions or deletions (indels) of nucleotides. Indels result in frameshift mutations or knockout in the targeted gene (Sonoda et al. 2006).

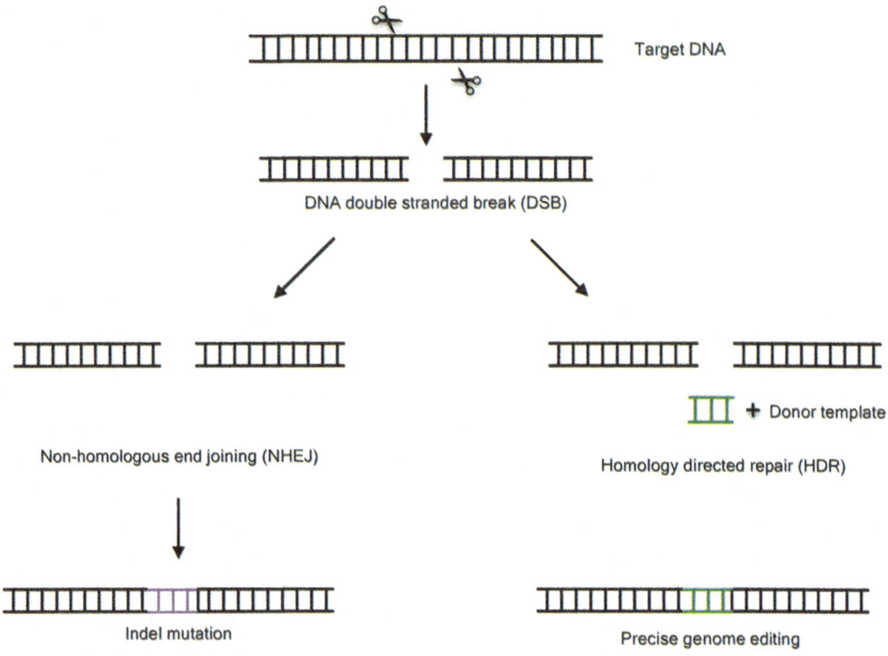

Fig. 4.1 Genome editing process upon creation of a double-strand break (DSB) by utilization of site-specific nucleases shown as scissors

The NHEJ consists of Ku-dependent or Ku-independent pathways. Among them, the dominant Ku-dependent pathway of the NHEJ repair mechanism is based on the DNA end protection factors known as Ku70/89 proteins, which often results in indels. The Ku-independent pathway, also known as the *microhomology-mediated end joining* (MMEJ), is used as the alternative and creates longer deletion regions at the DSB (Osakabe et al. 2014). On the contrary, HDR is known to be an error-free pathway that is used for the insertion or replacement of genes when there is a donor DNA template present. The reason for eukaryotic cells' dominancy on the NHEJ pathway is correlated with the fact that the NHEJ pathway can be used in all of the phases of the cell cycle whereas HDR can only work during the late S/G2 phase, which limits the occurrence (Sonoda et al. 2006).

So far, there are a few known nucleases that can be engineered to create a targeted DSB, which includes homing endonucleases, zinc finger nucleases (ZFNs), transcription activator-like effector nucleases (TALENs), and clustered regularly interspaced short palindromic repeat/Cas (CRISPR/Cas) systems. All the mentioned systems are currently used and have been demonstrated among various model plants as well as crops. However, the recent CRISPR system has been the game-changer by making genome editing more easy, feasible, and accessible. Also, one important factor that sets CRISPR apart from the other systems is that it utilizes an RNA guidance system, whereas the other methods depend on protein engineering for target DNA recognition (Belhaj et al. 2013).

4.2.1 Zinc Finger Nucleases (ZFNs)

Zinc finger nucleases (ZFNs) are one of the first discovered genome engineering tools, which simply work by engineering a chimeric protein consisting of two different domains. One of them is DNA-binding domain composed of a zinc finger motif, while the other is the endonuclease FokI, a dimeric type IIS restriction enzyme, which cleaves the DNA (Urnov et al. 2010). Although ZFNs are recognized as an effective genome editing tool due to its high specificity for the target sequence, the disadvantages such as difficulty of design, long process of engineering, high cost, and low efficiency have reduced its popularity (Ramirez et al. 2008).

Even though the modular construction of ZFNs is complicated, a lot of plant gene modification applications were performed in tobacco (Wright et al. 2005), *Arabidopsis* (Petolino et al. 2010), and crops including canola (Gupta et al. 2012), soybean (Curtin et al. 2011), and maize (Ainley et al. 2013). ZFN applications were used to establish both abiotic and biotic stress resistance. In tobacco, herbicide resistance was performed by Townsend et al. in 2009, and bialaphos-resistant maize was achieved by Shukla et al. (2009). Osakabe et al. (2010) promoted a new phenotype of *Arabidopsis* with abscisic acid (ABA) insensitivity.

4.2.2 Transcription Activator-Like Effector Nucleases (TALENs)

Transcription activator-like effectors (TALEs) are effector proteins that are originally found in a phytopathogen, *Xanthomas* genus, which transfers these TALEs into plant cells after infection. They take a role in regulating specific gene expressions by binding to the promoter regions of the plant genome (Doyle et al. 2013). The center of DNA-binding region of TALEs has a modular architecture, and it is compromised of the tandem repeats. These TALE tandem repeats can be easily reprogrammable, and thus enable the use of TALEs as a genome editing tool (Mak et al. 2012). Even though the TALEN system creates a big diversity for binding specificity, it must be engineered specifically for each target. In addition, the necessity of simultaneous binding of two TAL monomers for a successful DSB creates a big challenge to this system (Nemudryi et al. 2014). Also, TALENs generally consist of 959–1500 amino acids per pair, which makes it a very-large-sized protein to be delivered into the plant cells (Zaman et al. 2019).

The efficiency of TALENs in animals and human cell lines is higher compared with plants (Joung and Sander 2013). However, some successful TALEN studies can be seen in plants (Sun et al. 2016). In TALEN studies, mutation induction through NHEJ was mostly used and loss of function was created at targeted sites (Joung and Sander 2013). Rice cultivars resistant to bacterial blight were produced by utilizing TALEN-mutagenized OsSWEET14 gene (Li et al. 2012). Later, a similar strategy to develop blight-resistant plants was done by using custom engineering of designer TALEs (Li et al. 2013a).

4.2.3 CRISPR/Cas Systems

(CRISPR)/Cas9 system is recognized as a milestone for the genome editing technology among different organisms since its discovery in 2012 (Jinek et al. 2012). The CRISPR system is originally found in bacterial cells as a part of their immune mechanism against phase and viruses. The firstly discovered and most commonly used CRISPR/Cas9 system is classified under class 2 as type II system, which was derived from *Streptococcus pyogenes*. CRISPR/Cas9 working mechanism requires two elements, the nuclease Cas9 and a guide RNA (gRNA). The nuclease protein Cas9 (CRISPR-associated protein 9) is composed of two lobes: the first being a large globular recognition (REC) lobe and the second a small nuclease lobe (NUC) (Belhaj et al. 2015; Gasiunas et al. 2012). The single guide RNA (sgRNA), also known as guide RNA (gRNA), is a chimeric molecule that consists of CRISPR RNA (crRNA) and *trans*-activating crRNA (tracrRNA). The gRNA contains a region to form a complex with Cas9 and a 20 bp homology region, which matches with the target DNA. An important requirement of the CRISPR/Cas9 system is protospacer adjacent motif region (PAM), a DNA sequence composed of 2–6 bp located on the upstream of the target DNA region (Jinek et al. 2012). After the gRNA/Cas9 complex is matched with the target, Cas9 nuclease creates a blunt-end cleavage at 3–4 bp upstream of PAM, which is followed by the intervention of cell's natural repair mechanisms resulting in a site-specific mutation, deletion, or integration (Sternberg et al. 2014). *S. pyogenes* Cas9 (SpCas9) recognizes only PAM regions consisting of the "NGG" sequence, which is seen as one of the important limitations of this system. To overcome this problem, many Cas9 variants, as well as Cas orthologues, have been explored and utilized as good alternatives to expand the target selection with distinctive PAM specifications (Anzalone et al. 2020; Kleinstiver et al. 2016; Zhang et al. 2018).

Off-target activity is one of the critical subjects in CRISPR/Cas systems (Zhang et al. 2015). In order to minimize possible off-targets, gRNA specificity must be carried out by selecting targets with minimal mismatches. Numerous bioinformatics tools have been created for optimizing the target selection process for plants (Kumar and Jain 2015). Alternatively, utilizing Cas9 variants or truncated gRNAs, as well as adjustment of CRISPR reagents' exposure levels, has shown to reduce off-targets (Osakabe et al. 2016; Zhang et al. 2018).

After CRISPR/Cas9 system, the discovery of type V (class 2) system that utilizes Cas12a nuclease has been one of the promising alternatives for genome editing tools, especially among plants (Bandyopadhyay et al. 2020; Zhang et al. 2018). Cas12a, formerly known as Cpf1, requires PAM sequence specificity with "TTTN" sequence, which eases AT-rich target region selection. Besides its different PAM requirements, Cas12a lacks the HNH domain and thus cleavage occurs only by the RuvC-like domain, which creates staggered ends with 5′ overhangs at the DSBs approximately 23 nucleotides distal of the PAM region. Also, it requires guidance only from crRNA, which shortens the guide RNA sequence and makes it more advantageous for multiplexing (Safari et al. 2019).

CRISPR/dCas9 system is another alternative genome editing tool, which is also used for functional genomics and system biology studies. In this system, three major components are used for transcriptional regulation: a complementary single guide RNA to the promoter region of a gene, catalytically inactive dCas9 protein and a transcriptional activator for CRISPR activation, and CRISPRa or repressor for CRISPR interference, CRISPRi, studies. These three components are bound to each other and target the promoter region, thus regulating the transcription of the downstream target gene by blocking RNA polymerase for inhibition or triggering transcription by activation. To silence or knock down gene expression, RNAi and CRISPRi function analogously with different principles and mechanisms (Boettcher and McManus 2015). RNAi method uses posttranscriptional mechanism by destroying transcribed mRNAs, while CRISPRi prevents DNA-level transcription. The field of functional genomics was revolutionized with the advancements of CRISPR/dCas9 technology since it is an efficient, less expensive, and simple tool for targeted gene repression and activation (Xu and Qi 2019).

On the other hand, CRISPRa and CRISPRi application with dCas9 is an alternative solution to overcome the limitations of GMO regulations since it does not make any change in genomic DNA. In addition, the multiplex genome editing property via application of multiple single guide RNAs highlights the CRISPR systems (Cong et al. 2013).

4.2.3.1 CRISPR/Cas Applications in Plants

CRISPR/Cas9 is a favored genome editing tool among plant studies as a result of its efficiency, simplicity, low cost, and allowing of faster genetic modification. In addition, the fact that CRISPR systems allow multiplex genome editing via the application of multiple single guide RNAs creates a big advantage (Cong et al. 2013). The potential of CRISPR systems is incomparable for sustainable agricultural development and crop breeding (Toda et al. 2019; Wurtzel et al. 2019).

Several impressive genetic modification achievements were provided by CRISPR/Cas9 including abiotic (drought, salt, cold) or biotic (bacterial, fungal, or viral pathogens) stress resistance, improved nutritional content, and herbicide resistance. Some examples might be given as thermosensitive genic male sterility which was accomplished in wheat (Okada et al. 2019) and maize (Li et al. 2017), herbicide resistance (Sun et al. 2016), pathogen resistance (Pyott et al. 2016), and improved nutritional properties in sorghum and wheat (Li et al. 2018; Zhang et al. 2018).

In potato plant, granule-bound starch synthase (GBSS) gene was knocked down via CRISPR/Cas9 to increase the starch quality in potatoes, which is a highly demanded commercial product (Andersson et al. 2017). In cucumber, knockdown of *elF4E gene* with CRISPR/Cas9, a translation initiation factor in eukaryotes, led to non-transgenic homozygotic mutant plants, which are resistant to *papaya ringspot mosaic virus* and *zucchini yellow mosaic virus* and immune to *cucumber vein yellowing virus* (Chandrasekaran et al. 2016). Genome editing with CRISPR/Cas9 has an everlasting potential to achieve disease management for plants with no detected resistance before, such as *tomato brown rugose fruit virus* and maize lethal necrosis (Garcia-Ruiz 2018; Luria et al. 2017; Wamaitha et al. 2018; see also chapter "Plant Molecular Farming of Antimicrobial Peptides for Plant Protection and Stress Tolerance").

4.2.4 The Role of Functional Genomics Studies

Functional genomics is evaluated as a powerful technique in terms of identification of gene functions and assessing cellular phenotypes arising from genome-wide perturbations. By collaborating with system biology studies, functional genomics is the growing field of science with the growth of bioinformatics tools and omics-based technologies. Using these advanced technologies, the roles of specific genes were able to be investigated by using tools including RNA interference (RNAi), TALENs, and ZFNs. In RNAi, the aim is inhibition of targeted protein translation by using a conserved eukaryotic machine of microRNAs (miRNAs) and small interfering RNAs (siRNAs) (Lee et al. 1993). Several crops were treated with RNAi, and tolerance against abiotic stress (Hanly et al. 2020) and biotic stress (Singh et al. 2020) was achieved. Although RNAi technology is quite popular, there are several limitations including difficulties in delivery, potential off-target effects or non-specificity, inconsistency, and incompleteness of knockdowns (Mamta and Rajam 2017).

Genome editing technologies have rapidly changed the face of biological research. With the rapid advancement of this technology, the use of genome editing applications has spread to many labs within several years of its initial development. As a result of this, genome editing has become more and more popular for functional genomic studies with its opportunities such as being a better alternative to RNAi by reducing the off-target effects (Boettcher and McManus 2015).

4.3 Plant Molecular Farming

The rapid developments in biotechnology and the increased world population have led to a great demand for industrial enzymes, biopolymers, food supplements, and therapeutic proteins such as monoclonal antibodies, growth factors, and cytokines. Recombinant protein technology has become a standard method for the bulk production of these compounds (Burnett and Burnett 2020; Thomas et al. 2002). The increase in the industrial demand for the low-cost and large-scale production of recombinant proteins has promoted the emergence of various host organisms such as bacteria, yeast, insects, mammalian cells, and plants (Gomes et al. 2016).

Plant molecular farming (PMF) is referred to as the utilization of genetically altered plants or plant cells/callus/tissues like biofactories for the large-scale production of economically important products. Although PMF has several concerns and limitations, plant-based production offers remarkable solutions for bottlenecks in the existing production systems in terms of yield, quality, and cost. In 1987, the first pharmaceutical protein, human growth hormone, was produced in transgenic sunflower and tobacco callus (Barta et al. 1986). This was followed by the production of immunoglobulin as the first recombinant antibody in tobacco (Hiatt et al. 1989). After a decade, avidin (egg protein) became the first protein produced for industrial purposes (Hood et al. 1997). In 2012, Protalix Biopharmaceutics developed the first commercial product produced from plant "Elelyso" for Gaucher's disease (Fox 2012). These developments have taken an initiative role in the usage of plants for valuable protein production.

The main steps in the production of compounds via molecular farming consist of (1) host plant selection, (2) plant tissue culture technique, (3) selection of the transformation strategy (stable or transient expression of the recombinant protein), and (4) downstream processes such as extraction and purification. Over the past few decades, numerous plant species have been introduced as a production platform including *Nicotiana tabacum* (tobacco), *Nicotiana benthamiana*, *Zea mays*, *Oryza sativa*, *Triticum aestivum*, *Solanum lycopersicum*, *Solanum tuberosum*, and *Daucus carota* (see also chapter "Molecular Farming of Pharmaceutical Proteins in Different

Crop Systems: A Way Forward"). Among these, *N. tabacum* and *N. benthamiana* are considered as the most favorable plant species for PMF due to their optimized transformation techniques, high biomass, and inedible plant properties (Spiegel et al. 2018). The production of the desired product can be performed in whole plants, organ cultures (i.e., hairy root cultures), callus cultures, or cell suspension cultures (Bourgaud et al. 2001; see also chapter "Tobacco Plants as a Versatile Host for the Expression of Glycoproteins").

4.3.1 Plants as Production Systems: Pros and Cons

Industrial enzymes, biopolymers, vaccines, and therapeutic proteins were produced so far in both prokaryotic and eukaryotic expression systems. Among host organisms, *Escherichia coli* was prominently used in recombinant protein production since it has a well-defined fermentation process and inexpensive culturing requirements (Schillberg et al. 2019). However, *E. coli*, along with other prokaryotic organisms, is not capable to perform significant cellular functions such as posttranslational modifications. Therefore, researchers have focused on especially Chinese hamster ovary (CHO) mammalian cells for protein production (Rader 2008). However, these cells require a long production period and expensive medium formulations. In addition, the scale-up process of mammalian cells can be challenging due to their specific requirements in terms of operational conditions in bioreactors. Also, contamination by pathogens such as viruses, bacteria, and prions causes a significant risk for the final product (Ma et al. 2003).

Plants have received a lot of interest in recent years as an alternative to expression systems including bacteria, yeast, and mammalian culture. The advantages of plant-based production systems have been summarized in Table 4.1. These advantages can be pointed out as low cost (particularly in high production quantities through effective biomanufacturing) and rapid, safe, efficient, and high-quality products (Chen and Davis 2016). Plants also eliminate some of the fundamental drawbacks of other expression systems, such as contamination risk and lack of posttranslational modifications. Plants are free of human pathogens and do not accumulate endotoxins. Plant cells also produce glycoproteins with N-glycan structures that only have minor differences compared with mammalian cells (Chen 2016). This enables the final product to have the appropriate posttranslational modifications and protein folding for manufacturing (see also Chaps. 1 and 2).

4.3.2 Plant Transformation Strategies

Plant molecular farming can be operated by two general strategies: stable and transient transformation. The transformation strategy generally depends on the purpose of the study in terms of cost, time, toxicity, and stability of the product.

Table 4.1 Comparison of commonly used host systems for recombinant protein production (adapted from Shanmugaraj et al. 2020)

Host system	Advantages	Disadvantages
Plant cell	Production of complex proteins Posttranslational modifications Similar glycosylation No contamination by human pathogens	Lower growth rate than prokaryotic cells
Bacteria	Easy to manipulate Rapid growth Easy scale-up	Improper protein folding Lack of posttranslational modifications Endotoxin accumulation
Yeast	Rapid growth Easy scale-up Proper posttranslational modifications	Hyperglycosylation
Mammalian cell	Proper protein folding Posttranslational modifications	High cost due to the expensive media compositions Limited scale-up capacity High risk of contamination

In stable transformation, *Agrobacterium*-mediated transformation is commonly used for delivery of the gene of interest to the host organism. Particle bombardment method is another established method that is widely used. These methods integrate the transferred gene into the genome, in which the offsprings have the possibility to express the desired product. So, stable transformation techniques are predominantly used for the long-term commercial production of recombinant proteins. However, optimization of transformation, selection of the transformed plants, and regeneration processes can take several months (approximately 6–18 months) depending on the species. Additionally, genetic instability and somaclonal variations can restrict the potential of stable transformation by creating variation in growth rates, undesired epigenetic changes, and transgene loss, which reduces the protein yield (Bhatia and Sharma 2015; Offringa et al. 1990). In such cases, cryopreservation has been regarded as a powerful strategy for the maintenance of genetic resources (Cho et al. 2007).

Nuclear or plastid transformation also allows stable expression of different proteins. The stable nuclear transformation has been considered as the most common transformation strategy in PMF. This method can also be used for the accumulation of proteins in the seeds of cereals for the durable storage without protein degradation (Horn et al. 2004; see also Chaps. 5 and 6). However, recombinant protein production in whole plants has limitations since outcrossing is a major concern for biosafety.

Alternatively, stable plastid transformation offers a remarkable solution for biosafety issues since chloroplasts are absent in pollens, while they are inherited through maternal tissues (Meyers et al. 2010). Also, a single plant cell contains 100 chloroplasts, each containing 100 copies of the chloroplast genome. Therefore, chloroplast transformation provides higher recombinant protein yield compared to nuclear transformation (Daniell et al. 2001). Until now, the highest recombinant protein concentration was observed in tobacco as 70% of total proteins via chloroplast transformation (Oey et al. 2009). Plastid transformation is frequently used in tobacco species, although there are studies conducted on several plants such as *Solanum melongena* and *Solanum lycopersicum* (Bock 2007; Singh et al. 2010; see also Chap. 9).

In recombinant protein production, transient expression is a valuable strategy in which the gene of interest does not integrate into the genome of the plant. This strategy can be performed via infiltration of *Agrobacterium* or viral vectors such as *potato virus X* (PVX), *tobacco mosaic virus* (TMV), and *cucumber mosaic virus* (CMV) (Lico et al. 2008). Transient expression systems are mostly preferable for pilot research since it provides a rapid expression (2–14 days depending on the host organism, product, and transformation method) of recombinant proteins. It can also be used for the production of toxic compounds in high concentrations such as alkaloids and tannins (see also chapter "Bioengineering of *Cannabis* Plants from Lab to the Field: Challenges and Opportunities"). However, the time to extract the protein must be well determined to maintain the stability of recombinant proteins.

Alternatively, plant cell suspension cultures have emerged for molecular farming. This system offers the dispersion of single-cell callus grown in liquid. Homogeneous liquid cultures can be obtained between fifth and tenth subcultures, which is more rapid than callus culture (Mustafa et al. 2011). Moreover, plant cell suspension culture enables reproducibility and a high rate of cell growth when compared with other plant tissue culture techniques (Muir and Hansch 1953). Since there is no need for the regeneration process, it accelerates protein production, which is a significant feature for PMF. The most remarkable feature of suspension culture for molecular farming is that it allows the large-scale production of valuable compounds, particularly in bioreactors. Plant cell suspension culture, which requires similar growth conditions to microbial cells, can be maintained in different types of bioreactors such as bubble column, membrane, wave, stirred-tank, and airlift bioreactors (Su et al. 2019). Besides, extraction and purification processes are easier and cost effective in plant cell suspension culture. These downstream processes can typically account for 80% of the total production cost (Roque et al. 2004; Schillberg et al. 2013). Furthermore, the target protein can be engineered to be secreted extracellularly into the culture medium, facilitating the workflow in downstream process (see also chapters "Production of Recombinant Proteins Using Plant Cell Suspension Cultures and Bioreactor Engineering: A Short Review" and "Scaling Up the Plant Molecular Farming via Bioprocessing of Plant Cell Suspension Culture").

Some examples include human granulocyte-colony-stimulating factor (hG-CSF), interferon-alpha 2b (IFN-α2b), and lignin, which have been successfully produced in *O. sativa*, *N. tabacum*, and *Forsythia koreana*, respectively (Hong et al. 2006; Kim et al. 2009; Tabar et al. 2012). Recombinant human glucocerebrosidase produced by Protalix from carrot cell suspension culture is one of the most well-known pharmaceuticals for the treatment of Gaucher's disease (Fox 2012).

4.3.3 *Plant Molecular Farming Applications*

Plant molecular farming applications provided plants as a platform to produce a lot of biological products such as vaccines, edible vaccines, therapeutic enzymes, therapeutic proteins, research enzymes, industrial enzymes, research reagents, feed additives, biopolymers, and biofuels. With this wide spectrum of biological products and potential candidates, PMF applications have gained importance and become more popular.

4.3.3.1 Therapeutic and Pharmaceutical Molecules

Antibodies or immunoglobulins can recognize specific antigens found on cells of foreign organisms and thus provide a defense mechanism against them. Different types of antibodies can be used in the diagnosis, prevention, as well as treatment of diseases. Plants are capable to produce specific antibodies after the transformation of the target DNA sequences. One of the most known antibodies produced in plants, avicidin, is an immunoglobulin that is used in the treatment of cancer. It was produced in transgenic corn by Monsanto. However, it was withdrawn from the market due to the side effects of the drug, which was not related to being plant based (Fischer and Buyel 2020). CaroRX, developed by Planet Biotechnology, is approved for use in the health sector in Europe and is actively used against oral bacterial infections in dental treatments (Ma et al. 1998). The ZMapp, an experimental drug, has been recognized as a milestone for the utilization of plants in antibody production. This drug has been developed for the Ebola virus by expressing three different chimeric monoclonal antibodies in tobacco. In the face of the Ebola crisis that exploded in Africa in 2014, the American Food and Drug Administration (FDA) allowed the emergency use of this drug for people with the disease, bypassing the clinical trial stages (Na et al. 2015; see also Chap. 7).

Tobacco is widely used in PMF applications; also, cereal crops are seen as good candidates for plant biofactories. 2F5 and 2F12 antibodies of *human immunodeficiency virus* (HIV) have been produced in tobacco (Floss et al. 2008; Sainsbury and Lomonossoff 2008). Brodzik et al. (2006) have produced BR55-2 antibody for human colorectal cancer in tobacco. In wheat and rice, the scFvT84.66 antibody was produced as a cancer tumor marker (Stöger et al. 2000). Likewise, 2G12 recombinant protein for HIV was produced in rice (Vamvaka et al. 2016).

4.3.3.2 Vaccines

Diseases such as HIV, hepatitis C, SARS, and influenza are still infecting many people at an increasing rate and threatening human health. Vaccines are used specifically to prevent the spread of such infectious diseases by protecting against a particular disease. The production of various pharmaceutical proteins used in the treatment or prevention of these diseases via PMF technologies has recently come into prominence. *N. tabacum* and *N. benthamiana* have been frequently encountered as model plants in plant-derived vaccines since they offer a rapid and high-volume production by using transient expression methods.

A variety of antigens for Ebola, Zika, and influenza have been engineered to be expressed in the nucleus and chloroplast of tobacco. In addition, protocol optimization was performed for recombinant protein production protocols for fruits and vegetables, including potatoes and tomatoes. Infectious bronchitis virus S1 glycoprotein was expressed in transgenic potatoes to confer protection against virus for chickens (Daniell et al. 2019). Leafy crops including clover, alfalfa, and lettuce were selected to succeed in the oral delivery of vaccine antigen. As a result of this, the elimination of purification and injection was targeted. Lettuce chloroplast was used to deliver a booster vaccine, which includes poliovirus capsid proteins expressed from lyophilized plant cells, and this induced neutralization antibodies to confer protection against all polio serotypes (Zhou et al. 2004; see also chapter "Plant Molecular Farming for Vaccine Development").

Cholera toxin B subunit (CTB) which was used for cholera was successfully expressed in tomato, tobacco, and rice (Daniell et al. 2001; Mishra et al. 2006; Nochi et al. 2007). For veterinary purposes, vaccines against Newcastle disease, avian influenza, enterotoxigenic *E. coli*, and foot-and-mouth disease were also produced in plants (Ling et al. 2010; see also chapter "Plant-Based Veterinary Vaccines"). In the transgenic pigeon pea (*Cajanus cajan*) plant, the hemagglutinin protein of rinderpest was produced (Satyavathi et al. 2003). Among carrots, potato, and tobacco, one or two of them were candidates for the production of the *Norwalk virus* capsid protein (Mason et al. 1996), L1 protein of *human papillomavirus* types 11 and 16 (Giorgi et al. 2010), and H5N1 pandemic vaccine candidate (D'Aoust et al. 2010).

The production of edible vaccines that activate the immune system has emerged as a valuable alternative to traditional vaccines. Edible plant tissues such as cereals, potato, lettuce, and various fruits are preferred in the development of edible vaccines. Since edible vaccines can be easily transported to different areas of the world without the need for a cold chain, it is considered a great advantage to overcome the transportation problem. On the other hand, some doubts arise due to the incomplete adjustment of the dose and the possibility of containing different amounts of proteins in the edible vaccines. The first example of plant-derived edible vaccines was produced via the transgenic potato producing an antigen against the hepatitis B in 2000 (Richter et al. 2000). Besides, studies were conducted against diseases such as *E. coli* heat-labile enterotoxin (LTB) and *Vibrio cholerae* toxin B subunit (CTB) in potato, corn, and tobacco (Nochi et al. 2007). To date, vaccines have been developed in plants against various diseases such as cholera, influenza, and SARS, but human clinical trials have not yet been completed. Plant-based vaccines have also been considered for the development of vaccines against the Covid-19 virus, which affected the world in 2020 and still persists (Lico et al. 2020; see also Chap. 12).

4.3.3.3 Biopolymers and Industrial Enzymes

Non-pharmaceutical plant-derived proteins have been on the market for a long time and generally consist of industrial enzymes such as trypsin, cellulase, laccase, and peroxidase (Basaran and Rodríguez-Cerezo 2008). For the production of this type of protein, whole plant, leaf, root, seed, or plant cell suspension culture techniques can be preferred according to the production strategy. It has been observed that long-term storage and higher yields are obtained when seeds are preferred for the industrial enzyme production (see also chapters "Molecular Farming of Industrial Enzymes: Products and Applications" and "Plant Molecular Farming for the Bulk Production of Industrial Enzymes").

In addition, plants are considered as a possible platform for biodegradable plastic-like substance production to be used in the biofuel industry. Biofuels arc fuels derived from biomass, and it is highly demanded as an alternative to fossil fuels. To date, copolymer production of polyhydroxybutyrate (PHB), cyanophycin, and polyhydroxyalkanoate (PHA) was expressed in transgenic plants (Conrad 2005; Matsumoto et al. 2009). In corn, polymer-degrading enzymes, exo-1,4 beta-glucanase, ligninase, cellulose, and hemicellulose are products derived from PMF to contribute to biofuel production (Park and Wi 2016).

4.4 Utilization of Genome Editing Tools for Plant Molecular Farming

The recombinant protein yield might be reduced due to RNA degradation or silencing (Brodersen and Voinnet 2006). Repression of genes involving small RNA processing is considered as a solution for higher recombinant protein production. Genome editing is an alternative tool for gene knockouts. Using CRISPR/Cas9 and TALENs, DCL2 and DCL3 genes that are used for small RNA processing have been repressed in *Glycine max* and *Medicago truncatula* (Curtin et al. 2018). Ludman et al. (2017) inactivated AGO2 using CRISPR/Cas9 in *N. benthamiana*, and higher expression levels were observed for green fluorescent protein, which suggested that CRISPR/Cas9 might be a convenient tool with its gene knockout application to increase recombinant protein production.

Phenolic compounds are required in the extraction of PMF products. Polyphenol oxidase (PPO) forms covalent complexes between proteins and phenols, which leads to protein precipitation, which is harmful to the output (Twyman et al. 2003). Using CRISPR/Cas9, the PPO gene was knocked out in potato tubers by 69% decrease in expression levels without other phenotypic effects (González et al. 2020). This makes CRISPR/Cas9 an alternative strategy in PMF applications to overcome the negative impacts of phenolic oxidation.

PMF products might also be impacted by flowering and senescence by influencing biomass quality and stem elongation. In tobacco, the FT5 gene was inactivated using CRISPR/Cas9, which leads to continuous accumulation of biomass under long-day conditions. In addition, since FT5 suppresses pollen and seed dispersal, it provides an advantage in biosafety issues (Schmidt et al. 2020). This revealed that CRISPR/Cas9 might be applied in tobacco for PMF applications to increase yield.

For manufacturing by-products such as building materials, whole-plant fiber processing is considered as a sustainable alternative (Revuelta-Aramburu et al. 2020). However, damaging metabolites should be removed from the plant waste to increase the acceptability of these materials. In previous studies, alkaloids and odorous compounds were removed from tobacco (Lin et al. 2016) and garlic, respectively (Mirondo and Barringer 2016). Genome editing applications of removing or inactivating the corresponding enzymes might be a good alternative for PMF studies of manufactured by-products.

Another strategy suggests that by reducing the endogenous storage protein accumulation, two- to tenfold increase in production of recombinant protein can be achieved (Takaiwa et al. 2017). In rice endosperm, an increase in the yield of recombinant cedar pollen allergen was achieved by the reduction of 13 kDa prolamins (Kawakatsu and Takaiwa 2012). Up to the present, RNAi was used in PMF applications for the suppression of endogenous seed storage proteins (Yuki et al. 2012). Recently, CRISPR system provided an alternative for targeting these genes in wheat (Sánchez-León et al. 2018), sorghum (Li et al. 2018), and Camelina (Lyzenga et al. 2019). For further PMF applications, endogenous storage protein depletion of seed crops might be developed via genome editing due to their stable phenotype (see also Chap. 3).

4.5 Conclusions

The usage of plants as a platform for valuable molecule production has always been evaluated as a system with high potential. However, producing target protein in a plant system was restricted because of the limitations of conventional transformation techniques, which results in random integration of transgene and copy number. This situation can cause undesirable knockdown of essential genes or random expression levels according to the position of integration. Moreover, a large class of mutant lines must be screened to find the most effective transgene integration, which is a time-consuming process. At this point, the plant production systems need an effective and more manageable way to integrate the transgene. Tools of targeted genome editing can be considered as a platform to compete with other systems. The fact that the studies of plant targeted genome editing are developing relatively slowly compared with other production systems like bacteria and yeast has created a big gap for plant production systems. Although different site-specific nucleases have been used in plants, the discovery of CRISPR-based systems has been a big milestone for plants, with much progress to be made.

As mentioned in the previous sections of this chapter, targeted genome editing of plants via CRISPR/Cas systems is a rapidly expanding area, including other wide range of applications via CRISPR tools such as base editing, activation, and repressor. However, in PMF, genome editing has not been a well-established method yet. The lack of development of genome editing applications for PMF has many aspects that need to be accomplished.

A considerable obstacle in the process of genome editing is the difficulties of plant transformation methods. The cell wall that is present in plants complicates the passage of genome editing reagents such as dsDNA, plasmids, RNPs, or RNA molecules by acting like a barrier (Zhang et al. 2018). The commonly used and well-established transformation methods like *Agrobacterium*-mediated transformation or biolistics should be optimized for each plant species to obtain higher transformation efficiencies. The development of new methods to overcome these limitations will surely accelerate the development of genome editing applications in the future, especially in the area of PMF (Keshavareddy et al. 2018; Nadakuduti and Enciso-Rodríguez 2021).

Another big obstacle is that the transgene insertion into the plant's genome via genome editing is not a well-established area. The knock-in studies utilizing HDR pathway have been demonstrated in model plants as well as crop plants like *N. benthamiana*, *O. sativa*, and maize; however, the frequency of integration was low (Begemann et al. 2017; Li et al. 2013b; Svitashev et al. 2016; Xu et al. 2017). The lack of development and low efficiency of knock-in studies are associated with the fact that plants generally utilize NHEJ mechanism over HDR for the repair of a DSB (Osakabe et al. 2010). Targeted knock-in via HDR pathway shows a low frequency of 0.01–0.1% compared with random integration events (Terada et al. 2007). In general, the studies indicate that to increase HDR repair frequency, factors such as the donor type and length can be optimized as well as the position of homology arms (Davis and Maizels 2014; Zhang et al. 2017). It is known that the donor template must be present at the time of the DSB; thus, the synchronization of delivery must be adjusted accordingly. The application of dual gRNA has been shown to increase HDR frequency (Zhao et al. 2016). One CRISPR/Cas9 study increased knock-in efficiency by creating direct recruitment of donor DNA to the target site by placing the target sequence both at the upstream and downstream of the donor DNA fragment, which was liberated by Cas9 nucleases (Zhang et al. 2017). Another study developed transgenic lines of *A. thaliana* plants expressing Cas9 under germ line-specific promoters, which showed an increase in heritable gene targeting frequency (Miki et al. 2018). Schiermeyer et al. (2019) have demonstrated a study with tobacco BY-2 cells by using the HDR repair mechanism to insert up to 20 kb DNA sequences. In another approach, inhibition of NHEJ repair mechanism either genetically or chemically has been suggested to improve HDR frequency. For this strategy, LIG4 (DNA ligase 4) and Ku70 mutant plants were developed in which the outcome was plants with suppressed NHEJ repair pathways (Endo et al. 2016; Qi et al. 2013). In some situations, PMF requires multigene constructs for the biosynthesis of proteins. Since the insertion of a large construct is difficult via targeted genome editing, it may not be the best approach to produce valuable products. However, using genome editing tools for transgene stacking has been suggested as an alternative approach (Buyel et al. 2021). Overall, it is suggested that the development of efficient knock-in strategies, especially via CRISPR, is crucial for both plant-based research including PMF.

References

Ainley WM, Sastry-Dent L, Welter ME, Murray MG, Zeitler B et al (2013) Trait stacking via targeted genome editing. Plant Biotechnol J 11:1126–1134

Andersson M, Turesson H, Nicolia A, Falt AS, Samuelsson M et al (2017) Efficient targeted multiallelic mutagenesis in tetraploid potato (*Solanum tuberosum*) by transient CRISPR-Cas9 expression in protoplasts. Plant Cell Rep 36:117–128

Anzalone AV, Koblan LW, Liu DR (2020) Genome editing with CRISPR–Cas nucleases, base editors, transposases and prime editors. Nat Biotechnol 38(7):824–844

Azegami T, Yuki Y, Kiyono H (2020) Plant-based mucosal vaccine delivery systems. In: Versteeg-Buschman L (ed) Mucosal vaccines. Academic, Boston, MA, pp 357–370

Bandyopadhyay A, Kancharla N, Javalkote VS, Dasgupta S, Brutnell TP (2020) CRISPR-Cas12a (Cpf1): a versatile tool in the plant genome editing tool box for agricultural advancement. Front Plant Sci 11:1589

Barampuram S, Zhang ZJ (2011) Recent advances in plant transformation. In: Birchler JA (ed) Plant chromosome engineering: methods and protocols, Methods in molecular biology. Springer Science+Business Media, pp 1–35

Barta A, Sommergruber K, Thompson D, Hartmuth K, Matzke MA et al (1986) The expression of a nopaline synthase—human growth hormone chimeric gene in transformed tobacco and sunflower callus tissue. Plant Mol Biol 6(5):347–357

Basaran P, Rodríguez-Cerezo E (2008) Plant molecular farming: opportunities and challenges. Crit Rev Biotechnol 28(3):153–172

Begemann MB, Gray BN, January E, Gordon GC, He Y et al (2017) Precise insertion and guided editing of higher plant genomes using Cpf1 CRISPR nucleases. Sci Rep 7(1):1–6

Belhaj K, Chaparro-Garcia A, Kamoun S, Nekrasov V (2013) Plant genome editing made easy: targeted mutagenesis in model and crop plants using the CRISPR/Cas system. Plant Methods 9:1–10

Belhaj K, Chaparro-Garcia A, Kamoun S, Patron NJ, Nekrasov V (2015) Editing plant genomes with CRISPR/Cas9. Curr Opin Biotechnol 32:76–84

Bevan MW, Chilton MD (1982) Multiple transcripts of T-DNA detected in nopaline crown gall tumors. J Mol Appl Genet 1(6):539–546

Bhatia S, Sharma K (2015) Technical glitches in micropropagation. In: Modern applications of plant biotechnology in pharmaceutical sciences. Academic, Cambridge, MA, pp 393–404

Bock R (2007) Plastid biotechnology: prospects for herbicide and insect resistance, metabolic engineering and molecular farming. Curr Opin Biotechnol 18(2):100–106

Boettcher M, McManus MT (2015) Choosing the right tool for the job: RNAi, TALEN, or CRISPR. Mol Cell 58(4):575–585

Bourgaud F, Gravot A, Milesi S, Gontier E (2001) Production of plant secondary metabolites: a historical perspective. Plant Sci 161(5):839–851

Brodersen P, Voinnet O (2006) The diversity of RNA silencing pathways in plants. Trends Genet 22(5):268–280

Brodzik R, Glogowska M, Bandurska K, Okulicz M, Deka D et al (2006) Plant-derived anti-Lewis Y mAb exhibits biological activities for efficient immunotherapy against human cancer cells. Proc Natl Acad Sci U S A 103(23):8804–8809

Burnett MJ, Burnett AC (2020) Therapeutic recombinant protein production in plants: challenges and opportunities. Plants People Planet 2(2):121–132

Buyel JF (2019) Plant molecular farming–integration and exploitation of side streams to achieve sustainable biomanufacturing. Front Plant Sci 9:1893

Buyel JF, Stöger E, Bortesi L (2021) Targeted genome editing of plants and plant cells for biomanufacturing. Transgenic Res 30(4):401–426

Chandrasekaran J, Brumin M, Wolf D, Leibman D, Klap C, Pearlsman M, Sherman A, Arazi T, Gal-On A (2016) Development of broad virus resistance in non-transgenic cucumber using CRISPR/Cas9 technology. Mol Plant Pathol 17(7):1140–1153

Chen Q (2016) Glycoengineering of plants yields glycoproteins with polysialylation and other defined N-glycoforms. Proc Natl Acad Sci U S A 113(34):9404–9406

Chen Q, Davis KR (2016) The potential of plants as a system for the development and production of human biologics. F1000Research 5:F1000 Faculty Rev-912

Cho JS, Hong SM, Joo SY, Yoo JS, Kim DI (2007) Cryopreservation of transgenic rice suspension cells producing recombinant hCTLA4Ig. Appl Microbiol Biotechnol 73(6):1470–1476

Cohen SN, Chang AC, Boyer HW, Helling RB (1973) Construction of biologically functional bacterial plasmids in vitro. Proc Natl Acad Sci U S A 70(11):3240–3244

Cong L, Ran FA, Cox D, Lin S, Barretto R, Habib N, Hsu PD, Wu X, Jiang W, Marraffini LA, Zhang F (2013) Multiplex genome engineering using CRISPR/Cas systems. Science 339(6121):819–823

Conrad U (2005) Polymers from plants to develop biodegradable plastics. Trends Plant Sci 10(11):511–512

Curtin SJ, Zhang F, Sander JD, Haun WJ, Starker C et al (2011) Targeted mutagenesis of duplicated genes in soybean with zinc-finger nucleases. Plant Physiol 156(2):466–473

Curtin SJ, Xiong Y, Michno JM, Campbell BW, Stec AO et al (2018) Crispr/Cas9 and TALENs generate heritable mutations for genes involved in small RNA processing of *Glycine max* and *Medicago truncatula*. Plant Biotechnol J 16(6):1125–1137

D'Aoust MA, Couture MMJ, Charland N, Trépanier S, Landry N, Ors F, Vézina LP (2010) The production of hemagglutinin-based virus-like particles in plants: a rapid, efficient and safe response to pandemic influenza. Plant Biotechnol J 8(5):607–619

Daniell H, Lee SB, Panchal T, Wiebe PO (2001) Expression of the native cholera toxin B subunit gene and assembly as functional oligomers in transgenic tobacco chloroplasts. J Mol Biol 311(5):1001–1009

Daniell H, Rai V, Xiao Y (2019) Cold chain and virus-free oral polio booster vaccine made in lettuce chloroplasts confers protection against all three poliovirus serotypes. Plant Biotechnol J 17(7):1357–1368

Davis L, Maizels N (2014) Homology-directed repair of DNA nicks via pathways distinct from canonical double-strand break repair. Proc Natl Acad Sci U S A 111(10):E924–E932

Doyle EL, Stoddard BL, Voytas DF, Bogdanove AJ (2013) TAL effectors: highly adaptable phytobacterial virulence factors and readily engineered DNA-targeting proteins. Trends Cell Biol 23(8):390–398

Endo M, Mikami M, Toki S (2016) Biallelic gene targeting in rice. Plant Physiol 170(2):667–677

Fischer R, Buyel JF (2020) Molecular farming–the slope of enlightenment. Biotechnol Adv 40:107519

Fischer R, Schillberg S, Buyel FJ, Twyman R (2013) Commercial aspects of pharmaceutical protein production in plants. Curr Pharm Des 19(31):5471–5477

Floss DM, Sack M, Stadlmann J, Rademacher T, Scheller J et al (2008) Biochemical and functional characterization of anti-HIV antibody–ELP fusion proteins from transgenic plants. Plant Biotechnol J 6(4):379–391

Fox JL (2012) First plant-made biologic approved. Nat Biotechnol 30(6):472–473

Fraley RT (1983) Liposome-mediated delivery of tobacco mosaic virus RNA into petunia protoplast. Plant Mol Biol 2(1):5–14

Garcia-Ruiz H (2018) Susceptibility genes to plant viruses. Viruses 10:484

Gasiunas G, Barrangou R, Horvath P, Siksnys V (2012) Cas9–crRNA ribonucleoprotein complex mediates specific DNA cleavage for adaptive immunity in bacteria. Proc Natl Acad Sci U S A 109(39):E2579–E2586

Giorgi C, Franconi R, Rybicki EP (2010) Human papillomavirus vaccines in plant. Expert Rev Vaccines 9(8):913–924

Gomes AR, Byregowda SM, Veeregowda BM, Balamurugan V (2016) An overview of heterologous expression host systems for the production of recombinant proteins. Adv Anim Vet Sci 4(7):346–356

González MN, Massa GA, Andersson M, Turesson H, Olsson N et al (2020) Reduced enzymatic browning in potato tubers by specific editing of a polyphenol oxidase gene via ribonucleoprotein complexes delivery of the CRISPR/Cas9 system. Front Plant Sci 10:1649

Gupta M, DeKelver RC, Palta A, Clifford C, Gopalan S et al (2012) Transcriptional activation of *Brassica napus* beta-ketoacyl-ACP synthase II with an engineered zinc finger protein transcription factor. Plant Biotechnol J 10(7):783–791

Hanly A, Karagiannis J, Lu QSM, Tian L, Hannoufa A (2020) Characterization of the role of SPL9 in drought stress tolerance in *Medicago sativa*. Int J Mol Sci 21(17):6003

Herrera-Estrella L, De Block M, Messens EHJP, Hernalsteens JP, Van Montagu M, Schell J (1983) Chimeric genes as dominant selectable markers in plant cells. EMBO J 2(6):987–995

Hiatt A, Caffferkey R, Bowdish K (1989) Production of antibodies in transgenic plants. Nature 342(6245):76–78

Hong SY, Kwon TH, Jang YS, Kim SH, Yang MS (2006) Production of bioactive human granulocyte-colony stimulating factor in transgenic rice cell suspension cultures. Protein Expr Purif 47(1):68–73

Hood EE, Witcher DR, Maddock S, Meyer T, Baszczynski C, Bailey M, Howard JA (1997) Commercial production of avidin from transgenic maize: characterization of transformant, production, processing, extraction and purification. Mol Breed 3(4):291–306

Horn ME, Woodard SL, Howard JA (2004) Plant molecular farming: systems and products. Plant Cell Rep 22(10):711–720

Jinek M, Chylinski K, Fonfara I, Hauer M, Doudna JA, Charpentier E (2012) A programmable dual-RNA–guided DNA endonuclease in adaptive bacterial immunity. Science 337(6096):816–821

Joung JK, Sander JD (2013) TALENs: a widely applicable technology for targeted genome editing. Nat Rev Mol Cell Biol 14(1):49–55

Kawakatsu T, Takaiwa F (2012) Reduction of 13 kD prolamins increases recombinant protein yield and recovery rate in rice endosperm. Plant Signal Behav 7(11):1402–1403

Keshavareddy G, Kumar ARV, Ramu VS (2018) Methods of plant transformation-a review. Int J Curr Microbiol Appl Sci 7(7):2656–2668

Kim HJ, Ono E, Morimoto K, Yamagaki T, Okazawa A, Kobayashi A, Satake H (2009) Metabolic engineering of lignan biosynthesis in *Forsythia* cell culture. Plant Cell Physiol 50(12):2200–2209

Klein TM, Wolf ED, Wu R, Sanford JC (1987) High-velocity microprojectiles for delivering nucleic acids into living cells. Nature 327(6117):70–73

Kleinstiver BP, Pattanayak V, Prew MS, Tsai SQ, Nguyen NT, Zheng Z, Joung JK (2016) High-fidelity CRISPR–Cas9 nucleases with no detectable genome-wide off-target effects. Nature 529(7587):490–495

Kumar V, Jain M (2015) The CRISPR–Cas system for plant genome editing: advances and opportunities. J Exp Bot 66(1):47–57

Lee RC, Feinbaum RL, Ambros V (1993) The *C. elegans* heterochronic gene lin-4 encodes small RNAs with antisense complementarity to lin-14. Cell 75(5):843–854

Li T, Liu B, Spalding MH, Weeks DP, Yang B (2012) High-efficiency TALEN-based gene editing produces disease-resistant rice. Nat Biotechnol 30(5):390–392

Li T, Huang S, Zhou J, Yang B (2013a) Designer TAL effectors induce disease susceptibility and resistance to *Xanthomonas oryzae* pv. Oryzae in rice. Mol Plant 6(3):781–789

Li JF, Norville JE, Aach J, McCormack M, Zhang D et al (2013b) Multiplex and homologous recombination–mediated genome editing in *Arabidopsis* and *Nicotiana benthamiana* using guide RNA and Cas9. Nat Biotechnol 31(8):688–691

Li J, Zhang H, Si X, Tian Y, Chen K et al (2017) Generation of thermosensitive male-sterile maize by targeted knockout of the ZmTMS5 gene. J Genet Genomics 44(9):465–468

Li A, Jia S, Yobi A, Ge Z, Sato SJ et al (2018) Editing of an alpha-kafirin gene family increases, digestibility and protein quality in sorghum. Plant Physiol 177(4):1425–1438

Lico C, Chen Q, Santi L (2008) Viral vectors for production of recombinant proteins in plants. J Cell Physiol 216(2):366–377

Lico C, Santi L, Baschieri S, Noris E, Marusic C et al (2020) Plant molecular farming as a strategy against COVID-19–the Italian perspective. Front Plant Sci 11:609910

Lin S, Zhang X, Song S, Hayat K, Eric K, Majeed H (2016) Tobacco alkaloids reduction by casings added/enzymatic hydrolysis treatments assessed through PLSR analysis. Regul Toxicol Pharmacol 75:27–34

Ling HY, Pelosi A, Walmsley AM (2010) Current status of plant-made vaccines for veterinary purposes. Expert Rev Vaccines 9(8):971–982

Ludman M, Burgyán J, Fátyol K (2017) Crispr/Cas9 mediated inactivation of Argonaute 2 reveals its differential involvement in antiviral responses. Sci Rep 7(1):1–12

Luria N, Smith E, Reingold V, Bekelman I, Lapidot M et al (2017) A new Israeli *Tobamovirus* isolate infects tomato plants harboring *Tm-22* resistance genes. PLoS One 12:e0170429

Lyzenga WJ, Harrington M, Bekkaoui D, Wigness M, Hegedus DD, Rozwadowski KL (2019) CRISPR/Cas9 editing of three CRUCIFERIN C homologues alters the seed protein profile in *Camelina sativa*. BMC Plant Biol 19(1):1–16

Ma JKC, Hikmat BY, Wycoff K, Vine ND, Chargelegue D, Yu L, Lehner T (1998) Characterization of a recombinant plant monoclonal secretory antibody and preventive immunotherapy in humans. Nat Med 4(5):601–606

Ma JK, Drake PM, Christou P (2003) The production of recombinant pharmaceutical proteins in plants. Nat Rev Genet 4(10):794–805

Mak ANS, Bradley P, Cernadas RA, Bogdanove AJ, Stoddard BL (2012) The crystal structure of TAL effector PthXo1 bound to its DNA target. Science 335(6069):716–719

Mamta B, Rajam MV (2017) RNAi technology: a new platform for crop pest control. Physiol Mol Biol Plants 23(3):487–501

Mason HS, Ball JM, Shi JJ, Jiang X, Estes MK, Arntzen CJ (1996) Expression of Norwalk virus capsid protein in transgenic tobacco and potato and its oral immunogenicity in mice. Proc Natl Acad Sci U S A 93(11):5335–5340

Mathur V, Javid L, Kulshrestha S, Mandal A, Reddy AA (2017) World cultivation of genetically modified crops: opportunities and risks. In: Lichtfouse E (ed) Sustainable agriculture reviews. Springer International Publishing, pp 45–87

Matsumoto KI, Murata T, Nagao R, Nomura CT, Arai S et al (2009) Production of short-chain-length/medium-chain-length polyhydroxyalkanoate (PHA) copolymer in the plastid of *Arabidopsis thaliana* using an engineered 3-ketoacyl-acyl carrier protein synthase III. Biomacromolecules 10(4):686–690

Meyers B, Zaltsman A, Lacroix B, Kozlovsky SV, Krichevsky A (2010) Nuclear and plastid genetic engineering of plants: comparison of opportunities and challenges. Biotechnol Adv 28(6):747–756

Miki D, Zhang W, Zeng W, Feng Z, Zhu JK (2018) CRISPR/Cas9-mediated gene targeting in *Arabidopsis* using sequential transformation. Nat Commun 9(1):1–9

Mirondo R, Barringer S (2016) Deodorization of garlic breath by foods, and the role of polyphenol oxidase and phenolic compounds. J Food Sci 81(10):C2425–C2430

Mishra S, Yadav DK, Tuli R (2006) Ubiquitin fusion enhances cholera toxin B subunit expression in transgenic plants and the plant-expressed protein binds GM1 receptors more efficiently. J Biotechnol 127(1):95–108

Muir RM, Hansch C (1953) On the mechanism of action of growth regulators. Plant Physiol 28(2):218

Mustafa NR, De Winter W, Van Iren F, Verpoorte R (2011) Initiation, growth and cryopreservation of plant cell suspension cultures. Nat Protoc 6(6):715–742

Na W, Park N, Yeom M, Song D (2015) Ebola outbreak in Western Africa 2014: what is going on with Ebola virus? Clin Exp Vaccine Res 4(1):17–22

Nadakuduti SS, Enciso-Rodríguez F (2021) Advances in genome editing with CRISPR systems and transformation technologies for plant DNA manipulation. Front Plant Sci 11:2267

Nemudryi AA, Valetdinova KR, Medvedev SP, Zakian SM (2014) Talen and CRISPR/Cas genome editing systems: tools of discovery. Acta Nat 6(3):19–40

Nochi T, Takagi H, Yuki Y, Yang L, Masumura T et al (2007) Rice-based mucosal vaccine as a global strategy for cold-chain-and needle-free vaccination. Proc Natl Acad Sci U S A 104(26):10986–10991

Oey M, Lohse M, Kreikemeyer B, Bock R (2009) Exhaustion of the chloroplast protein synthesis capacity by massive expression of a highly stable protein antibiotic. Plant J 57(3):436–445

Offringa R, De Groot MJ, Haagsman HJ, Does MP, Van Den Elzen PJ, Hooykaas PJ (1990) Extrachromosomal homologous recombination and gene targeting in plant cells after *Agrobacterium*-mediated transformation. EMBO J 9(10):3077–3084

Okada A, Arndell T, Borisjuk N, Sharma N, Watson-Haigh NS et al (2019) CRISPR/Cas9-mediated knockout of Ms1 enables the rapid generation of male-sterile hexaploid wheat lines for use in hybrid seed production. Plant Biotechnol J 17:1905–1913

Osakabe K, Osakabe Y, Toki S (2010) Site-directed mutagenesis in *Arabidopsis* using custom-designed zinc finger nucleases. Proc Natl Acad Sci U S A 107(26):12034–12039

Osakabe Y, Osakabe K, Shinozaki K, Tran LSP (2014) Response of plants to water stress. Front Plant Sci 5:86

Osakabe Y, Watanabe T, Sugano SS, Ueta R, Ishihara R, Shinozaki K, Osakabe K (2016) Optimization of CRISPR/Cas9 genome editing to modify abiotic stress responses in plants. Sci Rep 6(1):1–10

Park KY, Wi SJ (2016) Potential of plants to produce recombinant protein products. J Plant Biol 59(6):559–568

Petolino JF, Worden A, Curlee K, Connell J, Strange Moynahan TL, Larsen C, Russell S (2010) Zinc finger nuclease-mediated transgene deletion. Plant Mol Biol 73(6):617–628

Pyott DE, Sheehan E, Molnar A (2016) Engineering of CRISPR/Cas9-mediated potyvirus resistance in transgene-free *Arabidopsis* plants. Mol Plant Pathol 17(8):1276–1288

Qi Y, Zhang Y, Zhang F, Baller JA, Cleland SC et al (2013) Increasing frequencies of site-specific mutagenesis and gene targeting in *Arabidopsis* by manipulating DNA repair pathways. Genome Res 23(3):547–554

Rader RA (2008) Expression systems for process and product improvement. BioProcess Int 6(4):4–9

Ramirez CL, Foley JE, Wright DA, Müller-Lerch F, Rahman SH et al (2008) Unexpected failure rates for modular assembly of engineered zinc fingers. Nat Methods 5(5):374–375

Revuelta-Aramburu M, Verdú-Vázquez A, Gil-López T, Morales-Polo C (2020) Environmental analysis of the use of plant fiber blocks in building construction. Sci Total Environ 725:138495

Richter LJ, Thanavala Y, Arntzen CJ, Mason HS (2000) Production of hepatitis B surface antigen in transgenic plants for oral immunization. Nat Biotechnol 18(11):1167–1171

Roque ACA, Lowe CR, Taipa MÂ (2004) Antibodies and genetically engineered related molecules: production and purification. Biotechnol Prog 20(3):639–654

Safari F, Zare K, Negahdaripour M, Barekati-Mowahed M, Ghasemi Y (2019) CRISPR Cpf1 proteins: structure, function and implications for genome editing. Cell Biosci 9(1):1–21

Sainsbury F, Lomonossoff GP (2008) Extremely high-level and rapid transient protein production in plants without the use of viral replication. Plant Physiol 148(3):1212–1218

Sánchez-León S, Gil-Humanes J, Ozuna CV, Giménez MJ, Sousa C, Voytas DF, Barro F (2018) Low-gluten, nontransgenic wheat engineered with CRISPR/Cas9. Plant Biotechnol J 16(4):902–910

Satyavathi V, Prasad V, Shaila M, Sita LG (2003) Expression of hemagglutinin protein of Rinderpest virus in transgenic pigeon pea [*Cajanus cajan* (L.) Millsp.] plants. Plant Cell Rep 21(7):651–658

Schiermeyer A, Schneider K, Kirchhoff J, Schmelter T, Koch N et al (2019) Targeted insertion of large DNA sequences by homology-directed repair or non-homologous end joining in engineered tobacco BY-2 cells using designed zinc finger nucleases. Plant Direct 3(7):e00153

Schillberg S, Raven N, Fischer R, Twyman MR, Schiermeyer A (2013) Molecular farming of pharmaceutical proteins using plant suspension cell and tissue cultures. Curr Pharm Des 19(31):5531–5542

Schillberg S, Raven N, Spiegel H, Rasche S, Buntru M (2019) Critical analysis of the commercial potential of plants for the production of recombinant proteins. Front Plant Sci 10:720

Schmidt FJ, Zimmermann MM, Wiedmann DR, Lichtenauer S, Grundmann L et al (2020) The major floral promoter NtFT5 in tobacco (*Nicotiana tabacum*) is a promising target for crop improvement. Front Plant Sci 10:1666

Shanmugaraj B, Bulaon ICJ, Phoolcharoen W (2020) Plant molecular farming: a viable platform for recombinant biopharmaceutical production. Plants 9(7):842

Shukla VK, Doyon Y, Miller JC, DeKelver RC, Moehle EA et al (2009) Precise genome modification in the crop species *Zea mays* using zinc-finger nucleases. Nature 459:437–441

Singh AK, Verma SS, Bansal KC (2010) Plastid transformation in eggplant (*Solanum melongena* L.). Transgenic Res 19(1):113–119

Singh S, Gupta M, Pandher S, Kaur G, Goel N, Rathore P (2020) Author Correction: Using de novo transcriptome assembly and analysis to study RNAi in *Phenacoccus solenopsis* Tinsley (Hemiptera: Pseudococcidae). Sci Rep 10(1):1–1

Sonoda E, Hochegger H, Saberi A, Taniguchi Y, Takeda S (2006) Differential usage of non-homologous end-joining and homologous recombination in double strand break repair. DNA Repair 5(9–10):1021–1029

Spiegel H, Stöger E, Twyman RM, Buyel JF (2018) Current status and perspectives of the molecular farming landscape. In: Kermode AR, Jiang L (eds) Molecular pharming: applications, challenges and emerging areas. Wiley, pp 1–23

Sternberg SH, Redding S, Jinek M, Greene EC, Doudna JA (2014) DNA interrogation by the CRISPR RNA-guided endonuclease Cas9. Nature 507(7490):62–67

Stöger E, Vaquero C, Torres E, Sack M, Nicholson L, Drossard J, Williams S, Keen D, Perrin Y, Christou P, Fischer R (2000) Cereal crops as viable production and storage systems for pharmaceutical scFv antibodies. Plant Mol Biol 42(4):583–590

Su R, Sujarani M, Shalini P, Prabhu N (2019) A review on bioreactor technology assisted plant suspension culture. Asian J Biotechnol Bioresour Technol 5:1–13

Sun Y, Zhang X, Wu C, He Y, Ma Y, Hou H, Guo X, Du W, Zhao Y, Xia L (2016) Engineering herbicide-resistant rice plants through CRISPR/Cas9-mediated homologous recombination of acetolactate synthase. Mol Plant 9(4):628–631

Svitashev S, Schwartz C, Lenderts B, Young JK, Mark Cigan A (2016) Genome editing in maize directed by CRISPR–Cas9 ribonucleoprotein complexes. Nat Commun 7(1):1–7

Tabar MS, Solouki M, Tohidfar M, Sadeghizadeh M (2012) Expression of human granulocyte-colony stimulating factor ('hG-CSF') gene in tobacco (*Nicotiana tabacum*). Aust J Crop Sci 6(1):135–140

Takaiwa F, Wakasa Y, Hayashi S, Kawakatsu T (2017) An overview on the strategies to exploit rice endosperm as production platform for biopharmaceuticals. Plant Sci 263:201–209

Terada R, Johzuka-Hisatomi Y, Saitoh M, Asao H, Iida S (2007) Gene targeting by homologous recombination as a biotechnological tool for rice functional genomics. Plant Physiol 144(2):846–856

Thomas B, Van Deynze A, Bradford K (2002) Production of therapeutic proteins in plants. UCANR Publications, Oakland, pp 1–12

Toda E, Koiso N, Takebayashi A, Ichikawa M, Kiba T, Osakabe K, Osakabe Y, Sakakibara H, Kato N, Okamoto T (2019) An efficient DNA- and selectable-marker-free genome-editing system using zygotes in rice. Nat Plants 5(4):363–368

Townsend JA, Wright DA, Winfrey RJ, Fu F, Maeder ML, Joung JK, Voytas DF (2009) High-frequency modification of plant genes using engineered zinc-finger nucleases. Nature 459:442–445

Twyman RM, Stoger E, Schillberg S, Christou P, Fischer R (2003) Molecular farming in plants: host systems and expression technology. Trends Biotechnol 21(12):570–578

Urnov FD, Rebar EJ, Holmes MC, Zhang HS, Gregory PD (2010) Genome editing with engineered zinc finger nucleases. Nat Rev Genet 11(9):636–646

Vamvaka E, Twyman RM, Murad AM, Melnik S, Te AYH, Arcalis E, Altmann F, Stoger E, Rech E, Ma JKC, Christou P (2016) Rice endosperm produces an underglycosylated and potent form of the HIV-neutralizing monoclonal antibody 2G12. Plant Biotechnol J 14:97–108

Wamaitha MJ, Nigam D, Maina S, Stomeo F, Wangai A et al (2018) Metagenomic analysis of viruses associated with maize lethal necrosis in Kenya. Virol J 15(1):1–19

Wright DA, Townsend JA, Winfrey RJ Jr, Irwin PA, Rajagopal J et al (2005) High-frequency homologous recombination in plants mediated by zinc-finger nucleases. Plant J 44(4):693–705

Wurtzel ET, Vickers CE, Hanson AD, Millar AH, Cooper M et al (2019) Revolutionizing agriculture with synthetic biology. Nat Plants 5(12):1207–1210

Xu X, Qi LS (2019) A CRISPR–dCas toolbox for genetic engineering and synthetic biology. J Mol Biol 431(1):34–47

Xu R, Qin R, Li H, Li D, Li L, Wei P, Yang J (2017) Generation of targeted mutant rice using a CRISPR-Cpf1 system. Plant Biotechnol J 15(6):713–717

Ye X, Al-Babili S, Kloti A, Zhang J, Lucca P, Beyer P, Potrykus I (2000) Engineering the provitamin A (β-carotene) biosynthetic pathway into (carotenoid-free) rice endosperm. Science 287(5451):303–305

Yuki Y, Mejima M, Kurokawa S, Hiroiwa T, Kong IG et al (2012) RNAi suppression of rice endogenous storage proteins enhances the production of rice-based *Botulinum neurotoxin* type A vaccine. Vaccine 30(28):4160–4166

Zaman QU, Li C, Cheng H, Hu Q (2019) Genome editing opens a new era of genetic improvement in polyploid crops. Crop J 7(2):141–150

Zerbini FM, da Silva FN, Patricia G, Urquiza C, Basso MF (2014) Transgenic plants. In: Borem A, Fritsche-Neto R (eds) Biotechnology and plant breeding: applications and approaches for developing improved cultivars. Elsevier, pp 179–199

Zhang XH, Tee LY, Wang XG, Huang QS, Yang SH (2015) Off-target effects in CRISPR/Cas9-mediated genome engineering. Mol Ther Nucleic Acids 4:e264

Zhang JP, Li XL, Li GH, Chen W, Arakaki C et al (2017) Efficient precise knockin with a double cut HDR donor after CRISPR/Cas9-mediated double-stranded DNA cleavage. Genome Biol 18(1):1–18

Zhang Y, Massel K, Godwin ID, Gao C (2018) Applications and potential of genome editing in crop improvement. Genome Biol 19:210

Zhao Y, Zhang C, Liu W, Gao W, Liu C et al (2016) An alternative strategy for targeted gene replacement in plants using a dual-sgRNA/Cas9 design. Sci Rep 6(1):1–11

Zhou JY, Cheng LQ, Zheng XJ, Wu JX, Shang SB, Wang JY, Chen JG (2004) Generation of the transgenic potato expressing full-length spike protein of infectious bronchitis virus. J Biotechnol 111(2):121–130

Chapter 5
Seed-Based Production System for Molecular Farming

Fumio Takaiwa

Abstract Seed is an ideal bioreactor for the expression of high-value recombinant proteins. High production can be achieved by taking advantage of established seed-specific promoters and targeting signal. When they are produced in seeds and are deposited in the seed-specific subcellular compartments, large amounts (quantities) of expressed recombinant proteins are stably accumulated without degradation even if stocked at ambient temperature for several years. Furthermore, when non-purified seed-made pharmaceuticals such as vaccines or antibodies are orally administered, immune reaction can be more efficiently induced in mucosal and systemic manners without expensive downstream processing as compared to the purified naked ones. This is mainly attributed to the fact that they can be protected from harsh conditions in gastrointestinal tract by bioencapsulation of two barriers of cell wall and protein body. Taken together, seed-based manufacturing system provides not only scalable, robust, and cost-effective production of various recombinant proteins, but also safe, convenient, and effective delivery of biopharmaceuticals and nutraceuticals to gastrointestinal tract.

Keywords Seed-specific promoter · Subcellular localization · ER stress · Plant-made pharmaceuticals · Protein body · Seed storage proteins

5.1 Introduction

Seeds are inherently natural reservoir of nutrients (proteins, carbohydrates, and lipids) for germinating seedling as well as human and livestock. High amounts of seed proteins are synthesized and accumulated during seed maturation. Leguminous

F. Takaiwa (✉)
Soul Signal Institute, Shiraoi, Hokkaido, Japan

C. Kole et al. (eds.), *Tools & Techniques of Plant Molecular Farming*, Concepts and Strategies in Plant Sciences, https://doi.org/10.1007/978-981-99-4859-8_5

seeds accumulate proteins at a level of 20–40% of seed weight, whereas the protein content of cereal grains is 7–15%. Majority of these seed proteins are stably deposited as seed storage proteins (SSPs) in the specialized intracellular compartments referred to as protein bodies (PBs). Such high protein contents in various seeds are in remarkable contrast with 2% or less of green tissues and tubers. Therefore, in view of such high protein contents and their stability, seeds have got a lot of attention as one of the ideal bioreactors for production of recombinant proteins to advance molecular farming. Seed production platform has several advantages over the conventional mammalian and bacterial production platforms in terms of low-cost production, high scalability (easy control of production scale), high productivity, high safety (non-contamination of human pathogens), high stability for several years at ambient temperature, and easy administration through oral rout without purification (Stöger et al. 2005; Lau and Sun 2009; Boothe et al. 2010; Sabalza et al. 2013; Wakasa and Takaiwa 2013; Takaiwa et al. 2017). Numerous industrial enzymes, pharmaceutical proteins (vaccines, antibodies, and cytokines), and nutraceuticals (bioactive peptides) have been mainly produced in cereal and leguminous seeds. The first commercialized plant-made recombinant proteins were avidin and trypsin, which were produced in transgenic maize grains by ProdiGene (Hood et al. 1997). Aprotinin, β-glucosidase, and several industrial proteins/enzymes (β-1,4-endoglucanase, cellobiohydrolase, manganese peroxidase, and amylopullulanase) have also been marketed. Other cereal grains are also utilized as production hosts. Human lactoferrin and human lysozyme are produced for the treatment of acute diarrhea and dehydration in transgenic rice cultivated in fields by Ventria Bioscience (Huang et al. 2002; Nandi et al. 2005). Human serum albumin (HAS) produced in transgenic rice seeds is widely used as a pathogen-free mammalian cell culture supplement (He et al. 2011). Over 30 types of human growth hormone and cytokines are produced in transgenic barley grains by ORF Genetics in Iceland, and their purified ones are commercialized for diagnostic, research, and cosmetic applications (Magnusdottir et al. 2013).

As shown in Table 5.1, plant-based production platform is categorized into at least five groups based on expression styles (stable expression in nuclear or plastid genomes vs. transient expression based on agroinfiltration or plant virus infection), production sites of recombinant proteins in plant (seeds, vegetative tissues, chloroplasts), and propagation style (plant tissues vs. in vitro culture cells such as cell suspension cultures and hairy root cultures). Individual production platform has advantages and disadvantages regarding operation cost, production yields, and production timescale (see also Chaps. 1 and 2).

Table 5.1 Comparison of plant production systems

Factor	Seed production	Vegetative production	Chloroplast production	In vitro culture production	Transient expression
Scalability	Excellent	Excellent	Excellent	Equipment limitations	Greenhouse size limitation
Infrastructure (cost)	Field (low cost)	Field (low cost)	Field (low cost)	Equipment (expensive)	Greenhouse (expensive)
Timescale for production	Months	Months	Months	Weeks	Weeks (within 4 weeks)
Consumption	Edible/purified	Edible/purified	Edible/purified	Edible/purified	Purified
Glycosylation	Some glycosylation	Some glycosylation	No glycosylation	Some glycosylation	Some glycosylation
Use in food and feed	Possible	Limited	Limited	No	No
Social acceptance	Necessary	Necessary	No need	No need	No need
Product stability	High	Low	Low	Low	Low
Storage form	Dried/ambient	Desiccated/frozen	Desiccated/frozen	Desiccated/frozen	Desiccated/frozen
Storage and shipping cost	Low	High	High	High	High
Host used for expression	Broad	Broad	Limited (several species)	Limited	Limited (mainly tobacco)
Expression level	High	Low	Very high	Middle	High

5.2 Seed-Based Production System

5.2.1 Selection of Production Hosts for Seed-Based Platform

The choice of host plant is a critical factor for the success of the plant molecular pharming approach. Host plant suitable for seed-based production platform has to be determined in view of a broad range of criteria including the nature of the target recombinant protein, biomass yield, ease of transformation, posttranslational modifications, scale-up of production, maintenance costs, span of production cycles, and downstream processing requirements (Twyman et al. 2003; Sabalza et al. 2013).

It is important to note that recombinant proteins expressed in seeds are remarkably stable at ambient temperature. This is associated with the physiological conditions of mature seed such as dormancy, low water content, and relatively high content of protease inhibitors. Actually, desiccation after maturation decreases activities of proteases in seed cells, resulting in protection of synthesized recombinant proteins from degradation. Therefore, the proteins produced in seeds allow long-term storage for at least 3 years at room temperature without detectable loss of protein activity or degradation (proteolytic cleavage). Moreover, unlike soybean, wheat, tobacco, and peanut containing food allergens or toxic compounds, many cereal seeds including rice and maize are generally regarded as safe (GRAS), leading to reduction of cost required for downstream processing and allowance of oral intake (see also Chap. 6).

Maize, rice, barley, and wheat seeds offer good production platforms for high-value recombinant proteins such as pharmaceutical proteins (see also chapter "Molecular Farming of Pharmaceutical Proteins in Different Crop Systems: A Way Forward"). Especially, maize seed is an excellent bioreactor because its grain size is larger than other cereal grins. Furthermore, maize not only has the highest biomass yield in hectare at the lowest production costs among food crops, but also its transformation is relatively easy. One major drawback is that it is a cross-pollinating species, which leads to substantial risk of gene flow. Similarly, rice has several advantages in terms of high production yield, ease of transformation, capacity of rapid scale-up, and self-pollinating trait, reducing the probability of horizontal gene flow. Drawback of rice is that seed protein content (7%) is lower than other cereal crops, and its production cost is expensive (see also chapter "The Use of Rice Seed as Bioreactor"). Barley is also a self-pollinating crop. Its seed protein content (15%) is the highest of cereal crops, but its grain yield is lower than maize or rice. Barley is less widely cultivated and is more difficult to transform than them. Biomass yield of wheat grain is the lowest of many cereals, and transformation is also difficult and protein expression level is relatively low. This is related to the fact that wheat genome size (about 17 Gb) is about 7–40-fold larger than that of rice (about 390 Mb) or maize (about 2.4 Mb) and its genome structure is complex due to polyploidy (allohexaploid). Therefore, wheat may not be suitable for production of recombinant proteins, although it is a self-pollinating crop like barley or rice (Table 5.2).

Table 5.2 Properties of crop seeds used as production platform of pharmaceutical proteins

Species	2S albumin	7S globulin	11–12S globulin	Prolamin	Storage tissue	Protein/ grain WT	Biomass (kg/ha)	Pollination
Maize		Globulin-1 and -2 (embryo)		Zein (70%): α^*, β^+, γ^+, δ^*	Endosperm	10	870	Cross-pollinating
Rice	α-Amylase inhibitor	Reg 1, Reg 2 (embryo)	Glutelin (60–80%); GluA, GluB, GluC, GluD	Prolamin (10–20%): $16K^+$, $14K^+$, $13K^*$, $10K^+$, 26 kDa α-globulin (5%)	Endosperm	8	730	Self-pollinating
Wheat			Triticin (5%)	Glutelin: HMW (13%), LMW (37%): B^+, C^+, D^*/ gliadin (50%): α/β^+, γ^+, ω^* $^+$ S rich 70–80%, * S poor 10–15%	Endosperm	12	270	Self-pollinating
Barley				Hordein: B^+, C^*, D, γ^+ $^+$S rich 80% (B; 70%), *S poor 10–20%	Endosperm	13	310	Self-pollinating
Oat			12S globulin (70–85%)	Avenin (15–20%)	Endosperm	15	340	Self-pollinating
Soybean		β-Conglycinin (20–30%): α, α′, β (Cys less)	Glycinin (40%): $Gy1^+$, 2^+, 3^+ (Cys rich) Gly4*, 5* (Cys poor)		Cotyledon	40	260	Self-pollinating
Pea		Vicilin, convicilin (40%)	Legumin (60%): LegA, E, J, S		Cotyledon	25	250	Self-pollinating
Common bean	Arcelin-5 (40%)	Phaseolin (50%)			Cotyledon	25	240	Self-pollinating
Broad bean		Vicilin (10%)	Legumin (80%) type A, type B Leg4		Cotyledon	29	110	Self-pollinating

(continued)

Table 5.2 (continued)

Species	2S albumin	7S globulin	11–12S globulin	Prolamin	Storage tissue	Protein/grain WT	Biomass (kg/ha)	Pollination
Rapeseed	Napin (20%)		Cruciferin (60%)		Cotyledon	22	150	Cross-pollinating
Arabidopsis	Albumin (10%)		Cruciferin (90%) A, B, C		Cotyledon	20	No data	Self-pollinating
Tobacco			12S globulin		Endosperm	No data	No data	Self-pollinating

* S poor, $^{+}$ S rich

Legume seeds have high production system for endogenous seed proteins because they are rich in proteins ranging from 20% to 40% of total seed weight (see also chapter "Legume Seed: A Useful Platform for the Production of Medical Proteins/Peptide"). Soybean and pea are self-pollinating crops and thus a low risk of contamination of pollens. However, annual grain yields (biomass of seed) are lower than maize or rice. Transformation of these leguminous crops is more difficult than *Arabidopsis* or tobacco. Expression levels of recombinant proteins are also lower than rice. On the other hand, *Arabidopsis* or tobacco has been exploited as model plants for production of recombinant proteins in dicot seeds (see also chapter "Tobacco Plants as a Versatile Host for the Expression of Glycoproteins"). This is due to merits of high transformation efficiency, high expression level, and short regeneration time (life cycle).

Oil crops (rapeseed, sunflower, safflower) offer another inexpensive platform for the expression of recombinant proteins. They have been produced as fusion proteins with the oleosin protein localized in surfaces of oil bodies observed in oil crop seeds.

5.2.2 Properties of Seed Proteins Accumulated in Seeds Used for Production Hosts

Plant seeds are divided into monocotyledonous seed (cereal seed), endospermic dicotyledonous seed (tobacco, castor bean), and non-endospermic dicotyledonous seed (legumes, cruciferous). Monocot seeds predominantly accumulate seed proteins in endosperm at levels of 7–15% of seed weight, whereas dicot seed proteins are stored in cotyledon and embryo at levels of 20–40%. Seed proteins are divided into water-soluble 2S albumin, salt-soluble globulin, alcohol-soluble prolamin, and dilute acidic/alkaline-soluble glutelin based on the property of their solubility according to the Osborne's classification. Globulin is the major seed storage protein (SSPs) of dicots, while prolamins are predominant in many cereal grains except for rice (glutelins) or oat (globulin) (Shewry et al. 1995; Kawakatsu and Takaiwa 2017). They are deposited in two types of PBs different in origin. One is ER-derived PB, and the other is protein storage vacuole (PSV).

Globulins and 2S albumins in non-endospermic dicotyledonous seeds are deposited in PSVs of cotyledon and embryo, whereas their endosperm tissue is absorbed by the cotyledon during the seed development. Globulin SSPs are categorized into two groups based on their sedimentation coefficients: 11–12S globulins (soybean glycinin, pea legumin, cruciferin of rapeseed, and *Arabidopsis*) consisting of hexametric subunits with molecular weights (MW) of about 300–450 kDa and 7S globulins (soybean β-conglycinin, pea vicilin, common bean phaseolin) composed of trimeric subunits with MW of about 150–200 kDa that lack cysteine residues (Table 5.2). Tobacco seed of endospermic dicot mainly accumulates 12S globulin in endosperm tissue. 2S albumins have MW of 10–15 kDa. They are synthesized as preproteins, which are posttranslationally processed to give rise to large (8 kDa) and small (4 kDa) subunits linked by two interchain disulfide bonds.

Trafficking and deposition of these dicotyledonous SSPs start with the translocation via N-terminal signal peptide and subsequent assembly in ER lumen. Upon trafficking to the Golgi apparatus, the SSPs aggregate in the cisternae and ultimately bud off from a *trans*-Golgi as dense vesicles (DVs). These DVs fuse with other different DVs to become prevacuolar compartments (PVCs)/multivesicular bodies (MVBs), from which SSPs are delivered to PSVs. Some SSPs are packaged in precursor-accumulating (PAC) vesicles and directly delivered from ER to PSVs in a Golgi-independent manner (Zheng et al. 2022). 11–12S globulins are posttranslationally cleaved with Asn-specific vacuolar processing enzyme (cysteine protease) into mature acidic (35–40 kDa) and basic (20 kDa) subunits linked by a disulfide bond and then assembled to hexametric molecules in PSVs. On the other hand, subunit compositions of 7S globulins vary considerably, mainly because of differences in the heterogeneity in the size of the subunits that comprise the trimeric molecules and extent of posttranslational processing (proteolysis and glycosylation). Vicilin subunits of pea are initially synthesized as precursor with MW of 47 and 50 kDa, and then posttranslationally proteolyzed and glycosylated, giving rise to mature subunits with MW between 12,500 and 33,000. By contrast, common bean phaseolin and soybean β-conglycinin (α, α′, and β) differ from pea vicilin, in which glycosylation is more extensive but proteolysis does not occur. Trimeric subunits with different MW are assembled as 7S globulin in PSVs (Shewry et al. 1995).

On the other hand, prolamins are the major seed protein components of most cereal grains (Shewry and Halford 2002). Prolamins of the Triticeae species (wheat, barley, and rye) are divided into sulfur-rich, sulfur-poor, and high-molecular-weight groups, whereas those of the panicoid cereals (maize, sorghum, and millet) are composed of four groups (α, β, γ, and δ). Rice prolamins are also classified into 16 kDa, 14 kDa, 13 kDa, and 10 kDa groups. In the panicoid cereals and rice, these prolamins are synthesized on the rough ER and form an insoluble matrix (high aggregates) by disulfide bonds and hydrophobic interactions after translocation into ER lumen, which are then budded from the ER as discrete round-shaped organelle referred to as ER-derived PBs in the cytoplasm. Rice ER-derived PB called PB-I consists of 10 kDa prolamin core, 13 kDa prolamin inner layer, 14 kDa and 16 kDa prolamin middle layer, and 13 kDa prolamin outer layer (Saito et al. 2012). In the maize PB, major Cys-poor α-zein (19 kDa and 22 kDa) comprises most of the core with δ-zein (18 kDa and 10 kDa), which is surrounded by a thin layer of β-zein (15 kDa) and γ-zein (50 kDa, 27 kDa, and 16 kDa) (Lending and Larkins 1989). By contrast, wheat glutens (polymeric glutenin and monomeric gliadins) and barley hordeins are first deposited in ER-derived PBs and subsequently transported to PSV by autophagy-like system or are transported through endomembrane system to PSVs via Golgi apparatus (Levanony et al. 1992). That is, there are two trafficking routes to the PSVs, Golgi-mediated and direct pathway route bypassing the Golgi apparatus. It should be noted that the same individual protein is trafficked to PSVs by either pathway depending on the development stage: Golgi-dependent traffic at early stage and direct transport at later stage (Tosi et al. 2009). On the other hand, rice glutelins and oat 12S globulin homologous to dicot 11–12S globulins (about 35% homology) and rice 26 kDa α-globulin that are expressed in the endosperm are transported from the cisternal ER to the PSVs via the Golgi apparatus or directly transported from ER to the PSVs via precursor-accumulating (PAC) vesicles (Yamagata et al. 1982; Takahashi et al. 2005). Rice α-globulin is stored in the peripheral matrix region that surrounds the inner crystalloid glutelin-localizing region of PSVs. In the maize endosperm, two globulins (legumin-1 and 18 kDa α-globulin) sharing high homology to rice α-globulin are found in PSVs. On the other hand, the minor 7S globulin (maize globulin 1 and globulin 2, rice Reg 1 and Reg 2) is expressed in the embryo and aleurone layer of maize and rice seeds, which are deposited in PSVs.

The endosperm and cotyledon act as a storage compartment preserving nutrient resources for germinating embryo. The endosperm in cereal grain accounts for 70–90% of grain weight, whereas cotyledon of dicot seed comprises about 60% of seed weight. These tissues are filled with PBs, starch granules (amyloplasts), and oil bodies.

5.3 Expression in Seed Production System

5.3.1 Promoters

Expression of recombinant protein is primely determined by the transcription level. To maximize the transcript level in seed, reasonable strategy is to utilize the seed-specific promoter derived from the major SSP genes in host plant used as bioreactor. Seed-specific promoters of the SSP genes have strong promoter activities to direct the expression in seed without having any influence on vegetative growth of host even though detrimental recombinant products were produced. Most of the SSPs are usually coded by multigene family. It is known that individual member is differentially expressed in the temporal or spatial specific manner during seed maturation.

In rice seeds, at least 15 rice seed promoters were isolated and characterized by fusing to GUS reporter gene in stable transgenic rice (Wu et al. 1998a; Qu and Takaiwa 2004; Qu et al. 2008). Rice seed protein genes are differently expressed in spatial and temporal specific manner (Qu and Takaiwa 2004). Glutelin is the most major SSP accounting for about 60% of total seed proteins, which is coded by 15 multigenes (Kawakatsu et al. 2008). Glutelin promoters (*GluA-1, GluA-2, GluA-3, GluB-1*, *GluB-2,* and *GluB-4*) predominantly direct the expression in subaleurone layers. Rice prolamin promoters (10 kDa, 13 kDa, 14 kDa, and 16 kDa prolamin genes) confer expression in aleurone and subaleurone layers. Glutelin *GluC* and *GluD* promoters direct expression throughout the whole endosperm. The expression of 26 kDa globulin and 14–16 kDa allergen genes is restricted to inner starchy endosperm. 18 kDa Oleosin and embryo globulin promoters exhibit the specific expression in embryo and aleurone layer. It was demonstrated that promoters from the

GluA-2 (Gt1), *GluB-1*, and *GluB-4* glutelins, 26 kDa globulin, and 10 kDa and 16 kDa prolamins exhibited much higher activities than the others. Furthermore, it was shown that 10 kDa and 16 kDa prolamin genes and *GluD* glutelin gene are initially expressed during seed development, which are followed by expression of most glutelins and 14 kDa prolamin and then by 13 kDa prolamin.

Based on these findings, glutelin (*GluA2(Gt1)*, *GluB-1*, *GluB-4*), 26 kDa globulin, 18 kDa oleosin, and prolamin (10 kDa and 16 kDa) promoters, which exhibit difference in spatial and temporal expression pattern, have been utilized for the expression of recombinant proteins on a case-by-case basis. A toolbox of seed-specific promoters with different spatial and temporal properties is valuable for seed-based production of recombinant proteins depending on the purpose.

5.3.2 Regulatory Mechanisms of Seed-Specific Expression

Seed-specific expression is determined by the combinatory interaction of several *cis*-regulatory elements in the promoter. Individual *cis*-regulatory element is specifically recognized by cognate seed-specific transcription factor (TF). It has been known that seed-specific expression observed in the SSP genes is regulated by an ensemble of common *cis*-elements consisting of GCN4-like motif (G/A)TGA(G/C)TCA(T/C) or ACGT motif ((C/T)ACGTAG) interacting with basic leucine zipper (bZIP) TF, RY motif (CATGCA(C/T)) with alternation of purine and pyrimidine nucleotides interacting with B3 domain TF, prolamin (P)-box (AAAG) interacting with DNA-binding one finger (DOF) class of zinc finger TF, and AACA-like motif ((C/T)AACAA(A/C)) interacting with R2R3MYB TF, irrespective of the difference in expression site between monocot (endosperm) and dicot (embryo and cotyledon) seeds (Vicente-Carbajosa and Carbonero 2005; Kawakatsu and Takaiwa 2010; Yang et al. 2023). This finding indicates that seed-specific regulation mechanism is evolutionarily conserved in various SSP genes of higher plants. Therefore, given that recombinant proteins were expressed under the control of the selected seed-specific promoter, the inherent spatiotemporal specificity observed in such promoter is expected to be fundamentally retained even if expressed in any plants. As shown in Table 5.3, orthologue genes coding for these bZIP, prolamin-box factor (PBF), MYB, and B3 TFs, which bind to *cis*-elements involved in seed-specific expression, have been reported in various crops. For example, the GCN4 motif is recognized by bZIP family TF, including maize O2, rice RISBZ1, wheat SPA, and barley BLZ1. As other novel bZIP TFs, maize OHP1, OHP2, and bZIP22, rice RITA1 and REB, wheat SHP, and barley BLZ2 have been identified to be implicated in seed-specific expression of SSPs.

Rice SSP genes contain all or some of the following four types of *cis*-elements: GCN4 motif (TGA(G/C)TCA), prolamin-box (P-box) (TGTAAAG), AACA motif (AACAAAC), and ACGT motif (ACGTG) (Washida et al. 1999; Wu et al. 2000). In many cereal prolamin genes encoding wheat glutenin and gliadin, barley hordein, maize zein, and rye secalin, the highly conserved "endosperm box" consisting of the

Table 5.3 Transcription factors involved in seed-specific expression and their *cis*-elements

TF	Binding sequence	Maize	Rice	Wheat	Barley	*Arabidopsis*
bZIP	TGAGTCA (GCN4 like) or ACGTA (O2 like)	O2	RISBZ1	SPA	BLZ2	bZIP10, bZIP25
		OHP1, OHP2	REB, RITA, RISBZ2-5	SHP	BLZ1	
	CACGTG (G-box)	ZmbZIP22				bZIP67
	ACGTGG/T(ABRE)		TRAB1, OsABI5	TaABI5		AREB3, ABI5
Dof	TGTAAAG	MPBF1	RPBF	WPBF	BPBF (SAD)	
R2R3MYB (GAMYB)	AACAAAC		OsMYB5	TaGAMYB	HvGAMYB (HvMYB3)	
R1MYB	GATA				HvMYB	
B3 domain	CATGCA (RY motif)	ZmFUSCA3	OsFUSCA3	TaFUSCA3	HvFUSCA3	FUSCA3
		ZmABI9, VP1	OsVP1			ABI3
NAC	CACGCAA	ZmNAC128/130	OsNAC20/26	TANAC019		
MADS	CATGT	ZmMADS47				

GCN4 motif and P-box has been characterized to be implicated in the endosperm-specific expression (Agarwal et al. 2011).

It has been demonstrated in stable transgenic rice seeds that the GCN4 motif acts as a key regulatory element for determining the aleurone- and subaleurone-specific expression, since the trimer of the GCN4 motif fused to the core (−46) promoter of CaMV 35S could only lead to expression in the outer endosperm of seed and its mutation resulted in the complete suppression (Wu et al. 1998b). On the other hand, each of the AACA motif, P-box, or ACGT motif did not confer seed-specific expression, but was instead mainly involved in the quantitative regulation of SSP gene expression (Wu et al. 2000). Combinatorial interaction of these *cis*-elements (presence or absence of some *cis*-elements and their arrangements) in the 5′ flanking region may play a crucial role in determining promoter activity and spatial expression pattern (Wu et al. 2000). Enhancement of promoter activity and the desired spatial expression could be obtained by arrangement of *cis*-elements.

Maize Opaque2 (O2) like bZIP and prolamin-box-binding factor (PBF) DOF TFs activate the transcription of many cereal SSP genes via binding to the GCN4 motif and P-box, respectively (Yamamoto et al. 2006; Kawakatsu et al. 2009). Furthermore, it is important to note that promoter activities are more highly activated by the combination of O2-like bZIP and PBF than the individual one, indicating the synergistic or combinatory interaction of these transcription factors (Yamamoto et al. 2006; Zhang et al. 2016). They have also been reported to be involved in the regulation of many starch, lipid, and amino acid metabolic pathway genes (Wang et al. 2013a, b; Zhang et al. 2016). Such protein and starch levels in seed are known to be influenced by fluctuations in nitrogen (N) availability. It has recently been suggested that PBF contributes to carbohydrate and nitrogen assimilation in developing endosperm through an N-dependent mechanism (Ning et al. 2023). Notably, temporal and spatial expression of seed proteins is dependent on the specificity of these transcription factors because they are regulated in endosperm- and maturation-specific manner (Vicente-Carbajosa et al. 1997; Onodera et al. 2001). Recently, NAC and MADS-box TFs have also been shown to be implicated in seed-specific regulation of some SSP genes in maize, wheat, and rice (Qiao et al. 2016; Zhang et al. 2019; Gao et al. 2021; Wang et al. 2020). Seed-specific expression of rice albumin (16 kDa allergen) genes is regulated by NAC20/26, which was activated by interaction with RPBF (Wu et al. 2023). It is noteworthy that maize seed O11 TF with a basic helix-loop-helix (bHLH) domain acts as the central hub of regulatory network for maize endosperm development and nutrient metabolism, since it directly regulates the expression of genes encoding key transcription factors (O2 and PBF) involved in nutrient metabolism (Feng et al. 2018). Moreover, the recently identified maize ZmABI19 TF with B3 domain is demonstrated to be involved in regulating the seed development and grain filling by direct binding to the promoters of various TFs (O2, O11, PBF1, bZIP22, NAC130) via the RY motif (Yang et al. 2021). Furthermore, this ZmABI19 TF is implicated not only in the development of BETL (unique endosperm cell layer) by transactivating the expression of Sweet4c sugar transporter involved in the nutrient transport, but also in the regulation of viviparous-1 (VP1), which plays a role in the cross talk between endosperm and embryo through responding to nutrient status of endosperm and scutulum. ZmABI19 also regulates auxin-responsive genes in early seed development. It is interesting to note that the ZmABI19 and O11 interact with each other and antagonistically regulate the expression of O2.

Regarding the embryo-specific expression of SSP genes in dicot plants, it has been reported that the B3 domain TFs, FUSCA3 (FUS3), ABA INSENSITIVE3 (ABI3), and LEAFY COTYLEDON2 (LEC2), strongly activate the expression of many seed maturation-specific genes including SSP genes in a cooperative and synergistic manner with other TFs (Kagaya et al. 2005; Santos-Mendoza et al. 2008; Verdier and Thompson 2008). Especially, LEC1 protein, heme activator protein 3 (HAP3) family CCAAT box-binding factor, has been demonstrated to be involved in regulating the expression of SSP genes in conjunction with ABI3 and FUS3 TFs, since loss-of-function mutation of LEC1 downregulated the expressions of ABI3 and FUS3 and furthermore reduced the accumulation of various SSPs. LEC1 also physically interacts with these TFs to regulate a variety of seed developmental processes. Taken together, it is concluded that LEC1 acts as a central regulator controlling the seed distinct development processes (Jo et al. 2020). Deletion or mutation of RY element in "legumin box" conserved in the promoters of various dicot SSP genes such as 11S globulins (pea legumin, soybean glycinin, and rape seed cruciferin), 7S globulins (soybean β-conglycinin), and 2S napin has been reported to result in loss or significant reduction of seed-specific expression. RY motif has been demonstrated to be recognized by ABI3 and FUS3 containing the B3 domains. This motif is inevitable for seed-specific expression, but not sufficient for full activity. Other *cis* elements including G-box (CACGTG), ABA-responsive element (ACGTG(G/T)(A/C)), AACA motif, or CAAT box have been reported to be involved in seed-specific expression of many dicot SSP genes as the combinatory element(s) with the RY motif. For example, in the gene expression of *Arabidopsis* At2S1 albumin and CRU3 12S globulin, the G-box in their promoters is bound by O2-like TFs (AtbZIP10/AtbZIP25) and was highly transactivated by the combination with ABI3 TF (Lara et al. 2003). Furthermore, synergistic activation between other bZIP TF (ABI5) recognizing the ABA-responsive element and the ABI3 TF was also reported to be implicated in the expression of some seed maturation genes (Nakamura et al. 2001; Carles et al. 2002).

5.3.3 Seed-Specific Promoters Used for Expression of High-Value Recombinant Proteins

For production of recombinant proteins in cereal seeds, rice glutelin promoter (*Gt1 (GluA-2), GluB-1,* and *GluB-4*), maize zein promoter (27 kDa γ-zein or 19 kDa α-zein), and barley hordein (B hordein) promoter have been mainly utilized. Notably, although maize embryo occupies only 10–12% of the seed volume, embryo-specific *globulin-1* promoter has been preferentially used for the expression in maize grain because the produced recombinant protein stably accumulates in embryo. Rice 18 kDa oleosin promoter conferred high expression in embryo and aleurone layer.

By contrast, when high-value recombinant proteins have been produced in dicot plants, seed-specific promoters from soybean glycinin (A1aB1b (*G1*)) and β-conglycinin (α′ subunit), pea legumin (*Leg4*), broad bean unknown seed protein (*USP*), common bean arcelin-5 and β-phaseolin, and rapeseed cruciferin genes have been utilized in transgenic soybean, pea, tobacco, and *Arabidopsis* seeds. Arcelin-5 and β-phaseolin account for about 40% and 50% of total seed proteins in common bean seed, so that these promoters have been frequently used for the production of pharmaceutical proteins in *Arabidopsis* and tobacco seeds. These promoters boosted the accumulation of a single-chain Fv (scFv) up to 36.5% of the total soluble protein (TSP), which was in remarkable contrast with about 1% directed by CaMV 35S promoter (De Jaeger et al. 2002).

Constitutive promoters such as enhanced CaMV 35S promoter (2×CaMV35S), maize ubiquitin (*Ubi*) promoter, and rice actin promoter with the first intron in the 5′ untranslated region have also been used for the production of recombinant proteins. Although recombinant proteins can be produced at relatively high levels in seeds, these constitutive promoters sometimes have a detrimental effect on vegetative growth and seed production yield by their expression in various tissues and sometimes induce gene silencing.

5.4 Translational and Posttranslational Regulation

Transcript levels are not always correlated with the accumulated protein amounts. To improve production yield of recombinant protein, it is important to stabilize the mRNA and to optimize translation because mRNA stability generally parallels with translation product (Koziel et al. 1996). At present, it is known that eukaryotic cells have mRNA quality control system referred to as “mRNA surveillance” to monitor aberrant mRNAs causing premature stop codons, nonstop codon, or structures to inhibit translation elongation. To remove such aberrant mRNAs on the ribosome during translation, three degradation processes such as nonsense-mediated decay (NMD), no-go decay (NGD), and nonstop decay (NSD) are known to involve (Isken and Maquat 2007; Shoemaker and Green 2012). Furthermore, protein synthesis by the ribosomes is regulated by ribosome-associated quality control pathway, which senses translational stalling induced by insufficient amounts of particular amino acid or tRNA and faulty mRNA (Brandman and Hegde 2016).

5.4.1 5′ UTR (Leader Sequence)

It has been reported that addition of monocot *Ubi1*, *Adh1 Sh1*, *and Act1* introns increased the production of the reporter gene in transgenic rice seeds. Such 5′ UTR intron acts as transcription enhancer affecting gene expression as well as its translation. However, *Ubi1* and *Adh1* introns did not always have any influence on the accumulation level of fungal laccase gene in transgenic maize grain (Hood et al. 2003). Furthermore, several plant virus 5′ noncoding leader sequences (tobacco mosaic virus (TMV)-Ω, TMV coat and brome mosaic virus (BMV) coat, tobacco etch virus, maize dwarf mosaic virus) have also been used for the purpose of improving the rate of translation as a translational enhancer. However, such plant viral 5′ UTR worked poorly in monocots, as opposed to high translational enhancement observed in dicots. Actually, addition of TMV-Ω or TMV coat sequence as leader sequence in the 5′ UTR did not have any effect on the production of the sweet protein brazzein in transgenic maize seeds, when expressed under the control of the embryo-specific globulin-1 (*Glb-1*) promoter (Lamphear et al. 2005). These results indicate that the effect of a 5′ UTR may be different between dicots and monocots. Furthermore, extensive predicted mRNA secondary structures that might hinder translation should be avoided.

5.4.2 Translation Initiation

The coding DNA sequence is known to have efficient control elements for translation initiation. Therefore, in order to boost the translation, the sequence surrounding the initiation codon should be modified to fit the consensus initiation sequence for dicot (A(A/C)aAUGGC) and monocot ((A/G)(A/C)cAUGGC) (Joshi et al. 1997). It is important to note that a purine at −3 position and a GC at +4 and +5 are highly conserved in plant mRNAs, whereas U in position −1 should be avoided. Alanine is highly observed to follow the N-terminal methionine in high-expressing plant genes.

5.4.3 Codon Usage

Codon usage of the target recombinant protein gene is highly related to the yield of produced protein. That is, difference in codon usage affects an influence on various steps of protein production such as RNA processing, protein translation, translation fidelity, and protein folding (Webster et al. 2017). Especially, suppressive effect is caused by the rare codons and expression-limiting regulatory elements in the expressed sequence (Jackson et al. 2014). Therefore, codon optimization is an inevitable process to improve the accumulation level of target recombinant protein. Various computer programs have been developed to aid the gene sequence optimization process. In order to boost the yield of recombinant protein in seed by codon usage optimization, rare codons in recombinant protein gene have to be replaced with synonymous counterparts used preferentially in major SSP genes of the host plant to avoid ribosomal pausing and attenuation of translation rates. That is, yield of recombinant protein is expected to be improved by maximizing codon preference or harmonizing codon preference. Codon harmonization requires information about codon usage and enlargement of available tRNA pool size because there is a strong correlation between the frequency of codon usage and the presence of their cognate tRNAs.

It is important to take account of that preferred codon usage that is considerably different not only between monocot and dicots, but also among various tissues in the same plant (Camiolo et al. 2012). Such difference may be overcome by optimizing the codons according to the tissues of host plant in which foreign recombinant protein genes are expressed. For example, it should be noted that G or C bias at the wobble position in many genes expressed in monocot vegetative tissues reaches 64% or more, whereas those of seed protein genes are not so high. Furthermore, undesirable regulatory elements within the coding sequence, such as the A/U-rich mRNA destabilizing element (AUUUA)-like sequences, polyadenylation signal (AAUAAA), and consensus intron splice sites, should be avoided from the coding sequence of the produced recombinant protein.

5.4.4 Stop Codon and 3′ Untranslated Region (UTR)

As stop codon in plants, UAA and UGA are preferentially utilized in plants. The avoidance of C and the preference for an A in the +1 position should be selected as the base immediately following the stop codon. The 3′ UTR has been implicated in determining the stability or instability of an mRNA. Multiple polyadenylation signals (AAUAAA) are present in the 3′ UTR located downstream of the translation stop codon. The poly(A) tail is known to play an important role in determining transcript stability. By contrast, some A/U-rich sequence elements which destabilize the mRNA have been identified in the 3′ UTR regions (Rothnie 1996; Gutiérrez et al. 1999). Thus, these sequences must be avoided. It was demonstrated in transgenic rice seed that use of the rice glutelin *GluB-1* 3′ UTR containing multiple polyadenylation signals enhanced accumulation levels of modified mite allergen Der f 2 by about fourfold, when compared to a nopaline synthase terminator frequently utilized as the terminator (Yang et al. 2009). This result indicated that the choice of suitable 3′ UTR can significantly contribute to high mRNA stability and improvement of recombinant production yield. Potato proteinase inhibitor II terminator has also been reported to increase translation product protein and utilized frequently as transcriptional terminator for production of recombinant proteins (Hood et al. 2007).

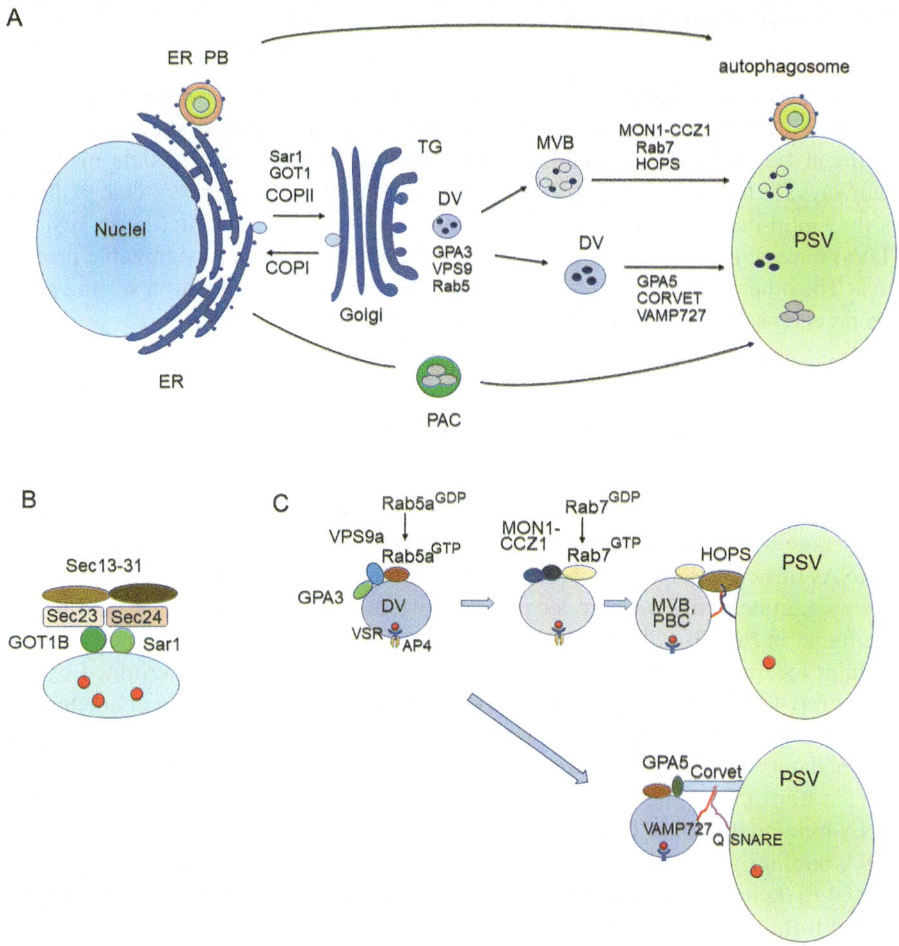

Fig. 5.1 Schematic overview of trafficking pathways of SSPs in seed cells. (**a**) SSPs produced as secretory proteins are first transported from the ER to the Golgi apparatus via the putative COPII machinery and then targeted to PSVs via either the direct DV-to-PSV route or the indirect DV-to-PSV route through MVB/PVC as intermediate compartments. Some SSPs such as cereal prolamins are directly synthesized and accumulated in the ER, followed by budding off as ER-derived PBs. Parts of these PBs are then transported to the PSVs as autophagosomes. Some SSPs are directly transported through PAC vesicles from the ER to the PSVs, bypassing the Golgi (Ren et al. 2022). (**b**) COPII vesicle formation. Sec23–Sec24 and Sec13–Sec31 heterodimers are formed as outer coat of COP vesicles by action of GOT1B and Sar1-GTP. (**c**) DV and MVE/PVC biogenesis and their fusion with PSV. In the DV-mediated post-Golgi trafficking pathway, GPA3 recruits VPS9a to activate Rab5a. The activated Rab5a-GTP recruits its effector GPA5, and GPA5 interacts with the CORVET and VAMP727 containing SNARE complex to PSV fusion. In the case of indirect DV-to-PVC trafficking pathway, Rab5a-GTP recruits the MON1-CCZ1 complex. The MON-CCZ1 acts as the GEF of Rab7-type GTPase to activate Rab7. HOPS is employed as its tethering complex to PSV fusion

5.5 Trafficking Process

Understanding of endomembrane trafficking process is an essential step to efficiently deliver the expressed recombinant protein to the suitable subcellular compartment for stable accumulation. There are two main distinct vesicle-mediated pathways for the trafficking of SSPs from the ER to PSVs (Fig. 5.1a). One pathway is the Golgi-dependent aggregation sorting route via plant-specific dense vesicles (DVs) that are enclosed by a single membrane but devoid of a recognizable protein coat. The other pathway is the direct ER-to-PSV insoluble aggregation sorting route via precursor-accumulating (PAC) vesicles (Vitale and Hinz 2005).

5.5.1 mRNA-Based Trafficking Mechanism

Localization of mRNAs at the subcellular level is an essential mechanism for specific protein targeting and local control of protein synthesis (Tian et al. 2020). mRNAs of rice SSPs are asymmetrically distributed in the subcellular manner. For example, glutelin mRNAs are localized at cisternal ER, whereas prolamin mRNAs are found in PB-ER (Chou et al. 2019a; Tian et al. 2020). Such differences in subcellular localizations of mRNAs have been demonstrated to be determined by zip codes (*cis*-localization elements) on the mRNA. The zip codes were characterized to be located at the coding region and the 3′ UTR of glutelin and prolamin mRNAs (Hamada et al. 2003; Washida et al. 2009). They contain the information required for the transport of the corresponding mRNAs, which are recognized by specific RNA-binding proteins (RBPs). Typical RBP possesses one or more conserved RNA-binding motifs. Furthermore, mRNAs have been demonstrated to be transported to their destination sites by co-opting several membrane trafficking factors, which form a ribonucleoprotein (RNP) complex. At present, several cytoskeleton-associated RBPs involved in SSP mRNA localization have been isolated and characterized (Crofts et al. 2010; Yang et al. 2014; Chou et al. 2019b). Some of the RBPs have been identified to play various roles in RNA metabolism including RNA transport and localization, since loss-of-function mutant of RBP resulted in mislocalization of mRNAs or formation of abnormal PBs due to reduced RNA binding and/or interruption of protein-protein interaction (Tian et al. 2018, 2019). In the transportation process of mRNAs on microtubules or actin filaments to the target site, RNP complexes assembled by multiple cytoskeletal associated RBPs are remodeled by changing protein factors to target the RNA to its proper location. Several mRNAs encoding proteins implicated in endomembrane trafficking such as Rab5, Golgi transporter, or guanine exchange factor (GET) have been detected in moving RNA-transporting particles. Loss of function of these endomembrane trafficking factors has been reported to result in mislocalization of seed mRNAs in rice. Especially, mutation of protein factors involved in the ER-to-Golgi trafficking pathway such as Golgi transporter 1 (GOTB1) gave rise to mislocalization of prolamin and globulin mRNAs, whereas loss of function of protein factors involved in Golgi-to-PSV trafficking pathway (Rab5a and GEF) disrupted the proper localization of glutelin mRNA (Fukuda et al. 2011, 2013, 2016; Yang et al. 2018). These evidences indicate that the association between mRNA transport particles and endomembrane is indispensable to localize the mRNA to the proper site. Co-transport of mRNAs with membranous compartments is a common mechanism of mRNA-based trafficking in higher eukaryotes, thus indicating that the transport and localization of SSP mRNAs depend on the membrane vesicular transport process.

5.5.2 *ER-to-Golgi Trafficking*

Once proteins are properly folded and packaged (assembled) in the ER, they are exported from the ER to other organelles through a vesicle-mediated membrane trafficking system. Golgi-dependent pathway requires two transport processes: ER-to-Golgi and Golgi-to-PSV (Vitale and Hinz 2005; Zheng et al. 2022; Ren et al. 2022). Coat protein complex II (COPII) mediates the first step of anterograde transport of newly synthesized proteins from the ER to Golgi apparatus. A group of evolutionally conserved proteins (Sar1, Sec23, Sec24, Sec13, and Sec31) constitute the basic COPII coat machinery (Fig. 5.1b). In the process of COPII vesicle formation, GOT1B participates in COPII coat formation at the ER exit sites via interaction with Sec23. The heterodimer of Sec23/24 is recruited by GOT1B and Sar1-GTP (activated by Sec12) cooperatively to form the pre-budding complex in which cargos are loaded. Then the heterodimeric of Sec13/31 is recruited to form the outer coat of the COPII vesicles before vesicle budding and subsequent fusion with the *cis*-Golgi apparatus. When Sar1 and GOT1B expressions were impaired by knockdown or knockout, export of storage proteins from the ER was significantly depressed, resulting in the retention of transported proteins in the ER lumen and appearance of novel ER-bounded protein bodies (intracisternal granules) (Tian et al. 2013; Fukuda et al. 2016; Wang et al. 2016). These results suggest that Sar 1 GTPase and GOT1B play an essential role in the formation of COPII vesicles for ER-to-Golgi traffic (Tian et al. 2013). Similar observation was reported in the *Arabidopsis* maigo mutants, in which functions of the tethering factors involved in protein export from ER were abolished by knockout mutation (Li et al. 2006).

5.5.3 *Golgi-to-PSV Trafficking*

Dense vesicles (DVs) involved in PSV trafficking are unique to plants, which deliver the cargo proteins including SSPs to PSVs. The formation of DVs is initiated from the *cis*-Golgi, and then DVs traverse Golgi apparatus and bud off from the *trans*-Golgi network (TGN). The TGN acts as a sorting station where various proteins are sorted and directed to target site (subcellular compartment) according to the post-Golgi transport pathways (Shimizu and Uemura 2022). DVs directly fuse with PSVs, or first fuse with multivesicular bodies (MVBs)/prevacuolar compartments (PVCs), which are followed by fusion with PSVs (Fig. 5.1). That is, SSPs in DVs are targeted to PSVs either via the direct DV-to-PSV route or via the indirect DV-to-PSV route through the MVB/PVC as intermediate compartments. In this process, vacuolar sorting receptors (VSRs) or receptor homology region-transmembrane domain-RING-H2 proteins (RMRs) in the DVs facilitate the concentration and aggregation of soluble storage proteins. Vacuolar sorting is mediated by specific protein-protein interactions between the sorting receptors and the cargo proteins. Once VSRs bind to cargo proteins via their target signals termed vacuolar sorting determinants (VSDs), the receptor-ligand complex is sorted by heterotetrametric adaptor protein (AP) complexes that mediate intracellular membrane trafficking along endocytic and secretory transport pathways (Shimizu and Uemura 2022). AP complexes recognize the cytosolic domain of VSRs, which are required for the formation of transport vesicles and cargo sorting in all eukaryotes. APs are known to play critical roles in protein sorting among various post-Golgi pathways through recognizing specific cargo protein signal. Five types of AP (AP-1 to AP-5) have been identified in eukaryotic cells, each of which is composed of three large subunits, one middle subunit, and one small subunit. Individual AP complex is implicated in distinct trafficking pathways. AP-4 is identified as the receptor-mediated vacuolar sorting proteins that have participated in the targeting of SSPs from the TGN to PSV. Mutations in the AP-4 led to abnormal trafficking of 12S globulin precursor in *Arabidopsis* seed (Fuji et al. 2016).

The Rab family of small GTPases and their common guanine exchange factor (GEF) are involved in specifying the vesicular trafficking and the R-soluble *N*-ethylmaleimide-sensitive factor attachment protein receptor (R-SNARE) complex for mediating membrane fusion between Golgi and PSV post-Golgi compartments. Loss-of-function mutation of Rab5a encoding a small GTPase indicated that it is implicated in the intracellular transport of SSPs from Golgi to the PSVs (Wang et al. 2010; Fukuda et al. 2011). Furthermore, loss-of-function mutation of its GEF, vacuolar protein sorting 9a (VPS9a), also disrupted the transport of SSPs, resulting in the formation of large dilated paramural bodies (Fukuda et al. 2013). These findings indicated that Rab5a-VPS9a cooperatively functions in regulating post-Golgi trafficking of SSPs to the PSVs in rice endosperm (Wang et al. 2010; Liu et al. 2013). Moreover, it was shown that the GPA3 (plant-specific kelch-repeat protein) functions as an adaptor protein to recruit VPS9a, which forms a regulatory complex with Rab5a by direct interaction with VPS9a (Fig. 5.1c). Actually, mutation of rice GPA3 caused mistargeting of DVs containing SSPs and formation of a new structure named the paramural body in apoplast (Ren et al. 2014).

Rab5 is a core regulator of storage protein trafficking from the *trans*-Golgi network (TGN) to PSV and participates in different post-Golgi transport pathways by employing different downstream effectors, such as GPA5 or MONENSIN SENSITIVITY1 (MON1)-CALCIUM CAFFEINE ZINC1 (CCZ1) protein complex. The CCZ1 interacts with MON1 to form dimeric complex (MON1-CCZ1), which functions as the effector of Rab5 and serves as the GEF for Rab7-type

GTPases, leading to Rab5-to-Rab7 conversion. Activation of Rab7 by MON1-CCZ1 is essential for membrane fusion between MVB/PVC and PSV. Mutation in either MON1 or CCZ1 resulted in mistargeting of SSPs to the apoplast space (Pan et al. 2021). MVB/PVCs acting as intermediate compartment from DVs to PSVs are formed by fusion of DVs with small vesicles, which are then delivered to PSVs by the homotypic fusion and vacuolar protein sorting (HOPS) (Ebine et al. 2014; Minamino and Ueda 2019). HOPS as the tethering complex interacts with Rab7 and R-SNARE, VAMP71, which likely mediates homotypic vacuole fusion (Fig. 5.1c). Rab5- and Rab7-dependent pathway modulated by the MON1-CCZ1 complex is conserved in various organisms including yeast and animals. The VAMP71-containing SNARE complex and HOPS participate in the regulation of membrane fusion between the MVB/PVC and PSV. The R-SNARE on the transport intermediate forms trans-SNARE complex with the Q-SNAREs (SYP22, VTI11, SYP51) on the target membrane. The tight binding of SNARE proteins leads to fusion of membranes of the transport intermediate and the target compartment (Zhang et al. 2021; Ito and Uemura 2022).

On the other hand, it has recently been demonstrated that loss-of-function mutation of GPA5 (plant-unique phox-homology domain-containing protein), a homolog of ENDOSOMAL RAB EFFECTOR WITH PX-DOMAIN (EREX), resulted in discharge of its cargo protein into the apoplast due to defective directional targeting from DV to PSV (Ren et al. 2020). GPA5 acts in concert with Rab5a and its GEF VPS9a to regulate DV-mediated post-Golgi trafficking to PSVs. Furthermore, GPA5 physically interacts with a class C core vacuole/endosome tethering (CORVENT) and plant seed-specific vesicle-associated membrane protein 727 (VAMP727)-containing SNARE complex required for the membrane tethering and fusion of DVs with PSVs. Rab5a and GPA3 are dually localized to both the Golgi/TGN and DVs (Fukuda et al. 2011; Ren et al. 2014), whereas GPA5 is only localized to DVs. Thus, in the DV-mediated post-Golgi traffic process to PSVs, GPA3 first recruits VPS9a to activate Rab5a, and the activated Rab5a recruits its effector GPA5 onto DVs and then interacts with the CORVET- and VAMP727-containing R-SNARE complexes to execute direct fusion of DVs with PSVs (Fig. 5.1c). This route is plant specific and is dependent on the Rab5 only. It is interesting to note that plants use both evolutionarily conserved machinery (such as Rab5, CORVET, and SNARE) and plant-unique factors (GPA3, GPA5, and VAMP727) to mediate storage protein transport. Understanding of trafficking mechanisms from ER to PSV will provide insight into exploitation about effective delivery system of massive amounts of recombinant proteins to PSVs.

5.6 Modification to Humanized Glycosylation

Plants attach β-1,2-xylose and α-1,3-fucose residues to the recombinant proteins containing the N glycosylation-specific sequence (Asn-X-Ser/Thr), when they are synthesized as secretory proteins and transported through Golgi apparatus. In contrast, N glycosylation sites of mammalian glycoproteins such as antibodies are modified with α-1,6-fucose, β-1,4-galactose and sialic acid residues (Gomord et al. 2010). Difference in such N glycosylation patterns is highly associated with pharmacokinetics and immunogenicities of pharmaceutical proteins. Furthermore, such plant-specific N-glycans have been reported to be sometimes immunogenic (Garcia-Casado et al. 1996). Therefore, glycoengineering strategies have been developed to eliminate plant-specific N-glycans and to enable the incorporation of human-type N-glycans with sialic acid into glycoproteins for pharmaceutical use. For this purpose, plant-specific glycosyltransferases have been knocked down by RNA interference (RNAi)-mediated suppression or knocked out by mutagenesis, homologous recombination, or genome editing using CRISPR/Cas9, while protein sialylation has been incorporated by coordinated expression of multiple human genes encoding the entire mammalian pathway for sialic acid synthesis, transport, and transfer (Castilho et al. 2010; Montero and Steinkellner 2018; Liu and Timko 2022; see also Chaps. 3 and 4).

N-glycosylation of secretory proteins is a major posttranslational modification in eukaryotes, which starts in the ER by transfer of an oligosaccharide precursor N-glycan $Glc_3Man_9GlcNAc_2$ from a dolichol lipid carrier onto specific Asn residues constitutive of the N-glycosylation consensus sequence Asn-X-Ser/Thr (X is any amino acid except Pro) in the translating recombinant glycoproteins by the action of oligosaccharyltransferase (OST) complex. N-glycosylation has influence on folding, trafficking, protein interactions, and efficacy of subunit vaccines and other biologics.

Targeting of recombinant proteins to either ER-derived PB or PSV leads to a difference in posttranslational modification by glycosylation. Glycoproteins deposited in ER-PBs are modified with high-mannose-type N-glycans in common between plants and mammals, whereas those transported to PSV through Golgi apparatus possess plant-specific complex-type glycosylation by the N-glycan-processing steps mediated by Golgi-resident enzymes. Plant complex N-glycans uniquely contain core α-1,3-fucose (Fuc) and β-1,2-xylose (Xyl) modification and terminal Lewis a (Le^a) epitopes with β-1,3-galactose and α-1,4-fucose linked to terminal *N*-acetylglucosamine (GlcNAc). When targeted to plasma membrane, apoplast, or vacuole, truncated paucimannosidic-type glycans are generated (Strasser 2016; Nagashima et al. 2018). These are in contrast with the human complex N-glycans modified with α-1,6-fucose and β-1,4-galactose and often sialylation, and furthermore with different epitopes, such as Lewis x, *N*-acetyllactosamine (LacNc), and *N,N*′-di-*N*-acetyllactosediamine (LaDiNAc). It is important to note that maturation of complex N-glycans is not uniform. That is, complex N-glycan structure is variant among plant species or tissues as well as development processes. For example, monocot seeds (rice or maize) are devoid of or very poor in the terminal Lewis a epitope compared to dicot seeds (peanut, pea, or mung bean). Furthermore, N-glycosylation level and its pattern in seed are also lower and simpler than those in vegetative tissues. Plant-specific complex N-glycan structures (α-1,3-fucose and β-1,2-xylose residues, Lewis a epitope) have been reported to be allergenic in several mammals, when exploited as therapeutic proteins for parenteral administration

(Gomord et al. 2010). Moreover, plant-derived glycoprotein yet lacks terminal sialic acids and galactose residues. Such difference in the posttranslational glycosylation results in the alternation of biological functions, such as pharmacokinetics of a biopharmaceutical product (half-life stability in the blood) and immunogenicity in humans. Therefore, humanized N-glycosylation has been proceeded by knockout of plant-specific complex N-glycans and knock-in of heterogeneous glycosyltransferases for the production of galactosylated and sialylated N-glycans by replacement of pathways involved in the synthesis, transport, and transfer of the glycosylating residues with human counterparts and addition of sialic acid residues. Up to date, recombinant proteins have been reported to have homologous humanlike sialylated glycosylation profiles by introducing multiple human enzyme genes (α-1,6-fucosyltransferase (FUT8), β-1,4-galactosyltransferase (β-1,4-GalT), and sialyltransferase) involved in the modification of α-1,6-fucose, β-1,4-galactose, and sialic acid (Kallolimath et al. 2016; Montero and Steinkellner 2018). Especially, β-1,4-galactosylation of N-glycan chain has been known to be a prerequisite for the addition of terminal sialic acid residues, since β-1,4-galactose residues serve as the acceptor substrate for the sialic acid residue *N*-acetylneuraminic acid (Neu5Ac).

At present, plant-specific β-1,2-xylose and α-1,3-fucose residues on glycoproteins are completely removed in *N. benthamiana* plants, tobacco BY-2 cells, and rice culture cells by using CRISPR/Cas9-mediated multiplex XylT and FucT knockout mutation technology (Hanania et al. 2017; Mercx et al. 2017; Jansing et al. 2019; Jung et al. 2021). However, it should be noted that plant-specific Lewis a (Le^a) epitope was detected in rice cell line even after the complete deletion of α-1,3-fucose and β-1,2-xylose residues. This indicates that the double knockout of both the OsXylT and OsFucT genes was not sufficient to remove the plant-specific complex-type N-glycans on glycoproteins (Jung et al. 2021). Therefore, elimination of Le^a epitopes could be achieved by knockout of the two genes encoding β-1,3-galactosyltransferase gene (β-1,3-GalT) and α-1,4-fucosyltransferase (α-1,4-FucT) (Parsons et al. 2012). Furthermore, in order to reduce paucimannosidic-type glycans (two terminal GlcNAc residues), genes encoding the β-*N*-acetylhexosaminidases (HEXOs) were knocked down in *A. thaliana* and *N. benthamiana* (Liebminger et al. 2011; Shin et al. 2017). By the way, it should be noted that elimination of the FuxT and XylT by knockout or knock-in of mammalian glycosyltransferase led to reduced viability (growth, reproduction), stress response, and appearance of phenotype abnormalities in some plants (Strasser 2014; Kaulfürst-Soboll et al. 2021).

Plant and mammalian N-glycans differ in the degree of branching. Plant N-glycans are biantennary, while mammalian ones are tri- and tetra-antennary. This is attributed to the lack of some genes encoding branching enzymes (*N*-acetylglycosyltransferases, i.e., GnTIII–GtTV) (Castilho and Steinkellner 2012; Montero and Steinkellner 2018). Various bisected and multi-antennary structures can be generated by introducing the *N*-acetylglucosaminyltransferases (GnTs) that are absent in plants (Frey et al. 2009; Karg et al. 2010; Nagels et al. 2011).

Introduction of β-1,4-GalT and enzymes involved in the human sialylation pathway results in β-1,4-galactosylation followed by terminal sialylation to multi-antennary structures, resulting in the generation of fully sialylated human-type

N-glycans (Liu and Timko 2022). Double XylT/FucT knockout line of *N. benthamiana* and a moss triple knockout line of XylT, FucT, and GalT3 were used as bioreactor to produce fully sialylated mammalian glycoproteins by stably expressing seven mammalian genes (Kallolimath et al. 2016; Bohlender et al. 2020).

On the other hand, O-glycosylation on Ser/Thr of secretory proteins is fundamentally different between plants and humans. In humans, mucin-type O-glycosylation is the attachment of a single *N*-acetylgalactosamine (GalNAc) residue to Ser/Thr amino acids (GalNAcα-Ser/Thr), which is further modified by the stepwise addition of different monosaccharides. By contrast, plant O-glycosylation is characterized by the attachment of arabinogalactan polysaccharide or arabino-oligosaccharides. In this plant O-glycosylation process, proline residues are converted to hydroxyproline in the ER, and then arabinose residues are attached in the Golgi body. This conversion is initiated by the action of the prolyl-4-hydroxylase (P4H) family, which is followed by decoration with arabinose or arabinogalactan residues (Strasser 2016). Therefore, to humanize plant O-linked glycans, the P4H genes have been knocked out to prevent this conversion (Schoberer and Strasser 2018). Next, implementation of humanized mucin-type O-glycosylation was achieved in ΔXT/FT knockdown *N. benthamiana* plants by overexpressing the human GalNAc transferase 2 (GalNAcT2) gene, resulting in O-GalNAc synthesis on recombinant glycoproteins (Dicker et al. 2016; Montero and Steinkellner 2018). This O-linked GalNAc was elongated with β-1,3-galactose by overexpression of β-1,3-galactosyltransferase (C1GalT1) and then sialylated with α-2,3- and α-2,6-sialyltransferases to generate sialylated O-glycans (Dicker et al. 2016).

5.7 Production of Recombinant Proteins in Seed

5.7.1 Subcellular Localization of Recombinant Protein Expressed as Secretory Protein

Accumulation levels, stability, and posttranslational modifications of recombinant proteins in transgenic seeds are highly influenced by the intracellular sorting and localization (Khan et al. 2012; Hofbauer and Stoger 2013). Therefore, exploring subcellular compartment suitable for deposition of the recombinant protein in the selected production host represents a major issue to maximize production level. In many cases, much higher yields of recombinant proteins can be generally obtained by targeting to secretory pathway using endomembrane system compared to the cytosol. This is attributed to an oxidizing environment, few proteases, and sufficient amounts of chaperones required for folding and assembly that are equipped in ER. Moreover, many pharmaceutical proteins are posttranslationally modified by glycosylation and thereby have to be synthesized as secretory proteins using endomembrane system via ER/Golgi tracking route. When the recombinant proteins are produced as secretory proteins, ER, PB, PSV, and amyloplast (starch granule) are

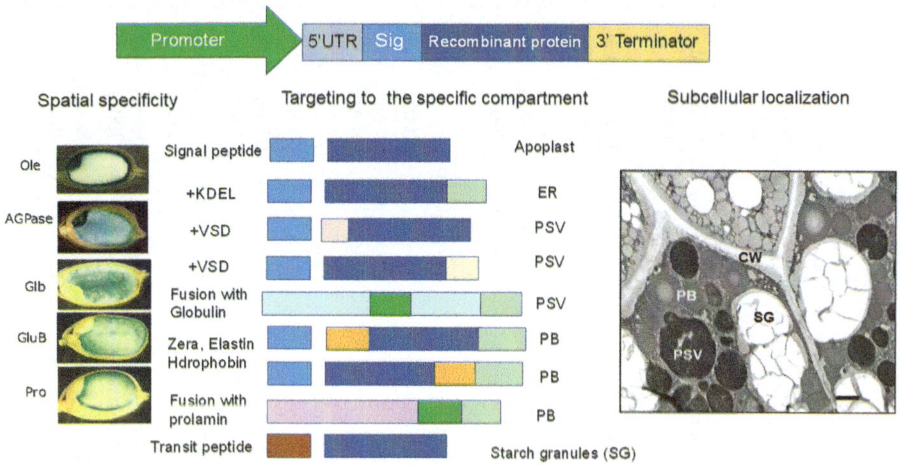

Fig. 5.2 Subcellular targeting of recombinant proteins in seed cells. Recombinant proteins with an N-terminal signal peptide (SP) enter the cell secretory pathway in ER and then travel through the Golgi system to be secreted in the apoplast, or directed to the PSV, if a vacuolar sorting determinant is present in the protein sequence. Proteins having ER retention signal (KDEL tetrapeptide) at the C-termini are retained in ER by retrieval function. When Zera, elastin-like polypeptide (ELP), or fungal hydrophobin is fused to the N- or C-termini of recombinant proteins, they are incorporated into PBs by their aggregation property. Bioactive peptides are highly produced as fusion proteins with SSPs such as globulins or prolamins to be deposited into PSVs or PBs. To target starch granules (SG), transit peptide is attached at the N-termini of recombinant proteins

expected to be suitable deposition compartments because they are inherent storage organs of reserves (sources of nitrogen and carbohydrates during seed germination) and provide ample deposition space. On the other hand, physicochemical property of the targeted recombinant protein has to be also taken into account. Especially, when recombinant proteins detrimental to plant cells such as industrial processing enzymes are produced in seeds, they have to be stored outside the cell by secretion, such as periplasm (apoplast) between plasm membrane and cell wall (Fig. 5.2).

When recombinant proteins are expressed as secretory protein by attaching the signal peptide at their N-termini, they are generally secreted into extracellular space (apoplast) via the default pathway (bulk flow). When fungal laccase and cellobiose 1 were expressed in maize grains under the control of the embryo globulin 1 promoter, they were predominantly deposited in apoplast as expected (Hood et al. 2003, 2007; see also chapters "Molecular Farming of Industrial Enzymes: Products and Applications" and "Plant Molecular Farming for the Bulk Production of Industrial Enzymes"). Many recombinant proteins (HA78 mAb, HA78 scFv, 2G12 mAb, human α-L-iduronidase, etc.) expressed in *Arabidopsis* seeds were also transported to apoplast (Loos et al. 2011a, b; Downing et al. 2006). By contrast, when *Aspergillus* phytase was expressed in the endosperm of rice, maize, or wheat seeds, it was deposited in PSVs and ER-derived PBs (Drakakaki et al. 2006; Arcalis et al. 2004, 2010). This result was in contrast with that the same *Aspergillus* phytase was secreted to apoplast of rice and

Table 5.4 Production yields and subcellular locations of recombinant proteins produced in various transgenic seeds

Host plant	Recombinant protein	Application	Promoter, 5′ UTR	Signal	Localization	Production level (yield)	Reference
Maize	Avidin	Research reagent	Maize ubiquitin, ubiquitin intron	No signal	Apoplast, embryo (57%)	5.7% of TSP (2.3% of TSP)	Hood et al. (1997)
	Aprotinin	Research reagent (protease inhibitor)	Maize ubiquitin, ubiquitin intron	No signal	Apoplast	0.069% of TSP	Zhong et al. (1999)
	Glucuronidase (GUS)	Model protein	Maize ubiquitin, ubiquitin intron	No signal	NA	0.7% of TSP	Kusnadi et al. (1998)
	Bovine trypsin	Protease	Embryo globulin 1	BAA SP	NA	58 mg/kg Grain (3% of TSP)	Woodard et al. (2003)
	Fungal laccase	Industrial enzyme	Embryo globulin 1	BAA SP	Cell wall (apoplast)	0.8% of TSP	Hood et al. (2003)
	β-1,4 Endoglucanase	Industrial enzyme	Embryo globulin 1	BAA SP, KDEL	ER	17.9% of TSP	Hood et al. (2007)
	β-1,4 Endoglucanase	Industrial enzyme	Embryo globulin 1	BAA SP, vacuole signal	Vacuole	16% of TSP	Hood et al. (2007)
	Manganese peroxidase	Industrial enzyme	Embryo globulin 1	BAA SP	NA	15% of TSP	Clough et al. (2006)
			CaMV 35S	BAA SP	NA	3% of TSP	Clough et al. (2006)
	Cellobiohydrolase 1	Industrial enzyme	Embryo globulin 1	BAA SP	Cell wall (apoplast)	17.8% of TSP	Hood et al. (2007)
	Cellobiohydrolase 1	Industrial enzyme	Embryo globulin 1	BAA SP, KDEL	ER	16.3% of TSP	Hood et al. (2007)

Host plant	Recombinant protein	Application	Promoter, 5′ UTR	Signal	Localization	Production level (yield)	Reference
	Cellobiohydrolase 1	Industrial enzyme	19 kDa Azein	BAA SP	NA	9% of TSP (0.225 g/kg seed)	Requesens et al. (2019)
	Aspergillus phytase	Industrial enzyme	Glutelin Gt1	Murine Ig k chain SP	shift from PSV to ER-derived PB	NA	Arcalis et al. (2010)
	Type 2 brazzein	Sweetener	Embryo globulin 1	BAA SP	NA	4% of TSP	Lamphear et al. (2005)
	2G12 mAb	HIV	Glutelin Gt1	Gt1 SP, KDEL	ER-derived PB	75 μg/g seed (5.7% of TSP)	Rademacher et al. (2008)
	LTB (heat-labile enterotoxin B subunit)	Diarrhea (enterotoxigenic *E. coli*)	27 kDa γ-Zein	γ-Zein SP	Starch granule	NA	Chikwamba et al. (2003)
	LTB (heat-labile enterotoxin B subunit)	Diarrhea (enterotoxigenic *E. coli*)	27 kDa γ-Zein	BAA SP, KDEL	NA	3.7% of TSP	Chikwamba et al. (2002)
	CTB (cholera toxin B subunit)	Diarrhea (chorera)	27 kDa γ-Zein, TEV	No signal	NA	0.0014% of TSP	Karaman et al. (2012)
	S protein of transmissible gastroenteritis virus (TGEV)	Transmissible gastroenteritis virus	Maize ubiquitin, ubiquitin intron	BAA SP	Apoplast	13 mg/kg Grain 2%	Lamphear et al. (2004)
	G protein of rabies virus	Rabies virus	Maize ubiquitin, ubiquitin intron	No signal	NA	25 μg/g Grain	Loza-Rubio et al. (2012)

(continued)

Table 5.4 (continued)

Host plant	Recombinant protein	Application	Promoter, 5′ UTR	Signal	Localization	Production level (yield)	Reference
	Surface glycoprotein F of Newcastle disease virus	Newcastle disease virus	Maize ubiquitin, ubiquitin intron	No signal	NA	3% of TSP	Guerrero-Andrade et al. (2006)
	Hepatitis B surface antigen	Hepatitis B virus	Embryo globulin 1	BAA SP	NA	0.51% of TSP	Hayden et al. (2012)
	H3N2 nucleoprotein (NP)	Influenza virus	27 kDa γ-Zein	γ-zein SP	NA	70 μg/g Seed (0.06% TSP)	Nahampun et al. (2015)
	Rotavirus VP6	Rotavirus	27 kDa γ-zein	γ-Zein SP, KDEL	NA	3.5 μg/mg Seed	Feng et al. (2017)
Wheat	Human serum albumin	Cell culture additive	Rice glutelin Gt1	Murine Ig k chain SP, KDEL	Prolamin-aggregate (PB) within vacuole	NA	Arcalis et al. (2004)
	Aspergillus phytase	Industrial enzyme	Rice glutelin GT1	Murine Ig k chain SP	PSV and cytoplasmic PB	NA	Arcalis et al. (2004)
	Pea legumin	Seed storage enzyme	Rice glutelin Gt1	Murine Ig k chain SP	Inclusion body at the peripheral of prolamin body	NA	Arcalis et al. (2004)
	scFvT84,66	Cancer (tumor marker)	Maize ubiquitin LPH	SP, KDEL	NA	1.4 mg/g Seed	Stöger et al. (2000)
Barley	Collagen type I (Cla I (full), 45 kDa)	Structural component	Rice glutelin GluB-1	Basic chitinase SP, KDEL	NA	140 mg/kg Seed	Eskelin et al. (2009)
	Fungal xylanase	Industrial enzyme	Rice glutelin GluB1, barley hordein Hor2-4	No signal	NA	127 AU/h grain, 22 AU/h grain	Patel et al. (2000)

Host plant	Recombinant protein	Application	Promoter, 5′ UTR	Signal	Localization	Production level (yield)	Reference
	β-Glucanase	Industrial enzyme	Barley D-hordein (Hor3-1)	Hor3-1 SP	NA	54 μg/mg Soluble protein	Horvath et al. (2000)
	2G12 mAb	HIV	Oat globulin 1	LegB4 SP, KDEL	ER-derived PB	160 μg/g Seed (0.4% of TSP)	Hensel et al. (2015)
	Peptide LL35 (human cathelicidin)	Antimicrobial peptide	Barley B1 hordein	ZmCKX1 SP, KDEL	ER-derived PB	550 μg/kg Seed	Holásková et al. (2018)
Rice	Human serum albumin	Cell culture additive	Glutelin Gt1	Gt1 SP	NA	2.75 g/kg Seed (0.58% TSP)	He et al. (2011)
	Human lactoferrin	Gastrointestinal infection	Glutelin Gt1	Gt1 SP	NA	4.5–5.5 mg/g Seed	Nandi et al. (2005)
	Human lysozyme	Gastrointestinal infection	Glutelin Gt1	Gt1 SP	PB-II	350 μg/mg TSP	Yang et al. (2003)
	Aspergillus phytase	Industrial enzyme	Glutelin Gt1	Murine Ig k chain SP	PB-I, PB-II	ND	Drakakaki et al. (2006)
			CaMV 35S	Murine Ig k chain SP	Apoplast (calli, leaf) PB-I/PB-II (seed)	0.5% of TEP	Drakakaki et al. (2006)
	7Crp T cell epitope	Japanese cedar pollinosis	Glutelin GluB-1	GluB-1 SP, KDEL signal	PB-I	50–60 μg/grain	Takagi et al. (2005a)
	6Chao T cell epitope	Cypress pollinosis	Glutelin GluB-1	GluB-1 SP, KDEL signal	PB-I	12 μg/grain	Takaiwa and Yang (2014)

(continued)

Table 5.4 (continued)

Host plant	Recombinant protein	Application	Promoter, 5′ UTR	Signal	Localization	Production level (yield)	Reference
	Mite allergen Der f 2 (Cys less)	Mite allergy	Glutelin GluB-1	GluB-1 SP, KDEL signal	Derf2 body	15–30 μg/grain	Yang et al. (2012a)
	Mite allergen Der p1 (45–145)	Mite allergy	Glutelin GluB-1	GluB-1 SP, KDEL	PB-I	90 μg/grain	Suzuki et al. (2011a)
	Mite allergen Der p 1 (Cys/Ala)	Mite allergy	Glutelin GluB-1	GluB-1 SP, KDEL	PB-I	58 μg/seed	Yang et al. (2008)
	Shuffled Cry j 2	Japanese cedar pollinosis	Glutelin GluB-1 or-4	Signal peptide, KDEL	PB-I	13 μg/grain	Suzuki et al. (2011b)
	Cry j 1 fragment fused to glutelin	Japanese cedar pollinosis	Glutelin GluB-1	GluB-1 SP	PB-I	80–220 μg/grain	Yang et al. (2007a)
	Shuffled Bet v 1 (TPC7)	Birch pollinosis	Glutelin GluB-1	GluB-1 SP, KDEL	TPC7 body	207 μg/grain	Wang et al. (2013a)
	2×β-amyloid (1–42)	Alzheimer	Glutelin GluB-1	GluB-1 SP, KDEL	Modified PB-I	8 μg/seed	Oono et al. (2010)
	Surface glycoprotein Newcastle disease	Newcastle disease	Glutelin Gt1	No signal	NA	2.5–5.5 μg/g	Yang et al. (2007b)
	PreS1 fused to HBV surface antigen (SS1)	Hepatitis B virus	Glutelin GluB-4	GluB-1 SP	Cytoplasm	31.5 μg/g	Qian et al. (2008)
	IBDV VP2	Infectious bursal disease	Glutelin Gt1	No signal	NA	4.5% of TSP	Wu et al. (2007)
	CTB (cholera toxin B subunit)	Diarrhea (cholera)	Glutelin GluB-1	GluB-1 SP, KDEL	PB-I, PB-II	30 μg/grain	Nochi et al. (2007)
	As16 fused to CTB	Roundworm	Glutelin GluB-1	GluB-1 SP, KDEL	NA	50 μg/g	Matsumoto et al. (2009)

Host plant	Recombinant protein	Application	Promoter, 5′ UTR	Signal	Localization	Production level (yield)	Reference
	Human bFBG	Cytokine	Glutelin Gt13a	Gt13a SP	NA	185 mg/kg grain	An et al. (2013)
	Human GM-CSF	Cytokine	Glutelin Gt1	Gt1 SP	NA	1.3% of TSP	Sardana et al. (2007)
	Human IL-10	Cytokine	Glutelin GluB-1	GluB-1 SP, KDEL	Aberrant ER-derived PB	220 μg/grain	Yang et al. (2012b)
	Mouse IL-4	Cytokine	Glutelin GluB-1	GluB-1 SP, KDEL	ER-derived PB	0.43 mg/g seed	Fujiwara et al. (2016)
	Mouse TGF-β	Cytokine	Glutelin GluB-1	GluB-1 SP, KDEL	Aberrant ER-derived PB	110 μg/grain	Takaiwa et al. (2016)
	Human TGF-β	Cytokine	26 kDa Globulin	Glb SP, KDEL	ER granules	452 μg/grain	Takaiwa et al. (2021)
	Mouse IL-6	Cytokine	Glutelin GluB1	GluB-1 SP, KDEL	PB-I	0.16 mg/g seed	Fujiwara et al. (2016)
	scFv T84.66	Cancer (tumor marker)	Ubiquitin promoter	mAbH SP, KDEL	PB-I, PSV	30 μg/g seed	Torres et al. (2001)
	Guy's 13 secretory IgA	*Dental caries*	*Ubiquitin*	Native SP, KDEL	ER-derived PB, PSV (assemble)	NA	Nicholson et al. (2005)
	2G12 mAb	HIV	Glutelin Gt1, ubi intron, TEV	Native SP	PSV, (ER-derived PB)	34–42 μg/g dry seed	Vamvaka et al. (2016a)
	Griffithsin (HIV antiviral lectin)	HIV	Maize zein	Amylase3A SP	PSV	223 μg/g dry seed	Vamvaka et al. (2016b)
	18 × Novokinin	Hypertension	Glutelin GluB-1	GluB-1 SP, KDEL	Nucleolus	85 μg/g seed	Wakasa et al. (2011a)

(continued)

Table 5.4 (continued)

Host plant	Recombinant protein	Application	Promoter, 5′ UTR	Signal	Localization	Production level (yield)	Reference
	GLP-1 fused to 26 kDa globulin	Type 2 diabetes	26 kDa Globulin	Globulin fusion	PSV	20–50 μg/grain	Sugita et al. (2005)
	Lactostatin fused to glutelin	Hypercholesterolemia	Glutelin GluB-1	Glutelin fusion	PSV	1.6 mg/g grain	Wakasa et al. (2006)
	APL6 (collagen 2) fused to glutelin	Arthritis	Glutelin GluB-1	Glutelin fusion	PSV	7–24 mg/g grain	Iizuka et al. (2014)
	APL12 (GPI) fused to glutelin	Arthritis	Glutelin GluB-1	Glutelin fusion	PSV	3.8 mg/g grain	Hirota et al. (2017)
	APL12 (GPI) fused to prolamin	Arthritis	Prolamin	Prolamin fusion	PB-I	200 μg/seed	Takaiwa et al. (2018)
	Novokinin fused to glutelins	Hypertension	Glutelin	Glutelin fusion	PSV	470 μg/g seed	Yang et al. (2006)
	Human insulin-like growth factor	Type 2 diabetes	Glutelin Gt1 3a	Gt1 SP	PSV, ER	6.8% of TSP	Xie et al. (2008)
Arabidopsis	MBP10 scFv-Fc	Maltose-binding protein	β-Phaseolin, arc5-1 5′ UTR	2S2 SP, KDEL	ER-derived PB, apoplast (35–40%)	12.4% of TSP	Van Droogenbroeck et al. (2007)
	EHF34 scFv-Fc	Hantaan virus				9.0% of TSP	
	HA78 scFv-Fc	Hepatitis A virus				7.1% of TSP	
	scFvG4	Model protein	β-Phaseolin, arc5-1 5′ UTR	2S2 SP, KDEL	NA	36.5% of TSP	De Jaeger et al. (2002)
	PRRSV glycoprotein	Porcine reproductive and respiratory syndrome virus	β-Phaseolin TMV ω	2S2 SP, KDEL	NA	2.7% of TSP	Piron et al. (2014)
	Glucocerebrosidase	Gaucher's disease	Arcelin-5-1	arc5-1 SP	PSV	0.1% of TSP	He et al. (2012)

Host plant	Recombinant protein	Application	Promoter, 5′ UTR	Signal	Localization	Production level (yield)	Reference
	HA78 mAb	Hepatitis A virus	β-Phaseolin	2S2 SP	Apoplast, Golgi dense vesicles	8.8 μg/mg seed	Loos et al. (2011a)
	HA78 scFv-Fc	Hepatitis A virus	β-Phaseolin	2S2 SP	Apoplast (Golgi vesicle)	8.0 μg/mg seed	Loos et al. (2011b)
				2S2 SP, KDEL	ER-derived vesicles, (PSV)	3.9 μg/mg seed	
	2G12 mAb	HIV	β-Phaseolin	2S2 SP	Apoplast (Golgi vesicle)	3.6 μg/g seed	Loos et al. (2011a)
				2S2 SP, KDEL	PSV, (ER)	3.0 μg/mg seed	
			CaMV 35S	Native SP	Apoplast (EM, EN)	NA	Arcalis et al. (2013)
				Native SP, KDEL	PSV (EM)	NA	
	2G12 scFv-Fc	HIV	β-Phaseolin	2S2 SP	ER-derived vesicles, nuclear envelop	0.8 μg/mg seed	Loos et al. (2011b)
				2S2 SP, KDEL	ER-derived vesicle, nuclear envelop	0.8 μg/mg seed	
	VHHC-IgA	Enterotoxigenic *E. coli*	β-Phaseolin, ω leader 5′ UTR	2S2 SP, KDEL	NA	3% of TSP	Virdi et al. (2013)
	GAD67/65mut	Type 1 diabetes	β-Phaseolin, ω leader 5′ UTR	2S2 SP, KDEL	PSV (ER to PSV)	4.5 mg/g	Morandini et al. (2011)
	Mouse hIL-10	Cytokine	β-Phaseolin, ω leader 5′ UTR	2S2 SP, KDEL	ER-like membrane	0.3 mg/g seed (0.41% of TSP)	Morandini et al. (2011)

(continued)

Table 5.4 (continued)

Host plant	Recombinant protein	Application	Promoter, 5′ UTR	Signal	Localization	Production level (yield)	Reference
	Proinsulin	Type 1 diabetes	Arcelin 5-1	2S2 SP, KDEL	PSV	4.8 μg/g (00054% of TSP)	Morandini et al. (2011)
	Human α-L-iduronidase	Lysosomal storage disorder	Arcelin-5-1	Arcelin-5-1 SP	Apoplast	1.8% of TSP	Downing et al. (2006)
Pea	scFv T84.66	Cancer (tumor marker)	Pea legumin A, ω leader 5′ UTR	LPH SP, KDEL	Pea cotyledon	9 μg/g	Perrin et al. (2000)
	scFv ABA Ab	ABA	USP	Signal	ER	2% of TSP	Saalbach et al. (2001)
	Capsid protein VP60 fused to CTB	Rabbit hemorrhagic disease virus	CaMV35S	SP, KDEL	NA	0.001% of TSP	Mikschofsky et al. (2009)
Soybean	Human basic fibroblast growth factor	Cytokine	Glycinin A1aB1b (G1)	G1 SP	PSV	2.3% of TSP	Ding et al. (2006)
	Human coagulation factor IX (hFIX)	Blood clotting (zymogen)	7S globulin α′ subunit	α-Coixin SP	PSV	0.23% of TSP (0.8 g/kg seed)	Cunha et al. (2011a, b)
	Human growth hormone	Growth hormone	7S globulin α′ subunit	α-Coixin SP	PSV	2.9% of TSP	Cunha et al. (2011a, b)
	Novokinin fused to 7S α′ subunit	Hypertension	7S globulin α′ subunit	α′ subunit SP	PSV	0.1–0.2% of TSP (1–2 μg/mg seed)	Nishizawa et al. (2008)
	Human epidermal growth factor	Growth factor	Glycinin A1aB1b	Chitinase SP	NA	129 μg/g seed	He et al. (2016)

Host plant	Recombinant protein	Application	Promoter, 5′ UTR	Signal	Localization	Production level (yield)	Reference
	Cyanovirin N (microbiocide)	HIV	7S globulin α′ subunit	α′ Subunit SP	PSV	350 mg/kg seed	O'Keefe et al. (2015)
	LTB (heat-labile toxin B subunit)	Diarrhea (enterotoxigenic *E. coli*)	Glycinin A1aB1b	Chitinase SP, KDEL	PSV	2.4% of TSP	Moravec et al. (2007)
Tobacco	Human growth hormone	Growth hormone	r-Kafirin promoter	Kafirin SP	NA	0.16% of TSP	Leite et al. (2000)
	Aspergillus phytase	Industrial enzyme	2×CaMV35S	Murine Ig k SP	Apoplast/PSV (EM), PSV/apoplast (EN), apoplast (leaves)	NA	Arcalis et al. (2013)
	Human cytomegalovirus glycoprotein B (HCMV)	Human cytomegalovirus	Glutelin Gt3	Gt3 SP	PSV	70–146 ng/mg protein	Wright et al. (2001)
	14D9 mAb	Catalytic antibody	7S globulin α′ subunit, TEV	HCMSP	PSV (matrix)	0.06% of TSP	Petruccelli et al. (2006)
				HCMSP, KDEL	PSV, apoplast	1.08% of TSP	
			CaMV 35S, TEV	HCMSP	Apoplast (leaf), PSV (seed)	2.9% (leaf), 0.21% of TSP (seed)	
				HCMSP, KDEL	ER (leaf), PSV, apoplast (seed)	5.2% (leaf), 0.41% of TSP (seed)	

(continued)

Table 5.4 (continued)

Host plant	Recombinant protein	Application	Promoter, 5′ UTR	Signal	Localization	Production level (yield)	Reference
	HBsAg mAb	Hepatitis B virus	Phaseolin, arc-5I 5′ UTR	2S2 SP, KDEL	PSV	6.5 mg/g seed	Hernadez-Velazquez et al. (2015)
	HBsAg scFv	Hepatitis B virus	CaMV 35S, ω leader 5′ UTR	Sporamin, vacuole TS	PSV, apoplast	0.2% of TSP	Ramirez et al. (2001)
	2G12 mAb	HIV	CaMV 35S	Native SP	Apoplast/PSV (EM), PSV/ apoplast (EN), apoplast (leaves)	NA	Arcalis et al. (2013)
				Native SP, KDEL	PSV (EM, EN), ER (leaves)	NA	
	HA H5N1	Avian influenza virus	Phaseolin	SP, KDEL	NA	3 mg/g seed	Ceballo et al. (2017)
	Human granulocyte-colony-stimulating factor	Growth factor	CaMV 35S	Ubiquitin + phaseolin SP	Apoplast (leaves), PSV (seed)	2.5 mg/g TSP (leaves), 1.3 mg/g TSP (seed)	Tian and Sun (2011)
	Human complementary factor 5a	Sepsis	2×CaMV35S, ω leader 5′ UTR	SP, KDEL, AFVY vacuole SP	ER or Vacuole (seed)	10.6 μg/g leaf 35.8 μg/g seed	Nausch et al. (2012)

NA not analyzed, *BAA SP* barley alpha-amylase signal peptide, *2S2 SP* signal peptide of the *Arabidopsis* 2S2 SSP gene, *HCMSP* heavy-chain murine signal peptide, *LPH* heavy-chain leader peptide from the TMV, *TEV* tobacco etch virus leader sequence, *EM* embryo, *EN* endosperm tobacco leaf cells, when directed by the CaMV 35S promoter (Drakakaki et al. 2006; Arcalis et al. 2013). It should be noted that it was deposited to PSVs and apoplast in these transgenic seeds. Furtheremore, some antibodies (14D9 mAb and HBsAg mAb) were transported to PSV in tobacco seeds (Ramirez et al. 2001; Petruccelli et al. 2006). Targeting to PSVs was also observed in transgenic rice and tobacco seeds, when 2G12 mAb, griffithsin, and human cytomegalovirus glycoprotein B were specifically expressed in their endosperm tissues (Wright et al. 2001; Vamvaka et al. 2016a, b). On the other hand, when *Aspergillus* phytase was expressed in seeds and leaves of *Medicago*, it was efficiently secreted to the apoplast as expected (Abranches et al. 2008). Thus, these results suggest that the deposition of recombinant protein is highly influenced by the expressed seed tissue used as host. It is surprising that the LT-B expressed in maize endosperm was unexpectedly detected in amyloplasts as intracellular localization site, although the LT-B with the γ-zein N-terminal signal peptide was expressed under the control of the γ-zein endosperm-specific promoter (Chikwamba et al. 2003) (Table 5.4).

On the other hand, when the recombinant proteins were expressed in seeds by attaching the signal peptide and KDEL ER retention tag at their N- and C-termini, various recombinant proteins (7Crp peptide, shuffled Cry j 1 or 2, Der p 1 (45–145), shuffled Bet v 1 (TPC7), CTB, TGF-β, and 2×β-amyloid) were transported to ER-derived PBs as expected (Takagi et al. 2005a; Nochi et al. 2007; Oono et al. 2010; Suzuki et al. 2011a, b; Yang et al. 2012b; Wang et al. 2013a; Wakasa et al. 2013; Takaiwa et al. 2016, 2021). The ER-derived PB structures were sometimes distorted depending on their production levels. Aberrant small ER-derived granules, new bodies, or giant bodies were formed depending on the individual physicochemical property of the expressed recombinant protein. It is pointed out that such ER-derived novel bodies are morphologically similar to Russell bodies found in mammalian cells acting as self-protection against overloaded transport-incompetent proteins (Arcalis et al. 2019). By contrast, human serum albumin was deposited in PSVs in a form of aggregates when expressed in wheat endosperm (Arcalis et al. 2004). Despite the presence of ER retention signal, vacuolar targeting or secretion to apoplast was observed for the production of 14D mAb and HBs mAb, when expressed in tobacco seeds (Petruccelli et al. 2006; Hernández-Velázquez et al. 2015). 2G12 mAb was also secreted to apoplast in *Arabidopsis* seeds. It should be noted that MBP10 single-chain variable fragment (scFv)-Fc Ab bearing the N-terminal signal peptide and C-terminal KDEL tag was localized in apoplast as well as ER-derived PBs in *Arabidopsis* seeds. This finding suggests that substantial amounts of scFv product may lead to direct transportation from ER to the periplasm space bypassing the Golgi apparatus. This may be caused by ER saturation effect due to high amount of accumulation of scFv. Furthermore, 18 × novokinin was unexpectedly sequestered to nucleolus in the endosperm cells of transgenic rice (Wakasa et al. 2011a). This may be explained by the property that the 18 × novokinin sequence has a property to function as nuclei target signal, since the GFP reporter fused to the 18 × novokinin was observed to be transported to nuclei (Table 5.4).

It has been known in rice and maize seeds that incorporation of the expressed recombinant protein into ER-derived PBs is attributed to the intermolecular interaction between the recombinant protein and endogenous Cys-rich prolamins by disulfide bond formation via free Cys residues (Takaiwa et al. 2009; Peters et al. 2013). The C-terminal attachment of the KDEL ER retention tag generally resulted in two- to tenfold enhancement of accumulation level as compared with the constructs lacking this signal (Takagi et al. 2005a; Conrad and Fiedler 1998). This is associated with the evidence that ER is a favorable subcellular compartment for deposition of expressed proteins. Furthermore, when four polypeptide components (heavy chain, light chain, secretory chain, and linker chain) constituting the secretory IgA against *Streptococcus mutants* surface antigen were co-expressed by attaching the N-terminal signal peptide and C-terminal ER retention tag in rice seeds under the control of the maize ubiquitin promoter, the assembled chains were targeted to PSVs and the non-assembled ones were retained in ER-derived PBs (Nicholson et al. 2005). This result suggested that the correctly assembled secretory IgA may be allowed to be transported to PSVs (Nicholson et al. 2005). Similar observation was done by co-expression of soybean glycinin A1aB1b and A3B4 subunits in transgenic rice seeds. Assembly of these two types of glycinins enhanced their accumulation levels and facilitated the transportation of these subunits into PSVs, when compared to the transgenic rice seeds, in which individual glycinin was independently expressed (Takaiwa et al. 2008). Furthermore, it is interesting to note that deposition site is altered during seed maturation process. When 2G12 mAb or fungal phytase was produced in maize endosperm, their products were predominantly localized within the PSVs in young endosperm cells. This intracellular localization was shown to be progressively altered to ER-derived PBs along seed maturation (Arcalis et al. 2010; Drakakaki et al. 2006). These findings suggest that seed developmental stages are implicated in determining the intracellular trafficking routes.

5.7.2 *Targeting to PSV*

PSVs are unique to plants, and they function as the main deposition sites of SSPs in dicots and some monocots (rice and oat). In order to deposit the recombinant protein in PSV, one strategy is to utilize the fusion with the globulin SSPs (soybean globulins or rice glutelins), which are transported to PSVs. 11–12S globulins or rice glutelins have four variable (flexible) regions in their molecules, which are corresponding to exposed regions in their tertiary structures. Especially, the C-terminal region of acidic subunit is highly variable. Various small-size polypeptides or bioactive peptides have been introduced into the C-terminal highly variable region of rice glutelins or soybean 11S glycinins by substitution insertion (Yang et al. 2007a; Maruyama et al. 2014). When expressed under the seed-specific promoter, modified glutelins or glycinins containing the desired sequences were highly produced and deposited in PSVs. After most of them were transported to PSVs via Golgi apparatus, they were posttranslationally processed into mature acidic and basic subunits (Table 5.4).

The other strategy is to use the vacuolar sorting signals. For transport of cargo proteins to PSVs or lytic vacuoles, vacuolar sorting determinant (VSD) in the cargo amino acid sequence and vacuolar sorting receptor (VSR) have been known to be involved. The VSDs are grouped into sequence-specific vacuolar sorting signal (ssVSS), C-terminal vacuolar sorting signal (ctVSS), and physical structure/conformation VSDs (psVSDs) (Neuhause and Rogers 1998; Vitale and Hinz 2005). The ssVSSs were found out in barley aleurone, sweet potato sporamin, and castor bean ricin and 2S albumin. The conserved Asn-Pro-Ile-Arg (NPIR) motif is identified as a core signal sequence. The ssVSSs have been shown to interact with the BP-80 family sorting receptors. The ctVSSs are rich in hydrophobic residues, which are localized within ten amino acids from the C-terminal end of various SSPs such as 11S and 7S soybean globulins, common bean phaseolin, barley lectin, and rice glutelin. Proteins carrying the ctVSS are demonstrated to be targeted to PSVs via DVs or PAC vesicles by aggregation-based sorting. These ctVSSs do not have any consensus sequence (Zhang et al. 2021). Given that the ctVSS signals were linked to recombinant proteins, they would be targeted to PSV. Actually, when the sequences containing the ctVSSs derived from soybean 7S subunit or 11S A1aB1b glycinin were attached to the C-terminus of GFP reporter and then their localization was examined in soybean maturing seeds by transient expression, they were demonstrated to be targeted to PSVs as expected (Nishizawa et al. 2003; Maruyama et al. 2006).

The VSDs in cargo protein are recognized by two types of vacuolar sorting receptors, BP-80/VSR family and receptor homology region-transmembrane domain-RING-H2 (RMR) protein family. Vacuolar sorting is known to be mediated by specific protein-protein interactions between the sorting receptors and the cargo proteins, which is implicated in trafficking to PSVs. The BP-80/VSR family recognizes both ssVSDs and ctVSDs in cargo proteins, while the RMRs only interact with the ctVSDs. Vacuolar sorting is known to be mediated by specific protein-protein interaction between the sorting receptors and the cargo proteins, which is implicated in traafficking to PSV. RMRs involved in the vacuolar delivery of SSPs are observed to be localized not only in PSVs but also in Golgi apparatus, TGN, and MVEs, coinciding with the distribution of SSPs.

Up to date, either efficient targeting of recombinant protein to PSV by using the VSD or enhancement of recombinant protein yield by VSD-mediated targeting has not yet been reported. Further studies on targeting mechanism from ER to PSV will be required.

5.7.3 Protein Fusion Strategy in Seed-Based Production for Recombinant Protein

Maize SSP 27 kDa γ-zein plays an important role in initiating and organizing the PBs. N-terminal region of γ-zein called Zera is composed of eight repeats of the hexapeptide VHLPPP and seven cysteine residues, which is implicated in ER retention and interchain disulfide bond formation, respectively. When a variety of recombinant proteins were fused to this Zera sequence, formation of ER-derived PB-like structures was induced in eukaryotic cells through aggregation (oligomerization) via disulfide bonds within ER (Llop-Tous et al. 2010). Fusion to this Zera sequence gave rise to enhanced accumulation and stability as compared with the fusion to the KDEL ER retention signal. When the Zera was fused to common bean phaseolin SSP, the fusion protein zeolin accumulated to levels in excess of 3.5% TSP (Mainieri et al. 2004).

Formation of PB-like structure via accretion to ER has been observed to be induced by fusing the elastin-like polypeptides (ELPs) or fungal hydrophobins (HFBs) to recombinant proteins in various tissues, which are involved in the protection of proteins from degradation (Conley et al. 2011). ELPs comprise numerous repeats of a short peptide such as VPGXG, which is reminiscent of the structure of the mammalian connective tissue protein elastin. The repeated domains allow self-assembly, which confers stability to the fusion partners by packing them. When scFv antibody was C-terminally fused to 100 repeats of ELP (VPGVG) and then expressed in tobacco seeds under the control of seed-specific *Leg4* or *USP* promoter, accumulation of the scFv/ELP fusion was enhanced to 40-fold higher level than the scFv only, resulting in production at a concentration of about 25% of TSP (Scheller et al. 2006).

It is generally difficult to stably and highly accumulate the small-size bioactive peptide in transgenic seeds. Then, reporter proteins such as GFP or major SSPs have been used as fusion partner or carrier to stabilize the peptide or small-size protein.

When amyloid β-peptide with a length of 42 amino acids (Aβ42) was linked to GFP reporter, the GFP/Aβ42 fusion product stably accumulated at a level of 400 μg/g grain in transgenic rice (Yoshida et al. 2011). When three tandem repeats of Aβ (4–10) peptide were inserted into three flexible regions of soybean A1aB1b glycinin and were then expressed in soybean seeds deficient in major native SSPs, the engineered glycinin accumulated at about 0.5% of TSP (Maruyama et al. 2014). T cell epitope peptides derived from Japanese cedar pollen allergens (Cry j 1 and Cry j 2) or analogue peptide ligands derived from autoantigens (type II collagen, glucose-6-phosphate-isomerase, M3 muscarinic acetylcholine receptor) causing rheumatoid arthritis and Sjögren's syndrome could also be accumulated well in transgenic rice seeds by inserting into highly variable regions of rice SSP glutelins or by attaching to the C-termini of rice prolamins (Takagi et al. 2005b; Iizuka et al. 2014; Hirota et al. 2017; Takaiwa et al. 2018; Takaiwa 2023). These results indicate that small-size peptide can be stably and highly accumulated as fusion with SSPs. These transgenic seeds can be utilized as oral peptide vaccine (see also chapter "Plant Molecular Farming for Vaccine Development").

Many bioactive peptides derived from enzymatic hydrolysis of food proteins have been reported to be effective as health-promoting agents for lifestyle-related diseases (Udenigwe and Aluko 2012). In order to confer new function to seed, these bioactive peptides such as glucagon-like peptide (GLP-1) (Sugita et al. 2005; Yasuda et al. 2006), hypocholesterolemic lactostatin peptide (IIKPW) (Wakasa et al. 2006), and an antihypertension novokinin peptide (RPLKPW), (Yang et al.

2006) have been accumulated in rice seeds as part of SSPs by expression as fusion protein with glutelin or 11–12S globulin in a similar manner as endogenous SSPs. These fusion proteins were highly produced and transported to PSVs and then assembled like endogenous SSPs in transgenic rice seeds, indicating that they can be utilized as a potential carrier for the production of heterogeneous bioactive peptide(s). Furthermore, soybean α′ subunit of 7S globulin has also been utilized as a carrier to express tandem repeats of novokinin peptide, hypocholesterolemic LPYPR peptide, and memory-enhancing rubiscolin (YPLDLF). The modified 7S α′ subunits containing functional peptides were expressed up to 0.2% of the total seed protein in the best transgenic lines (Yamada et al. 2008; Nishizawa et al. 2008).

It has been investigated how much length of polypeptide can be acceptable as fusion carrier by substitutive insertion into highly variable region of glutelin. Polypeptide with a length of about 150 amino acids could be inserted into the highly variable region of glutelin GluA-2 acidic subunit and be stably and highly accumulated in transgenic rice seeds (Yang et al. 2007a). Its highest accumulation level reached about 15% of total seed protein. However, this fusion product could not be transported to PSV-like native glutelin and was retained as aggregates in ER lumen through intermolecular interaction with Cys-rich prolamins via disulfide bonds.

Ubiquitin, C-terminal region of rice chaperone luminal binding protein (BiP1) (256 amino acids), and cholera toxin B subunit (CTB) have also been utilized as a fusion carrier or partner. Ubiquitin has been developed as a fusion partner for enhancing the accumulation of recombinant proteins. Ubiquitin fusion to human granulocyte-colony-stimulating factor (hG-CSF) resulted in sevenfold and twofold enrichment than that without fusion partner in leaves and seeds of tobacco, respectively (Tian and Sun 2011).

When the C-terminal region of BiP1 was also used as a fusion partner of the human insulin-like growth factor 1 (IGF-1), this fusion product accumulated at a level of 6.8% of TSP in transgenic rice seed under the control of the glutelin *Gt13a* promoter containing its signal peptide (Xie et al. 2008). Oral administration of this unprocessed grain reduced blood glucose level by enhancing islet cell survival and increasing insulin secretion in mice diabetes model.

Many antigens have been expressed by fusing to the CTB acting as a mucosal adjuvant. This is due to the fact that the CTB facilitates transportation of conjugated antigen to mucosal immune tissues through the binding to GM1 ganglioside receptor present on mucosal epithelium. Therefore, the CTB was fused to the hybrid T cell epitope peptide (7Crp) derived from Japanese cedar pollen allergens and expressed as 7Crp/CTB fusion protein in transgenic rice seed. When these transgenic rice seeds were orally administered to model mice, oral immune tolerance activity (suppression of specific IgE level and sneezing number) was activated to about 50-fold compared to the control rice containing 7Crp alone (Takagi et al. 2008). Furthermore, oral administration of transgenic rice seeds containing the CTB/16 kDa antigen from *Ascaris suum* (As16) resulted in the induction of antigen-specific IgG and high protection from pathogen attack (depression of lungworm burdens) after challenge with lungworm eggs (Matsumoto et al. 2009). When human rotavirus VP7(HRVVP7)/CTB and HRVVP7 antigens were expressed in *Arabidopsis* seeds and then TSPs from these transgenic seeds were orally administered to mice, expressions of antigen-specific systemic IgG and mucosal IgA antibodies was more highly upregulated by the HRVVP7/CTB fusion protein as compared to the HRVVP7 only (Li et al. 2018). Mice immunized with HRVVP/CTB were protected from the challenge with virulent rotavirus.

Oil contents of sunflower, safflower, and rapeseeds are about 50%, 30%, and 40%, respectively, which are in contrast with 2–5% of many cereal seeds. Oil bodies are the organelles that originate from the ER and function to store seed oils. The oil body membrane contains high abundant proteins termed as oleosin. Recombinant proteins are correctly targeted to oil bodies in seeds when expressed as translational fusions with oleosins. The oil bodies and associated proteins can be easily separated from the majority of other seed cell components by flotation centrifugation.

Human growth hormone and human insulin have been expressed as fusion proteins by linking to oleosins in *Arabidopsis* and safflower seeds, which were deposited in oil bodies (Boothe et al. 2010). When they were purified by cost-effective oil body-mediated processing, their functions were demonstrated to be bioequivalent with the existing commercial products. Furthermore, the recombinant apolipoprotein A1 Milano was also expressed as fusion with oleosin in safflower seeds under the control of phaseolin promoter (Nykiforuk et al. 2011). ApoAI Milano accumulated at the level of 7.8 g/kg seed. The purified ApoAI Milano was functionally equivalent to the marketed one.

5.7.4 Enhancement of Production Yields of Recombinant Proteins by Reduction of Endogenous SSPs

Improvement of production yields of recombinant proteins is always required for practical purpose in molecular farming. It has been reported that reduction of endogenous SSPs by either knockout by mutations and genome editing or knockdown by RNA interference (RNAi) gave rise to upregulation of production yields of the foreign recombinant proteins in rice, maize, barley, and soybean seeds (Tada et al. 2003; Schmidt and Herman 2008; Shigemitsu et al. 2012; Yuki et al. 2012, 2013; Yang et al. 2012a, b; Takaiwa et al. 2021; Panting et al. 2021). This effect has been explained by the following several reasons: (1) supply of ample vacant deposition space for the foreign recombinant protein by the reduction of endogenous SSPs; (2) alleviation of competition with the endogenous SSPs in the transcription, translation, and posttranslation processes in terms of scramble for transcription factors, tRNAs, amino acids, and trafficking; (3) mitigation of ER load causing the ER stress; and (4) induction of proteome rebalancing to compensate for reduced nitrogen levels (Takaiwa 2013b; Herman 2014; Wu and Messing 2014). It is well known that seed has a property to compensate for the reduced SSPs by enhancing the production of other seed proteins in order to maintain the constant protein content (nitrogen and sulfate level) required for germinating seedling (Kawakatsu et al.

2010; Schmidt et al. 2011; Holding 2014; Wu and Messing 2014). Thus, shortage of endogenous SSPs is presumed to be recovered by enhanced accumulation of foreign recombinant protein(s) in maturing seed to keep the homeostasis of seed protein (nitrogen) contents. Taken together, combination of high expression of the target recombinant proteins under the control of the strong seed-specific promoter and simultaneous suppression of endogenous SSPs is expected to offer powerful production tool for boosting the accumulation levels of the desired recombinant proteins.

ER quality control within ER lumen may be one of the critical steps determining the accumulation level. Unfolded protein response which influences recombinant protein production may be alleviated by reduction of ER load through the suppression of endogenous SSP production. As a result, production of foreign recombinant protein can be increased in hosts in which the expression of endogenous SSPs is depressed. Especially, this influence on production yield is further enhanced by reduction of SSPs stored at the same deposition compartment as the targeting site of recombinant protein. Production yields of several cytokines (hIL-10, IFN-γ, mIL-6, hTGF-β) (Yang et al. 2012b; Takaiwa 2013b; Takaiwa et al. 2021), antigens (botulinum neurotoxin, CTB, shuffled Cry j 1 and Cry j 2, 7Crp peptide) (Yuki et al. 2012, 2013; Entesari et al. 2018; Takaiwa et al. 2019), antibodies (human norovirus VHH) (Tokuhara et al. 2013; Sasou et al. 2021), and human serum albumin (HSA) (Pang et al. 2020) were significantly enhanced two- to tenfold by reduction of endogenous SSP production (Table 5.5). Interestingly, expression level of HSA gene in the knocked-in line integrated at the seed storage protein locus (glutelin GluA1) was much higher than that of the random integration lines (Pang et al. 2020). Furthermore, it is important to note that rebalancing of proteome is highly influenced by the sulfur level of amino acids included in the expressed recombinant proteins as well as the nitrogen level. When methionine-rich 10 kDa δ-zein was expressed in the soybean seed of 7S Cys-less β-conglycinin knockdown line, the produced maize zein level could not be enhanced (Kim et al. 2014). By contrast, when these transgenic plants were grown in sulfur-rich medium, accumulation level of the 10 kDa δ-zein increased 3–16-fold. This result suggests that sulfur availability rather than proteome rebalancing was crucial for the enriched accumulation of heterologous methionine-rich proteins in soybean seeds. Such rebalancing for sulfate storage between cysteine and methionine was also observed in knocked-down Cys-rich zein transgenic maize (Wu and Messing 2014). When expression of Cys-rich γ- and β-zeins were depressed by RNAi-mediated suppression, accumulation of the methionine-rich δ-zein was increased (Wu et al. 2012). Conversely, when the methionine-rich protein was overexpressed, synthesis of Cys-rich proteins was reduced, indicating the presence of rebalancing mechanism between Met-rich and Cys-rich proteins regarding the sulfur storage in seeds.

It is well known that reduction of some major SSPs or nitrogen supply by fertilizer results in the enrichment of Cys-poor SSPs rather than Cys-rich SSPs. Furthermore, it has been sometimes observed that inherent intracellular localization of recombinant proteins was altered by the suppression of endogenous SSPs. For example, when CTB, botulinum neurotoxin, or norovirus HVV antibody was highly

Table 5.5 Enhancing effect on accumulation levels of recombinant proteins in seeds by suppression of endogenous seed storage proteins

Species	Product	Promoter and signal	Suppression	Expression level	Localization	Reference
Rice	Soybean glycinin A1aB1b	Glutelin GluB-1 + SP	a123 or LGC1 (low glutelin mutant)	2.4 mg/g seed (2.3-fold)	PSV (PB-II)	Tada et al. (2003)
	Soybean β-conglycinin α′	Glutelin GluB-2	a123 + RNAi: GluA, 13 kDa	>20 mg/g seed	PSV (PB-II)	Cabanos et al. (2014)
	Human growth hormone	10 kDa prolamin + SP	RNAi: GluA, 13 kDa	NA	PSV (PB-II)	Shigemitsu et al. (2012)
	Human IL-10 + KDEL	Glutelin GluB-1 + SP	RNAi: 16 kDa, 13 kDa, 10 kDa	219 μg/grain (threefold)	IL-10 body	Yang et al. (2012b)
	Mouse IL-6 + KDEL	Glutelin GluB-1 + SP	RNAi: 16 kDa, 13 kDa, 10 kDa	2.4-fold	NA	Takaiwa (2013a)
	Mouse IFN-γ + KDEL	Glutelin GluB-1 + SP	RNAi: 16 kDa, 13 kDa, 10 kDa	1.6-fold	NA	Takaiwa (2013a)
	Mouse 2×TGF-β + KDEL	Glutelin GluB-1 + SP	RNAi: 16 kDa, 13 kDa, 10 kDa	110 μg/grain (twofold)	PB-I	Takaiwa et al. (2016)
	Human TGF-β + KDEL	26 kDa globulin + SP	RNAi: GluA, B, D, 16 kDa, 13 kDa	430 μg/grain (tenfold)	ER granules	Takaiwa et al. (2021)
	CTB + KDEL	13 kDa prolamin + 10 kDa SP	RNAi: GluA, 13 kDa	2.35 mg/g grain (sixfold)	Apoplast, cytoplasm	Kurokawa et al. (2014)
	Botulinum neurotoxin type A	10 kDa prolamin + SP	RNAi: GluA, 13 kDa	100 μg/grain (tenfold)	Cytoplasm	Yuki et al. (2012)
	Shuffled Cry j 1, Cry j 2 + KDEL	Glutelin GluB-4	RNAi: 16 kDa, 13 kDa, 10 kDa	120–140 μg/grain (twofold)	ER granules	Takaiwa et al. (2019)
	7Crp + KDEL	Glutelin GluB-1	RNAi: GluA, B, 13 kDa	110 μg/grain (fourfold)	ER granules	Entesari et al. (2018)
	Human serum albumin	Glutelin Gt1	Site-specific integration into GluA1 locus	5.1 mg/g seed (fivefold)	NA	Pang et al. (2020)
	Anti-Rota VHH antibody	13 kDa prolamin + SP	RNAi: GluA, 13 kDa	170 μg/seed	PB-I, PB-II	Tokuhara et al. (2013)

Species	Product	Promoter and signal	Suppression	Expression level	Localization	Reference
Soybean	β-Amyloid fused to A1aB1b	Glycinin A1aB1b	JQ line: glycine less, β-conglycinin less mutant	>11% of PSP (sixfold)	PSV, ER-derived PB	Maruyama et al. (2014)
	10 kDa δ zein	Glycinin	RNAi: β-conglycinin	3–16-fold enhancement in sulfur-rich condition	ER-derived PB	Kim et al. (2014)
Barley	Epidermal growth factor	B hordein, oat globulin	Risø 56, HorB mutant	Twofold	NA	Panting et al. (2021)

NA not analyzed, *SP* signal peptide

produced in transgenic rice seeds by RNAi-mediated suppression of endogenous SSPs, their main subcellular localizations were changed to cell wall and cytoplasm from ER-derived PBs (Yuki et al. 2012, 2013; Kurokawa et al. 2014; Sasou et al. 2021). Moreover, when most of the SSPs (glutelins and prolamins) were severely decreased by RNAi-mediated suppression in transgenic rice seeds, it is notable that recombinant protein (hTGF-β) was deposited as a major protein in ER-derived vesicles and its accumulation level was remarkably increased up to the level of more than 30% of total seed proteins (Takaiwa et al. 2021).

Rebalancing between storage proteins and oil (fatty acids) has also been observed in oilseeds (rapeseed and sunflower seed). When the SSPs' expression was knocked down by RNAi-mediated suppression, oil content was significantly increased (Rolletschek et al. 2020). Furthermore, cross talk (compensatory rebalancing) among prolamin members in the same ER-derived PBs or between 12S cruciferin and 2S napin within the same PSVs has also been observed in rice, *Arabidopsis*, and rapeseeds (Kohno-Murase et al. 1994; Goossens et al. 1999; Takaiwa et al. 2018). In *Arabidopsis* seeds, reduction of endogenous 2S napin by RNAi gave rise to remarkable enhancement of common bean arcelin-5-1 SSP to the level of more than 24% of total seed proteins (Goossens et al. 1999).

5.8 Protein Quality Control of Recombinant Proteins

SSPs are synthesized as secretory proteins on rough ER. After co-translational cleavage of the N-terminal signal peptide, SSPs are translocated into ER lumen. Before transportation to the destined target site, qualities of synthesized secretory proteins have to be checked in the ER lumen according to ER quality control (ERQC) system to maintain ER homeostasis whether they are properly folded, assembled, and glycosylated (Takaiwa et al. 2017). Many chaperons and folding enzymes such as luminal binding proteins (BiPs), protein disulfide isomerases (PDIs), calnexin (CNX)/calreticulin (CRT), and co-chaperons (DnaJ) are involved in these processes. When recombinant proteins are highly produced as secretory proteins in seeds, expression of protein-folding chaperons is induced in order to control the quality of synthesized secretory proteins, which is designated as unfolded protein response (UPR) (Ron and Walter 2007; Hetz 2012). On the other hand, given that the imbalance occurred between the ER-folding capacity and the burden of incoming proteins and the level of misfolded proteins overwhelmed the ERQC system, ER stress and ER-associated degradation (ERAD) would be elicited as a result of accumulation of unfolded and misfolded proteins. Furthermore, severe or prolonged ER stress gives rise to cell death as well as damage of ER or cell organelle. In higher plants, the UPR is mediated by two ER transmembrane sensors (UPR signaling pathways) to alleviate the ER stress, activating the orthologues of mammalian activating transcription factor 6 (ATF6) as a sensing protein and inositol-requiring protein 1 (IRE1) as a transducer of the UPR (Liu and Howell 2016). ATF6 is translocated from the ER to the Golgi through an interaction with the coat protein

II (COPII) complex, where it is digested by site-1 proteases (S1P) and site-2 proteases (S2P) to get full function. In *Arabidopsis* and rice, AtbZIP17, AtbZIP28, and AtIRE1a,1b/AtbZIP60 or OsbZIP39, OsbZIP60, and OsIRE1/OsbZIP50 have been identified, respectively (Iwata and Koizumi 2012; Liu and Howell 2016). By the way, orthologue of protein kinase RNA-like ER kinase (PERK) in mammals, which is involved in the attenuation of general protein synthesis through phosphorylation of eukaryotic translation initiator 2α (eIF2α), has not been identified in plants.

BiP is an ER-resident HSP70 family protein and serves as a stress sensor that activates the UPR cascade. Upon ER stress, IRE1 associates in dimer or oligomers after BiP release and activates its RNase domain by autophosphorylation of the kinase domain. Furthermore, IRE1 mediates unconventional cytoplasmic splicing of AtbZIP60u or OsbZIP50u mRNA by cleavage of two specific points on the double stem-loop-forming structure, resulting in a frameshift that is accompanied by the production of an activated transcription factor containing nuclear localization signal (Deng et al. 2011; Nagashima et al. 2011; Hayashi et al. 2013). This process leads to upregulation of many UPR target genes to enhance the protein-folding capacity in the ER lumen and ER-associated degradation of misfolded proteins. Furthermore, when ER stress is severe, IRE1 degrades the ER membrane-localized mRNAs through its RNase activity according to the regulated IRE1-dependent decay (RIDD) pathway (Hallien and Weissman 2006; Mishiba et al. 2013; Hayashi et al. 2016). That is, RIDD is implicated in quantitative regulation to reduce the burden by secretory proteins imported into the ER lumen through degradation of ER-localized mRNAs. On the other hand, *Arabidopsis* AtbZIP17 and AtbZIP28 (Liu et al. 2007; Iwata et al. 2017) or rice OsbZIP39 and OsbZIP60 (Takahashi et al. 2012) transcription factors, orthologues of ATF6 transcription factors residing in the ER membrane, are translocated to the Golgi apparatus by interaction with Sar 1 through coat protein II (COPII) formation upon sensing ER stress. In the Golgi apparatus, they are cleaved proteolytically twice, firstly by the luminal S1P and then in the intramembrane S2P, to release the cytosolic effector portions of these transcription factors, which then enter into the nucleus and then are involved in the activation of UPR target genes. The expression of various UPR genes is known to be regulated by binding to the *cis*-elements (pUPREII: GATGACGCGTAC, pUPRE: ATTGGTCCACGTCATC, ER stress response element 1(ERSE1): CCAAT-N10-CACG) of their promoters not only by the AtbZIP60s or OsbZIP50s, but also by the AtbZIP17 and AtbZIP28 or OsbZIP39 and OsbZIP60 (Hayashi and Takaiwa 2013).

High levels of recombinant protein production may impose a heavy load on the cellular homeostasis. Presence of unfolded recombinant proteins in the ER lumen results in a disturbance of trafficking of secretory proteins. When secretory proteins fail to fold correctly, their dislocation from the ER to cytoplasm is followed by ubiquitination and final degradation by the proteasome. That is, the misfolded proteins are subjected to polyubiquitination in the ER and then degraded by the 26S proteasome in the cytoplasm according to the ER-associated degradation (ERAD) pathway (Liu and Li 2014; Strasser 2018). ERAD system is known to involve four steps: recognition, ubiquitination, dislocation, and degradation. Based on the different subcellular localization of misfolded domain of the ERAD substrate (ER lumen,

ER membrane, and cytoplasm), three different ERAD pathways are known to be implicated in the elimination of misfolded proteins, although their molecular mechanisms are poorly understood in plants. For example, when misfolded proteins are aggregated in ER lumen of seed, expression of ERAD-related genes is induced to remove misfolded proteins from ER. One of the key members connecting ubiquitination of misfolded proteins by the Hrd1 ubiquitin ligase complex is Hard3. This is based on the finding that aggregation of Cys-rich prolamins was observed to be induced in ER lumen without dislocalization, when OsHard3 was specifically suppressed in maturing seed by RNAi (Ohta and Takaiwa 2015). Expression of genes associated with the ERAD machinery including homologs of EDEMs (MNS4/5), OS9 (EBS6/OS9), Hrd1, Hrd3/Sel1L (EBS5/Hrd3A), and Derlin-1 was activated by this process.

The ER stress levels were primarily controlled by the physicochemical properties of the expressed protein rather than the expressed level. When some peptides such as mGLP-1, β-amyloid, and cathelicidin were expressed in transgenic rice seeds as secretory products by ligating N-terminal signal peptide and C-terminal ER retention tag, ER stress was induced through the activation of IRE1/OsbZIP50 pathway, resulting in the pronounced high production of various chaperons (BiPs, PDILs), irrespective of low levels of recombinant products (Oono et al. 2010; Wakasa et al. 2012). In these transgenic rice seeds, accumulation levels of SSPs and starch were severely reduced and grain phenotype was distorted (opaque phenotype). Furthermore, when hIL-7 or alpha-antitrypsin was produced, expressions of ERAD- or programmed cell death (PCD)-related genes were upregulated (Kudo et al. 2013; Zhang et al. 2012). Interestingly, when some recombinant proteins such as hIL-10 and house dust mite allergen Der f 2 derivative were expressed as secretory proteins by attaching the N-terminal signal peptide and C-terminal KDEL ER retention signal, they mainly self-aggregated in ER lumen and budded from ER by forming abnormal ER-derived granules with the size of 0.2–0.3 μm or distorted ER-derived PBs (Yang et al. 2012b). Occurrence of such abnormal ER-PBs in transgenic rice seeds may play a role in alleviating the toxic effects from improperly folded proteins by transporting the insoluble material from the ER to physically independent noble PB bodies like Russell body (Arcalis et al. 2019).

Production of sufficient amounts of chaperons and folding enzymes and UPR signaling control may improve the production yield of recombinant proteins, since the protein-folding capacity is regarded as one of the important bottlenecks in the production of secreted proteins. However, trial of improvement of recombinant protein production by chaperone engineering or URP engineering has not been performed so much in plant cells. To date, there are only a few reports on the effect of production level of recombinant proteins by modulating chaperone proteins in higher plants. Extreme overexpression of BiP1 resulted in the generation of transgenic rice seeds, which displayed an opaque phenotype with floury and shrunken phenotype like ER stress (Yasuda et al. 2009). By contrast, PDIL1-1 overexpression had no significant effect on grain phenotype, almost the same to the wild type, although knockout of this gene (esp2 mutant) leads to morphological changes in PB-Is due to aggregation of proglutelin and Cys-rich prolamins via disulfide bonds

(Takemoto et al. 2002). Overexpression of the BiP1 altered the morphology of intracellular structure, resulting in generation of abnormal ER-PBs. Furthermore, when the BiP1 was either overexpressed or suppressed, the accumulation levels of SSPs and starch contents were significantly reduced in most lines. However, it should be noted that the total seed protein level was definitely increased in transgenic rice seed containing slightly higher level of BiP than the normal one, when many independent transgenic rice seeds containing different amounts of BiP1 were used for analysis (Wakasa et al. 2013). Therefore, taking it into account that judicious modification of chaperons would lead to enhanced production of recombinant proteins, it might be critical to finely tune the expression level of chaperons or folding enzymes without affecting grain phenotype or grain biomass yield. Recently, it has been demonstrated in tobacco transient expression system that the production levels of some viral glycoproteins such as HIV gp140 were significantly improved by co-expression with human calreticulin chaperone protein (Margolin et al. 2020). These results indicate that appropriate tuning of chaperone protein expression levels to alleviate ER stress against overexpressed recombinant protein is important to improve their production levels.

5.9 Efficacy of Seed-Made Recombinant Proteins

5.9.1 Vaccines

Antigens produced in seeds can be administered as edible (mucosal) vaccine without purification, thus allowing very cost-effective, safer, and painless treatment (Peters and Stoger 2011; Takaiwa 2011; Azegami et al. 2015). Furthermore, vaccine produced in seed is remarkably stable at ambient temperature, so that there is no need for cold chain for storage and transportation (Nochi et al. 2007). That is, seed-based oral vaccine cut down the cost required for the storage and transportation. Furthermore, expensive downstream purification process can be eliminated, providing the cost-effective and convenient immune therapy.

The oral route of antigen administration is safe and convenient. However, when purified antigen is used, more than 1000-fold amount of antigen is usually required to get a similar level of effect to that obtained by the parental administration (Neutra and Kozlowski 2006; Streatfield 2006). This is due to the fact that the purified antigen is sensitive to harsh conditions (digestive enzymes and low pH) in gastrointestinal tract. However, when antigens are deposited within ER-derived PBs or PSVs of seeds and orally administered without purification, they exhibit resistance to harsh conditions, thus resulting in an effective delivery to gut-associated lymphoid tissues (GALT) without severe degradation (Takagi et al. 2010; Takaiwa 2013a). This is attributed to the fact that antigens are bioencapsulated by double barriers of PBs and cell wall in plant cell (Takaiwa et al. 2015; Schwestka and Stoger 2021).

PBs containing antigens were shown to be taken up by M cell-mediated transcytosis in Peyer's patches of GALT (Nochi et al. 2007; Takaiwa et al. 2015).

T cell epitope peptides and hypoallergic derivatives derived from allergens act as safe tolerogens for the treatment of various allergy diseases, since side effects can be avoided due to lack or destruction of the tertiary structure required for recognition by the specific IgE by fragmentation, shuffling, or site-directed mutagenesis of Cys residues involved in disulfide bonds. When such artificial antigens were produced as secretory proteins in transgenic rice seeds under the control of the endosperm-specific promoter, they could highly and stably accumulate in PBs (Takaiwa et al. 2007). When these transgenic seeds were orally administered to mice model and then challenged with the causing antigens, not only antigen-specific T cell proliferative activities but also allergen-specific IgG and IgE levels were significantly downregulated compared to those fed non-transgenic seeds. Furthermore, expression levels of Th2-type cytokine (IL-4, IL-5, and IL-13) and histamine were depressed and clinical symptoms such as sneezing number were alleviated, indicating that antigen-specific immune tolerance was induced by the oral administration of transgenic rice seeds containing shuffled JCP Cry j 1 and Cry j 2, shuffled Bet v 1 (TPC7), fragmented Der p1, and mutagenized Der f 2 (Suzuki et al. 2011a; Yang et al. 2012a; Wakasa et al. 2013; Fukuda et al. 2018; Ishida et al. 2021). Recently, clinical study using transgenic rice seeds containing the hybrid T cell epitopes (7Crp peptide for JC pollinosis) was achieved for JC allergy patients. Allergen-specific T cell responses were significantly reduced by oral intake of transgenic rice in a dose-dependent manner. However, oral intake of transgenic rice for 20 weeks showed that neither medication score nor QOL symptom scores for allergic rhinitis were improved during the JC pollen season (Endo et al. 2021).

Cholera toxin B subunit (CTB), *E. coli* heat-labile enterotoxin B subunit (LTB), hepatitis B surface, Norwalk virus coat protein antigen, rotavirus capsid protein VP6, and infectious bursal disease virus (IBDV) V2 have been produced in various transgenic seeds (Chikwamba et al. 2003; Nochi et al. 2007; Wu et al. 2007; Hayden et al. 2012; Feng et al. 2017). When these seed-based subunit vaccines were orally administered, protective immune reactions have been observed to be induced by production of antigen-specific mucosal SIgA and systemic IgG Abs in experimental animal models. For example, oral administration of MucoRice-CTB powder to model mice and macaque induced protective immunity comprising antigen-specific systemic and mucosal Ab immune responses (Nochi et al. 2007, 2009). Interestingly, this vaccine also exhibited cross-reactivity with the LTB in piglets challenged with enterotoxigenic *E. coli* (ETEC) and protected them after its oral administration (Takeyama et al. 2015). Up to date, several plant-based vaccine candidates have been advanced into early-phase human clinical trials (Sethi et al. 2021). MucoRice-CTB is the first non-purified seed-based vaccine, which has recently been moved to phase I clinical trial. It was demonstrated that intake of good manufacturing practice (GMP)-compliant MucoRice-CTB induced neutralizing antibodies against diarrheal toxins in a gut microbiota-dependent manner, without major adverse events (Yuki et al. 2021).

5.9.2 Antibodies

Passive immunotherapy through oral route of seed-based antibody is one of the most promising applications (Juarez et al. 2016). The 2G12 monoclonal antibody (mAb) is known to neutralize human immunodeficiency virus (HIV)-1 by binding with high affinity to a cluster of high-mannose oligosaccharides on the envelope glycoprotein 120 (gp120). This 2G12 mAb has been produced in the endosperm of rice and maize seeds (Rademacher et al. 2008; Vamvaka et al. 2016a, b). Crude extracts from the transgenic rice and maize seeds containing the 2G12 mAb exhibited almost equivalent HIV-binding activities to the purified one. However, the purified 2G12 mAb with mammalian glycans produced in CHO cells exhibited nearly 14 times and 4 times higher potential in HIV neutralization activities than those in rice or maize grain, respectively, indicating that the difference in glycosylation mode is greatly responsible for assembly and virus neutralization activity of antibody. These findings suggest the potentiality of cost-effective oral passive immunotherapy using non-purified seed-based antibody, given that glycosylation of 2G12 mAb could be humanized in plant production system.

Llama variable domain of the heavy chain of heavy antibody (VHH) composed of heavy-chain dimers is heat- and acid-stable protein with the MW of 15 kDa (one-tenth of IgG), thus allowing for oral administration against intestinal infectious diseases (van der Linden et al. 1999). The variable domain of a llama VHH, which specifically binds to human norovirus, was expressed in transgenic rice seeds. Oral administration of this transgenic seeds to model mice was demonstrated to protect diarrhea caused by rotavirus infection through the neutralizing activity via specific VHH antibody (Tokuhara et al. 2013). Recently, monomeric VHH (7C6) against GII.4 norovirus and heterodimeric VHH (7C6-1E4) against GII.4 and GII.17 noroviruses have been highly expressed at the levels of 0.54% and 0.28% of seed weight by RNAi-mediated suppression of SSPs in transgenic rice seeds (Sasou et al. 2021). These seed-based VHH antibodies blocked the invasion of norovirus into human intestinal epithelium cells and exhibited high neutralizing activities against noroviruses. Furthermore, neutralizing activity of the MucoRice VHH (7C6-1E4) was shown to be retained even after heat treatment at 90 °C for 20 min.

The secretory IgA (SIgA) antibody is also a reasonable choice for oral passive immunotherapy, since secretory IgA (SIgA) specifically neutralizes viruses and prevents bacterial colonization at the mucosal surface. Then, anti-enterotoxigenic *E. coli* (ETEC) antibody, in which Fc part of the porcine IgA was fused to the VHH against ETEC, was expressed in *Arabidopsis* and soybean seeds. It was shown that the VHH-IgA antibody was produced at a level of 0.2% of their seed weights. Oral administration of these transgenic seeds was demonstrated to protect the diarrhea of ETEC-challenged piglets (Virdi et al. 2013, 2019; see also Chap. 7).

5.9.3 *Cytokines*

Cytokines are a family of signaling polypeptides involved in intercellular interactions in the process of immune response as well as the regulation of a number of normal physiological functions. Various cytokines have been produced in monocot and dicot seeds. GM-CSF is a major growth factor for neutrophilic granulocytes and monocytes. In the transgenic rice and tobacco seeds, recombinant GM-CSF was produced at the levels of 1.3% and 0.03% of the TSP under the control of the rice glutelin *GT1* and *GT3* endosperm-specific promoters, respectively (Sardana et al. 2007). The biological activity of the purified recombinant GM-CSF was equivalent to the existing commercialized product expressed in CHO.

IL-10 is a cytokine that plays an important role in numerous inflammatory and immunoregulatory reactions. Human IL-10 (hIL-10) has been produced in transgenic rice and *Arabidopsis* seeds (Fujiwara et al. 2010; Morandini et al. 2011). Its production level could be enhanced by proteome rebalancing via RNAi-mediated suppression of endogenous prolamins (Yang et al. 2012b). The purified IL-10 comprised only noncovalent dimers and showed higher activity than the commercial IL-10. Furthermore, it had very low endotoxin contamination (Fujiwara et al. 2010). The mIL-4 and mIL-6 were also produced at the levels of 0.43 mg/g and 0.16 mg/g grain in transgenic rice seeds, respectively (Fujiwara et al. 2016). These purified cytokines exhibited high activity. Furthermore, other several cytokines such as IL-7, TGF-β, and IFN-γ were also highly expressed and produced in rice seeds (Kudo et al. 2013; Takaiwa et al. 2016, 2021). On the other hand, several human cytokines have also been produced in barley seeds and are marketed now. It is interesting to note that oral administration of transgenic rice seeds expressing hIL-10 or hGM-CSF reduced the severity of colitis developed in IL-10-deficient mice and increased the number of leukocytes in non-leukopenic and leukopenic mice (Ning et al. 2008; Takagi et al. 2015).

5.10 Prospective

Seed-based vaccines or antibodies are highly stable at ambient temperature and allow oral administration, when expressed in cereal and leguminous crops. This is attributed to the findings that such seed-made biopharmaceuticals are mainly deposited in PBs of seed cells, which exhibit high resistance to harsh conditions and digestive enzymes in gastrointestinal tract (Takaiwa et al. 2015; Kwon and Daniell 2015). Oral delivery of such proteins is stable and convenient and furthermore does not demand sterile injection conditions. Furthermore, they are very cost effective due to no need for cold-chain and downstream processing.

In the case of seed-based vaccines, orally administrated vaccines efficiently elicit the systemic and mucosal immune responses. This can be attributed to the fact that antigens bioencapsulated by two barriers of PBs and cell wall in seed cell can be delivered to mucosal immune tissues in GALT without degradation. After uptake via M cells in Peyer's patch, they are processed by antigen presentation cells such as dendric cells and macrophages and then presented to naïve T cells in Peyer's patch and/or mesenteric lymphoid nodes, resulting in the induction of pathogen-specific immune reaction or allergen-specific immune tolerance. Production of protective antigen-specific antibodies (sIgA and IgG) is induced against infection diseases, whereas allergen-specific immune tolerance is induced by deletion or energy of allergen-specific T cells, induction of allergen-specific regulatory T cells (Tr1, Th3, and iFoxP3^{+}CD25^{+}), or allergen-specific blocking antibodies (IgG4 and IgA).

Up to date, there are only a few plant-made pharmaceuticals which have achieved the practical use, although a lot of clinical trials have been ongoing (Sethi et al. 2021; Stander et al. 2022). The first one is the USDA-approved poultry vaccine against Newcastle disease virus, and the other is the first FDA-approved recombinant protein "Elelyso" (taliglucerase alfa) used as an enzyme replacement therapy for the treatment of Gaucher's disease (Shaaltiel et al. 2007; Liew and Hair-Bejo 2015). They were produced in tobacco and carrot cell suspension cultures, respectively. On the other hand, an experimental drug ZMapp, an anti-Ebola antibody cocktail of three chimeric monoclonal antibodies produced by transient expression in tobacco plants (*Nicotiana benthamiana*), was used for treatment of patients during the Ebola outbreak (Olinger et al. 2012). In 2022, plant-based vaccine against COVID-19 expressed transiently in tobacco leaves was officially approved for human use in Canada (Stander et al. 2022; see also Chap. 12).

Production of pharmaceutical proteins in transgenic plants used for preclinical and clinical trials has to be conducted in accordance with good manufacturing practice (GMP) guidelines (Fischer et al. 2012; see also chapter "Production of Pharmaceutical Proteins in Genetically Modified Seeds"). Manufacture processes for plant-made pharmaceutical proteins using whole plant have been first approved for the 2G12 mAb expressed in transgenic tobacco leaves, which were cultivated in a specialized containment greenhouse (Ma et al. 2015). On the other hand, non-processed intact seeds expressing recombinant protein, which were cultivated in open fields or conventional greenhouses, cannot be accepted as pharmaceutical formulation because of quantitative variation of active ingredients included in individual seed. Some kinds of processing (partial purified formulation) are inevitably required to reduce the variations among seeds. Processing through powdering of polished grains or preparation of crude PBs is bare minimum rule to control the quality standards required as pharmaceuticals (Wakasa et al. 2015; Yuki et al. 2021).

Plant cells are now approved as a cost-effective production platform of protein drugs from the Food and Drug Administration (FDA) when cultivated in GMP facilities. At present, transgenic rice expressing MucoRice-CTB used in phase I clinical trial was cultivated in a GMP-grade closed clean hydroponic harvesting facility (approved GMP facility) to minimize variations in expression level and to offer high quality (Kashima et al. 2016). Powder of polished grains was utilized as drug formulation without purification. Similarly, transgenic carrot cells expressing human glucocerebrosidase for Gaucher's disease have been developed as non-purified pharmaceutical products aiming at oral administration (phase II), which were propagated by in vitro suspension culture in GMP facilities (Shaaltiel et al. 2015). Taken together, seed-based pharmaceutical production is expected to provide a promising strategy for exploitation of the cost-effective oral pharmaceuticals.

References

Abranches R, Arcalis E, Marcel S, Altmann F, Ribeiro-Pedro M, Rodriguez J, Stoger E (2008) Functional specialization of Medicago truncatula leaves and seeds does not affect the subcellular localization of a recombinant protein. Planta 227:649–658

Agarwal, F., S. Kapoor. And A.K. Tyagi. 2011. Transcription factors regulating the progression of monocot and dicot seed development. Bioessays 33:189–202

An N, Ou J, Jiang D, Zhang L, Liu J, Fu K, Dai Y, Yang D (2013) Expression of a functional recombinant human basic fibroblast growth factor from transgenic rice seeds. Int J Mol Sci 14:3556–3567

Arcalis E, Marcel S, Altmann F, Kolarich D, Drakakaki G, Fischer R, Christou P, Stoger E (2004) Unexpected deposition patterns of recombinant proteins in post-endoplasmic reticulum compartments of wheat endosperm. Plant Physiol 136:3457–3466

Arcalis E, Stadlmann J, Marcel S, Drakakaki G, Winter V, Rodriguez J, Fischer R, Altmann F, Stoger E (2010) The changing fate of a secretory glycoprotein in developing maize endosperm. Plant Physiol 153:693–702

Arcalis E, Stadlmann J, Rademacher T, Marcel S, Sack M, Altmann F, Stoger E (2013) Plant species and organ influence the structure and subcellular localization of recombinant glycoproteins. Plant Mol Biol 83:105–117

Arcalis E, Ibl V, Hilscher J, Rademacher T, Avesani L, Morandini F, Bortesi L, Pezzotti M, Vitale A, Pum D, De Meyer T, Depicker A, Stoger E (2019) Russell-like bodies in plant seeds share common features with prolamin bodies and occur upon recombinant protein production. Front Plant Sci 10:777

Azegami T, Itoh H, Kiyono H, Yuki Y (2015) Novel transgenic rice-based vaccines. Arch Immunol Ther Exp 63:87–99

Azzoni AR, Kusnadi AR, Miranda EA, Nikolov ZL (2002) Recombinant aprotinin produced in transgenic corn seed: extraction and purification studies. Biotechnol Bioeng 80:268–276

Bohlender LL, Parsons J, Hoernstein SNW, Rempfer C, Ruiz-Molina N, Lorenz T, Rodríguez Jahnke F, Figl R, Fode B, Altmann F, Reski R, Decker EL (2020) Stable protein sialylation in Physcomitrella. Front Plant Sci 11:610032

Boothe J, Nykiforuk C, Shen Y, Zaplachinski S, Szarka S, Kuhlman P, Murray E, Morck D, Moloney MM (2010) Seed-based expression systems for plant molecular farming. Plant Biotechnol J 8:588–606

Brandman O, Hegde RS (2016) Ribosome-associated protein quality control. Nat Struct Mol Biol 23:7–14

Cabanos C, Kato N, Amari Y, Fujiwara K, Ohno T, Shimizu K, Goto T, Shimada M, Kuroda M, Masuda T, Takaiwa F, Utsumi S, S. Nagaoka S, Maruyama N. (2014) Development of a novel transgenic rice with hypocholesterolemic activity via high-level accumulation of the α' subunit of soybean β-conglycinin. Transgenic Res 23:609–620

Camiolo S, Farina L, Porceddu A (2012) The relation of codon bias to tissue-specific gene expression in Arabidopsis thaliana. Genetics 192:641–649

Carles C, Bies-Etheve N, Aspart L, Léon-Kloosterziel KM, Koornneef M, Echeverria M, Delseny M (2002) Regulation of Arabidopsis thaliana Em genes: role of ABI5. Plant J 30:373–283

Castilho A, Steinkellner H (2012) Glyco-engineering in plants to produce human-like N-glycan structures. Biotechnol J 7:1088–1098

Castilho A, Strasser R, Stadlmann J, Grass J, Jez J, Gattinger P, Kunert R, Quendler H, Pabst M, Leonard R, Altmann F, Steinkellner H (2010) In planta protein sialylation through overexpression of the respective mammalian pathway. J Biol Chem 285:15923–15930

Ceballo Y, Tiel K, López A, Cabrera G, Pérez M, Ramos O, Rosabal Y, Montero C, Menassa R, Depicker A, Hernández A (2017) High accumulation in tobacco seeds of hemagglutinin antigen from avian (H5N1) influenza. Transgenic Res 26:775–789

Chikwamba R, Cunnick J, Hathaway D, McMurray J, Mason H, Wang K (2002) A functional antigen in a practical crop: LT-B producing maize protects mice against *Escherichia coli* heat labile enterotoxin (LT) and cholera toxin (CT). Transgenic Res 11:479–493

Chikwamba RK, Scott MP, Mejía LB, Mason HS, Wang K (2003) Localization of a bacterial protein in starch granules of transgenic maize kernels. Proc Natl Acad Sci U S A 100:11127–11132

Chou, H.L., L. Tian, T. Kumamaru, S. Hamada. And T.W. Okita. 2017. Multifunctional RNA binding protein OsTudor-SN in storage protein mRNA transport and localization. Plant Physiol 175:1608–1623

Chou HL, Tian L, Fukuda M, Kumamaru T, Okita TW (2019a) The role of RNA-binding protein OsTudor-SN in post-transcriptional regulation of seed storage proteins and endosperm development. Plant Cell Physiol 60:2193–2205

Chou HL, Tian L, Washida H, Fukuda M, Kumamaru T, Okita TW (2019b) The rice storage protein mRNAs as a model system for RNA localization in higher plants. Plant Sci 284:203–211

Clough RC, Pappu K, Thompson K, Beifuss K, Lane J, Delaney DE, Harkey R, Drees C, Howard JA, Hood EE (2006) Manganese peroxidase from the white-rot fungus Phanerochaete chrysosporium is enzymatically active and accumulates to high levels in transgenic maize seed. Plant Biotechnol J 4:53–62

Conley AJ, Joensuu JJ, Richman A, Menassa R (2011) Protein body-inducing fusions for high-level production and purification of recombinant proteins in plants. Plant Biotechnol J 9:419–433

Conrad U, Fiedler U (1998) Compartment-specific accumulation of recombinant immunoglobulins in plant cells: an essential tool for antibody production and immunomodulation of physiological functions and pathogen activity. Plant Mol Biol 38:101–109

Crofts AJ, Crofts N, Whitelegge JP, Okita TW (2010) Isolation and identification of cytoskeleton-associated prolamin mRNA binding proteins from developing rice seeds. Planta 231:1261–1276

Cunha NB, Murad AM, Cipriano TM, Araújo AC, Aragão FJ, Leite A, Vianna GR, McPhee TR, Souza GH, Waters MJ, Rech EL (2011a) Expression of functional recombinant human growth hormone in transgenic soybean seeds. Transgenic Res 20:811–826

Cunha NB, Murad AM, Ramos GL, Maranhão AQ, Brígido MM, Araújo AC, Lacorte C, Aragão FJ, Covas DT, Fontes AM, Souza GH, Vianna GR, Rech EL (2011b) Accumulation of functional recombinant human coagulation factor IX in transgenic soybean seeds. Transgenic Res 20:841–855

De Jaeger G, Scheffer S, Jacobs A, Zambre M, Zobell O, Goossens A, Depicker A, Angenon G (2002) Boosting heterologous protein production in transgenic dicotyledonous seeds using Phaseolus vulgaris regulatory sequences. Nat Biotechnol 20:1265–1268

Deng Y, Humbert S, Liu JX, Srivastava R, Rothstein SJ, Howell SH (2011) Heat induces the splicing by IRE1 of a mRNA encoding a transcription factor involved in the unfolded protein response in Arabidopsis. Proc Natl Acad Sci U S A 108:7247–7252

Dicker M, Tschofen M, Maresch D, König J, Juarez P, Orzaez D, Altmann F, Steinkellner H, Strasser R (2016) Transient glyco-engineering to produce recombinant IgA1 with defined N- and O-glycans in plants. Front Plant Sci 7:18

Ding SH, Huang LY, Wang YD, Sun HC, Xiang ZH (2006) High-level expression of basic fibroblast growth factor in transgenic soybean seeds and characterization of its biological activity. Biotechnol Lett 28:869–875

Downing WL, Galpin JD, Clemens S, Lauzon SM, Samuels AL, Pidkowich MS, Clarke LA, Kermode AR (2006) Synthesis of enzymatically active human alpha-L-iduronidase in Arabidopsis cgl (complex glycan-deficient) seeds. Plant Biotechnol J 4:169–181

Drakakaki G, Marcel S, Arcalis E, Altmann F, Gonzalez-Melendi P, Fischer R, Christou P, Stoger E (2006) The intracellular fate of a recombinant protein is tissue dependent. Plant Physiol 141:578–586

Ebine K, Inoue T, Ito J, Ito E, Uemura T, Goh T, Abe H, Sato K, Nakano A, Ueda T (2014) Plant vacuolar trafficking occurs through distinctly regulated pathways. Curr Biol 24:1375–1382

Endo T, Asaka D, Nakayama T, Saito S, Kodama H, Mitsuyoshi R, Takaishi S, Sugimoto N, Omae S, Takagi H, Wakasa Y, Ozawa K, Takano M, Takaiwa F, Kojima H, Saito S (2021) Immunological and symptomatic effects of oral intake of transgenic rice containing 7 linked major T-cell epitopes from Japanese cedar pollen allergens. Int Arch Allergy Immunol 182:109–119

Entesari M, Wakasa Y, Zanjani BM, Takaiwa F (2018) Change in subcellular localization of over-expressed vaccine peptide in rice endosperm cell that is caused by suppression of endogenous seed storage proteins. Plant Cell Tissue Org Cult 133:275–287

Eskelin K, Ritala A, Suntio T, Blumer S, Holkeri H, Wahlström EH, Baez J, Mäkinen K, Maria NA (2009) Production of a recombinant full-length collagen type I alpha-1 and of a 45-kDa collagen type I alpha-1 fragment in barley seeds. Plant Biotechnol J 7:657–672

Feng H, Li X, Song W, Duan M, Chen H, Wang T, Dong J (2017) Oral administration of a seed-based bivalent rotavirus vaccine containing VP6 and NSP4 induces specific immune responses in mice. Front Plant Sci 8:910

Feng F, Qi W, Lv Y, Yan S, Xu L, Yang W, Yuan Y, Chen Y, Zhao H, Song R (2018) OPAQUE11 is a central hub of the regulatory network for maize endosperm development and nutrient metabolism. Plant Cell 30:375–396

Fischer R, Schillberg S, Hellwig S, Twyman RM, Drossard J (2012) GMP issues for recombinant plant-derived pharmaceutical proteins. Biotechnol Adv 30:434–439

Frey AD, Karg SR, Kallio PT (2009) Expression of rat beta(1,4)-*N*-acetylglucosaminyltransferase III in *Nicotiana tabacum* remodels the plant-specific *N*-glycosylation. Plant Biotechnol J 7:33–48

Fuji K, Shirakawa M, Shimono Y, Kunieda T, Fukao Y, Koumoto Y, Takahashi H, Hara-Nishimura I, Shimada T (2016) The adaptor complex AP-4 regulates vacuolar protein sorting at the trans-Golgi network by interacting with VACUOLAR SORTING RECEPTOR1. Plant Physiol 170:211–219

Fujiwara Y, Aiki Y, Yang L, Takaiwa K, Kosaka A, Tsuji NM, Shiraki K, Sekikawa K (2010) Extraction and purification of human interleukin-10 from transgenic rice seeds. Protein Expr Purif 72:125–130

Fujiwara Y, Yang L, Takaiwa F, Sekikawa K (2016) Expression and purification of recombinant mouse interleukin-4 and -6 from transgenic rice seeds. Mol Biotechnol 58:223–231

Fukuda M, Satoh-Cruz M, Wen L, Crofts AJ, Sugino A, Washida H, Okita TW, Ogawa M, Kawagoe Y, Maeshima M, Kumamaru T (2011) The small GTPase Rab5a is essential for intracellular transport of proglutelin from the Golgi apparatus to the protein storage vacuole and endosomal membrane organization in developing rice endosperm. Plant Physiol 157:632–644

Fukuda M, Wen L, Satoh-Cruz M, Kawagoe Y, Nagamura Y, Okita TW, Washida H, Sugino A, Ishino S, Ishino Y, Ogawa M, Sunada M, Ueda T, Kumamaru T (2013) A guanine nucleotide exchange factor for Rab5 proteins is essential for intracellular transport of the proglutelin from the Golgi apparatus to the protein storage vacuole in rice endosperm. Plant Physiol 162:663–674

Fukuda K, Ishida W, Harada Y, Wakasa Y, Takagi H, Takaiwa F, Fukushima A (2015) Prevention of allergic conjunctivitis in mice by a rice-based edible vaccine containing modified Japanese cedar pollen allergens. Br J Ophthalmol 99:705–709

Fukuda M, Kawagoe Y, Murakami T, Washida H, Sugino A, Nagamine A, Okita TW, Ogawa M, Kumamaru T (2016) The dual roles of the Golgi transport 1 (GOT1B): RNA localization to the cortical endoplasmic reticulum and the export of proglutelin and α-globulin from the cortical ER to the Golgi. Plant Cell Physiol 57:2380–2391

Fukuda K, Ishida W, Harada Y, Wakasa Y, Takagi H, Takaiwa F, Fukushima A (2018) Efficacy of oral immunotherapy with a rice-based edible vaccine containing hypoallergenic Japanese cedar pollen allergens for treatment of established allergic conjunctivitis in mice. Allergol Int 67:119–123

Gao Y, An K, Guo W, Chen Y, Zhang R, Zhang X, Chang S, Rossi V, Jin F, Cao X, Xin M, Peng H, Hu Z, Guo W, Du J, Ni Z, Sun Q, Yao Y (2021) The endosperm-specific transcription factor TaNAC019 regulates glutenin and starch accumulation and its elite allele improves wheat grain quality. Plant Cell 33:603–622

Garcia-Casado G, Sanchez-Monge R, Chrispeels MJ, Armentia A, Salcedo G, Gomez L (1996) Role of complex asparagine-linked glycans in the allergenicity of plant glycoproteins. Glycobiology 6:471–477

Gil-Humanes J, Pistón F, Shewry PR, Tosi P, Barro F (2011) Suppression of gliadins results in altered protein body morphology in wheat. J Exp Bot 62:4203–4213

Gomord V, Fitchette AC, Menu-Bouaouiche L, Saint-Jore-Dupas C, Plasson C, Michaud D, Faye L (2010) Plant-specific glycosylation patterns in the context of therapeutic protein production. Plant Biotechnol J 8:564–587

Goossens A, Van Montagu M, Angenon G (1999) Co-introduction of an antisense gene for an endogenous seed storage protein can increase expression of a transgene in Arabidopsis thaliana seeds. FEBS Lett 456:160–164

Guerrero-Andrade O, Loza-Rubio E, Olivera-Flores T, Fehervari-Bone T, Gomez-Lim MA (2006) Expression of the Newcastle disease virus fusion protein in transgenic maize and immunological studies. Transgenic Res 15:455–463

Gustafsson C, Govindarajan S, Minshull J (2004) Codon bias and heterologous protein expression. Trends Biotechnol 22:346–353

Gutiérrez RA, MacIntosh GC, Green PJ (1999) Current perspectives on mRNA stability in plants: multiple levels and mechanisms of control. Trends Plant Sci 4:429–438

Hallien J, Weissman JS (2006) Decay of endoplasmic reticulum-localized mRNAs during the unfolded protein response. Science 313:104–107

Hamada S, Ishiyama K, Sakulsingharoj C, Choi SB, Wu Y, Wang C, Singh S, Kawai N, Messing J, Okita TW (2003) Dual regulated RNA transport pathways to the cortical region in developing rice endosperm. Plant Cell 15:2265–2272

Hanania U, Ariel T, Tekoah Y, Fux L, Sheva M, Gubbay Y, Weiss M, Oz D, Azulay Y, Turbovski A, Forster Y, Shaaltiel Y (2017) Establishment of a tobacco BY2 cell line devoid of plant-specific xylose and fucose as a platform for the production of biotherapeutic proteins. Plant Biotechnol J 15:1120–1129

Hayashi S, Takaiwa F (2013) The plant-unique cis-element that mediates signaling from multiple endoplasm reticulum stress sensors. Plant Signal Behav 8:e24316

Hayashi S, Wakasa Y, Takahashi H, Kawakatsu T, Takaiwa F (2012) Signal transduction by IRE1-mediated splicing of bZIP50 and other stress sensors in the endoplasmic reticulum stress response of rice. Plant J 69:946–956

Hayashi S, Takahashi H, Wakasa Y, Kawakatsu T, Takaiwa F (2013) Identification of a cis-element that mediates multiple pathways of the endoplasmic reticulum stress response in rice. Plant J 74:248–257

Hayashi S, Wakasa Y, Ozawa K, Takaiwa F (2016) Characterization of IRE1 ribonuclease-mediated mRNA decay in plants using transient expression analyses in rice protoplasts. New Phytol 210:1259–1268

Hayden CA, Egelkrout EM, Moscoso AM, Enrique C, Keener TK, Jimenez-Flores R, Wong JC, Howard JA (2012) Production of highly concentrated, heat-stable hepatitis B surface antigen in maize. Plant Biotechnol J 10:979–984

He Y, Ning T, Xie T, Qiu Q, Zhang L, Sun Y, Jiang D, Fu K, Yin F, Zhang W, Shen L, Wang H, Li J, Lin Q, Sun Y, Li H, Zhu Y, Yang D (2011) Large-scale production of functional human serum albumin from transgenic rice seeds. Proc Natl Acad Sci U S A 108:19078–19083

He X, Galpin JD, Tropak MB, Mahuran D, Haselhorst T, von Itzstein M, Kolarich D, Packer NH, Miao Y, Jiang L, Grabowski GA, Clarke LA, Kermode AR (2012) Production of active human glucocerebrosidase in seeds of *Arabidopsis thaliana* complex-glycan-deficient (cgl) plants. Glycobiology 22:492–503
He Y, Schmidt MA, Erwin C, Guo J, Sun R, Pendarvis K, Warner BW, Herman EM (2016) Transgenic soybean production of bioactive human epidermal growth factor (EGF). PLoS One 11:e0157034
Hensel G, Floss DM, Arcalis E, Sack M, Melnik S, Altmann F, Rutten T, Kumlehn J, Stoger E, Conrad U (2015) Transgenic production of an anti HIV antibody in the barley endosperm. PLoS One 10:e0140476
Herman EM (2014) Soybean seed proteome rebalancing. Front Plant Sci 2014(5):437
Hernández-Velázquez A, López-Quesada A, Ceballo-Cámara Y, Cabrera-Herrera G, Tiel-González K, Mirabal-Ortega L, Pérez-Martínez M, Pérez-Castillo R, Rosabal-Ayán YO, Ramos-González OG, Enríquez-Obregón G, Depicker A, Pujol-Ferrer M (2015) Tobacco seeds as efficient production platform for a biologically active anti-HBsAg monoclonal antibody. Transgenic Res. 24:897-909
Hetz C (2012) The unfolded protein response: controlling cell fate decisions under ER stress and beyond. Nat Rev Mol Cell Biol 13:89–102
Hirota T, Tsuboi H, Iizuka-Koga M, Takahashi H, Asashima H, Yokosawa M, Kondo Y, Ohta M, Wakasa Y, Matsumoto I, Takaiwa F, Sumida T (2017) Suppression of glucose-6-phosphate-isomerase induced arthritis by oral administration of transgenic rice seeds expressing altered peptide ligands of glucose-6-phosphate-isomerase. Mod Rheumatol 27:457–465
Hofbauer A, Stoger E (2013) Subcellular accumulation and modification of pharmaceutical proteins in different plant tissues. Curr Pharm Des 19:5495–5502
Holásková E, Galuszka P, Mičúchová A, Šebela M, Öz MT, Frébort I (2018) Molecular farming in barley: development of a novel production platform to produce human antimicrobial peptide LL-37. Biotechnol J 13:e1700628
Holding DR (2014) Recent advances in the study of prolamin storage protein organization and function. Front Plant Sci 5:276
Hood EE, Witcher DR, Maddock S, Meyer T, Baszczynski C, Bailey M, Flynn P, Register J, Marshall L, Bond D, Kullisek E, Kusnadi A, Evangelista R, Nikolov Z, Wooge C, Mehigh RJ, Hernan R, Kappel WK, Ritland D, Li C-P, Howard JA (1997) Commercial production of avidin from transgenic maize: characterization of transformant, production, processing, extraction and purification. Mol Breed 3:291–306
Hood EE, Bailey MR, Beifuss K, Magallanes-Lundback M, Horn ME, Callaway E, Drees C, Delaney DE, Clough R, Howard JA (2003) Criteria for high-level expression of a fungal laccase gene in transgenic maize. Plant Biotechnol J 1:129–140
Hood EE, Love R, Lane J, Bray J, Clough R, Pappu K, Drees C, Hood KR, Yoon S, Ahmad A, Howard JA (2007) Subcellular targeting is a key condition for high-level accumulation of cellulase protein in transgenic maize seed. Plant Biotechnol J 5:709–719
Horvath H, Huang JT, Wong O, Kohl E, Okita T, Kannangara CG, Von Wettstein D (2000) The production of recombinant proteins in transgenic barley grains. Proc Natl Acad Sci U S A 97:1914–1919
Huang J, Nandi S, Wu L, Yalda D, Bartley G, Rodriguez R, Lonnerdal B, Huang N (2002) Expression of natural antimicrobial human lysozyme in rice grain. Mol Breed 10:83–94
Iizuka M, Wakasa Y, Tsuboi H, Asashima H, Hirota T, Kondo Y, Matsumoto I, Takaiwa F, Sumida T (2014) Suppression of collagen-induced arthritis by oral administration of transgenic rice seeds expressing altered peptide ligands of type II collagen. Plant Biotechnol J 12:1143–1152
Ishida W, Kishimoto T, Takaiwa F, Fukuda K (2021) Prophylactic and therapeutic effects of oral immunotherapy on birch pollen-induced allergic conjunctivitis in mice with a rice-based edible vaccine expressing a hypoallergenic birch pollen allergen. Cell 10:3361
Isken O, Maquat LE (2007) Quality control of eukaryotic mRNA: safeguarding cells from abnormal mRNA function. Genes Dev 21:1833–1856

Ito E, Uemura T (2022) RAB GTPases and SNAREs at the trans-Golgi network in plants. J Plant Res 135:389–403

Iwata Y, Koizumi N (2012) Plant transducers of the endoplasmic reticulum unfolded protein response. Trends Plant Sci 17:720–727

Iwata Y, Ashida M, Hasegawa C, Tabara K, Mishiba KI, Koizumi N (2017) Activation of the Arabidopsis membrane-bound transcription factor bZIP28 is mediated by site-2 protease, but not site-1 protease. Plant J 91:408–415

Jackson MA, Sternes PR, Mudge SR, Graham MW, Birch RG (2014) Design rules for efficient transgene expression in plants. Plant Biotechnol J 12:925–933

Jansing J, Sack M, Augustine SM, Fischer R, Bortesi L (2019) CRISPR/Cas9-mediated knockout of six glycosyltransferase genes in Nicotiana benthamiana for the production of recombinant proteins lacking β-1,2-xylose and core α-1,3-fucose. Plant Biotechnol J 17:350–361

Jo L, Pelletier JM, Hsu SW, Baden R, Goldberg RB, Harada JJ (2020) Combinatorial interactions of the LEC1 transcription factor specify diverse developmental programs during soybean seed development. Proc Natl Acad Sci U S A 117:1223–1232

Joshi CP, Zhou H, Huang X, Chiang VL (1997) Context sequences of translation initiation codon in plants. Plant Mol Biol 35:993–1001

Juarez P, Virdi V, Depicker A, Orzaez D (2016) Biomanufacturing of protective antibodies and other therapeutics in edible plant tissues for oral applications. Plant Biotechnol J 14:1791–1799

Jung JW, Shin JH, Lee WK, Begum H, Min CH, Jang MH, Oh HB, Yang MS, Kim SR (2021) Inactivation of the β(1,2)-xylosyltransferase and the α(1,3)-fucosyltransferase gene in rice (Oryza sativa) by multiplex CRISPR/Cas9 strategy. Plant Cell Rep 40:1025–1035

Kagaya Y, Toyoshima R, Okuda R, Usui H, Yamamoto A, Hattori T (2005) LEAFY COTYLEDON1 controls seed storage protein genes through its regulation of FUSCA3 and ABSCISIC ACID INSENSITIVE3. Plant Cell Physiol 46:399–406

Kallolimath S, Castilho A, Strasser R, Grünwald-Gruber C, Altmann F, Strubl S, Galuska CE, Zlatina K, Galuska SP, Werner S, Thiesler H, Werneburg S, Hildebrandt H, Gerardy-Schahn R, Steinkellner H (2016) Engineering of complex protein sialylation in plants. Proc Natl Acad Sci U S A 113:9498–9503

Karaman S, Unnick J, Wang K (2012) Expression of the cholera toxin B subunit (CT-B) in maize seeds and a combined mucosal treatment against cholera and traveler's diarrhea. Plant Cell Rep 31:527–537

Karg SR, Frey AD, Kallio PT (2010) Reduction of *N*-linked xylose and fucose by expression of rat β1,4-*N*-acetylglucosaminyltransferase III in tobacco BY-2 cells depends on Golgi enzyme localization domain and genetic elements used for expression. J Biotechnol 146:54–65

Kashima K, Yuki Y, Mejima M, Kurokawa S, Suzuki Y, Minakawa S, Takeyama N, Fukuyama Y, Azegami T, Tanimoto T, Kuroda M, Tamura M, Gomi Y, Kiyono H (2016) Good manufacturing practices production of a purification-free oral cholera vaccine expressed in transgenic rice plants. Plant Cell Rep 35:667–679

Kaulfürst-Soboll H, Mertens-Beer M, Brehler R, Albert M, von Schaewen A (2021) Complex *N*-glycans are important for normal fruit ripening and seed development in tomato. Front Plant Sci 12:635962

Kawaguchi R, Bailey-Serres J (2002) Regulation of translational initiation in plants. Curr Opin Plant Biol 5:460–465

Kawakatsu T, Takaiwa F (2010) Cereal seed storage protein synthesis: fundamental processes for recombinant protein production in cereal grains. Plant Biotechnol J 8:939–953

Kawakatsu T, Takaiwa F (2017) In: Thomas B, Murray BG, Murphy DJ (eds) Encyclopedia of applied plant sciences, vol 2. Academic, Waltham, MA, pp 100–105

Kawakatsu T, Yamamoto MP, Hirose S, Yano M, Takaiwa F (2008) Characterization of a new rice glutelin gene GluD-1 expressed in the starchy endosperm. J Exp Bot 59:4233–4245

Kawakatsu T, Yamamoto MP, Touno SM, Yasuda H, Takaiwa F (2009) Compensation and interaction between RISBZ1 and RPBF during grain filling in rice. Plant J 59:908–920

Kawakatsu T, Hirose S, Yasuda H, Takaiwa F (2010) Reducing rice seed storage protein accumulation leads to changes in nutrient quality and storage organelle formation. Plant Physiol 154:1842–1854

Khan I, Twyman RM, Arcalis E, Stoger E (2012) Using storage organelles for the accumulation and encapsulation of recombinant proteins. Biotechnol J 7:1099–1108

Kim WS, Jez JM, Krishnan HB (2014) Effects of proteome rebalancing and sulfur nutrition on the accumulation of methionine rich delta-zein in transgenic soybeans. Front Plant Sci 5:633

Kohno-Murase J, Murase M, Ichikawa H, Imamura J (1994) Effects of an antisense napin gene on seed storage compounds in transgenic Brassica napus seeds. Plant Mol Biol 26:1115–1124

Koziel MG, Cadine NB, Desai N (1996) Optimizing expression of transgenes with an emphasis on post-transcriptional events. Plant Mol Biol 32:393–405

Kriechbaum R, Ziaee E, Grünwald-Gruber C, Buscaill P, van der Hoorn RAL, Castilho A (2020) BGAL1 depletion boosts the level of β-galactosylation of N- and O-glycans in N. benthamiana. Plant Biotechnol J 18:1537–1549

Kudo K, Ohta M, Yang L, Wakasa Y, Takahashi S, Takaiwa F (2013) ER stress response induced by the production of human IL-7 in rice endosperm cells. Plant Mol Biol 81:461–475

Kurokawa S, Kuroda M, Mejima M, Nakamura R, Takahashi Y, Sagara H, Takeyama N, Satoh S, Kiyono H, Teshima R, Masumura T, Yuki Y (2014) RNAi-mediated suppression of endogenous storage proteins leads to a change in localization of overexpressed cholera toxin B-subunit and the allergen protein RAG2 in rice seeds. Plant Cell Rep 33(1):75–87

Kusnadi AR, Hood EE, Witcher DR, Howard JA, Nikolov ZL (1998) Production and purification of two recombinant proteins from transgenic corn. Biotechnol Prog 14:149–155

Kwon KC, Daniell H (2015) Low-cost delivery of protein drugs bioencapsulated in plant cells. Plant Biotechnol J 13:1017–1022

Lamphear BJ, Jilka JM, Kesl L, Welter M, Howard JA, Streatfield SJ (2004) A corn-based delivery system for animal vaccines: an oral transmissible gastroenteritis virus vaccine boosts lactogenic immunity in swine. Vaccine 22:2420–2424

Lamphear BJ, Barker DK, Brooks CA, Delaney DE, Lane JR, Beifuss K, Love R, Thompson K, Mayor J, Clough R, Harkey R, Poage M, Drees C, Horn ME, Streatfield SJ, Nikolov Z, Woodard SL, Hood EE, Jilka JM, Howard JA (2005) Expression of the sweet protein brazzein in maize for production of a new commercial sweetener. Plant Biotechnol J 3:103–114

Lara P, Oñate-Sánchez L, Abraham Z, Ferrándiz C, Díaz I, Carbonero P, Vicente-Carbajosa J (2003) Synergistic activation of seed storage protein gene expression in Arabidopsis by ABI3 and two bZIPs related to OPAQUE2. J Biol Chem 278:21003–21011

Lau OS, Sun SSM (2009) Plant seeds as bioreactors for recombinant protein production. Biotechnol Adv 27:1015–1022

Leite A, Kemper EL, de Silva MJ, Luchessi AD, Siloto RMP, Bonaccorsi ED, El-Dorry HF, Arruda P (2000) Expression of correctly processed human growth hormone in seeds of transgenic tobacco plants. Mol Breed 6:47–53

Lending CR, Larkins BAL (1989) Changes in the zein composition of protein bodies during maize endosperm development. Plant Cell 1:1011–1023

Levanony H, Rubin R, Altschuler Y, Galili G (1992) Evidence for a novel route of wheat storage proteins to vacuoles. J Cell Biol 119(5):1117–1128

Li L, Shimada T, Takahashi H, Ueda H, Fukao Y, Kondo M, Nishimura M, Hara-Nishimura I (2006) MAIGO2 is involved in exit of seed storage proteins from the endoplasmic reticulum in Arabidopsis thaliana. Plant Cell 18:3535–3547

Li Y, Guan L, Liu X, Liu W, Yang J, Zhang X, Wang F, Guo Y, Li H, Li X (2018) Oral immunization with rotavirus VP7-CTB fusion expressed in transgenic Arabidopsis thaliana induces antigen-specific IgA and IgG and passive protection in mice. Exp Ther Med 15:4866–4874

Liebminger E, Veit C, Pabst M, Batoux M, Zipfel C, Altmann F et al (2011) Beta-*N*-acetylhexosaminidases HEXO1 and HEXO3 are responsible for the formation of paucimannosidic *N*-glycans in *Arabidopsis thaliana*. J Biol Chem 286:10793–10802

Liew PS, Hair-Bejo M (2015) Farming of plant-based veterinary vaccines and their applications for disease prevention in animals. Adv Virol 2015:936940

Liu JX, Howell SH (2016) Managing the protein folding demands in the endoplasmic reticulum of plants. New Phytol 211:418–428

Liu Y, Li J (2014) Endoplasmic reticulum-mediated protein quality control in Arabidopsis. Front Plant Sci 5:162

Liu H, Timko MP (2022) Improving protein quantity and quality-the next level of plant molecular farming. Int J Mol Sci 23:1326

Liu JX, Srivastava R, Che P, Howell SH (2007) An endoplasmic reticulum stress response in Arabidopsis is mediated by proteolytic processing and nuclear relocation of a membrane-associated transcription factor, bZIP28. Plant Cell 19:4111–4119

Liu F, Ren Y, Wang Y, Peng C, Zhou K, Lv J, Guo X, Zhang X, Zhong M, Zhao S, Jiang L, Wang H, Bao Y, Wan J (2013) OsVPS9A functions cooperatively with OsRAB5A to regulate post-Golgi dense vesicle-mediated storage protein trafficking to the protein storage vacuole in rice endosperm cells. Mol Plant 6:1918–1932

Llop-Tous I, Madurga S, Giralt E, Marzabal P, Torrent M, Ludevid MD (2010) Relevant elements of a maize gamma-zein domain involved in protein body biogenesis. J Biol Chem 285:35633–35644

Loos A, Van Droogenbroeck B, Hillmer S, Grass J, Kunert R, Cao J, Robinson DG, Depicker A, Steinkellner H (2011a) Production of monoclonal antibodies with a controlled N-glycosylation pattern in seeds of *Arabidopsis thaliana*. Plant Biotechnol J 9:179–192

Loos A, Van Droogenbroeck B, Hillmer S, Grass J, Pabst M, Castilho A, Kunert R, Liang M, Arcalis E, Robinson DG (2011b) Expression of antibody fragments with a controlled N-glycosylation pattern and induction of endoplasmic reticulum-derived vesicles in seeds of *Arabidopsis*. Plant Physiol 155:2036–2048

Loza-Rubio E, Rojas-Anaya E, Lopez J, Olivera-Flores MT, Gomez-Lim M, Tapia-Perez G (2012) Induction of a protective immune response to rabies virus in sheep after oral immunization with transgenic maize, expressing the rabies virus glycoprotein. Vaccine 30:5551–5557

Ma JK, Drossard J, Lewis D, Altmann F, Boyle J, Christou P, Cole T, Dale P, van Dolleweerd CJ, Isitt V, Katinger D, Lobedan M, Mertens H, Paul MJ, Rademacher T, Sack M, Hundleby PA, Stiegler G, Stoger E, Twyman RM, Vcelar B, Fischer R (2015) Regulatory approval and a first-in-human phase I clinical trial of a monoclonal antibody produced in transgenic tobacco plants. Plant Biotechnol J 13:1106–1120

Magnusdottir A, Vidarsson H, Björnsson JM, Örvar BL (2013) Barley grains for the production of endotoxin-free growth factors. Trends Biotechnol 31:572–580

Mainieri D, Rossi M, Archinti M, Bellucci M, De Marchis F, Vavassori S, Pompa A, Arcioni S, Vitale A (2004) Zeolin. A new recombinant storage protein constructed using maize gamma-zein and bean phaseolin. Plant Physiol 136:3447–3456

Margolin E, Oh YJ, Verbeek M, Naude J, Ponndorf D, Meshcheriakova YA, Peyret H, van Diepen MT, Chapman R, Meyers AE, Lomonossoff GP, Matoba N, Williamson AL, Rybicki EP (2020) Co-expression of human calreticulin significantly improves the production of HIV gp140 and other viral glycoproteins in plants. Plant Biotechnol J 18:2109–2117

Maruyama N, Mun LC, Tatsuhara M, Sawada M, Ishimoto M, Utsumi S (2006) Multiple vacuolar sorting determinants exist in soybean 11S globulin. Plant Cell 18:1253–1273

Maruyama N, Fujiwara K, Yokoyama K, Cabanos C, Hasegawa H, Takagi K, Nishizawa K, Uki Y, Kawarabayashi T, Shouji M, Ishimoto M, Terakawa T (2014) Stable accumulation of seed storage proteins containing vaccine peptides in transgenic soybean seeds. J Biosci Bioeng 118:441–447

Matsumoto Y, Suzuki S, Nozoye T, Yamakawa T, Takashima Y, Arakawa T, Tsuji N, Takaiwa F, Hayashi Y (2009) Oral immunogenicity and protective efficacy in mice of transgenic rice plants producing a vaccine candidate antigen (As16) of Ascaris suum fused with cholera toxin B subunit. Transgenic Res 18:185–192

Mercx S, Smargiasso N, Chaumont F, De Pauw E, Boutry M, Navarre C (2017) Inactivation of the β(1,2)-xylosyltransferase and the α(1,3)-fucosyltransferase genes in Nicotiana tabacum BY-2 cells by a multiplex CRISPR/Cas9 strategy results in glycoproteins without plant-specific glycans. Front Plant Sci 27(8):403

Mikschofsky H, Schirrmeier H, Keil GM, Lange B, Polowick PL, Keller W, Broer I (2009) Pea-derived vaccines demonstrate high immunogenicity and protection in rabbits against rabbit haemorrhagic disease virus. Plant Biotechnol J 7:537–549

Minamino N, Ueda T (2019) RAB GTPases and their effectors in plant endosomal transport. Curr Opin Plant Biol 52:61–68

Mishiba K, Nagashima Y, Suzuki E, Hayashi N, Ogata Y, Shimada Y, Koizumi N (2013) Defects in IRE1 enhance cell death and fail to degrade mRNAs encoding secretory pathway proteins in the Arabidopsis unfolded protein response. Proc Natl Acad Sci U S A 110:5713–5718

Montero-Morales L, Steinkellner H (2018) Advanced plant-based glycan engineering. Front Bioeng Biotechnol 6:81

Morandini F, Avesani L, Bortesi L, Van Droogenbroeck B, De Wilde K, Arcalis E, Bazzoni F, Santi L, Brozzetti A, Falorni A, Stoger E, Depicker A, Pezzotti M (2011) Non-food/feed seeds as biofactories for the high-yield production of recombinant pharmaceuticals. Plant Biotechnol J 9:911–921

Moravec T, Schmidt MA, Herman EM, Woodford-Thomas I (2007) Production of *Escherichia coli* heat labile toxin (LT) B subunit in soybean seed and analysis of its immunogenicity as an oral vaccine. Vaccine 25:1647–1657

Nagashima Y, Mishiba K, Suzuki E, Shimada Y, Iwata Y, Koizumi N (2011) Arabidopsis IRE1 catalyses unconventional splicing of bZIP60 mRNA to produce the active transcription factor. Sci Rep 1:29

Nagashima Y, von Schaewen A, Koiwa H (2018) Function of N-glycosylation in plants. Plant Sci 274:70–79

Nagels B, Van Damme EJM, Pabst M, Callewaert N, Weterings K (2011) Production of complex multi-antennary *N*-glycans in *Nicotiana benthamiana* plants. Plant Physiol 155:1103–1112

Nahampun HN, Bosworth B, Cunnick J, Mogler M, Wang K (2015) Expression of H3N2 nucleoprotein in maize seeds and immunogenicity in mice. Plant Cell Rep 34:969–980

Nakamura S, Lynch TJ, Finkelstein RR (2001) Physical interactions between ABA response loci of Arabidopsis. Plant J 26:627–635

Nandi, S., D. Yalda, S. Lu, Z. Nikolov, R. Misaki, K. Fujiyama. And N. Huang. 2005. Process development and economic evaluation of recombinant human lactoferrin expressed in rice grain. Transgenic Res 14:237–249

Nausch H, Mikschofsky H, Koslowski R, Meyer U, Broer I, Huckauf J (2012) Expression and subcellular targeting of human complement factor C5a in Nicotiana species. PLoS One 7:e53023

Neuhaus J, Rogers J (1998) Sorting of proteins to vacuoles in plant cells. Plant Mol Biol 38:127–144

Neutra MR, Kozlowski P (2006) Mucosal vaccines: the promise and the challenge. Nat Rev Immunol 6:148–158

Nicholson L, Gonzalez-Melendi P, van Dolleweerd C, Tuck H, Perrin Y, Ma JK, Fischer R, Christou P, Stoger E (2005) A recombinant multimeric immunoglobulin expressed in rice shows assembly-dependent subcellular localization in endosperm cells. Plant Biotechnol J 3:115–127

Ning T, Xie T, Qiu Q, Yang W, Zhou S, Zhou L, Zheng C, Zhu Y, Yang D (2008) Oral administration of recombinant human granulocyte-macrophage colony stimulating factor expressed in rice endosperm can increase leukocytes in mice. Biotechnol Lett 30:1679–1686

Ning L, Wang Y, Shi X, Zhou L, Ge M, Liang S, Wu Y, Zhang T, Zhao H (2023) Nitrogen-dependent binding of the transcription factor PBF1 contributes to the balance of protein and carbohydrate storage in maize endosperm. Plant Cell 35:409–434

Nishizawa K, Maruyama N, Satoh R, Fuchikami Y, Higasa T, Utsumi S (2003) A C-terminal sequence of soybean beta-conglycinin alpha′ subunit acts as a vacuolar sorting determinant in seed cells. Plant J 2003(34):647–659

Nishizawa K, Kita A, Doi C, Yamada Y, Ohinata K, Yoshikawa M, Ishimoto M (2008) Accumulation of the bioactive peptides, novokinin, LPYPR and rubiscolin, in seeds of genetically modified soybean. Biosci Biotechnol Biochem 72:3301–3305

Nochi T, Takagi H, Yuki Y, Yang L, Masumura T, Mejima M, Nakanishi U, Matsumura A, Uozumi A, Hiroi T, Morita S, Tanaka K, Takaiwa F, Kiyono H (2007) Rice-based mucosal vaccine as a global strategy for cold-chain- and needle-free vaccination. Proc Natl Acad Sci U S A 104:10986–10991

Nochi T, Yuki Y, Katakai Y, Shibata H, Tokuhara D, Mejima M, Kurokawa S, Takahashi Y, Nakanishi U, Ono F, Mimuro H, Sasakawa C, Takaiwa F, Terao K, Kiyono H (2009) A rice-based oral cholera vaccine induces macaque-specific systemic neutralizing antibodies but does not influence pre-existing intestinal immunity. J Immunol 183:6538–6544

Nykiforuk CL, Shen Y, Murray EW, Boothe JG, Busseuil D, Rhéaume E, Tardif JC, Reid A, Moloney MM (2011) Expression and recovery of biologically active recombinant Apolipoprotein AI(Milano) from transgenic safflower (Carthamus tinctorius) seeds. Plant Biotechnol J 9:250–263

O'Keefe BR, Murad AM, Vianna GR, Ramessar K, Saucedo CJ, Wilson J, Buckheit KW, da Cunha NB, Araujo AC, Lacorte CC, Madeira L, McMahon JB, Rech EL (2015) Engineering soya bean seeds as a scalable platform to produce cyanovirin-N, a non-ARV microbicide against HIV. Plant Biotechnol J 13:884–892

Ohta M, Takaiwa F (2015) OsHrd3 is necessary for maintaining the quality of endoplasmic reticulum-derived protein bodies in rice endosperm. J Exp Bot 66:4585–3593

Olinger GG, Pettitt J, Kim D, Working C, Bohorov O, Bratcher B, Hiatt E, Hume SD, Johnson AK, Morton J, Pauly M, Whaley KJ, Lear CM, Biggins JE, Scully C, Hensley L, Zeitlin L (2012) Delayed treatment of Ebola virus infection with plant-derived monoclonal antibodies provides protection in rhesus macaques. Proc Natl Acad Sci U S A 109:18030–18035

Onodera Y, Suzuki A, Wu CY, Washida H, Takaiwa F (2001) A rice functional transcriptional activator, RISBZ1, responsible for endosperm-specific expression of storage protein genes through GCN4 motif. J Biol Chem 276:14139–14152

Oono Y, Wakasa Y, Hirose S, Yang L, Sakuta C, Takaiwa F (2010) Analysis of ER stress in developing rice endosperm accumulating beta-amyloid peptide. Plant Biotechnol J 8:691–718

Pan T, Wang Y, Jing R, Wang Y, Wei Z, Zhang B, Lei C, Qi Y, Wang F, Bao X, Yan M, Zhang Y, Zhang P, Yu M, Wan G, Chen Y, Yang W, Zhu J, Zhu Y, Zhu S, Cheng Z, Zhang X, Jiang L, Ren Y, Wan J (2021) Post-Golgi trafficking of rice storage proteins requires the small GTPase Rab7 activation complex MON1-CCZ1. Plant Physiol 187:2174–2191

Pang J, Zhou J, Yang D (2020) Knock-in at GluA1 locus improves recombinant human serum albumin expression in rice grain. J Biotechnol 321:87–95

Panting M, Holme IB, Björnsson JM, Brinch-Pedersen H (2021) Modulation of barley (Hordeum vulgare L.) grain protein sink-source relations towards human epidermal growth factor instead of B-hordein storage protein. Mol Biotechnol 63:13–23

Parsons J, Altmann F, Arrenberg CK, Koprivova A, Beike AK, Stemmer C et al (2012) Moss-based production of asialo-erythropoietin devoid of Lewis A and other plant-typical carbohydrate determinants. Plant Biotechnol J 10:851–861

Patel M, Johnson JS, Brettell RIS, Jacobsen J, Xue GP (2000) Transgenic barley expressing a fungal xylanase gene in the endosperm of the developing grains. Mol Breed 6:113–123

Perrin Y, Vaquero C, Gerrard I, Sack M, Drossard J, Stoger E, Christou P, Fischer R (2000) Transgenic pea seeds as bioreactors for the production of a single-chain Fv fragment (scFV) antibody used in cancer diagnosis and therapy. Mol Breed 6:345–352

Peters J, Stoger E (2011) Transgenic crops for the production of recombinant vaccines and antimicrobial antibodies. Hum Vaccin 7:367–374

Peters J, Sabalza M, Ramessar K, Christou P, Capell T, Stöger E, Arcalís E (2013) Efficient recovery of recombinant proteins from cereal endosperm is affected by interaction with endogenous storage proteins. Biotechnol J 8:1203–1212

Petruccelli S, Otegui MS, Lareu F, Tran Dinh O, Fitchette AC, Circosta A, Rumbo M, Bardor M, Carcamo R, Gomord V (2006) A KDEL-tagged monoclonal antibody is efficiently retained in the endoplasmic reticulum in leaves, but is both partially secreted and sorted to protein storage vacuoles in seeds. Plant Biotechnol J 4:511–527

Piron R, De Koker S, De Paepe A, Goossens J, Grooten J, Nauwynck H, Depicker A (2014) Boosting in planta production of antigens derived from the porcine reproductive and respiratory syndrome virus (PRRSV) and subsequent evaluation of their immunogenicity. PLoS One 9:e91386

Qian B, Shen H, Liang W, Guo X, Zhang C, Wang Y, Li G, Wu A, Cao K, Zhang D (2008) Immunogenicity of recombinant hepatitis B virus surface antigen fused with preS1 epitopes expressed in rice seeds. Transgenic Res 17:621–631

Qiao Z, Qi W, Wang Q, Feng Y, Yang Q, Zhang N, Wang S, Tang Y, Song R (2016) ZmMADS47 regulates zein gene transcription through interaction with Opaque2. PLoS Genet 12:e1005991

Qu LQ, Takaiwa F (2004) Evaluation of tissue specificity and expression strength of rice seed component gene promoters in transgenic rice. Plant Biotechnol J 2:113–125

Qu LQ, Xing YP, Liu WX, Xu XP, Song YR (2008) Expression pattern and activity of six glutelin gene promoters in transgenic rice. J Exp Bot 59:2417–2424

Rademacher T, Sack M, Arcalis E, Stadlmann J, Balzer S, Altmann F, Quendler H, Stiegler G, Kunert R, Fischer R (2008) Recombinant antibody 2G12 produced in maize endosperm efficiently neutralizes HIV-1 and contains predominantly single GlcNAc glycans. Plant Biotechnol J 6:189–201

Ramirez N, Oramas P, Ayala M, Rodrıguez M, Perez M, Gavilondo J (2001) Expression and long-term stability of a recombinant single-chain Fv antibody fragment in transgenic Nicotiana tabacum seeds. Biotechnol Lett 23:47–49

Ren Y, Wang Y, Liu F, Zhou K, Ding Y, Zhou F, Wang Y, Liu K, Gan L, Ma W, Han X, Zhang X, Guo X, Wu F, Cheng Z, Wang J, Lei C, Lin Q, Jiang L, Wu C, Bao Y, Wang H, Wan J (2014) *GLUTELIN PRECURSOR ACCUMULATION3* encodes a regulator of post-Golgi vesicular traffic essential for vacuolar protein sorting in rice endosperm. Plant Cell 26:410–425

Ren Y, Wang Y, Pan T, Wang Y, Wang Y, Gan L, Wei Z, Wang F, Wu M, Jing R, Wang J, Wan G, Bao X, Zhang B, Zhang P, Zhang Y, Ji Y, Lei C, Zhang X, Cheng Z, Lin Q, Zhu S, Zhao Z, Wang J, Wu C, Qiu L, Wang H, Wan J (2020) *GPA5* encodes a Rab5a effector required for post-Golgi trafficking of rice storage proteins. Plant Cell 32:758–777

Ren Y, Wang Y, Zhang Y, Pan T, Duan E, Bao X, Zhu J, Teng X, Zhang P, Gu C, Dong H, Wang F, Wang Y, Bao Y, Wang Y, Wan J (2022) Endomembrane-mediated storage protein trafficking in plants: Golgi-dependent or Golgi-independent? FEBS Lett 596:2215–2230

Requesens DV, Romero MEG, Devaish SP, Chang Y-K, Flory A, Streatfield S, Ring R, Phillips C, Hood NC, Marbaniang CD, Howard J, Hood EE (2019) The maize α-zein promoter can be utilized as a strong inducer of cellulose enzyme expression in maize kernels. Transgenic Res 28:537–547

Rolletschek H, Schwender J, König C, Chapman KD, Romsdahl T, Lorenz C, Braun HP, Denolf P, Van Audenhove K, Munz E, Heinzel N, Ortleb S, Rutten T, McCorkle S, Borysyuk T, Guendel A, Shi H, Vander Auwermeulen M, Bourot S, Borisjuk L (2020) Cellular plasticity in response to suppression of storage proteins in the *Brassica napus* embryo. Plant Cell 32:2383–2401

Ron D, Walter P (2007) Signal integration in the endoplasmic reticulum unfolded protein response. Nat Rev Mol Cell Biol 8:519–529

Rothnie HM (1996) Plant mRNA 3′-end formation. Plant Mol Biol 32:43–61

Rubin R, Levanony H, Galili G (1992) Evidence for the presence of two different types of protein bodies in wheat endosperm. Plant Physiol 99:718–724

Saalbach I, Giersberg M, Conrad U (2001) High-level expression of a single-chain Fv fragment (scFv) antibody in transgenic pea seeds. J Plant Physiol 158:529–533

Sabalza M, Vamvaka E, Christou P, Capell T (2013) Seeds as a production system for molecular pharming applications: status and prospects. Curr Pharm Des 19:5543–5552

Saito Y, Shigemitsu T, Yamasaki R, Sasou A, Goto F, Kishida K, Kuroda M, Tanaka K, Morita S, Satoh S, Masumura T (2012) Formation mechanism of the internal structure of type I protein bodies in rice endosperm: relationship between the localization of prolamin species and the expression of individual genes. Plant J 70:1043–1055

Santos-Mendoza M, Dubreucq B, Baud S, Parcy F, Caboche M, Lepiniec L (2008) Deciphering gene regulatory networks that control seed development and maturation in Arabidopsis. Plant J 54:608–620

Sardana R, Dudani AK, Tackaberry E, Alli Z, Porter S, Rowlandson K, Ganz P, Altosaar I (2007) Biologically active human GM-CSF produced in the seeds of transgenic rice plants. Transgenic Res 16:713–721

Sasou A, Yuki Y, Kurokawa S, Sato S, Goda Y, Uchida M, Matsumoto N, Sagara H, Watanabe Y, Kuroda M, Sakon N, Sugiura K, Nakahashi-Ouchida R, Ushijima H, Fujihashi K, Kiyono H (2021) Development of antibody-fragment-producing rice for neutralization of human norovirus. Front Plant Sci 12:639953

Scheller J, Leps M, Conrad U (2006) Forcing single-chain variable fragment production in tobacco seeds by fusion to elastin-like polypeptides. Plant Biotechnol J 4:243–249

Schmidt MA, Herman EM (2008) Proteome rebalancing in soybean seeds can be exploited to enhance foreign protein accumulation. Plant Biotechnol J 6:832–842

Schmidt MA, Barbazuk WB, Sandford M, May G, Song Z, Zhou W, Nikolau BJ, Herman EM (2011) Silencing of soybean seed storage proteins results in a rebalanced protein composition preserving seed protein content without major collateral changes in the metabolome and transcriptome. Plant Physiol 156(1):330–345

Schoberer J, Strasser R (2018) Plant glyco-biotechnology. Semin Cell Dev Biol 80:133–141

Schwestka J, Stoger E (2021) Microparticles and nanoparticles from plants-the benefits of bioencapsulation. Vaccine 9:369

Sethi L, Kumari K, Dey N (2021) Engineering of plants for efficient production of therapeutics. Mol Biotechnol 63:1125–1137

Shaaltiel Y, Bartfeld D, Hashmueli S, Baum G, Brill-Almon E, Galili G, Dym O, Boldin-Adamsky SA, Silman I, Sussman JL, Futerman AH, Aviezer D (2007) Production of glucocerebrosidase with terminal mannose glycans for enzyme replacement therapy of Gaucher's disease using a plant cell system. Plant Biotechnol J 5:579–590

Shaaltiel Y, Gingis-Velitski S, Tzaban S, Fiks N, Tekoah Y, Aviezer D (2015) Plant-based oral delivery of β-glucocerebrosidase as an enzyme replacement therapy for Gaucher's disease. Plant Biotechnol J 13:1033–1040

Shewry PR, Halford NG (2002) Cereal seed storage proteins: structures, properties and role in grain utilization. J Exp Bot 370:947–958

Shewry PR, Napier JA, Tatham AS (1995) Seed storage proteins: structures and biosynthesis. Plant Cell 7:945–956

Shigemitsu T, Ozaki S, Saito Y, Kuroda M, Morita S, Satoh S, Masumura T (2012) Production of human growth hormone in transgenic rice seeds: co-introduction of RNA interference cassette for suppressing the gene expression of endogenous storage proteins. Plant Cell Rep 31:539–549

Shimizu Y, Uemura T (2022) The sorting of cargo proteins in the plant trans-Golgi network. Front Plant Sci 13:957995

Shin YJ, Castilho A, Dicker M, Sádio F, Vavra U, Grünwald-Gruber C, Kwon TH, Altmann F, Steinkellner H, Strasser R (2017) Reduced paucimannosidic N-glycan formation by suppression of a specific β-hexosaminidase from Nicotiana benthamiana. Plant Biotechnol J 15:197–206

Shoemaker CJ, Green R (2012) Translation drives mRNA quality control. Nat Struct Mol Biol 19:594–601

Stander J, Mbewana S, Meyers A (2022) Plant-derived human vaccines: recent development. BioDrugs 36:573–589

Stöger E, Vaquero C, Torres E, Sack M, Nicholson L, Drossard J, Williams S, Keen D, Perrin Y, Christou P, Fischer R (2000) Cereal crops as viable production and storage systems for pharmaceutical scFv antibodies. Plant Mol Biol 42:583–590
Stöger E, Ma JKC, Fischer R, Christou P (2005) Sowing the seeds of success: pharmaceutical proteins from plants. Curr Opin Biotechnol 16:167–173
Strasser R (2014) Biological significance of complex N-glycans in plants and their impact on plant physiology. Front Plant Sci 5:363
Strasser R (2016) Plant protein glycosylation. Glycobiology 26:926–939
Strasser R (2018) Protein quality control in the endoplasmic reticulum of plants. Annu Rev Plant Biol 69:147–172
Streatfield SJ (2006) Mucosal immunization using recombinant plant-based oral vaccines. Methods 38:150–157
Sugita K, Endo-Kasahara S, Tada Y, Yang L, Yasuda H, Hayashi Y, Jomori T, Ebinuma H, Takaiwa F, Sugita K (2005) Genetically modified rice seeds accumulating GLP-1 analogue stimulate insulin secretion from a mouse pancreatic beta-cell line. FEBS Lett 579:1085–1088
Suzuki K, Kaminuma O, Yang L, Takai T, Mori A, Umezu-Goto M, Ohtomo T, Ohmachi Y, Noda Y, Hirose S, Okumura K, Ogawa H, Takada K, Hirasawa M, Hiroi T, Takaiwa F (2011a) Prevention of allergic asthma by vaccination with transgenic rice seed expressing mite allergen: induction of allergen-specific oral tolerance without bystander suppression. Plant Biotechnol J 9:982–990
Suzuki K, Yang L, Takaiwa F (2011b) Transgenic rice accumulating modified cedar pollen allergen Cry j 2 derivatives. J Biosci Bioeng 113(2):249–251
Tada Y, Utsumi S, Takaiwa F (2003) Foreign gene products can be enhanced by introduction into low storage protein mutants. Plant Biotechnol J 1:411–422
Takagi H, Saito S, Yang L, Nagasaka S, Nishizawa N, Takaiwa F (2005a) Oral immunotherapy against a pollen allergy using a seed-based peptide vaccine. Plant Biotechnol J 3:521–533
Takagi H, Hiroi T, Yang L, Tada Y, Yuki Y, Takamura K, Ishimitsu R, Kawauchi H, Kiyono H, Takaiwa F (2005b) A rice-based edible vaccine expressing multiple T cell epitopes induces oral tolerance for inhibition of Th2-mediated IgE responses. Proc Natl Acad Sci U S A 102:17525–17530
Takagi H, Hiroi T, Yang L, Takamura K, Ishimitsu R, Kawauchi H, Takaiwa F (2008) Efficient induction of oral tolerance by fusing cholera toxin B subunit with allergen-specific T-cell epitopes accumulated in rice seed. Vaccine 26:6027–6030
Takagi H, Hiroi T, Hirose S, Yang L, Takaiwa F (2010) Rice seed ER-derived protein body as an efficient delivery vehicle for oral tolerogenic peptides. Peptides 31:1421–1425
Takagi H, Watanabe N, Hiroi T, Takaiwa F (2015) Efficacy of transgenic rice containing human interleukin-10 in experimental mouse models of colitis and pollen allergy. Plant Biotechnol 32:329–332
Takahashi H, Saito Y, Kitagawa T, Morita S, Masumura T, Tanaka K (2005) A novel vesicle derived directly from endoplasmic reticulum is involved in the transport of vacuolar storage proteins in rice endosperm. Plant Cell Physiol 46:245–249
Takahashi H, Kawakatsu T, Wakasa Y, Hayashi S, Takaiwa F (2012) A rice transmembrane bZIP transcription factor, OsbZIP39, regulates the endoplasmic reticulum stress response. Plant Cell Physiol 53:144–153
Takaiwa F (2011) Seed-based oral vaccines as allergen-specific immunotherapies. Hum Vaccin 7:357–366
Takaiwa F (2013a) Update on the use of transgenic rice seeds in oral immunotherapy. Immunotherapy 5:301–312
Takaiwa F (2013b) Increasing the production yield of recombinant protein in transgenic seeds by expanding the deposition space within the intracellular compartment. Bioengineered 4:136–139
Takaiwa F (2023) Influence on accumulation levels and subcellular localization of prolamins by fusion with the functional peptide in transgenic rice seeds. Mol Biotechnol. https://doi.org/10.1007/s12033-023-00666-6

Takaiwa F, Yang L (2014) Development of a rice-based peptide vaccine for Japanese cedar and cypress pollen allergies. Transgenic Res 23:573–584

Takaiwa F, Takagi H, Hirose S, Wakasa Y (2007) Endosperm tissue is good production platform for artificial recombinant proteins in transgenic rice. Plant Biotechnol J 5:84–92

Takaiwa F, Sakuta C, Choi SK, Tada Y, Motoyama T, Utsumi S (2008) Co-expression of soybean glycinins A1aB1b and A3B4 enhances their accumulation levels in transgenic rice seed. Plant Cell Physiol 49:1589–1599

Takaiwa F, Hirose S, Takagi H, Yang L, Wakasa Y (2009) Deposition of a recombinant peptide in ER-derived protein bodies by retention with cysteine-rich prolamins in transgenic rice seed. Planta 229:1147–1158

Takaiwa F, Wakasa Y, Takagi H, Hiroi T (2015) Rice seed for delivery of vaccines to gut mucosal immune tissues. Plant Biotechnol J 13:1041–1055

Takaiwa F, Yang L, Maruyama N, Wakasa Y, Ozawa K (2016) Deposition mode of transforming growth factor-β expressed in transgenic rice seed. Plant Cell Rep 35(12):2461–2473

Takaiwa F, Wakasa Y, Hayashi S, Kawakatsu T (2017) An overview on the strategies to exploit rice endosperm as production platform for biopharmaceuticals. Plant Sci 263:201–209

Takaiwa F, Yang L, Wakasa Y, Ozawa K (2018) Compensatory rebalancing of rice prolamins by production of recombinant prolamin/bioactive peptide fusion proteins within ER-derived protein bodies. Plant Cell Rep 37:209–223

Takaiwa F, Yang L, Takagi H, Maruyama N, Wakasa Y, Ozawa K, Hiroi T (2019) Development of rice-seed-based oral allergy vaccines containing hypoallergenic Japanese cedar pollen allergen derivatives for immunotherapy. J Agric Food Chem 67:13127–13138

Takaiwa F, Wakasa Y, Ozawa K, Sekikawa K (2021) Improvement of production yield and extraction efficacy of recombinant protein by high endosperm-specific expression along with simultaneous suppression of major seed storage proteins. Plant Sci 302:110692

Takemoto Y, Coughlan SJ, Okita TW, Satoh H, Ogawa M, Kumamaru T (2002) The rice mutant esp2 greatly accumulates the glutelin precursor and deletes the protein disulfide isomerase. Plant Physiol 128:1212–1222

Takeyama N, Yuki Y, Tokuhara D, Oroku K, Mejima M, Kurokawa S, Kuroda M, Kodama T, Nagai S, Ueda S, Kiyono H (2015) Oral rice-based vaccine induces passive and active immunity against enterotoxigenic E. coli-mediated diarrhea in pigs. Vaccine 33:5204–5211

Tian L, Sun SS (2011) Ubiquitin fusion expression and tissue-dependent targeting of hG-CSF in transgenic tobacco. BMC Biotechnol 11:91

Tian L, Dai LL, Yin ZJ, Fukuda M, Kumamaru T, Dong XB, Xu XP, Qu LQ (2013) Small GTPase Sar1 is crucial for proglutelin and α-globulin export from the endoplasmic reticulum in rice endosperm. J Exp Bot 64:2831–2845

Tian L, Chou HL, Zhang L, Hwang SK, Starkenburg SP, Doroshenk KA, Kumamaru T, Okita TW (2018) RNA-binding protein RBP-P is required for glutelin and prolamin mRNA localization in rice endosperm cells. Plant Cell 30:2529–2552

Tian L, Chou HL, Zhang L, Okita TW (2019) Targeted endoplasmic reticulum localization of storage protein mRNAs requires the RNA-binding protein RBP-L. Plant Physiol 179:1111–1131

Tian L, Chou HL, Fukuda M, Kumamaru T, Okita TW (2020) mRNA localization in plant cells. Plant Physiol 182:97–109

Tokuhara D, Álvarez B, Mejima M, Hiroiwa T, Takahashi Y, Kurokawa S, Kuroda M, Oyama M, Kozuka-Hata H, Nochi T, Sagara H, Aladin F, Marcotte H, Frenken LG, Iturriza-Gómara M, Kiyono H, Hammarström L, Yuki Y (2013) Rice-based oral antibody fragment prophylaxis and therapy against rotavirus infection. J Clin Invest 123:3829–3838

Torres E, Gonzalez-Melendi P, Stöger E, Shaw P, Twyman RM, Nicholson L, Vaquero C, Fischer R, Christou P, Perrin Y (2001) Native and artificial reticuloplasmins co-accumulate in distinct domains of the endoplasmic reticulum and in post-endoplasmic reticulum compartments. Plant Physiol 127:1212–1223

Tosi P, Parker M, Gritsch CS, Carzaniga R, Martin B, Shewry PR (2009) Trafficking of storage proteins in developing grain of wheat. J Exp Bot 60:979–991

Twyman RM, Stoger E, Schillberg S, Christou P, Fischer R (2003) Molecular farming in plants: host systems and expression technology. Trends Biotechnol 21:570–578
Udenigwe CC, Aluko RE (2012) Food protein-derived bioactive peptides: production, processing, and potential health benefits. J Food Sci 77:R11–R24
Vamvaka E, Twyman RM, Murad AM, Melnik S, Teh AY, Arcalis E, Altmann F, Stoger E, Rech E, Ma JK, Christou P, Capell T (2016a) Rice endosperm produces an underglycosylated and potent form of the HIV-neutralizing monoclonal antibody 2G12. Plant Biotechnol J 14:97–108
Vamvaka E, Arcalis E, Ramessar K, Evans A, O'Keefe BR, Shattock RJ, Medina V, Stöger E, Christou P, Capell T (2016b) Rice endosperm is cost-effective for the production of recombinant griffithsin with potent activity against HIV. Plant Biotechnol J 14:1427–1437
van der Linden RH, Frenken LG, de Geus B, Harmsen MM, Ruuls RC, Stok W, de Ron L, Wilson S, Davis P, Verrips CT (1999) Comparison of physical chemical properties of llama VHH antibody fragments and mouse monoclonal antibodies. Biochim Biophys Acta 1431:37–46
Van Droogenbroeck B, Cao J, Stadlmann J, Altmann F, Colanesi S, Hillmer S, Robinson DG, Van Lerberge E, Terryn N, Van Montagu M, Liang M, Depicker A, De Jaeger G (2007) Aberrant localization and under glycosylation of highly accumulating single-chain Fv-Fc antibodies in transgenic *Arabidopsis* seeds. Proc Natl Acad Sci U S A 104:1430–1435
Verdier J, Thompson RD (2008) Transcriptional regulation of storage protein synthesis during dicotyledon seed filling. Plant Cell Physiol 49:1263–1271
Vicente-Carbajosa J, Carbonero P (2005) Seed maturation: developing an intrusive phase to accomplish a quiescent state. Int J Dev Biol 49:645–651
Vicente-Carbajosa J, Moose SP, Parsons RL, Schmidt RJ (1997) A maize zinc-finger protein binds the prolamin box in zein gene promoters and interacts with the basic leucine zipper transcriptional activator Opaque2. Proc Natl Acad Sci U S A 94:7685–7690
Virdi V, Coddens A, De Buck S, Millet S, Goddeeris BM, Cox E, De Greve H, Depicker A (2013) Orally fed seeds producing designer IgAs protect weaned piglets against enterotoxigenic Escherichia coli infection. Proc Natl Acad Sci U S A 110:11809–11814
Virdi V, Palaci J, Laukens B, Ryckaert S, Cox E, Vanderbeke E, Depicker A, Callewaert N (2019) Yeast-secreted, dried and food-admixed monomeric IgA prevents gastrointestinal infection in a piglet model. Nat Biotechnol 37:527–530
Vitale A, Hinz G (2005) Sorting of proteins to protein vacuoles: how many mechanisms? Trends Plant Sci 10:316–323
Wakasa Y, Takaiwa F (2013) The use of rice seeds to produce human pharmaceuticals for oral therapy. Biotechnol J 8:1133–1143
Wakasa Y, Yasuda H, Takaiwa F (2006) High accumulation of bioactive peptide in transgenic rice seeds by expression of introduced multiple genes. Plant Biotechnol J 4:499–510
Wakasa Y, Zhao H, Hirose S, Yamauchi D, Yamada Y, Yang L, Ohinata K, Yoshikawa M, Takaiwa F (2011a) Antihypertensive activity of transgenic rice seed containing an 18-repeat novokinin peptide localized in the nucleolus of endosperm cells. Plant Biotechnol J 9:729–735
Wakasa Y, Yasuda H, Oono Y, Kawakatsu T, Hirose S, Takahashi H, Hayashi S, Yang L, Takaiwa F (2011b) Expression of ER quality control-related genes in response to changes in BiP1 levels in developing rice endosperm. Plant J 65:675–689
Wakasa Y, Hayashi S, Takaiwa F (2012) Expression of OsBiP4 and OsBiP5 is highly correlated with the endoplasmic reticulum stress response in rice. Planta 236:1519–1527
Wakasa Y, Takagi H, Hirose S, Yang L, Saeki M, Nishimura T, Kaminuma O, Hiroi T, Takaiwa F (2013) Oral immunotherapy with transgenic rice seed containing destructed Japanese cedar pollen allergens, Cry j 1 and Cry j 2, against Japanese cedar pollinosis. Plant Biotechnol J 11:66–76
Wakasa Y, Takagi H, Watanabe N, Kitamura N, Fujiwara Y, Ogo Y, Hayashi S, Yang L, Ohta M, Thet Tin WW, Sekikawa K, Takano M, Ozawa K, Hiroi T, Takaiwa F (2015) Concentrated protein body product derived from rice endosperm as an oral tolerogen for allergen-specific immunotherapy—a new mucosal vaccine formulation against Japanese cedar pollen allergy. PLoS One 10:e0120209

Wang Y, Ren Y, Liu X, Jiang L, Chen L, Han X, Jin M, Liu S, Liu F, Lv J, Zhou K, Su N, Bao Y, Wan J (2010) OsRab5a regulates endomembrane organization and storage protein trafficking in rice endosperm cells. Plant J 64:812–824

Wang S, Takahashi H, Kajiura H, Kawakatsu T, Fujiyama K, Takaiwa F (2013a) Transgenic rice seeds accumulating recombinant hypoallergenic birch pollen allergen Bet v 1 generate giant protein bodies. Plant Cell Physiol 54:917–933

Wang J-C, Xu H, Zhu Y, Liu Q-Q, Cai X-L (2013b) OsbZIP58, a basic leucine zipper transcription factor, regulates starch biosynthesis in rice endosperm. J Exp Bot 64:3453–3466

Wang Y, Liu F, Ren Y, Wang Y, Liu X, Long W, Wang D, Zhu J, Zhu X, Jing R, Wu M, Hao Y, Jiang L, Wang C, Wang H, Bao Y, Wan J (2016) GOLGI TRANSPORT 1B regulates protein export from endoplasmic reticulum in rice endosperm cells. Plant Cell 28:2850–2865

Wang J, Chen Z, Zhang Q, Meng S, Wei C (2020) The NAC transcription factors OsNAC20 and OsNAC26 regulate starch and storage protein synthesis. Plant Physiol 184:1775–1791

Washida H, Wu CY, Suzuki A, Yamanouchi U, Akihama T, Harada K, Takaiwa F (1999) Identification of cis-regulatory elements required for endosperm expression of the rice storage protein glutelin gene GluB-1. Plant Mol Biol 40:1–12

Washida H, Kaneko S, Crofts N, Sugino A, Wang C, Okita TW (2009) Identification of cis-localization elements that target glutelin RNAs to a specific subdomain of the cortical endoplasmic reticulum in rice endosperm cells. Plant Cell Physiol 50:1710–1714

Webster GR, The AY, Ma JK (2017) Synthetic gene design—the rationale for codon optimization and implications for molecular pharming in plants. Biotechnol Bioeng 114:492–502

Woodard SL, Mayor JM, Bailey MR, Barker DK, Love RT, Lane JR, Delaney DE, McComas-Wagner JM, Mallubhotla HD, Hood EE, Dangott LJ, Tichy SE, Howard JA (2003) Maize (Zea mays)-derived bovine trypsin: characterization of the first large-scale, commercial protein product from transgenic plants. Biotechnol Appl Biochem 38:123–130

Wright KE, Prior F, Sardana R, Altosaar I, Dudani AK, Ganz PR, Tackaberry ES (2001) Sorting of glycoprotein B from human cytomegalovirus to protein storage vesicles in seeds of transgenic tobacco. Transgenic Res 10:177–181

Wu Y, Messing J (2014) Proteome balancing of the maize seed for higher nutritional value. Front Plant Sci 5:240

Wu CY, Adachi T, Hatano T, Washida H, Suzuki A, Takaiwa F (1998a) Promoters of rice seed storage protein genes direct endosperm-specific gene expression in transgenic rice. Plant Cell Physiol 39:885–889

Wu CY, Suzuki A, Washida H, Takaiwa F (1998b) The GCN4 motif in a rice glutelin gene is essential for endosperm-specific gene expression and is activated by Opaque-2 in transgenic rice plants. Plant J 14:673–683

Wu CY, Washida H, Onodera Y, Harada K, Takaiwa F (2000) Quantitative nature of the Prolamin-box, ACGT and AACA motifs in a rice glutelin gene promoter: minimal cis-element requirements for endosperm-specific gene expression. Plant J 23:415–421

Wu J, Yu L, Li L, Hu J, Zhou J, Zhou X (2007) Oral immunization with transgenic rice seeds expressing VP2 protein of infectious bursal disease virus induces protective immune responses in chickens. Plant Biotechnol J 5:570–578

Wu Y, Wang W, Messing J (2012) Balancing of sulfur storage in maize seed. BMC Plant Biol 12:77

Wu MW, Liu J, Bai X, Chen WQ, Ren Y, Liu JL, Chen MM, Zhao H, Yao X, Zhang JD, Wan J, Liu CM (2023) Transcription factors NAC20 and NAC26 interact with RPBF to activate albumin accumulations in rice endosperm. Plant Biotechnol J 21(5):890–892

Xie T, Qiu Q, Zhang W, Ning T, Yang W, Zheng C, Wang C, Zhu Y, Yang D (2008) A biologically active rhIGF-1 fusion accumulated in transgenic rice seeds can reduce blood glucose in diabetic mice via oral delivery. Peptides 29:1862–1870

Yamada Y, Nishizawa K, Yokoo M, Zhao H, Onishi K, Teraishi M, Utsumi S, Ishimoto M, Yoshikawa Y (2008) Anti-hypertensive activity of genetically modified soybean seeds accumulating novokinin. Peptides 29:331–337

Yamagata, H., T. Sugimoto, K. Tanaka. And Z. Kasai. 1982. Biosynthesis of storage proteins in developing rice seeds Plant Physiol 70:1094–1100

Yamamoto MP, Onodera Y, Touno SM, Takaiwa F (2006) Synergism between RPBF Dof and RISBZ1 bZIP activators in the regulation of rice seed expression genes. Plant Physiol 141:1694–1707

Yang D, Guo F, Liu B, Huang N, Watkins SC (2003) Expression and localization of human lysozyme in the endosperm of transgenic rice. Planta 216:597–603

Yang, L., Y. Tada, M.P. Yamamoto, H. Zhao, M. Yoshikawa. And F. Takaiwa. 2006. A transgenic rice seed accumulating an anti-hypertensive peptide reduces the blood pressure of spontaneously hypertensive rats FEBS Lett 580:3315–3320

Yang L, Suzuki K, Hirose S, Wakasa Y, Takaiwa F (2007a) Development of transgenic rice seed accumulating a major Japanese cedar pollen allergen (Cry j 1) structurally disrupted for oral immunotherapy. Plant Biotechnol J 5:815–826

Yang Z, Liu Q, Pan Z, Yu HX, Jiao XA (2007b) Expression of the fusion glycoprotein of Newcastle disease virus in transgenic rice and its immunogenicity in mice. Vaccine 25:591–598

Yang L, Kajiura H, Suzuki K, Hirose S, Fujiyama K, Takaiwa F (2008) Generation of a transgenic rice seed-based edible vaccine against house dust mite allergy. Biochem Biophys Res Commun 365:334–339

Yang, L., Y. Wakasa, T. Kawakatsu. And F. Takaiwa. 2009. The 3′-untranslated region of rice glutelin GluB-1 affects accumulation of heterologous protein in transgenic rice Biotechnol Lett 31:1625–1631

Yang L, Hirose S, Suzuki K, Hiroi T, Takaiwa F (2012a) Expression of hypoallergenic Der f 2 derivatives with altered intramolecular disulphide bonds induces the formation of novel ER-derived protein bodies in transgenic rice seeds. J Exp Bot 63:2947–2959

Yang L, Hirose S, Takahashi H, Kawakatsu T, Takaiwa F (2012b) Recombinant protein yield in rice seed is enhanced by specific suppression of endogenous seed proteins at the same deposit site. Plant Biotechnol J 10:1035–1045

Yang Y, Crofts AJ, Crofts N, Okita TW (2014) Multiple RNA binding protein complexes interact with the rice prolamin RNA cis-localization zipcode sequences. Plant Physiol 164:1271–1282

Yang Y, Chou HL, Crofts AJ, Zhang L, Tian L, Washida H, Fukuda M, Kumamaru T, Oviedo OJ, Starkenburg SR, Okita TW (2018) Selective sets of mRNAs localize to extracellular paramural bodies in a rice glup6 mutant. J Exp Bot 69:5045–5058

Yang T, Guo L, Ji C, Wang H, Wang J, Zheng X, Xiao Q, Wu Y (2021) The B3 domain-containing transcription factor ZmABI19 coordinates expression of key factors required for maize seed development and grain filling. Plant Cell 33:104–128

Yang T, Wu X, Wang W, Wu Y (2023) Regulation of seed storage protein synthesis in monocot and dicot plants: a comparative review. Mol Plant 16:145–167

Yasuda H, Tada Y, Hayashi Y, Jomori T, Takaiwa F (2005) Expression of the small peptide GLP-1 in transgenic plants. Transgenic Res 14:677–684

Yasuda H, Hayashi Y, Jomori T, Takaiwa F (2006) The correlation between expression and localization of a foreign gene product in rice endosperm. Plant Cell Physiol 47:756–763

Yasuda H, Hirose S, Kawakatsu T, Wakasa Y, Takaiwa F (2009) Overexpression of BiP has inhibitory effects on the accumulation of seed storage proteins in endosperm cells of rice. Plant Cell Physiol 50:1532–1543

Yoshida T, Kimura E, Koike S, Nojima J, Futai E, Sasagawa N, Watanabe Y, Ishiura S (2011) Transgenic rice expressing amyloid beta-peptide for oral immunization. Int J Biol Sci 7:301–307

Yuki Y, Mejima M, Kurokawa S, Hiroiwa T, Kong IG, Kuroda M, Takahashi Y, Nochi T, Tokuhara D, Kohda T, Kozaki S, Kiyono H (2012) RNAi suppression of rice endogenous storage proteins enhances the production of rice-based botulinum neurotoxin type A vaccine. Vaccine 30:4160–4166

Yuki Y, Mejima M, Kurokawa S, Hiroiwa T, Takahashi Y, Tokuhara D, Nochi T, Katakai Y, Kuroda M, Takeyama N, Kashima K, Abe M, Chen Y, Nakanishi U, Masumura T, Takeuchi Y,

Kozuka-Hata H, Shibata H, Oyama M, Tanaka K, Kiyono H (2013) Induction of toxin-specific neutralizing immunity by molecularly uniform rice-based oral cholera toxin B subunit vaccine without plant-associated sugar modification. Plant Biotechnol J 11:799–808

Yuki Y, Nojima M, Hosono H, Tanaka H, Kimura Y, Satoh T, Imoto S, Uematsu S, Kurokawa S, Kashima K, Mejima M, Ouchida-Nakahashi R, Uchida Y, Murui T, Fijihashi K, Yoshikawa N, Nagamura F, Kiyono H (2021) Oral MucoRice-CTB vaccine for safety and microbiota-dependent immunogenicity in humans: a phase 1 randomised trial. Lancet Microbe 2:E429–E440

Zhang L, Shi J, Jiang D, Stupak J, Ou J, Qiu Q, An N, Li J, Yang D (2012) Expression and characterization of recombinant human alpha-antitrypsin in transgenic rice seed. J Biotechnol 164:300–308

Zhang Z, Yang J, Wu Y (2015) Transcriptional regulation of zein gene expression in maize through the additive and synergistic action of Opaque2, Prolamin-box binding factor, and O2 heterodimerizing proteins. Plant Cell 27:1162–1172

Zhang Z, Zheng X, Yang J, Messing J, Wu Y (2016) Maize endosperm-specific transcription factors O2 and PBF network the regulation of protein and starch synthesis. Proc Natl Acad Sci U S A 113:10842–10847

Zhang Z, Dong J, Ji C, Wu Y, Messing J (2019) NAC-type transcription factors regulate accumulation of starch and protein in maize seeds. Proc Natl Acad Sci U S A 116:11223–11228

Zhang X, Li H, Lu H, Hwang I (2021) The trafficking machinery of lytic and protein storage vacuoles: how much is shared and how much is distinct? J Exp Bot 72:3504–3512

Zheng P, Zheng C, Otegui MS, Li F (2022) Endomembrane mediated-trafficking of seed storage proteins: from Arabidopsis to cereal crops. J Exp Bot 73:1312–1326

Zhong GY, Peterson D, Delaney DE, Bailey M, Wicher DR, Register JC, Bond D, Li CP, Marshall L, Kulisek E, Ritland D, Meyer T, Hood EE, Howard JA (1999) Commercial production of aprotinin in transgenic maize seeds. Mol Breed 5:345–356

Chapter 6
Seed-Based Production of Recombinant Proteins

Cristiano Lacorte, Amanda Lopes Ferreira, Aline Melro Murad, Nicolau Brito da Cunha, and Patricia Valle Pinheiro

Abstract The use of plants to produce recombinant proteins has become a promising alternative to current expression systems based on microorganisms and cell cultures. Producing recombinant proteins in plants presents advantages such as reduced upstream costs, lower probability of infection by mammals' pathogens, and easy scale-up production. Furthermore, plants can make posttranslational modifications and express large and complex proteins. Seeds stand out for their intrinsic characteristics among the plant tissues that can be used for recombinant protein production. In seeds, recombinant proteins are stored in protein storage vacuoles in the endosperm cells. They can be stably maintained for 4–6 years at room temperature without significant loss of protein and biological activity. Progress in methods for genome editing, alteration of the glycosylation pattern of the recombinant proteins produced, and use of protein fusions to aid downstream processes are some trends that shall contribute further to make molecular farming an attractive option for recombinant protein production. This chapter discusses the potential of seeds as a platform for producing recombinant proteins, across the scale-up of the production systems, basic approaches for the purification of recombinant proteins from plant cells, and biosafety issues.

C. Lacorte (✉)
Embrapa Recursos Genéticos e Biotecnologia, Brasília, Brazil
e-mail: cristiano.lacorte@embrapa.br

A. L. Ferreira · P. V. Pinheiro
Embrapa Arroz e Feijão, Santo Antônio de Goiás, Brazil
e-mail: patricia.pinheiro@embrapa.br

A. M. Murad
Universidade Federal de Santa Catarina, Florianópolis, Brazil

N. B. da Cunha
Universidade de Brasília, Brasília, Brazil

C. Kole et al. (eds.), *Tools & Techniques of Plant Molecular Farming*, Concepts and Strategies in Plant Sciences, https://doi.org/10.1007/978-981-99-4859-8_6

Keywords Plant molecular farming · Heterologous protein production · Foreign gene expression · Glycosylation · Fusion proteins · Recombinant protein purification · Biosafety

Abbreviations

AA	Amino acids
APHIS	Animal and Plant Health Inspection Service
ELP	Elastin-like polypeptides
ER	Endoplasmic reticulum
FDA	Food and Drug Administration
GFP	Green fluorescent protein
HIV	Human immunodeficiency virus
ITC	Inverse transition cycling
PB	Protein bodies
PSV	Protein storage vacuoles

6.1 Introduction

The production of high-value biopharmaceuticals and other recombinant proteins for diagnosis or industrial application is mainly based on microbial and cell cultures in large bioreactors (De Jaeger et al. 2002; Tschofen et al. 2016; Kesik-Brodacka 2018). However, research has demonstrated the expression of hundreds of different recombinant proteins in plants over the last decades. Some have reached commercial production, confirming the viability and potential of this approach (Schillberg et al. 2019; Liu and Timko 2022). The advantages of using plants as an alternative to the other, more established systems based on microbial and cell culture are the reduced costs, the lower probability of infection by pathogens capable of infecting humans and other mammals, and the easy scale-up production (Fischer and Buyel 2020; Schillberg and Finnern 2021). Biomass production from plants is more sustainable than bioreactor-based production once it requires soil, fertilizers, and water, instead of expensive complex culture media (Obembe et al. 2011). Besides, the technology for cultivation is largely available and demands less skilled staff (Fischer and Buyel 2020; Liu and Timko 2022).

Different plant-based systems have been described for the production of recombinant proteins (see also chapter "Molecular Farming of Pharmaceutical Proteins in Different Crop Systems: A Way Forward"). These include the transient expression in leaves and the stable integration of the foreign gene for the expression in seeds, leaves, fruits, tubers, roots, aquatic plants, moss, hairy root culture, and cell suspension culture (Xu et al. 2018). Each of these systems presents advantages and specific applications. However, to be competitive, a plant-based system must ensure a high-level expression of the recombinant protein. To this end, several factors that can increase both gene expression and protein accumulation in the cell should be considered. These include the type of tissue or cell; selection of strong promoters and enhancer elements; and features of the target protein, subcellular targeting, and posttranslational modifications (Streatfield 2007; Ghag et al. 2021). Next, the downstream processes must be cost effective, and the final product must be safe, sufficiently pure, and biologically active (Wilken and Nikolov 2012).

This chapter discusses factors involved in producing recombinant proteins using seeds as a platform. The advantages, limitations, and challenges of seed-based production systems are commented, as well as biosafety issues and general aspects associated with the purification of heterologous proteins from seeds.

6.2 Seeds as Bioreactors for Producing and Storing Recombinant Protein

Expressing a recombinant protein in seeds has some unique advantages. During development, endosperm cells are committed to producing and storing proteins and other nutrients (oils, starch, carbohydrates, etc.) to nurture the developing plantlet after germination (Li and Berger 2012). Thus, the target protein expressed in the seed finds a cellular environment that favors protein accumulation (Robinson et al. 2005; Khan et al. 2012).

Product yield and protein quality are critical factors for any recombinant protein production system. Yield involves the efficiency of biosynthesis (i.e., transcription and translation) and the stability of the recombinant protein in the cell (Chen et al. 2020; Liu and Timko 2022). Quality may include the correct assembly and posttranslational modifications, which are involved in protein turnover processes, impacting yield, and are likely necessary for the protein to retain its biological function (Vitale and Boston 2008; Thomas and Walmsley 2015; Strasser et al. 2021). This balance between translation and turnover will affect protein accumulation and involves features of the recombinant protein, the subcellular location where it is directed, and the metabolic burden associated with transcription and translation of this protein (Thomas and Walmsley 2015).

In leaves, very high transcriptional and translational levels of a heterologous gene can be achieved, particularly in transient expression assays (Fischer et al. 1999; Gleba et al. 2004; Pogue et al. 2010). This high expression of a foreign gene, nonessential, may impose a metabolic imbalance on the cell and activate endogenous protection mechanisms that can limit the expression and accumulation of the heterologous protein or direct the cell to apoptosis (Thomas and Walmsley 2015). In seeds, the transcription level may not be as high as in leaves. Still, it can be steadily maintained throughout the developing endosperm and embryo, reducing the metabolic burden on these cells (Boothe et al. 2010). That does not imply that a seed-based production system will produce any recombinant protein. According to the type of recombinant proteins, some may be poorly expressed while others may not even be detected. There will also be differences in the target protein level between different transgenic lines expressing the same gene (Streatfield 2007; Hood et al. 2012). In any case, transgenic cereals and legume grains have been shown to accumulate high amounts of different types of recombinant proteins at consistent levels throughout different generations and batches (Hudson et al. 2014; Mirzaee et al. 2022). Seed production may take a couple of months. Still, a recombinant protein therein is steadily accumulated and stably maintained throughout seed development and can be stored for 4–6 years after the harvest, allowing scale-up production (Oakes et al. 2009; Boothe et al. 2010).

Once the recombinant protein is stably maintained in the cell, the final yield and purity will also depend on the extraction and purification processes (Menkhaus et al. 2004; Janson 2011; Wilken and Nikolov 2012). Optimizing each of these aspects is an effort to make molecular farming an increasingly attractive alternative for recombinant protein production.

6.3 Setting a Seed-Based Platform for Recombinant Protein Production

The production of recombinant proteins in seeds implies that the foreign candidate gene is stably integrated into the genome of a transgenic plant and that this gene is expressed in the seed and inherited by the progeny. Hence, a key factor in setting a seed-based platform for recombinant protein production is the ability to transform the candidate crop. For many species, the transformation protocol may be lengthy and cumbersome, even intractable, in some cases. However, the years-long plant transformation experience has resulted in relatively efficient protocols for species with high-protein content seeds, including cereals and grain legumes, such as rice, maize, barley, pea, and soybean.

To assure that the candidate gene is expressed in the seed, the usual approach includes a seed-specific promoter in the expression cassette used for transformation (Fig. 6.1). Several seed-specific promoters have been identified and tested, both for monocots and dicots (Furtado et al. 2009; Joshi et al. 2015; Xu et al. 2016; Mirzaee et al. 2022). Promoter sequences of storage proteins, for example, follow a tissue-specific pattern and are strictly regulated in time during embryogenesis (Chen et al. 1989). Endosperm-specific promoters that have been used for transgene expression in cereals include the rice globulin, prolamin, glutelin GluB-4, *Gt13a* maize zein, and barley D hordein (Kawakatsu and Takaiwa 2010). In legume seeds, some endosperm-specific promoters tested are the soybean β-conglycinin α′ subunit, the pea legumin, and the arcelin and phaseolin, from common bean (Chen et al. 1989; De Jaeger et al. 2002; Mirzaee et al. 2022).

Other features that have also been shown to influence the expression levels in transgenic plants include the presence of introns, enhancer sequences, codon optimization, terminator sequence, and other 3′ flanking regions, such as scaffold matrix

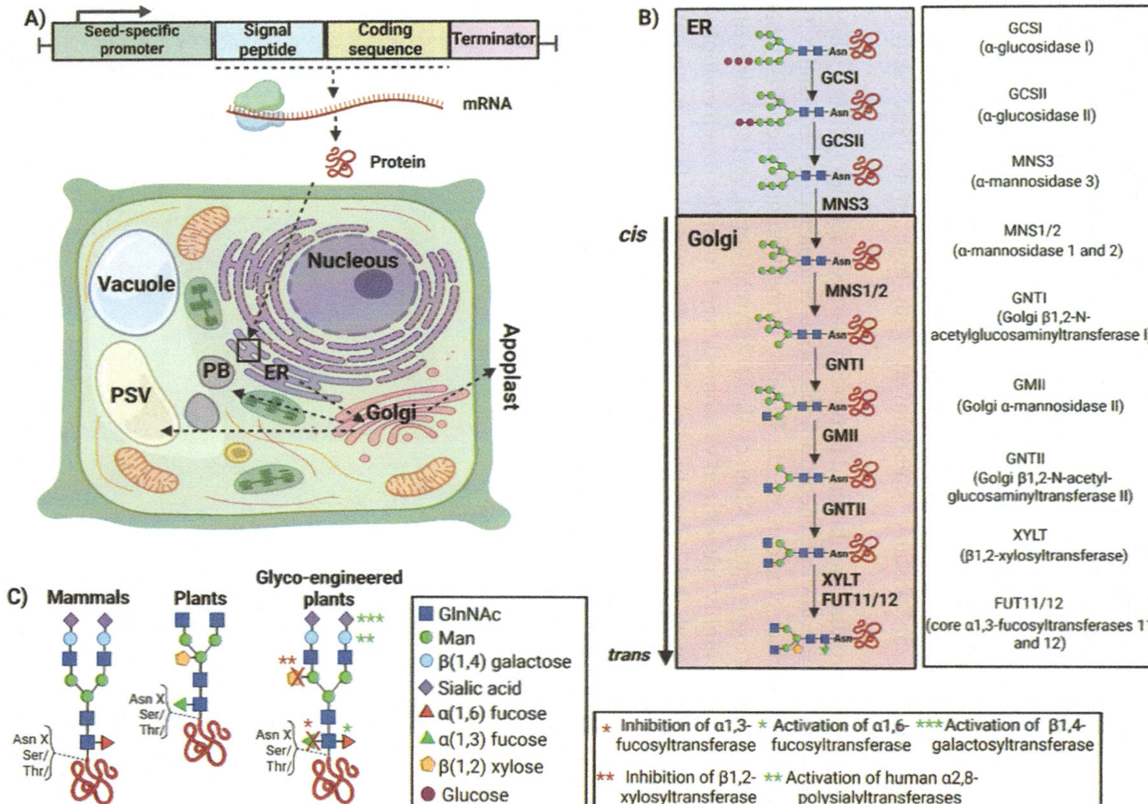

Fig. 6.1 Schematic overview of N-glycosylation in plants. (**a**) Following transcription in the nucleus, the protein is translated and enters the ER, where the glycosylation process begins. Next, the protein moves to the Golgi where it is further processed. The glycosylated protein may be secreted to the PSV, PB, or apoplast. (**b**) Glycosylation process in the ER and Golgi corresponding enzymes. (**c**) Differences in glycosylation pattern of plant and mammal proteins, and the approach for glycol-engineering protein in a transgenic plant

attachment regions (MARs) (Habibi et al. 2017; Webster et al. 2017; Diamos and Mason 2018).

Besides the promoter region, a cassette for protein expression in seeds includes a sequence for a signal peptide that will direct the protein to the endoplasmic reticulum (ER) and the secretory pathway (Fig. 6.1) (Arcalis et al. 2014). A signal peptide is a fragment of 20–30 amino acids present at the N- or C-terminal end of the target protein, which are recognized by specific complexes of RNA and proteins, called "signal recognition particles" (SRPs), that mediates the internalization of the target protein into the membranous organelles (Jolliffe et al. 2005; Robinson et al. 2005; Ashnest and Gendall 2018). Next, the signal peptide is cleaved, leaving the target protein at its intracellular destination (Bohnsack and Schleiff 2010). In the absence of signal peptides, the protein synthesized in free ribosomes accumulates in the cytoplasm, generally an unstable environment with high proteolytic activity (Obembe et al. 2011).

In seeds, storage proteins are directed to protein storage vacuoles (PSVs), which are derived from prevacuoles detached from the Golgi complex (Figs. 6.1 and 6.2). In *Poaceae*, which includes the cereals, the ER also forms protein bodies (PBs) that store mainly prolamin aggregates (Khan et al. 2012; Arcalis et al. 2014; Pedrazzini et al. 2016). PSVs are highly specialized vacuoles derived from the rough endoplasmic reticulum (Khan et al. 2012; Arcalis et al. 2014) and possibly from embryonic vacuoles (EVs), formed during seed development (Feeney et al. 2018). Their lumen has a pH close to neutral and practically no aminopeptidases, features that characterize them as a subcellular environment where protein degradation is minimal and an excellent target for addressing heterologous polypeptides (Takaiwa et al. 2007). In addition to providing a low oxidative environment, PSVs also harbor a high

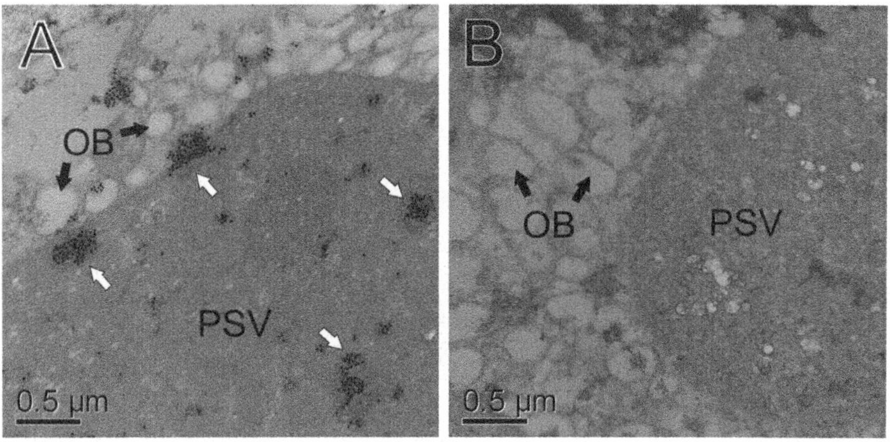

Fig. 6.2 Subcellular localization of recombinant FIX by immunocytochemistry in ultrathin sections of soybean cotyledons. (**a**) Immunogold-labeled FIX (white arrows) localizes to protein storage vacuoles (PSV) in transgenic soybean seeds. (**b**) Non-transgenic cotyledon. *OB* oil bodies. (Images: N.B. Cunha)

concentration of protease inhibitors, which increases their potential as a target for protein targeting aiming at increasing protein stability (Jolliffe et al. 2005; Oakes et al. 2009).

In cereal grains and legume seeds, the families of storage proteins represent the major part of the total seed protein content. For example, the family of glutelins in rice comprises 80% of the seed protein content; the glutelins, in wheat, 40%; and the zein, in maize, 60% (Kawakatsu and Takaiwa 2010). In soybeans, the globulins, glycinin, and β-conglycinin account for up to 80% of the total protein in the seed (Hudson et al. 2014). The expression of these storage proteins along the development of the seed is highly regulated and might constrain the accumulation of a heterologous protein.

A rebalancing of the seed storage protein of soybean was tested by partially suppressing the α/α′ subunit of β-conglycinin, resulting in the increased accumulation of glycinin, along with heterologous green fluorescent protein (GFP) regulated by glycinin promoter and terminator (Schmidt and Herman 2008). In a similar approach, Kim et al. (2014) found that the increase in the recombinant methionine-rich 11 kDa δ-zein in soybean was dependent on a sulfur-rich medium supplementation. In maize, Hood et al. (2012) crossed transgenic lines expressing a recombinant cellulase with high protein elite genotypes and selected for lines with higher content of cellulase. These approaches demonstrate that seed-based platforms have many possibilities to further optimize the production, both in yield and quality of recombinant proteins.

6.4 Posttranslational Modification in Plants and Its Relevance in Molecular Farming

As eukaryotic organisms, plants have the metabolic pathways for posttranslational modifications of proteins—glycosylation, acetylation, and phosphorylation, among others. These modifications occur in the ER and Golgi and are relevant for molecular farming (Fig. 6.1). It is estimated that 50–70% of human proteins are glycosylated (Walsh and Jefferis 2006). Likewise, about 50% of the biopharmaceuticals currently produced are represented by glycosylated proteins (Mizukami et al. 2018; Montero-Morales and Steinkellner 2018). Also, glycosylation and other posttranslational modifications are involved in protein stability and turnover, potentially impacting the accumulation and, hence, the recombinant protein's final yield (Thomas and Walmsley 2015; Varki 2017; Gupta and Shukla 2018).

The glycosylation pattern of proteins in plants, both N- and O-glycosylation, differs from that observed in insects, yeast, and animal cells. Glycosylated proteins from plants contain xylose, and arabinose residues (in *O*-glycans), which are not found in mammalian proteins. Moreover, plant *N*-glycans present α(1,3)-fucose, which is also present in mammalian cells but in β(1,6) linkage form. Plant glycans lack galactose and terminal sialic acids, which are present in mammalian glycoproteins (Fig. 6.1) (Strasser et al. 2021; Bohlender et al. 2022).

In many cases, these differences in the glycosylation pattern may not interfere with the biological activity or the functionality of the recombinant protein, particularly in non-pharma proteins. However, glycostructures can influence the pharmacokinetics, stability, and immunogenicity for biopharmaceuticals (Gupta and Shukla 2018; Bohlender et al. 2022). For example, the presence of nonhuman glycans, particularly the fucose and xylose residues, may cause allergies and immunogenic responses in humans (Montero-Morales and Steinkellner 2018), and non-sialylated glycoproteins are rapidly cleared from serum (Walsh and Jefferis 2006; Bohlender et al. 2022).

To circumvent these problems, *Nicotiana benthamiana* plants were glycol-engineered to present a more "humanlike" glycosylation pattern (Fig. 6.1). Transgenic *N. benthamiana* plants expressing β(1-4)-galactosyltransferase were successfully tested. The α(1,3)-fucosyltransferase and β(1,2)-xylosyltransferase genes were knocked out in the moss *Physcomitrella*, in transgenic *Arabidopsis thaliana*, and *N. benthamiana*, which produced *N*-glycans lacking xylose and fucose residues (Strasser et al. 2004). Next, *N. benthamiana* plants were modified to express the α-1,6-fucosyltransferase and the pathways involved in the biosynthesis, activation, transport, and transfer of Neu5Ac to terminal galactose of heterologous proteins (Fig. 6.1) (Castilho et al. 2010, 2011; Kallolimath et al. 2016).

Expression platforms that are able to produce proteins with extensively modified glycosylation patterns, as for *N. benthamiana*, for example, have not yet been developed for cereals or legume seeds. However, in one attempt to engineer the glycosylation pattern, Wang et al. (2017) expressed the human α-1,6-fucosyltransferase (FUT8) in rice, controlled by an endosperm-specific promoter. After crossing with a plant expressing recombinant human 1-antitrypsin, they confirmed the presence of α(1,6)-fucose and a reduction of β(1,3)-fucose both in the recombinant protein and in globulins. Vamvaka et al. (2016b) demonstrated that recombinant heavy chain of the HIV-neutralizing monoclonal antibody 2G12 expressed in rice seeds was predominantly non-glycosylated, potentially less immunogenic, and more potent in HIV-neutralization assays than the 2G12 antibodies produced in *Nicotiana tabacum*. Similar results were reported by Zhang et al. (2012), which found that approximately 70% of their rice-derived recombinant human a1-anti-α-trypsin was aglycosylated. Indeed, plants appear to tolerate alterations in the glycosylation pathways, not showing phenotypical alterations and being well suited for the production of glyco-engineered recombinant proteins. That flexibility was further demonstrated by the transient co-expression in *N. benthamiana* of specific glycosyltransferases allowing the production of the glycoproteins omega-1 and kappa-5 of *Schistosoma mansoni* containing the helminth-like glycosylation pattern (Wilbers et al. 2017; see also chapter "Tobacco Plants as a Versatile Host for the Expression of Glycoproteins").

New available technologies, such as targeted genome editing, could be used to efficiently knock down β1,2-xylose and core α1,3-fucose residues, for example. In any case, developing a glycol-engineered seed-based platform is a promising yet challenging process (Buyel et al. 2021; see also Chaps. 3 and 4).

6.5 Comparison of Current Seed-Based Platforms

6.5.1 Cereals

Cereals are among the most cultivated and consumed crops worldwide, and most of them are considered a staple food for many countries. Currently, the main products obtained through the production of recombinant proteins in commercially available seed-based platforms are amylase, peroxidase, and cellobiohydrolase I (maize); growth factors and cytokines (barley); and a variety of enzymes (lactoferrin, albumin, transferrin, and lysozyme) and growth factors in rice (Fischer and Buyel 2020; Mirzaee et al. 2022).

Maize was the first seed-based platform used to produce an industrial reagent, avidin, by ProdiGene Inc. (USA). The product was indistinguishable from its counterpart from hen egg white and presented a high yield (2.3% of extractable protein from seed, on average) (Hood et al. 1997; Fischer and Buyel 2020; Moon et al. 2020).

Because maize is a cross-pollinating crop, working with transgenic maize is challenging and requires strict biosafety protocols to be followed. Despite that, maize has several advantages as a seed-based platform for the production of recombinant proteins, such as larger grain size, high yield, and lower production cost compared to other cereals. Moreover, maize has a higher endosperm proportion; a set of specific promoters for the seeds, which can be used to drive the expression of the transgene alone or in combinations; and easy genetic transformation, with established protocols (Watson and Ramstad 1987; Hood et al. 1997; Witcher et al. 1998). As a result, maize seeds have been used as a platform for the production of industrial reagents, such as enzymes and cosmetics, and also pharmaceuticals, such as antibodies to treat human and animal diseases (Rademacher et al. 2008; Egelkrout et al. 2020) and vaccines (Nahampun et al. 2015).

On the contrary, rice and barley are self-pollinating crops, making the risk of undesirable gene flow very low. The rice transformation system is effective in most varieties cultivated worldwide. However, some genotypes may be more suitable for producing a high level of recombinant proteins. The amount of proteins in rice seeds is 7–15% of the total seed weight (Takaiwa et al. 2007). The productivity of recombinant protein in relation to the total weight of the rice seed has reached high levels, from 1% of the total seed weight to 20% of dry seed weight (Vamvaka et al. 2016a). However, in some cases, the high content of recombinant proteins in rice seeds results in grains with an impaired phenotype, indicating that further research is required (Kusaba et al. 2003; Tada et al. 2003; Wakasa and Takaiwa 2013). Besides that, the ease of processing and scale-up production by well-established cultivation systems favor the choice of rice as a platform for producing recombinant proteins.

The rice seed-based platform has been explored for the production of biopharmaceuticals, such as vaccines, growth factors, and antiviral proteins (Takagi et al. 2005; Xie et al. 2008; Vamvaka et al. 2016a). As one of the main staple foods globally, it would be useful to exploit the rice seed-based platform to deliver pharmaceuticals as food. Some studies have shown that the recombinant protein produced in rice seeds remains active in the seeds, even after processing them as a fine powder or crude extracts, for oral administration to mice or macaques (Takagi et al. 2005; Xie et al. 2008; Nochi et al. 2009). In one study, the oral administration of rice seeds expressing immunogens, processed as a fine powder, effectively inhibited allergy-associated immune responses in mice (Takagi et al. 2005). Another study developed a rice-based vaccine that expresses the B subunit of cholera toxin (CT), initially tested in mice with positive results. Subsequently, the vaccine was orally administered to macaques and induced CT-neutralizing IgG antibodies, confirming its effectiveness against cholera in nonhuman primates. This vaccine (MucoRice-CTB) has recently passed phase 1 human clinical trials (Nochi et al. 2009; Yuki et al. 2021).

Another interesting example was the use of crude extracts of rice seeds expressing a microbicide against human immunodeficiency virus (HIV) in cytotoxicity and antiviral assays with human cells (Vamvaka et al. 2016a). Results showed that the crude extracts had stronger binding activity to HIV than the wild-type rice seeds, similar to the purified protein, and were not toxic to human cell lines. Also, the crude extracts expressing the microbicide presented the same oligosaccharide-dependent binding properties as the same recombinant protein expressed in *Escherichia coli*. Altogether, these results show that it is possible to administer rice seeds expressing biopharmaceuticals without processing or with only minimal processing (see also chapter "The Use of Rice Seed as Bioreactor").

Barley is another cereal species that has been used as a seed-based platform for the production of recombinant proteins. The European regulatory agency (EFSA) has declared self-pollinating cereals such as wheat and barley as GRAS (generally recognized as safe) (Mirzaee et al. 2022; see also Chap. 14). In addition to being self-pollinating, barley has other characteristics of interest, such as its ability to regenerate, especially the cultivar Golden Promise, in which *transformation amenability* (TFA) alleles have been identified as responsible for its *Agrobacterium* transformation efficiency (Hisano and Sato 2016; Orman-Ligeza et al. 2020).

Other cereal seeds have been studied as a platform for producing heterologous proteins, however, on a smaller scale. For example, wheat seeds have been used to express TM-1 protein as an antigen to be used as an edible vaccine for chronic respiratory disease, a common disease in chickens, resulting in a significant level of protection (Shi et al. 2023).

6.5.2 Soybean

Among the plants that are candidates for seed-based stable accumulation of recombinant proteins, soybean plants, along with pea plants, present seeds with a high protein content (corresponding to approximately 40% of their weight) and, compared to other sources, represent a lower protein cost, due to their high seed yield (Mikschofsky and Broer 2012; Hudson et al. 2014; Vollmann 2016). Besides, recombinant proteins stored in soybean seeds were shown to remain stable and functional for long periods at room temperature (Oakes et al. 2009; Lobato Gómez et al. 2021). A study showed that the production of a functional subunit vaccine for *Staphylococcal enterotoxin* B in soybean seeds was stable over several soybean generations, and biochemically and immunologically similar to commercial recombinant forms (Hudson et al. 2014). The expression of cyanovirin-N, a lectin with antiviral activity, was demonstrated in soybean seeds at levels up to 10% of total soluble protein. Attempts to express this protein in transient assays in leaves of *N. benthamiana* were unsuccessful, demonstrating the potential of soybean as an alternative to express and accumulate this recombinant protein (O'Keefe et al. 2015).

From a regulatory point of view, it has a reduced risk of pollen contamination since soybean is largely self-pollinated (Paul and Ma 2011; Paul et al. 2011). Furthermore, soybean seeds accumulate proteins in PSV, resulting in optimal conditions for long-term storage of immunogenic and fully active recombinant proteins (Fig. 6.2) (Cunha et al. 2011).

Although soybean has not been commercially used for the production of recombinant proteins yet, numerous studies show the potential of this plant species for the production of pharmaceutical proteins, such as human growth hormone, proinsulin antiviral lectins, coagulation factor IX, antigens, antibodies, as well as non-pharma proteins (Moravec et al. 2007; Yamada et al. 2008; Cunha et al. 2011, 2014; Hudson et al. 2014; O'Keefe et al. 2015). Some of these studies have tested the potential of soybean expressing recombinant proteins when administered orally to mice, as seed extracts, with promising results (Moravec et al. 2007; Hudson et al. 2014). This indicates that soybean seeds could be formulated into edible products for oral delivery of pharmaceutical proteins (Adelakun et al. 2013).

Another important characteristic of this legume is its sensitivity to the photoperiod, expressed in the temporal modulation of the vegetative phase of its phenological cycle as a function of daily time and light intensity (O'Keefe et al. 2015). The production of seeds per plant can be greatly increased under controlled conditions in a greenhouse reaching up to 1000 seeds per plant. This increase in the production scale can be exploited by the molecular farming industry and is particularly relevant for biocontainment and more controlled cultivation conditions (Kantolic and Slafer 2007; see also chapter "Legume Seed: A Useful Platform for the Production of Medical Proteins/Peptide").

6.6 Scale-Up of Seed-Based Production Systems

The commercial success of large-scale seed molecular agriculture depends on technology, economics, and public acceptance (see also Chap. 15). Factors important for the biopharmaceutical industry include the expected reduction in costs and the indirect effects on the biopharmaceutical market (Twyman et al. 2003). Seed-based platforms can be used to produce recombinant proteins at a significantly lower cost compared to other systems such as microbial fermentation and cell cultures (Giddings 2001; Hood et al. 2002; Twyman et al. 2003). It is estimated that the production costs of recombinant proteins in maize, for example, will be threefold higher than that for the production of maize for food use (Mison and Curling 2000). Still, the savings in operational expenses can generate a considerable reduction in terms of capital investment (about 75–80%) and manufacture (50–60%) compared to microbial fermentation and cell culture production (Buyel 2019). In addition, the production of proteins that require lower purity, such as industrial enzymes and oral vaccines, is significantly reduced compared to other products that require expensive purification processes (Nikolov and Hammes 2002; see also chapters "Molecular Farming of Industrial Enzymes: Products and Applications" and "Plant Molecular Farming for Vaccine Development").

Field cultivation is the most reasonable option for the large-scale production of transgenic seeds. Although production in greenhouses may be feasible for transient transformations in short-cycle crops, generation time and space requirements would reduce the cost advantages of seed production (Boothe et al. 2010). The cost of production ($/g product) of the same recombinant protein produced in greenhouses is estimated as fivefold higher than that produced in an open-field system (Pogue et al. 2010). Besides, in terms of equipment for field production (from planting to harvesting), there is no difference between those used in conventional commodity crops. The existing technologies already meet this demand, enabling good scalability compared to other systems limited by the size of the culture reactor, for example. Furthermore, the long-term stability of the recombinant protein in the seeds allows the harvest to be decoupled from the purification process, generating greater flexibility and better stock management (Boothe et al. 2010).

Another important aspect of scale production is a high level of quality control to ensure high protein purity. Therefore, crop management must be carefully conducted, which is also necessary for biosafety reasons. Crops must provide high-quality seeds, which must present stable expression levels, and homogeneous products (e.g., glycosylation pattern and degradation levels) over generations and cultivation places to meet the established quality standards.

Finally, approximately 60–70% of the production cost of seed-based recombinant proteins is associated with downstream processing, so it is essential to develop techniques to improve it and make it increasingly efficient, reducing its cost (Dyr and Suttnar 1997). Therefore, understanding the conditions that affect the product's quantity and quality is essential to meet the market standards and competitiveness. An overview of the methods currently used for protein purification is presented in the following section.

6.7 Basic Approaches for the Purification of Recombinant Proteins from Plant Cells

The choice of an expression system should consider the particularities of the target protein to be purified. Expression based on the bacteria *Escherichia coli*, for example, could be advantageous in terms of production time and costs (Rosano and

Ceccarelli 2014; Lozano Terol et al. 2021). However, the lack of posttranslational modifications and contamination with bacterial endotoxins represent limitations for using this popular expression system (Sahdev et al. 2008). On the other hand, expression based on mammalian, yeast, and insect cells allows posttranslational modifications. However, these expression systems based on large bioreactors present a high upstream cost (Schillberg and Finnern 2021).

The production of biomass from plants requires soil and fertilizers, whereas bioreactor cultures demand complex media. That makes plant-based biomass production more sustainable than microbe and cell culture expression systems (Buyel et al. 2021), requiring fewer investments and demanding less specialized staff.

Downstream processes, independently of the expression system utilized, are critical in terms of cost and quality of the final product. It is estimated that downstream processing may represent 50–80% of the total costs, depending on the yield, recovery efficiency, and purity grade (Schillberg et al. 2019). Besides the costs, if purification of the recombinant protein is needed, one may be facing a technical challenge. Indeed, purification of a specific protein, either recombinant or endogenous, is generally not trivial, as each protein will have its specific physical characteristics. Plus, in the case of a recombinant protein, there will likely be no available protocol for purification from plant tissues. However, information on the purification method of the target protein from other expression systems may be helpful. Some factors do favor the downstream processing and purification of proteins from seeds since they are presented as a comparatively homogeneous starting material, have reduced water content, lack chlorophyll pigments and alkaloids, and confer high stability of the target protein stored in the seed storage vacuoles.

The source of the material to be processed needs to be handled carefully—plants need to be well cultivated so that the collected tissues are healthy. Next, some basic protein analyses are needed to confirm the presence of the heterologous protein and evaluate the expression level. These can be done from total extracts (i.e., tissues ground in extraction buffer), using well-established protocols for total protein quantification and detection, such as Western blot or ELISA. Finally, the amount of the recombinant protein can be expressed as a percentage of the total protein in the extract, or of the seed dry weight, as weight per mass of fresh or dry tissue (e.g., mg/g fresh leaves, µg/g dry seed weight).

The seed composition varies among species and requires specific extraction conditions regarding the contents of proteins, starch, oil, etc. For soybean seeds, for example, due to large amounts of oil (about 18–22% of seed weight), the material often requires homogenizing the seed meal with solvents, such as hexane. Furthermore, to ensure that the protein of interest maintains its stability and that its extraction process is efficient, it is essential that the extraction solutions present the appropriate pH and ionic strength (Robić et al. 2010). Hence, the extraction conditions consider pH, saline buffers, chaotropic agents/detergents, protease inhibitors, etc. (Janson 2011). Once extracted from the seeds, the target protein may be further purified by chromatography.

6.7.1 Non-affinity Absorption and Affinity Techniques to Purify Proteins

Proteins are made of amino acids (AAs) as basic building blocks assembled in a chain via amide bonds (peptide linkages). The 20 L-AAs found in proteins have four different ligand groups (an amino group, carboxyl group, hydrogen, and R-group). These groups and their interactions within the protein give unique biochemical characteristics and functions and influence their physiological and biological activities. The R-group of the 20 AA commonly found in proteins varies widely, especially their polarity at a biological pH (around pH 7.0), from polar and hydrophilic (water soluble) to nonpolar and hydrophobic (water insoluble) (Wu 2009). These physical properties can be exploited to aid the protein isolation and purification process.

In aqueous solutions, functional AA groups from folded proteins contribute significantly to the protein surface charge in a pH-dependent way. According to external pH, the overall charge may vary from positive (at low pH) to negative (at high pH). Therefore, separating a complex sample of proteins based on their surface charge helps purify a protein-rich sample with similar physicochemical characteristics in a reduced volume compared to the initial volume before separation (Bonner 2018). The purification may be optimized by using various techniques to exploit differences in the target protein's charge and biospecificity. Several successful cases in the literature explore these methodologies for efficient protein separation from seeds. For example, Zhang et al. (2012) purified the human alpha-antitrypsin protein expressed in transgenic rice seed using different anion exchange columns. In another work, cellobiohydrolase I was expressed in transgenic corn seeds and purified with ammonium sulfate precipitation (a fractionation technique also used to isolate proteins from a complex sample) together with both cationic and anionic exchange chromatography, yielding 63% of pure protein (Hood et al. 2014).

It is a common strategy to engineer the recombinant protein with affinity tags to facilitate affinity-based purification procedures. The most common examples found in the literature are proteins containing polyhistidine tails (6×His or 10×His) (Valdez-Ortiz et al. 2005). The sequence of six (or ten) consecutive histidine residues is currently one of the most used strategies worldwide to purify recombinant proteins for biochemical and structural studies.

Unlike ion exchange chromatography, affinity chromatography does not explore the physicochemical characteristics of proteins. The affinity chromatography technique is based precisely on the unique biospecificity of the protein engineered for isolation from a complex sample of proteins. Biospecificity involves the interaction between two immiscible phases, that is, the reversible interaction between a ligand (which can be a small molecule, enzymes, among others) immobilized on a resin (known as the stationary phase) and the recombinant target protein inserted into a solute (mobile phase) (Janson 2011). Menkhaus et al. (2004) compared various techniques, such as precipitation with polyethyleneimine cationic polyelectrolyte (PEI), anion/cation exchange, diafiltration (molecular exclusion), and immobilized metal ion affinity chromatography (IMAC) for the purification of histidine-tagged β-glucuronidase from transgenic pea seeds. They observed an increased recovery of pure protein and higher enzyme activity when utilizing affinity chromatography (Menkhaus et al. 2004).

In some cases, proteins from the host can be present as contaminants in samples purified by affinity purification. To remove these contaminants, denaturation and refolding steps of proteins are sometimes necessary. Fujiwara et al. (2010) observed the need for two steps of affinity purification combined with a denaturation step with 6 M guanidine to remove protein contaminants from rice seed in the purification of IL-4 and -6 cytokines (Fujiwara et al. 2016). Similar results were previously obtained by Fujiwara et al. (2010) during the purification of human interleukin-10 (IL-10), also expressed in rice seeds. These results demonstrate the efficiency of combining different chromatographic and purification techniques to remove contaminants and consequently increase the yield and purity of the recombinant proteins obtained (Fujiwara et al. 2010).

6.7.2 Chromatography-Free Protein Purification

Another approach for recombinant protein purification is based on fusion proteins (FPs). The idea is to exploit the unique properties of the fusion partner, allowing an increase in stability and facilitating the purification process (Viana et al. 2013; Ki and Pack 2020). Examples of these fusion partners include the synthetic peptide elastin-like polypeptides (ELPs) (Ciofani et al. 2014), γ-zein (Torrent et al. 2009), and hydrophobin (Lahtinen et al. 2008).

Derived from its soluble precursor, tropoelastin, elastin has a hydrophobic motif composed of a repeated sequence of hydrophobic amino acids alanine (Ala) and valine (Val), in addition to the presence of other residues in significant amounts, such as glycine (Gly) and proline (Pro) (Partridge et al. 1955). At temperatures below 25 °C, the protein remains soluble; however, when the temperature is raised to 37 °C, a precipitated protein known as coacervate is observed. This process is fully reversible upon returning the protein to room temperature (Urry et al. 1969). Based on these properties, synthetic peptides were developed, known as the elastin-like polypeptides (ELPs), composed of the canonical sequence of the pentapeptide (Val-Pro-Gly-Xaa-Gly)*n*.

When investigating the strategy of using FP in conjunction with scalable purification processes, Phan et al. (2014) observed an enhancement in expression levels (about tenfold higher) of the avian influenza virus (H5N1) hemagglutinin subtype 5 (H5) protein fused with ELP at the C-terminus in transgenic tobacco seeds. The enhanced accumulation of HA, which is the major antigen of the influenza virus, by ELP-FP resulted in high concentrations of the ELPylated target protein in the aqueous crude extract. Further purification was facilitated by using optimized processes involving inverse transition cycling (ITC). In comparison with another strategy, also using FP (fungal hydrophobin I—HFBI), the same authors observed that only

ELPylation was able to increase HA expression in seeds, resulting in high-purity protein (Phan et al. 2014), demonstrating the efficiency of using ELPylated proteins in the processes of expression and purification simply and inexpensively (Khan et al. 2012).

In contrast, in another study, Yang et al. (2021) showed that the γ-zein system, a member of the major prolamin storage family in maize, was more efficient for the accumulation of GFP in immature soybean seeds than the ELP system. In addition, the use of the γ-zein system provided a 3.9-fold increase in the accumulation of fused GFP in comparison with unfused GFP protein, demonstrating that the γ-zein system is a promising FP for future enhancement in the expression and purification of recombinant proteins in plants.

Although fusion proteins are generally non-immunogenic and biologically compatible, they may interfere with the activity of native proteins (Shamji et al. 2007; Viana et al. 2013). That implies the removal of the FP, which is done after purification by specific proteolytic enzymes that recognize cleavage sites placed at the junction of the target protein and the fusion partner. This additional step for recombinant protein purification may result in unspecific degradations of the target protein, reducing the final yield and impacting downstream costs (Tian and Sun 2011).

A promising alternative to proteolytic enzymes for the removal of FP is the use of inteins (Viana et al. 2013; Ki and Pack 2020). These proteins can catalyze their self-cleavage and, through amino acid substitution, can be regulated to cleave at either the N- or the C-terminus in response to reducing agents or changes in solution pH (Xu and Perler 1996; Perler 1998; Gillies et al. 2009). Therefore, the self-cleavage property of inteins can be applied to replace the traditional proteolytic cleavage. By fusing the intein (Eitag) with the ELP-FP, for example, the recombinant proteins can be purified by applying both ITC followed by autocatalysis by changing the pH of the solution. Tian and Sun (2011) explored the use of the ELP-intein system to increase the accumulation of the recombinant lectin fused with ELP in transgenic rice and tested the capacity of autocatalysis of intein after ELP extraction from seeds. Furthermore, the presence of Eitag + ELP did not alter the N-glycosylation patterns of the recombinant protein, demonstrating the potential application of the ELP-intein fusion system for the expression and purification of recombinant proteins in plants, especially in seeds.

6.8 A Brief Overview of Biosafety and Risk Assessment of Seed-Based Expression Systems

The technology for the production of recombinant proteins using these platforms is developing fast and focuses on two product lines: pharmaceuticals and non-pharmaceuticals. Despite intense research on developing biopharmaceutical production, most plant-based products currently available on the market belong to the non-pharma field, mainly because regulation processes are faster and less expensive for non-pharma products. These include products for the diagnosis, industrial reagents, and cosmetics, among others.

The biological safety assessment to produce recombinant proteins in seeds is an important issue. Biosafety involves several relevant issues, such as choice of plant platform, transgenic plants, field production, handling, harvesting, and transport. Therefore, one must consider plant biology from the perspective of productivity and how it impacts the environment, food security, and human health. Therefore, the best material based on technical aspects (e.g., seeds with better processing capacity, high protein content, stability, etc.) may not be the best choice if considering the regulatory issues of biosafety (Sparrow and Twyman 2009).

The use of seeds as a "bioreactor" has its risks of propagation in nature, contaminating non-transgenic plants, and being potentially hazardous to people and animals if used unintentionally as food and for insects and soil microorganisms (Lee et al. 2003). Most of the steps required to avoid mixing these seed-derived biopharmaceuticals in the food chain are relatively simple, such as meticulous planning and execution. The plants must be cultivated in an isolated area to avoid genetic and mechanical mixing of seeds containing biopharmaceuticals with those intended for food. Likewise, small-scale and large-scale field trials must be isolated from conventional practices with crops to avoid cross-pollination. Although these risks apply to all transgenic crops grown in the field, plants cultivated for molecular farming deserve special attention due to the nature of the recombinant proteins (i.e., biopharmaceuticals), consequently, with unpredictable, potentially hazardous outcomes (Basaran and Rodríguez-Cerezo 2008). Achieving an effective isolation level to avoid wind and insect pollination is challenging. If the plants are cultivated in confinement, the risks and threats to the environment would be reduced and would imply less strict regulatory issues.

Appropriate mitigation measures for recombinant protein-producing seeds will depend on several factors, including properties of the molecule, biology of the crop, and characteristics of the environment where it is being produced. Approaches of containment methods include identity preservation—using varieties that are visually distinct from traditional varieties (such as purple maize or black soybeans)—and use of marker genes, such as a fluorescent protein, barrier crops, and temporal barriers aiming to minimize undesirable crosses (Sparrow and Twyman 2009).

In general, regulatory guidelines for the production of recombinant proteins are similar across countries, but some specificities may apply. For example, in the United States, the production of biopharmaceuticals on transgenic plants is regulated by two agencies. The US Department of Agriculture (USDA), Animal and Plant Health Inspection Service (APHIS), focuses on the containment of these seed producers of protein. In contrast, the Food and Drug Administration (FDA) focuses on the manufacture of the drug or vaccine. APHIS reviews production license applications, assessing probable environmental impacts of these releases (Basaran and Rodríguez-Cerezo 2008). In the European Union, authorizations involve all member states and the European Commission (Breyer et al. 2009). In Brazil, GMO studies are only allowed in research institutions after authorization by the National Biosafety Technical Commission (CTNBio) (Mendonça-Hagler et al. 2008).

6.9 Conclusions and Perspectives

The demand for biopharmaceuticals continues to grow as new products are approved. To couple with this demand, a general trend for the production of recombinant proteins has been to increase yield and optimize upstream and downstream processes to reduce costs. In addition, bioequivalence is also very relevant for the biopharmaceutical industry, which pursues products that are as similar as possible to the original product.

As discussed in this chapter, several aspects of plant molecular farming are aligned with these demands. Producing recombinant protein in plants demands substantially lower costs for upstream processes as compared to the expression systems based on microorganisms and cell cultures. Concerning posttranslational modifications, efforts for glyco-engineering of plants have also achieved amazing progress. Although still restricted to model plants and transient assays using *N. benthamiana*, efforts toward a seed-based glyco-engineered platform are in progress. It is reasonable to consider its viability and availability before long.

As experience accumulates, methods for protein purification will become more efficient. These may compensate for limitations on the expression level, which can be very low, depending on the target protein. For seed-based expression, the final yield could be increased by applying more efficient extraction and purification, involving, for example, fusion proteins and fusion tags. Seed-based platforms, for their advantages in terms of protein content and long-term stability of the recombinant protein stored therein, offer great potential for new ideas to be implemented (see also Chap. 5). The progress witnessed in the last decades confirms the potential of molecular farming as an alternative system for expressing recombinant proteins and represents a field of opportunities.

References

Adelakun OE, Duodu KG, Buys E, Olanipekun BF (2013) Potential use of soybean flour (*Glycine max*) in food fortification. In: El-Shemy HA (ed) Soybean. Bio-active compounds. IntechOpen, Rijeka, Croatia, pp 513–520

Arcalis E, Ibl V, Peters J, Melnik S, Stoger E (2014) The dynamic behavior of storage organelles in developing cereal seeds and its impact on the production of recombinant proteins. Front Plant Sci 5:439

Ashnest JR, Gendall AR (2018) Trafficking to the seed protein storage vacuole. Funct Plant Biol 45:895–910

Basaran P, Rodríguez-Cerezo E (2008) Plant molecular farming: opportunities and challenges. Crit Rev Biotechnol 28:153–172

Bohlender LL, Parsons J, Hoernstein SNW, Bangert N, Rodriguez-Jahnke F et al (2022) Unexpected *Arabinosylation* after humanization of plant protein N-glycosylation. Front Bioeng Biotechnol 10:1–12

Bohnsack MT, Schleiff E (2010) The evolution of protein targeting and translocation systems. Biochim Biophys Acta 1803:1115–1130

Bonner P (2018) Protein purification, 2nd edn. Taylor & Francis, London, UK

Boothe J, Nykiforuk C, Shen Y, Zaplachinski S, Szarka S et al (2010) Seed-based expression systems for plant molecular farming. Plant Biotechnol J 8:588–606
Breyer D, Goossens M, Herman P, Sneyers M (2009) Biosafety considerations associated with molecular farming in genetically modified plants. J Med Plant Res 3:825–838
Buyel JF (2019) Plant molecular farming - integration and exploitation of side streams to achieve sustainable biomanufacturing. Front Plant Sci 9:1–17
Buyel JF, Stoger E, Bortesi L (2021) Targeted genome editing of plants and plant cells for biomanufacturing. Transgenic Res 30:401–426
Castilho A, Strasser R, Stadlmann J, Grass J, Jez J et al (2010) *In planta* protein sialylation through overexpression of the respective mammalian pathway. J Biol Chem 285:15923–15930
Castilho A, Gattinger P, Grass J, Jez J, Pabst M et al (2011) N-glycosylation engineering of plants for the biosynthesis of glycoproteins with bisected and branched complex N-glycans. Glycobiology 21:813–823
Chen ZL, Naito S, Nakamura I, Beachy RN (1989) Regulated expression of genes encoding soybean beta-conglycinins in transgenic plants. Dev Genet 10:112–122
Chen Q, Yu F, Xie Q (2020) Insights into endoplasmic reticulum-associated degradation in plants. New Phytol 226:345–350
Ciofani G, Genchi GG, Mattoli V, Mazzolai B, Bandiera A (2014) The potential of recombinant human elastin-like polypeptides for drug delivery. Expert Opin Drug Deliv 11:1507–1512
Cunha NB, Murad AM, Cipriano TM, Araújo ACG, Aragão FJL et al (2011) Expression of functional recombinant human growth hormone in transgenic soybean seeds. Transgenic Res 20:811–826
Cunha NB, Murad A, Vianna G, Rech E (2014) Recombinant biosynthesis of functional human growth hormone and coagulation factor IX in transgenic soybean seeds. BMC Proc 8:P112
De Jaeger G, Scheffer S, Jacobs A, Zambre M, Zobell O et al (2002) Boosting heterologous protein production in transgenic dicotyledonous seeds using *Phaseolus vulgaris* regulatory sequences. Nat Biotechnol 20:1265–1268
Diamos AG, Mason HS (2018) Chimeric 3′ flanking regions strongly enhance gene expression in plants. Plant Biotechnol J 16:1971–1982
Dyr JE, Suttnar J (1997) Separation used for purification of recombinant proteins. J Chromatogr B Biomed Sci Appl 699:383–401
Egelkrout E, Hayden C, Fake G, Keener T, Arruda P et al (2020) Oral delivery of maize-produced porcine epidemic diarrhea virus spike protein elicits neutralizing antibodies in pigs. Plant Cell Tissue Organ Cult 142:79–86
Feeney M, Kittelmann M, Menassa R, Hawes C, Frigerio L (2018) Protein storage vacuoles originate from remodeled preexisting vacuoles in *Arabidopsis thaliana*. Plant Physiol 177:241–254
Fischer R, Buyel JF (2020) Molecular farming - the slope of enlightenment. Biotechnol Adv 40:107519
Fischer R, Vaquero-Martin C, Sack M, Drossard J, Emans N et al (1999) Towards molecular farming in the future: transient protein expression in plants. Biotechnol Appl Biochem 30:113–116
Fujiwara Y, Aiki Y, Yang L, Takaiwa F, Kosaka A et al (2010) Extraction and purification of human interleukin-10 from transgenic rice seeds. Protein Expr Purif 72:125–130
Fujiwara Y, Yang L, Takaiwa F, Sekikawa K (2016) Expression and purification of recombinant mouse interleukin-4 and -6 from transgenic rice seeds. Mol Biotechnol 58:223–231
Furtado A, Henry RJ, Pellegrineschi A (2009) Analysis of promoters in transgenic barley and wheat. Plant Biotechnol J 7:240–253
Ghag SB, Adki VS, Ganapathi TR, Bapat VA (2021) Plant platforms for efficient heterologous protein production. Biotechnol Bioprocess Eng 26:546–567
Giddings G (2001) Transgenic plants as protein factories. Curr Opin Biotechnol 12:450–454
Gillies AR, Mahmoud RB, Wood DW (2009) PHB-intein-mediated protein purification strategy. Methods Mol Biol 498:173–183
Gleba Y, Marillonnet S, Klimyuk V (2004) Engineering viral expression vectors for plants: the 'full virus' and the 'deconstructed virus' strategies. Curr Opin Plant Biol 7:182–188

Gupta SK, Shukla P (2018) Glycosylation control technologies for recombinant therapeutic proteins. Appl Microbiol Biotechnol 102:10457–10468

Habibi P, Prado GS, Pelegrini PB, Hefferon KL, Soccol CR et al (2017) Optimization of inside and outside factors to improve recombinant protein yield in plant. Plant Cell Tissue Organ Cult 130:449–467

Hisano H, Sato K (2016) Genomic regions responsible for amenability to *Agrobacterium*-mediated transformation in barley. Sci Rep 6:37505

Hood EE, Witcher DR, Maddock S, Meyer T, Baszczynski C et al (1997) Commercial production of avidin from transgenic maize characterization of transformant, production, processing, extraction and purification. Mol Breed 3:291–306

Hood EE, Woodard SL, Horn ME (2002) Monoclonal antibody manufacturing in transgenic plants - myths and realities. Curr Opin Biotechnol 13:630–635

Hood EE, Devaiah SP, Fake G, Egelkrout E, Teoh KT et al (2012) Manipulating corn germplasm to increase recombinant protein accumulation. Plant Biotechnol J 10:20–30

Hood NC, Hood KR, Woodard SL, Devaiah SP, Jeoh T et al (2014) Purification and characterization of recombinant Cel7A from maize seed. Appl Biochem Biotechnol 174:2864–2874

Hudson LC, Garg R, Bost KL, Piller KJ (2014) Soybean seeds: a practical host for the production of functional subunit vaccines. Biomed Res Int 2014:340804–340804

Janson J-C (2011) Protein purification: principles, high resolution methods, and applications, 3rd edn. Wiley, Hoboken, NJ

Jolliffe NA, Craddock CP, Frigerio L (2005) Pathways for protein transport to seed storage vacuoles. Biochem Soc Trans 33:1016–1018

Joshi JB, Geetha S, Singh B, Kumar KK, Kokiladevi E et al (2015) A maize alpha-zein promoter drives an endosperm-specific expression of transgene in rice. Physiol Mol Biol Plants 21:35–42

Kallolimath S, Castilho A, Strasser R, Grünwald-Gruber C, Altmann F et al (2016) Engineering of complex protein sialylation in plants. Proc Natl Acad Sci U S A 113:9498–9503

Kantolic AG, Slafer GA (2007) Development and seed number in indeterminate soybean as affected by timing and duration of exposure to long photoperiods after flowering. Ann Bot 99:925–933

Kawakatsu T, Takaiwa F (2010) Cereal seed storage protein synthesis: fundamental processes for recombinant protein production in cereal grains. Plant Biotechnol J 8:939–953

Kesik-Brodacka M (2018) Progress in biopharmaceutical development. Biotechnol Appl Biochem 65:306–322

Khan I, Twyman RM, Arcalis E, Stoger E (2012) Using storage organelles for the accumulation and encapsulation of recombinant proteins. Biotechnol J 7:1099–1108

Ki MR, Pack SP (2020) Fusion tags to enhance heterologous protein expression. Appl Microbiol Biotechnol 104:2411–2425

Kim W-S, Jez JM, Krishnan HB (2014) Effects of proteome rebalancing and sulfur nutrition on the accumulation of methionine rich δ-zein in transgenic soybeans. Front Plant Sci 5:633

Kusaba M, Miyahara K, Iida S, Fukuoka H, Takano T et al (2003) Low glutelin content1: a dominant mutation that suppresses the glutelin multigene family via RNA silencing in rice. Plant Cell 15:1455–1467

Lahtinen T, Linder MB, Nakari-Setälä T, Oker-Blom C (2008) Hydrophobin (HFBI): a potential fusion partner for one-step purification of recombinant proteins from insect cells. Protein Expr Purif 59:18–24

Lee RWH, Pool AN, Ziauddin A, Lo RYC, Shewen PE et al (2003) Edible vaccine development: stability of *Mannheimia haemolytica* A1 leukotoxin 50 during post-harvest processing and storage of field-grown transgenic white clover. Mol Breed 11:259–266

Li J, Berger F (2012) Endosperm: food for humankind and fodder for scientific discoveries. New Phytol 195:290–305

Liu H, Timko MP (2022) Improving protein quantity and quality-the next level of plant molecular farming. Int J Mol Sci 23:1326

Lobato Gómez M, Huang X, Alvarez D, He W, Baysal C et al (2021) Contributions of the international plant science community to the fight against human infectious diseases – Part 1: Epidemic and pandemic diseases. Plant Biotechnol J 19:1901–1920

Lozano Terol G, Gallego-Jara J, Sola Martínez RA, Martínez Vivancos A, Cánovas Díaz M et al (2021) Impact of the expression system on recombinant protein production in *Escherichia coli* BL21. Front Microbiol 12:682001

Mendonça-Hagler L, Souza L, Aleixo L, Oda L (2008) Trends in biotechnology and biosafety in Brazil. Environ Biosaf Res 7:115–121

Menkhaus TJ, Pate C, Krech A, Glatz CE (2004) Recombinant protein purification from pea. Biotechnol Bioeng 86:108–114

Mikschofsky H, Broer I (2012) Feasibility of *Pisum sativum* as an expression system for pharmaceuticals. Transgenic Res 21:715–724

Mirzaee M, Osmani Z, Frébortová J, Frébort I (2022) Recent advances in molecular farming using monocot plants. Biotechnol Adv 58:107913

Mison D, Curling J (2000) The industrial production costs of recombinant therapeutic proteins expressed in transgenic corn. BioPharm 13:48–54

Mizukami A, Caron AL, Picanço-Castro V, Swiech K (2018) Platforms for recombinant therapeutic glycoprotein production. Methods Mol Biol 1674:1–14

Montero-Morales L, Steinkellner H (2018) Advanced plant-based glycan engineering. Front Bioeng Biotechnol 6:81

Moon K-B, Park J-S, Park Y-I, Song I-J, Lee H-J et al (2020) Development of systems for the production of plant-derived biopharmaceuticals. Plants 9:30

Moravec T, Schmidt MA, Herman EM, Woodford-Thomas T (2007) Production of *Escherichia coli* heat labile toxin (LT) B subunit in soybean seed and analysis of its immunogenicity as an oral vaccine. Vaccine 25:1647–1657

Nahampun HN, Bosworth B, Cunnick J, Mogler M, Wang K (2015) Expression of H3N2 nucleoprotein in maize seeds and immunogenicity in mice. Plant Cell Rep 34:969–980

Nikolov Z, Hammes D (2002) Production of recombinant proteins from transgenic crops. In: Hood EE, Howard JA (eds) Plants as factories for protein production. Springer, Dordrecht, Netherlands, pp 159–174

Nochi T, Yuki Y, Katakai Y, Shibata H, Tokuhara D et al (2009) A rice-based oral cholera vaccine induces macaque-specific systemic neutralizing antibodies but does not influence pre-existing intestinal immunity. J Immunol 183:6538–6544

O'Keefe BR, Murad AM, Vianna GR, Ramessar K, Saucedo CJ et al (2015) Engineering soya bean seeds as a scalable platform to produce cyanovirin-N, a non-ARV microbicide against HIV. Plant Biotechnol J 13:884–892

Oakes JL, Bost KL, Piller KJ (2009) Stability of a soybean seed-derived vaccine antigen following long-term storage, processing and transport in the absence of a cold chain. J Sci Food Agric 89:2191–2199

Obembe OO, Popoola JO, Leelavathi S, Reddy SV (2011) Advances in plant molecular farming. Biotechnol Adv 29:210–222

Orman-Ligeza B, Harwood W, Hedley PE, Hinchcliffe A, Macaulay M et al (2020) TRA1: a locus responsible for controlling *Agrobacterium*-mediated transformability in barley. Front Plant Sci 11:355

Partridge SM, Davis HF, Adair GS (1955) The chemistry of connective tissues. 2. Soluble proteins derived from partial hydrolysis of elastin. Biochem J 61:11–21

Paul M, Ma JKC (2011) Plant-made pharmaceuticals: leading products and production platforms. Biotechnol Appl Biochem 58:58–67

Paul M, van Dolleweerd C, Drake PMW, Reljic R, Thangaraj H et al (2011) Molecular Pharming: future targets and aspirations. Hum Vaccin 7:375–382

Pedrazzini E, Mainieri D, Marrano CA, Vitale A (2016) Where do protein bodies of cereal seeds come from? Front Plant Sci 7:1139

Perler FB (1998) Protein splicing of inteins and hedgehog autoproteolysis: structure, function, and evolution. Cell 92:1–4

Phan HT, Hause B, Hause G, Arcalis E, Stoger E et al (2014) Influence of elastin-like polypeptide and hydrophobin on recombinant hemagglutinin accumulations in transgenic tobacco plants. PLoS One 9:e99347

Pogue GP, Vojdani F, Palmer KE, Hiatt E, Hume S et al (2010) Production of pharmaceutical-grade recombinant aprotinin and a monoclonal antibody product using plant-based transient expression systems. Plant Biotechnol J 8:638–654

Rademacher T, Sack M, Arcalis E, Stadlmann J, Balzer S et al (2008) Recombinant antibody 2G12 produced in maize endosperm efficiently neutralizes HIV-1 and contains predominantly single-GlcNAc N-glycans. Plant Biotechnol J 6:189–201

Robić G, Farinas CS, Rech EL, Miranda EA (2010) Transgenic soybean seed as protein expression system: aqueous extraction of recombinant β-glucuronidase. Appl Biochem Biotechnol 160:1157–1167

Robinson DG, Oliviusson P, Hinz G (2005) Protein sorting to the storage vacuoles of plants: a critical appraisal. Traffic 6:615–625

Rosano GL, Ceccarelli EA (2014) Recombinant protein expression in *Escherichia coli*: advances and challenges. Front Microbiol 5:1–17

Sahdev S, Khattar SK, Saini KS (2008) Production of active eukaryotic proteins through bacterial expression systems: a review of the existing biotechnology strategies. Mol Cell Biochem 307:249–264

Schillberg S, Finnern R (2021) Plant molecular farming for the production of valuable proteins - critical evaluation of achievements and future challenges. J Plant Physiol 258–259:153359

Schillberg S, Raven N, Spiegel H, Rasche S, Buntru M (2019) Critical analysis of the commercial potential of plants for the production of recombinant proteins. Front Plant Sci 10:720

Schmidt MA, Herman EM (2008) Proteome rebalancing in soybean seeds can be exploited to enhance foreign protein accumulation. Plant Biotechnol J 6:832–842

Shamji MF, Betre H, Kraus VB, Chen J, Chilkoti A et al (2007) Development and characterization of a fusion protein between thermally responsive elastin-like polypeptide and interleukin-1 receptor antagonist: sustained release of a local anti-inflammatory therapeutic. Arthritis Rheum 56:3650–3661

Shi Y, Habibi P, Haq ANU, Saeed M, Gulghutay Amjad N et al (2023) Seed-based system for cost-effective production of vaccine against chronic respiratory disease in chickens. Mol Biotechnol 65:570–580

Sparrow PA, Twyman RM (2009) Biosafety, risk assessment and regulation of plant-made pharmaceuticals. Methods Mol Biol 483:341–353

Strasser R, Altmann F, Mach L, Glössl J, Steinkellner H (2004) Generation of *Arabidopsis thaliana* plants with complex N-glycans lacking β1,2-linked xylose and core α1,3-linked fucose. FEBS Lett 561:132–136

Strasser R, Seifert G, Doblin MS, Johnson KL, Ruprecht C et al (2021) Cracking the "sugar code": a snapshot of n- and o-glycosylation pathways and functions in plants cells. Front Plant Sci 12:640919

Streatfield SJ (2007) Approaches to achieve high-level heterologous protein production in plants. Plant Biotechnol J 5:2–15

Tada Y, Utsumi S, Takaiwa F (2003) Foreign gene products can be enhanced by introduction into low storage protein mutants. Plant Biotechnol J 1:411–422

Takagi H, Hiroi T, Yang L, Tada Y, Yuki Y et al (2005) A rice-based edible vaccine expressing multiple T cell epitopes induces oral tolerance for inhibition of Th2-mediated IgE responses. Proc Natl Acad Sci U S A 102:17525–17530

Takaiwa F, Takagi H, Hirose S, Wakasa Y (2007) Endosperm tissue is good production platform for artificial recombinant proteins in transgenic rice. Plant Biotechnol J 5:84–92

Thomas DR, Walmsley AM (2015) The effect of the unfolded protein response on the production of recombinant proteins in plants. Plant Cell Rep 34:179–187

Tian L, Sun SM (2011) A cost-effective ELP-intein coupling system for recombinant protein purification from plant production platform. PLoS One 6:e24183
Torrent M, Llompart B, Lasserre-Ramassamy S, Llop-Tous I, Bastida M et al (2009) Eukaryotic protein production in designed storage organelles. BMC Biol 7:5
Tschofen M, Knopp D, Hood E, Stoger E (2016) Plant molecular farming: much more than medicines. Annu Rev Anal Chem (Palo Alto, Calif) 9:271–294
Twyman RM, Stoger E, Schillberg S, Christou P, Fischer R (2003) Molecular farming in plants: host systems and expression technology. Trends Biotechnol 21:570–578
Urry DW, Starcher B, Partridge SM (1969) Coacervation of solubilized elastin effects a notable conformational change. Nature 222:795–796
Valdez-Ortiz A, Rascón-Cruz Q, Medina-Godoy S, Sinagawa-García SR, Valverde-González ME et al (2005) One-step purification and structural characterization of a recombinant His-tag 11S globulin expressed in transgenic tobacco. J Biotechnol 115:413–423
Vamvaka E, Arcalis E, Ramessar K, Evans A, O'Keefe BR et al (2016a) Rice endosperm is cost-effective for the production of recombinant griffithsin with potent activity against HIV. Plant Biotechnol J 14:1427–1437
Vamvaka E, Twyman RM, Murad AM, Melnik S, Teh AY et al (2016b) Rice endosperm produces an underglycosylated and potent form of the HIV-neutralizing monoclonal antibody 2G12. Plant Biotechnol J 14:97–108
Varki A (2017) Biological roles of glycans. Glycobiology 27:3–49
Viana FCJ, Dias CS, Franco LO, Lacorte C (2013) Heterologous production of peptides in plants: fusion proteins and beyond. Curr Protein Pept Sci 14:568–579
Vitale A, Boston RS (2008) Endoplasmic reticulum quality control and the unfolded protein response: insights from plants. Traffic 9:1581–1588
Vollmann J (2016) Soybean versus other food grain legumes: a critical appraisal of the united nations international year of pulses 2016. Die Bodenkultur 67:17–24
Wakasa Y, Takaiwa F (2013) The use of rice seeds to produce human pharmaceuticals for oral therapy. Biotechnol J 8:1133–1143
Walsh G, Jefferis R (2006) Post-translational modifications in the context of therapeutic proteins. Nat Biotechnol 24:1241–1252
Wang X, Jiang D, Shi J, Yang D (2017) Expression of alpha-1,6-fucosyltransferase (FUT8) in rice grain and immunogenicity evaluation of plant-specific glycans. J Biotechnol 242:111–121
Watson SA, Ramstad PE (1987) Corn: chemistry and technology. American Association of Cereal Chemists, St. Paul, USA
Webster GR, Teh AY, Ma JK (2017) Synthetic gene design-the rationale for codon optimization and implications for molecular pharming in plants. Biotechnol Bioeng 114:492–502
Wilbers RH, Westerhof LB, van Noort K, Obieglo K, Driessen NN et al (2017) Production and glyco-engineering of immunomodulatory helminth glycoproteins in plants. Sci Rep 10:45910
Wilken LR, Nikolov ZL (2012) Recovery and purification of plant-made recombinant proteins. Biotechnol Adv 30:419–433
Witcher DR, Hood EE, Peterson D, Bailey M, Bond D et al (1998) Commercial production of β-glucuronidase (GUS): a model system for the production of proteins in plants. Mol Breed 4:301–312
Wu G (2009) Amino acids: metabolism, functions, and nutrition. Amino Acids 37:1–17
Xie T, Qiu Q, Zhang W, Ning T, Yang W et al (2008) A biologically active rhIGF-1 fusion accumulated in transgenic rice seeds can reduce blood glucose in diabetic mice via oral delivery. Peptides 29:1862–1870
Xu MQ, Perler FB (1996) The mechanism of protein splicing and its modulation by mutation. EMBO J 15:5146–5153
Xu R, Li D, Li H, Li J, Yang Y et al (2016) Isolation of four rice seed-specific promoters and evaluation of endosperm activity. Plant Cell Tissue Organ Cult 128:125–132
Xu J, Towler M, Weathers PJ (2018) Platforms for plant-based protein production. In: Pavlov A, Bley T (eds) Bioprocessing of plant in vitro systems. Springer, Cham, Switzerland, pp 1–40

Yamada Y, Nishizawa K, Yokoo M, Zhao H, Onishi K et al (2008) Anti-hypertensive activity of genetically modified soybean seeds accumulating. Peptides 29:331–337
Yang J, Xun H, Niu L, He H, Cheng Y et al (2021) Elastin-like polypeptide and gamma-zein fusions significantly increase recombinant protein accumulation in soybean seeds. Transgenic Res 30:675–686
Yuki Y, Nojima M, Hosono O, Tanaka H, Kimura Y et al (2021) Oral MucoRice-CTB vaccine for safety and microbiota-dependent immunogenicity in humans: a phase 1 randomised trial. Lancet Microbe 2:e429–e440
Zhang L, Shi J, Jiang D, Stupak J, Ou J et al (2012) Expression and characterization of recombinant human alpha-antitrypsin in transgenic rice seed. J Biotechnol 164:300–308

Chapter 7
Plant-Based Antibody Manufacturing

Gregory P. Pogue, Kelsi Swope, Joseph Rininger, Lauren Schoukroun-Barnes, Josh Morton, Steve Hume, Krystal Hamorsky, Josh Fuqua, Joshua M. Royal, Michael H. Pauly, Max Brennan, Larry Zeitlin, Kevin Whaley, Sean Stevens, and Barry Bratcher

Abstract Plant-based manufacture of antibodies and other biologics is a rapidly growing area of interest to the biopharmaceutical industry. The use of plants can significantly accelerate biologics production, with relatively lower infrastructure costs, compared with mammalian cell-based manufacturing. Improvements in genomics, bioinformatics, and genome engineering tools have contributed to the generation of more efficient and productive plant strains. Novel technological approaches have streamlined expression, extraction, and purification of biologics from these plants and resulted in biologics, which have successfully been used in clinical trials and treatment of patients. In this chapter, we discuss the progress and remaining challenges for plant-based manufacturing of biologics, with a focus on the use of *Nicotiana benthamiana* for production of antibodies.

The original version of the chapter has been revised. A corrections to this chapter can be found at https://doi.org/10.1007/978-981-99-4859-8_16

G. P. Pogue (✉)
KBio, Inc., Owensboro, KY, USA

IC2 Institute, The University of Texas at Austin, Austin, TX, USA
e-mail: gregp@kbio.com

K. Swope · J. Morton · S. Hume · K. Hamorsky · J. Fuqua · J. M. Royal · S. Stevens · B. Bratcher
KBio, Inc., Owensboro, KY, USA
e-mail: kelsis@kbio.com; joshuam@kbio.com; steveh@kbio.com; krystalh@kbio.com; joshuaf@kbio.com; joshuar@kbio.com; seans@kbio.com; barryb@kbio.com

J. Rininger · L. Schoukroun-Barnes
Latham BioPharm Group, Inc., Elkridge, MD, USA
e-mail: jarininger@lathambiopharm.com; lschoukroun-barnes@lathambiopharm.com

M. H. Pauly · M. Brennan · L. Zeitlin · K. Whaley
ZabBio, Inc., San Diego, CA, USA
e-mail: michael.pauly@zabbio.com; max.brennan@zabbio.com; larry.zeitlin@zabbio.com; kevin.whaley@zabbio.com

C. Kole et al. (eds.), *Tools & Techniques of Plant Molecular Farming*, Concepts and Strategies in Plant Sciences, https://doi.org/10.1007/978-981-99-4859-8_7

Keywords Recombinant · Antibody · Plant-made pharmaceutical · Manufacturing · Biopharmaceutical · Glycosylation · *Nicotiana benthamiana*

Abbreviations

ADCC	Antibody-dependent cell-mediated cytotoxicity
AEX	Anion exchange chromatography
CAM-Cys	Carbamidomethyl-cysteine
CFU	Colony-forming unit
CHO	Chinese hamster ovary cells
COVID-19	SARS-CoV-2 virus
Cys	Cysteine
Da	Dalton
DF	Diafiltration
DP	Drug product
DS	Drug substance
DSP	Downstream process
ELISA	Enzyme-linked immunosorbent assay
EU	Enzyme units
Fc	Constant fragment
Fuc	Fucose
Gal	Galactose
GlcNAC	*N*-acetylglucosamine
GMP	Good manufacturing practices
HC	Heavy chain
HC-DNA	Host cell DNA
HCP	Host cell protein
HEK	Human embryonic kidney cells
HIV	Human immunodeficiency virus type 1
HSV	Herpes simplex virus
icIEF	Imaged capillary isoelectric focusing
IEX	Ion-exchange chromatography
IV	Intravenous
LC	Light chain
LC-MS	Liquid chromatography-mass spectrometry
Lys-C	Lys-C protease
mAb	Monoclonal antibody
Man	Mannose
MW	Molecular weight
NSO	Murine myeloma cells
PD	Pharmacodynamics
PK	Pharmacokinetics
PMP	Plant-made pharmaceutical
PVX	Potato virus X
SDS-PAGE	Sodium dodecyl sulfate-polyacrylamide gel electrophoresis
SEC-HPLC	Size-exchange-high-pressure liquid chromatography

TFF	Tangential flow filtration
TVCV	Turnip vein clearing virus
U.S.P.	U.S. Pharmacopeia
UF	Ultrafiltration
US FDA	U.S. Food and Drug Administration
USP	Upstream process
UV	Ultraviolet light wavelength
WHO	World Health Organization
Xyl	Xylose

7.1 Introduction

Since the first monoclonal antibody (mAb) was approved by the US FDA in 1986 (Lu et al. 2020), the global market has rapidly expanded to more than $188.18 billion by the end of 2022 (ReportLinker 2021). The high specificity of therapeutic antibodies combined with their ability to leverage the functions of the immune system provides powerful tools to target a variety of diseases. There are more than 300 antibody therapeutics currently marketed for anti-inflammatory (Antibody Society 2022) anti-infective, anticancer, and other indications. Antibody manufacturing has historically relied upon mammalian cell-based production using cell lines, such as Chinese hamster ovary (CHO), murine myeloma (NS0), or human embryonic kidney (HEK) cells (Moussavou et al. 2015). However, use of mammalian cells requires methods and infrastructure that can be cost prohibitive for many companies, which has created a large interest in alternative hosts for biologics manufacturing. Plant-based manufacturing is of growing interest due to key attributes including (Bio 2014):

1. Improved methods for extremely rapid and efficient expression of mAbs
2. Sophisticated engineering to reproduce humanlike posttranslational modifications necessary for antibody effector functions and half-life (Merlin et al. 2014; Sack et al. 2015; Chen and Davis 2016)
3. Reduced facility and production costs compared with mammalian cell culture methods, in part due to the ability to grow plants without the sterile conditions or bioreactors typical for mammalian cells (Buyel and Fischer 2012; Ecker et al. 2015; Ma et al. 2013; Tusé and McDonald 2014; Walwyn et al. 2015; Nandi et al. 2016; Alam et al. 2018; Mir-Artigues et al. 2019)
4. More rapid and less complex scale-up of plant-based production to meet clinical trial and market demands
5. Improved safety of plant-based production compared with mammalian systems, through avoidance of potential culture contamination by adventitious viral agents and mycoplasma (Barone et al. 2020; Gregersen 2008; Sack et al. 2015)

This chapter illustrates these key advantages while clarifying key features of plant-based manufacturing of mAbs using a case-based assessment of lot-to-lot consistency (Medicago 2022), homogeneity and consistency of protein N-glycosylation, stability of recombinant products, logistics surrounding production systems compared with mammalian-based systems, and opportunity for campaign manufacturing.

7.2 Background

7.2.1 *Examples of Plant-Based Pharmaceuticals*

Over the past 10 years, plants have been successfully used to produce protein-based vaccines and biotherapeutics (Table 7.1). The earliest example of a plant-made pharmaceutical (PMP) was Protalix's Elelyso, a recombinant enzyme for the treatment of Gaucher's disease, which is the first plant-expressed protein therapeutic to receive FDA approval. Protalix has three additional enzyme products in their development pipeline made using their plant-based manufacturing process, ProCellEx (Protalix Biotherapeutics, 2022). Medicago is another company that has demonstrated the ability to utilize plants to produce pharmaceutical products, including Covifenz, which was recently approved by Health Canada for the prevention of COVID-19 (Medicago 2022; see also Chap. 12). While these products are not yet approved for use by the US FDA, they demonstrate the ability of PMPs to progress into development as potential products for the prevention or treatment of human disease.

KBio and ZabBio have developed a variety of plant-produced monoclonal antibodies (mAbs) for Ebola, HSV, HIV, and contraception. KBio and collaborators have initiated trials in more than six indications (Table 7.2), including FDA fast-track approval of the ZMapp plant-derived trivalent mAb cocktail for Ebola and experimental use in Africa for the 2014–2016 outbreak (Centers for Disease Control and Prevention 2019).

7.2.2 *Overview of Manufacturing of Plant-Based Pharmaceuticals*

Plant-based biologics manufacturing typically starts with insertion of the sequence of interest into a plant virus-based expression vectors, which are transfected into *Agrobacterium* and cultured to provide material for plant infiltration. In parallel, plants are seeded and grown to an appropriate size for infiltration. A solution containing the *Agrobacterium* with the desired expression constructs is used to infiltrate

Table 7.1 Clinical stage companies with plant-based manufacturing

Company name	Platform
Protalix	Cell-based *Daucus carota* L. and *Nicotiana tabacum*
Ventria	*Oryza sativa* L.
Medicago	*Nicotiana benthamiana*
KBio/ZabBio	*Nicotiana benthamiana*
Bayer	*Nicotiana benthamiana*

Table 7.2 KBio plant-based biologics with varying regulatory interactions

Company	Product	Latest stage of regulatory interaction
KBio/ZabBio	Trivalent mAb cocktail	US Phase I/II
KBio/ZabBio	MB66	Phase I
KBio/ZabBio	ZB-06	Phase I
University of Louisville/KBio	Griffithsin	Phase I[a]
KBio	Quadrivalent influenza vaccine	Phase I
KBio	COVID-19 vaccine	Phase I

[a] To date, three separate phase I trials have been performed

plants under vacuum. Soon after infiltration, the plant tissues transformed by *Agrobacterium*-mediated insertion express the gene(s) of interest at high levels. After incubation for additional plant growth and expression and accumulation of the gene of interest, the plant biomass is harvested, followed by extraction, purification, and formulation of the antibody of interest. After purification, analytical tests for drug substance release are conducted to verify product identity, purity, safety, and other parameters as required to verify activity and potency (Table 7.4) (Swope et al. 2021). The bulk drug substance may then be stored for future use or directly transferred to drug product manufacturing. While the finer details of the plant-based manufacturing process may vary depending upon the specific biomolecule to be expressed, the general steps outlined are typical for most antibody therapeutics and are compared in Table 7.3 to mammalian cell-based manufacturing processes (Song et al. 2020).

7.2.3 Overview of Expression System Attributes

As shown in Table 7.3, there are a variety of differences between plant- and mammalian cell-based manufacturing processes. Each platform possesses a number positive and negative attributes, such as the following:

- **Yield**: Mammalian cells typically produce upwards of 6 g of antibody per liter of culture, with some published examples as high as 10 g/L. Plants have been shown to produce up to 2 g of antibody per kilogram of biomass. Although a strict comparison of protein production per cell is difficult to calculate, there is some evidence that the larger size of plant cells may allow for greater productivity compared with mammalian cells. Further enhancement of plant cell production through the reduction of intracellular vacuoles via alteration of cell osmolality and genome engineering of plant strains to increase protein expression on a per cell basis may be possible (Song et al. 2020).

Table 7.3 Differentiation of mammalian cell- and plant-based mAb/protein production systems

Characteristics	Mammalian cells (CHO, HEK)	Plants
Protein yield	Up to 10 g/L of culture[a] (LONZA 2020)	Up to 2 g/kg of biomass (Buyel et al. 2017)
Glycosylation	Mammalian N-glycosylation pattern (Buyel et al. 2017)	Plant or mammalian N-glycosylation pattern
Purification	Harvest Filtration Protein A antibody purification Low pH hold Intermediate polish Final polish Buffer exchange/concentration Analysis (Buyel et al. 2017)	Harvest Homogenization Filtration Protein A antibody purification Intermediate polish Final polish Buffer exchange/concentration Analysis (Buyel et al. 2017)
Safety	Typically requires viral clearance	Viral clearance step may not be required
Biological activity	Generally equivalent	
Production duration	90-day cell bank 14–28-day fed batch	4 (Diamos et al. 2020) to 14 (Hiatt and Pauly 2006) days from genes to full-length protein
Campaign manufacturing capability	Multiproduct, changeover is required, cannot turn over within the same day	Multiproduct, turnover is not product specific, within the same day

[a] Higher yields are possible with optimization of cellular culture processes

- **Glycosylation**: Plants exhibit glycan modifications distinct from mammals due to endoplasmic reticulum and Golgi processing, such as the inclusion of β1,2-linked xylose and α1,3-linked fucose (Sack et al. 2015). Interestingly, detailed evaluation of clinical safety and efficacy endpoints indicated that although anti-plant glycan antibodies are produced in the patient, these antibodies do not affect the safety or efficacy of taliglucerase alfa (Elelyso®). This study indicates that anti-plant glycan antibodies are not necessarily detrimental for the therapeutic use of plant-manufactured proteins (Tusé 2011; Rup et al. 2017). However, the use of a combination of genome engineering and transient expression techniques has successfully eliminated plant-specific glycosylation and demonstrated the ability to generate mammalian cell-specific glycosylation (see also Chaps. 3 and 4).
- **Purification**: Unlike mammalian cells, protein extraction from plant cells employs mechanical disruption followed by clarification. Once the clarified plant extract is obtained, the manufacturing process follows traditional GMP schemes similar to mammalian cell-based manufacturing (e.g., affinity, intermediate polish, final polish). However, unlike mammalian production, extended acidic pH treatment and purification to remove endogenous viruses are not required, which further streamlines plant-based manufacturing compared with mammalian cell processes.
- **Safety**: As mentioned above, mammalian cell culture carries the risk of viral contamination that is absent in plant-based manufacturing. Mammalian

Table 7.4 Bulk drug substance testing results for six *Nicotiana*-produced mAbs

mAb	mAb 1	mAb 2	mAb 3	mAb 4	mAb 5	mAb 6	mAb 7	mAb 8
Formulation	Formulation 1	Formulation 2	Formulation 2	Formulation 2	Formulation 1	Formulation 1	Formulation 3	Formulation 4
Appearance—particulates	Clear liquid, no visible particles	Clear liquid, no visible particles	Clear liquid, no visible particles	Clear liquid, no visible particles	Clear liquid, no visible particles	Clear liquid, no visible particles	Clear liquid, no visible particles	Clear liquid, no visible particles
Appearance—color	Matches fluid B	Without color	Without color	Without color	Without color	Without color	Without color	Without color
pH	6.50	6.00	6.00	6.00	6.50	6.50	6.00	5.50
Osmolality (mOsm/kg)	25.0	360.5	367.5	357.5	40.0	29.0	326	349
UV absorbance (mg/mL)	22.0	20.7	20.7	19.8	22.8	21.0	20.0	19.4
IEX	Matches standard	Matches standard	Matches standard	Matches standard	Matches standard	Matches standard	Matches standard	Matches standard
SEC-HPLC	100% Monomer	100% Monomer	100% Monomer	100% Monomer	100% Monomer	93% Monomer	100% Monomer	97% Monomer
SDS-PAGE NR	>99%	98%	99%	99%	>99%	>99%	97%	>99%
SDS-PAGE-R	>99%	98%	98%	98%	>99%	>99%	>99%	>99%
Endotoxin EU/mg	0.01	0.04	0.03	<0.03	2.93	0.54	0.02	<0.03
Bioburden (CFU/mL)	0	<1	<1	<1	0	0	0	0
Residual HCP (ng/mg)	<1.427	<15.6	<15.6	<15.6	<LOD	<LOD	<1.56	<16
Residual protein A (ng/mg)	<1.151	<10	<9	<2	0.50	0.50	<1.25	<0.41
HCDNA (pg/mg)	<0.309	<3.1	<6.2	<3.1	ND	ND	<0.34	<1

(continued)

Table 7.4 (continued)

mAb	mAb 1	mAb 2	mAb 3	mAb 4	mAb 5	mAb 6	mAb 7	mAb 8
Nicotine (μg/mL)	ND	<0.2	<0.2	<0.2	ND	ND	<2	<0.3
Heavy metals	ND	Meets specification	Meets specification	Meets specification	ND	ND	ND	ND

ND not detected

cell-based platforms have experienced 18 incidents of viral contamination since 1985 (Barone et al. 2020; Gregersen 2008; Sack et al. 2015). Due to the increased availability of plant-based manufacturing platforms, the World Health Organization (WHO) recently released (Medicago 2022), on October 12th, 2021, a guideline for the safe production and quality control of monoclonal antibodies for use in humans, which has specific areas focused on plant-based manufacturing considerations.

- **Biological Activity**: While there have been individual examples of protein activity being greater or lesser depending upon the cell type used for expression, there is no evidence of clear superiority of mammalian cells over plant cells. Several studies have demonstrated equivalent function of a variety of proteins expressed in each system, supporting the broad potential for plant-based production (Bardor et al. 2003; Dubald et al. 2009).
- **Production Timeline**: From sequence identification to a full-length protein generation, mammalian and plant manufacturing processes differ, ranging from 14 to 28 days or 4 to 14 days, respectively. However, this does not account for cell banking, which is required for mammalian cell manufacturing and causes an increase in time to GMP product by approximately 6 months. As will be shown in this chapter, the use of established Master Seed Banks for *Nicotiana* plants coupled with transient protein expression system allows gene to cGMP-ready production in as little as 3 months.

Campaign Manufacturing: Plant-based manufacturing requires growth spaces that are much less complex and costly than production using mammalian cell culture. Plant growth spaces do not require sterile conditions, and the environmental (light, temperature, humidity, etc.) controls are much less stringent than those necessary for mammalian cells. In addition, mammalian cells are typically grown using fed-batch approaches, where batches are typically a single continuous bioreactor process, whereas plant-based manufactured batches can include multiple infiltrations over a set period of time. For example, KBio's plant manufacturing campaign can consist of weekly production of 0.25–1 kg of product with current infrastructure, followed by lot blending to produce a final product for clinical testing or product launch.

7.3 Plant-Specific Expression System Characteristics

7.3.1 Manufacturing Upstream Process

In order to have a reproducible process, a controlled environment is required to ensure healthy and consistent plant growth. Thus, plants are grown indoors, without ground contact, to enable the greatest level of temperature, humidity, lighting, and other environmental factors and to limit the exposure of external elements and adventitious agents.

The upstream process (USP) in Fig. 7.1 begins with seeding to start the plant growth process. Seeds from a qualified seed bank are placed on top of soilless growth media and provided with water, fertilizer, and light to initiate germination. After the initial seeding, preparation of the *Agrobacterium tumefaciens* vectors used to express the homologous proteins begins. The gene of interest is inserted into a plant viral vector, such as *potato virus X* (PVX) or *turnip vein clearing virus* (TVCV), using typical molecular biological protocols (see also chapter "Plant Viral Vectors: Important Tools for Biologics Production"). In the case of antibodies, vectors are designed to separately express either a heavy or a light chain, using a codon-optimized version of the desired antibody sequence. The two vectors are transfected into the *Agrobacterium* via electroporation, and the resulting transformed *Agrobacterium* is grown in liquid culture medium to a volume appropriate for the number of plants to be infiltrated. The plants seeded and grown in parallel are

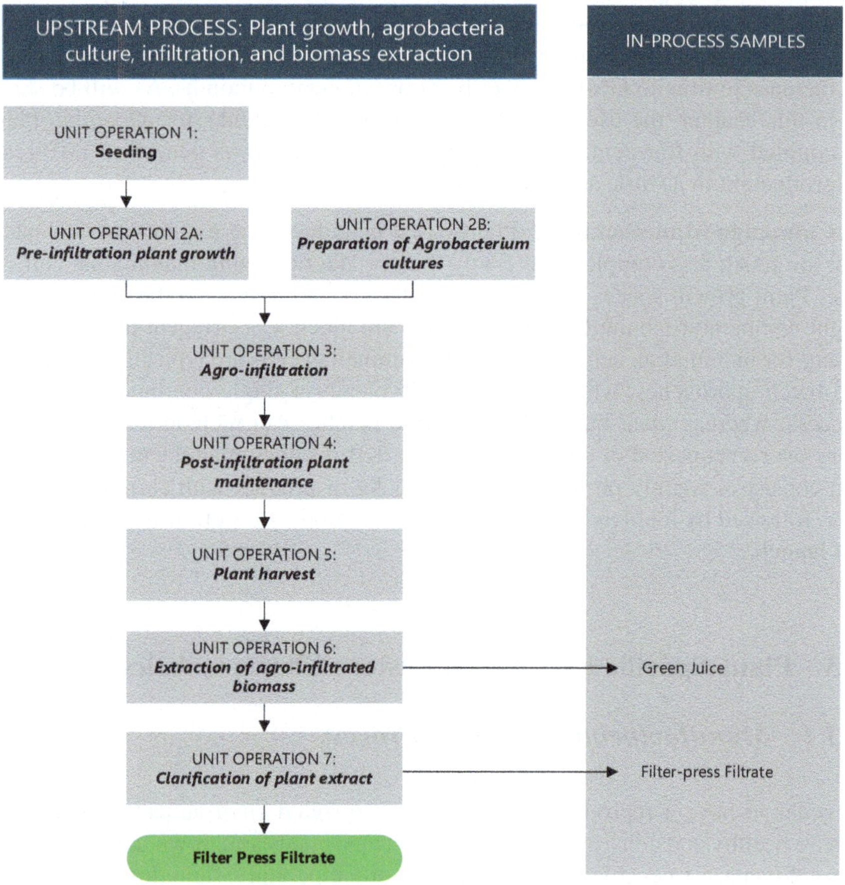

Fig. 7.1 Upstream plant-based manufacturing process for monoclonal antibody production (Swope et al. 2021)

inverted inside of a plant infiltration chamber filled with the *Agrobacterium* culture. A vacuum is applied while the aerial portions of the plants are submerged in culture media, which removes air from the stomata and into the solution. Upon release of the vacuum, the infiltration solution is forced into the plant cells through the stomata, taking the place of the previously expelled air. At the completion of infiltration, the plants are removed from the chamber, allowed to remain inverted to dry, set upright to recover, and grown for another 6–8 days to maximize antibody expression. The presence of cell walls and insoluble matrix materials in plants necessitates the substantially different purification methods compared with mammalian cells (Schillberg et al. 2019). Plants are harvested, and the biomass is extracted using a mechanical disintegrator to rupture the cells in the presence of solution containing antioxidants and chelating agents. The fiber and cellulose are then separated from the liquid using a hydraulic screw press. The liquid extract is collected, and clarification is done using either a plate and frame depth filter in combination with a filter aid or a continuous centrifugation. The filtrate and supernatant can be sampled at this point for initial analysis and further purification.

7.3.2 *Manufacturing Downstream Process*

Upon completion of the upstream process, the manufacturing process is continued within an ISO7 clean room suite, which constitutes the GMP process for plant-based manufacturing. Initial capture chromatography, polishing, and formulation of plant-based extract are substantially similar to that of mammalian cell-based processes (Schillberg et al. 2019).

The downstream process (DSP), shown in Fig. 7.2, begins with antibody capture using protein A affinity chromatography. The protein A-linked resin binds a conserved epitope in the Fc region of the antibody heavy chain and is common to most antibody purification processes. The KBio manufacturing process includes further purification by anion exchange (AEX) chromatography and multimodal chromatography to separate monomeric protein from other product-related impurities. Finally, the material is formulated according to application-specific needs using a tangential flow filtration (TFF) system. Excipients, such as polysorbate, may be added post-diafiltration to increase the stability of a molecule. This results in the final bulk drug substance product.

7.3.3 *Campaign Manufacturing*

Campaign manufacturing is defined as the manufacture of product batches in parallel or in temporal series. Plant-based campaign manufacturing may have advantages compared with mammalian cells including more efficient changeover between different products and lack of cleaning validation requirements, potentially allowing

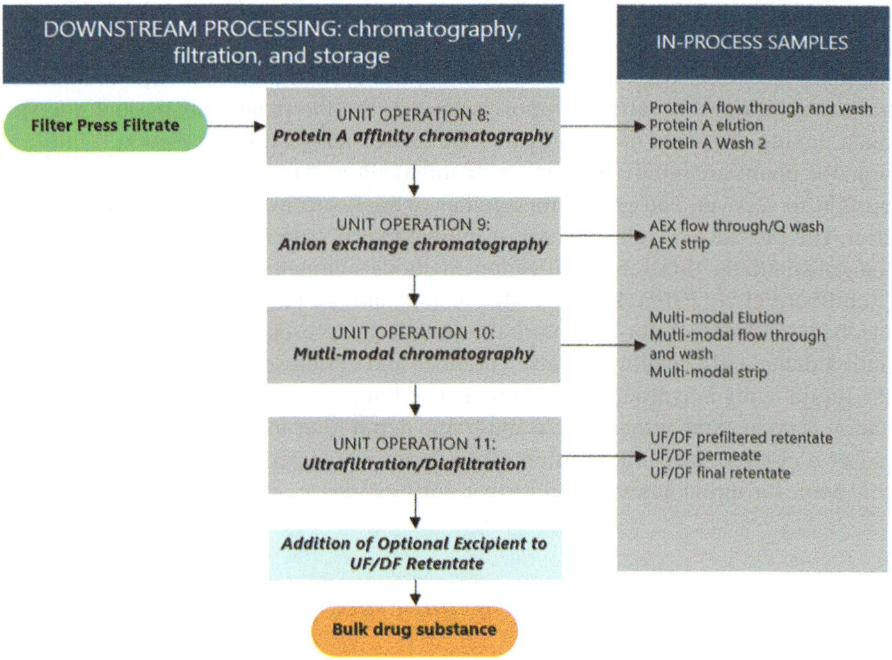

Fig. 7.2 Downstream manufacturing process for plant-based monoclonal antibody production

for more rapid production of more and different antibody products within the same facility.

Campaign manufacturing has been demonstrated with the production of DP2, a trivalent mAb cocktail, via the processes outlined in Figs. 7.1 and 7.2. Approximately 800 kg of biomass was grown and infiltrated with one of the three target antibodies (Strasser et al. 2008). Additional harvests were scheduled for lower expressing antibodies and were considered "sub-lots," which were pooled at unit operation 11: ultrafiltration/diafiltration (Fig. 7.2). This created individual antibody bulk drug substances, which were characterized individually as described in Sect. 7.3.1.

To further evaluate the lot-to-lot consistency of antibodies produced using their plant platform, KBio carried out a number of detailed tests for two DP2 lots, each comprised of two lots each of mAb2, mAb3, and mAb4 (Tables 7.4 and 7.6). Two tests illustrate the consistency in product manufacturing: glycan analysis and potency measurements using Biacore. Lot-to-lot consistency of a single antibody in the DP2 trivalent cocktail has been previously described.

KBio has demonstrated campaign manufacturing via the production of bulk drug substance for eight antibodies. Specific results for each antibody are presented in Table 7.4, detailing release test outcomes for the bulk drug substance. Key comparative outcomes include the absence of multimeric antibody aggregates and protein purity by various tests of >98%. Furthermore, process-related impurities such as endotoxin, host cell DNA, host cell protein, small molecules, and bioburden are uniformly low and meet specifications. All bulk drug substance lots passed release criteria as specified in the Certificate of Analysis (Table 7.4) and could be formulated for drug product release depending on the route of administration, infusion (formulation 2), or vaginal film (formulation 1) (Table 7.4). One difference of note is the osmolality, which is due to difference in buffer formulation and presence of sodium chloride for the infused forms of formulations 2, 3, and 4 (Table 7.4).

Critical to campaign manufacturing of products for clinical use are methods to effectively test lot-to-lot purity, potency, activity, and safety of products. The key analytical tests described here are generally applicable to all antibodies, with the exception of potency and activity testing that is molecule dependent. These tests are described in Table 7.4.

Several tests provide general information concerning the visible stability and key in-solution characteristics of the product. These include visible appearance, pH, and osmolality. Product quantity in formulation is measured by UV absorbance to determine protein content, reported in milligrams of product per milliliters of solution.

Product identity is established by ion-exchange chromatography (IEX), which allows separation of native proteins, in the absence of solvents, based on net surface charge. Each antibody exhibits a characteristic chromatogram that allows comparison with a characterized standard for confirmation of identity.

Purity is determined by several methods, including separation by reducing or nonreducing sodium dodecyl sulfate-polyacrylamide gel electrophoresis (SDS-PAGE) and Coomassie Brilliant Blue staining. These separations allow quantitation of antibody aggregates, free heavy and light chain, as well as other host cell protein impurities. Size-exclusion high-pressure liquid chromatography (SEC-HPLC) provides size-based separation of purified antibody product, fragments, and impurities for measurement using orthogonal testing approaches.

The safety of the antibody products is principally determined by measuring the amount of endotoxin and bioburden using standard U.S. Pharmacopeia (U.S.P.) methods. Key process impurities are also measured, including residual protein A from the purification process, host cell protein and DNA levels, and several small-molecule impurities. Potency testing is evaluated by individually developed enzyme-linked immunoassay (ELISA) methods, with specific ligands for each antibody. Criteria for acceptance of each test method are summarized in the Certificate of Analysis Table (Table 7.5).

7.3.4 Glycosylation Patterns

Glycosylation is the addition of sugar moieties to proteins as part of the posttranslational modification process of the endoplasmic reticulum and Golgi apparatus. Glycosylation plays an important role in processes such as protein stability and antibody-mediated immune effector functions. Glycosylation can be categorized based on the sugar-peptide bond and the oligosaccharide attached, which are often termed N-, O-, or C-linked depending upon the specific residue linkage (Gomord

Table 7.5 Model Certificate of Analysis for *Nicotiana*-produced mAb

Test parameters	Test method	Target specification
General	Appearance—particulates	Clear liquid, no visible particulates
General	Appearance—color	Without color
General	pH	6.5 ± 0.5
General	Osmolality mOsm/kg	10–100 mOsm/kg
Protein concentration	UV absorbance mg/mL	15–25 mg/mL
Identity	IEX	Chromatogram conforms to standard
Purity	SEC-HPLC	>90% monomer; <5% high-molecular-weight species
Purity	SDS-PAGE nonreducing	>90% main band
Purity	SDS-PAGE reducing	>95% monomer (sum of heavy and light chains)
Safety	Endotoxin EU/mg USP <85>	<5 EU/mg
Safety	Bioburden CFU/mL USP <61>	<1 CFU/mL
Impurities	Residual host cell protein ng/mg	<100 ng/mg
Impurities	Residual protein A ng/mg	<100 ng/mg
Impurities	Host cell DNA pg/mg	<10 pg/mg
Impurities	Nicotine μg/mL	Perform and report
Impurities	Heavy metals	Perform and report
Potency	Potency ELISA	100 ± 50% of reference

et al. 2010). N-linked glycosylation is one of the best characterized and critical modifications of antibodies; however, O-linked and other glycosylation types have been detected.

Glycosylation in the Fc region of antibody heavy chains can significantly affect the pharmacokinetics (PK) and pharmacodynamics (PD) of the molecule in vivo (Liu 2015). In addition, the presence or absence of specific glycans directs antibody binding to Fc receptors found on various cells in the body, generating specific cellular responses or "effector functions" (Liu 2015). For example, the lack of fucosylation in the Fc region leads to enhanced antibody-dependent cell-mediated cytotoxicity (ADCC), which can be highly desirable for therapeutic antibodies designed to induce and target cell clearance in oncology. Furthermore, aberrant glycosylation may result in poorly active or immunogenic products (Fig. 7.3).

Plant-specific protein glycosylation includes two glycans not found in mammalian cells, α1,3-fucose and β1,2-xylose, which can be immunogenic in patients (Bakker and Jan 2010). The specific elimination of these glycans can be accomplished by knockdown or knockout of their cognate xylose and fucosyl transferase genes (see Fig. 7.3). These genetic ablations have shown to reduce or eliminate the inclusion of plant-specific glycans and thus reduce the immunogenicity of plant-produced molecules in mammals. While these methods have reduced or eliminated

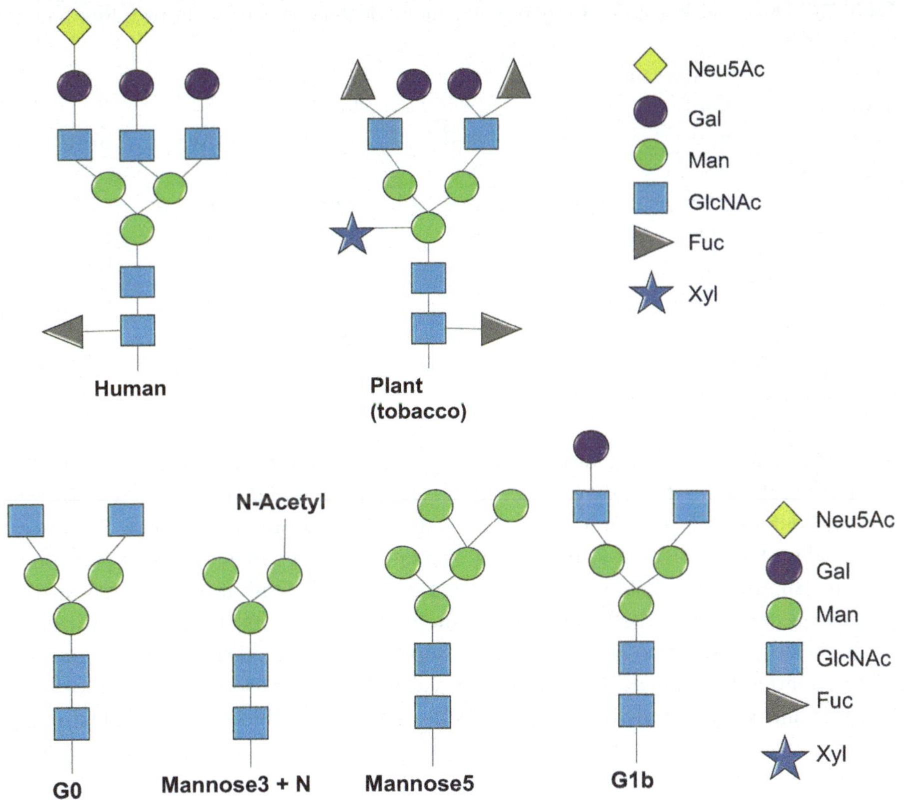

Fig. 7.3 Difference between native human glycans and those deriving from plants, such as *Nicotiana*

plant-specific glycosylation, they do not address the need for recreation of mammalian-specific glycosylation, which may be desirable for antibody function. The introduction of mammalian pathways for N-glycosylation (Gomord et al. 2010) has been demonstrated in plant cells (Ramírez-Alanis et al. 2018), including the improvement of target amino acid specificity (Larsen et al. 2020; Margolin et al. 2021).

KBio has previously used *Nicotiana benthamiana* transgenic lines with RNAi-based knockdown of fucosyl- and xylosyl-transferases. These seed lines exhibit very low accumulation of plant-specific xylose and fucose linkages, leaving the core human "G0" glycosylation patterns. Table 7.6 shows an example of characterization of N-linked glycosylation patterns present on each DS (antibody) component of DP2 for two separate DP lots. The primary glycoforms present on each antibody in both lots was the G0 form, or GlcNac(2)-Man(3)-GlcNac(2). mAb4 and mAb2 both show >90% abundance of this G0 form, whereas mAb3 shows >80% abundance. High-mannose structures, varying from Man5 to Man9, are also observed. Man9 is a non-processed endoplasmic reticulum (ER) glycan addition, and each truncated

Table 7.6 Quantitative analysis of glycan structures for three antibodies in two drug product (DP2) lots

	% Abundance			% Abundance		
Lot	Lot 1			Lot 2		
Antibody	mAb4	mAb3	mAb2	mAb4	mAb3	mAb2
Man3 + N	0.8	0.6	0.7	1.2	0.7	0.7
G0	92.5	83.6	93.9	90.3	80.7	91.1
Man5	BQL	0.6	BQL	1.1	0.8	BQL
G1b	1.1	0.7	0.6	1.2	0.7	0.6
Man7	0.6	1.9	0.6	0.7	2.3	0.7
Man8	1.8	5.9	2.2	1.3	6.7	2.7
Man9	1.9	4.8	2.0	1.2	5.4	2.5

BQL below level of quantitation

form represents incomplete enzymatic processing, generally thought to be affected by steric factors. mAb4 showed <5% high mannose abundance, and mAb2 showed <6% abundance of incompletely processed high-mannose forms. Lot 2 showed higher abundance, <16%, of various high-mannose chains. In each case, variation in glycan structure abundance varied less than 2% overall, with a high degree of glycosylation consistency between two distinct production lots of the three different antibodies. In addition to KBio, various groups have produced recombinant proteins using genetically modified *Nicotiana* lines exhibiting knockdown of the majority of fucosyl- and xylosyl-transferase activity that appears suitable for clinical testing.

7.3.5 *Analytical Analysis*

7.3.5.1 Additional Analytics

We further characterized two lots of a single antibody, mAb2, for the following properties:

1. **Tryptic peptide mapping**
2. **Free cysteine presence**
3. **LC-MS molecular weight determination**
4. **Isoelectric focusing profile**
5. **Disulfide mapping**

Each lot of mAb2 was subjected to reducing conditions and tryptic digestion, followed by LC-MS separation and molecular weight analysis. Weights were compared with deduced amino acid sequence from heavy and light chains for identity confirmation. Each lot showed >96% coverage of predicted tryptic peptides. Sequences confirmed deduced amino acid composition completely matched with the peptides obtained. Peptides lacking in analysis were held in common between both lots.

Using sequential Lys-C and trypsin digestion, non-reduced samples from each lot of mAb2 were analyzed by LC-MS for disulfide bond occupancy. All expected disulfide linkages were identified in each sample. Some expected disulfide-bonded peptides were found in multiple variants due to incomplete proteolytic cleavages during digestion. Scrambled disulfide-linked peptides are comparable to the most abundant of the possible corresponding non-scrambled disulfide-linked peptides. The relative abundance of each scrambled disulfide-bonded peptide was determined relative to the sum of the scrambled and the most abundant corresponding non-scrambled disulfide-linked peptides. Minor levels (~1.5%) of scrambled disulfide-linked peptides were observed.

To determine free cysteine composition, non-reduced samples from each lot of mAb2 were subjected to alkylation prior to sequential Lys-C and trypsin digestion. This treatment is expected to alkylate any free cysteines, resulting in a unique residue, carbamidomethyl-Cys (CAM-Cys), which is detected by LC-MS analysis. Less than a 2.7% difference in free cysteines was detected at each potential site when comparing both lots, showing high-fidelity protein folding in plant-expressed molecules.

The intact lots of mAb2 were subjected to LC-MS as non-reduced samples, to view molecular weight patterns of the intact drug substance. In Fig. 7.4, the full mass spectrum and zoomed views of the main peaks are shown for both protein lots.

The results show virtually identical major peak molecular weights for monomeric and dimeric forms of HC-LC heterodimers ranging from 144,429 to 155,788 Da and 298,967 to 297,757 Da, respectively. Variation in glycoforms correlates with the observed molecular weight differences. Some free heavy chain at MW 73,657–75,167 Da is also observed. (Note that the detection of free chains is not possible using SDS-PAGE and Coomassie Brilliant Blue staining.)

Charge heterogeneity in two lots of mAb2 was determined by imaged capillary isoelectric focusing (icIEF) and was performed using a ProteinSimple iCE3 system. Major charged forms were virtually identical with primary peaks identified at p*I*'s of 8.3, 8.4, and 8.5. Some variations in minor forms between p*I* of 8.1 and 8.28 were observed. Overall, these data show high conservation in protein charge of different drug substance lots (Fig. 7.5).

Taken together, these results show a high degree of consistency between lots 1 and 2 of mAb2 manufactured at KBio including glycosylation, amino acid composition, disulfide mapping, intact protein molecular weight, and surface charge retention. Thus, the plant production systems are comparable to mammalian cell-based platforms in their ability to produce consistent lots of therapeutic molecules (Liu et al. 2016; Sifniotis et al. 2019).

7.4 Transition from Drug Substance to Final DP

Bulk drug product (DP) is produced from the bulk drug substance (DS) and takes into account concentration, delivery, formulation, and application. In the case of DP2, three individual antibodies (drug substances: mAb2, mAb3, mAb4, Table 7.3)

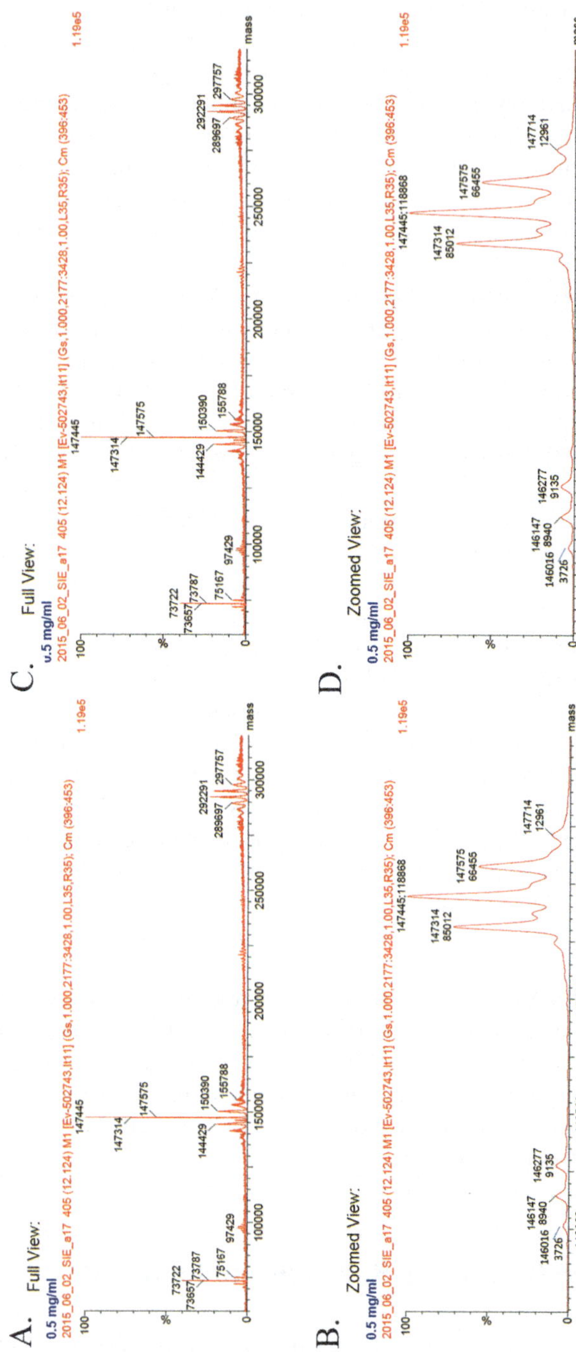

Fig. 7.4 Intact protein molecular weight of mAb2 by LC-MS, full scan in Panels **a** and **c**, and zoomed view in Panels **b** and **d**

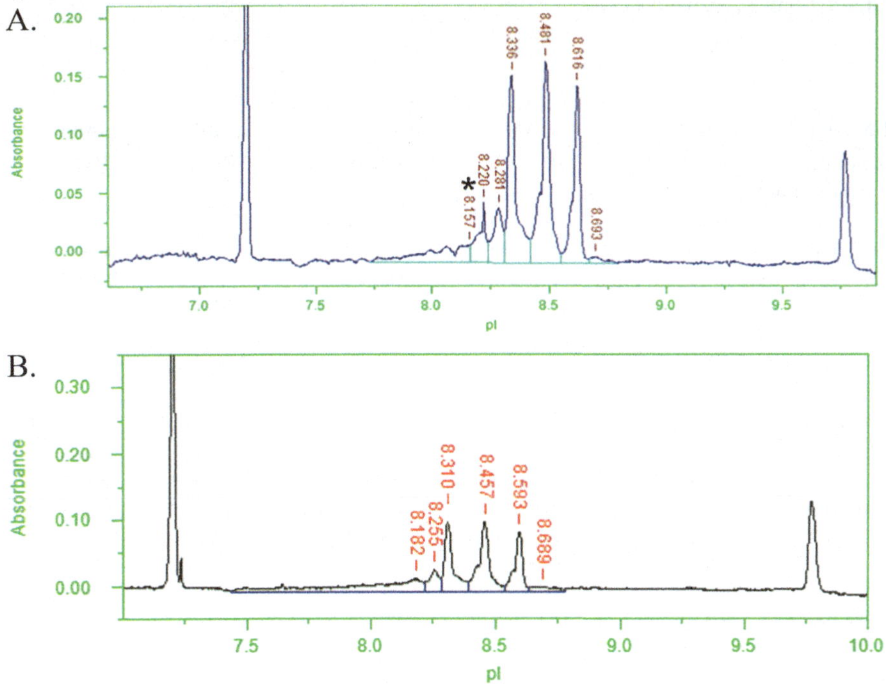

Fig. 7.5 icICF analysis of mAb2 drug substance lots

with the same formulation were compounded into one trivalent cocktail (drug product) containing equal ratios of each antibody. In the case of DP1, a single antibody (drug substance: mAb1, Table 7.3) was ultraconcentrated and formulated into a polymer-based dissolvable vaginal film (drug product) as a human contraceptive. In the case of DP3, two antibodies (drug substances: mAb5, mAb6, Table 7.4) were each ultraconcentrated and then combined into a polymer-based dissolvable vaginal film delivering equal amounts of each anti-sexually transmitted disease antibody (drug product). DP1 and DP3 followed the same formulation into a vaginal film. DP4 and DP5 are each made up of a single antibody (drug substances: mAb7 and mAb8, respectively, Table 7.3). For DP4 and DP5, bulk drug substance is stored at ultralow temperature for long-term storage. Upon need, bulk DS is removed from storage, distributed into vials for administration, and held at common storage temperature as drug product. The drug products included in Table 7.7 are examples produced by KBio demonstrating the flexibility of formulation and route of delivery of plant-produced antibodies. In addition to intravenous administration and dissolvable vaginal films, other viable routes for delivery of PMPs under evaluation include intramuscular (Medicago, Covifenz) and oral (Ventria, VEN120) (Politch et al. 2021). KBio has carried out a series of stability studies of the drug products in Table 7.7, demonstrating stability over 12 and in some cases in excess of 12 months.

Table 7.7 DP quality characteristics

DP name	DP1	DP2	DP3	DP4	DP5
DS name	mAb1	mAb2, mAb3, and mAb4	mAb5 and mAb6	mAb7	mAb8
Route of administration	Vaginal film	IV	Vaginal film	IV	IV
Appearance—particulates	Homogenous, opaque, square piece of film	Clear liquid, no visible particles	Homogenous, opaque, square piece of film	Clear liquid, no visible particles	Clear liquid, no visible particles
pH	6.0	5.8	5.3	6.0	5.50
Osmolality (mOsm/kg)	42	367	45	321	348
UV absorbance	22 mg/film	23 mg/mL	20 mg/film	20 mg/mL	19 mg/mL
Identity by IEX	Matches standard	Matches standard	Matches standard	Matches standard	Matches standard
SEC-HPLC	96% Monomer, 4% aggregate	97.5% Monomer, 1.3% aggregate	96% Monomer, 2% aggregate	100% Monomer	97% Monomer, 1% aggregate
SDS-PAGE (NR)	>99%	>99%	>99%	>99%	>99%
SDS-PAGE (R)	>99%	>99%	>99%	>99%	>99%
Endotoxin	<26 EU/film	0.21 EU/mg	<26 EU/film	0.03 EU/mg	<0.3 EU/mg
Bioburden (CFU/mL)	<1	<1	<1	<1	<1[a]

[a] Based on DS results. DP5 was tested for sterility and passed

Despite slight increased aggregation over time, the overall purity continues to fall within the acceptable ranges for DP1 vaginal films after 26 months stored at 2–8 °C. DP3 vaginal films were also found to be stable at 2–8 °C, but stability testing was concluded after 12 months. DP2 trivalent cocktail was found to be stable for 24 months stored at −20 °C. DP4, a single antibody therapeutic, was stable for 6 months at 2–8 °C, and DP5, also a single antibody therapeutic, was found to be stable for 18 months at −20 °C as shown in Table 7.8. Further, the high consistency from lot to lot (shown here and in Swope et al. (2022)) allows for multiple lot blending and other approaches to ensure that sufficient drug is available for both development and clinical testing.

The consistency in manufacturing conditions and the high quality of product emerging from the KBio system allow for dramatic truncation in development times for biotherapeutic products (Table 7.9), as shown by the production of the ZMapp trivalent antibody product (Swope et al. 2022). In response to the West Africa Ebola outbreak that began in December 2013, three candidate antibodies were obtained, and within 1 month, small quantities of each antibody were produced to support a pilot study in nonhuman primates. Initial experiments demonstrated a cocktail of three antibodies that protected rhesus macaques from lethal challenge when

Table 7.8 Drug product stability

Product	Final time point (months)	Storage temperature (°C)	Initial % monomer	Initial % aggregate	Final % monomer	Final % aggregate
DP1	24	2–8	96	4	96	4
DP2	26	−20	99	1	97	2
DP3	12	2–8	96	2	94	3
DP4	6	2–8	100	0	98	2
DP5	18	−20	97	<1	95	2

Note: Some stability programs are ongoing

Table 7.9 KBio-manufactured antibodies in preclinical and clinical phases

Monoclonal antibody	Final drug product formulation	Purpose	Phase	Clinical trial ID
DP1	Vaginal film	Contraception	Clinical phase 1	NCT04731818
DP2	I.V. delivery	Ebola treatment	Clinical phase 1/2	NCT02363322
DP3	Vaginal film	HSV treatment	Clinical phase 1	NCT02579083
DP4	I.V. delivery	SARS-CoV-19	GLP Pharm/ Tox	N/A
DP5	I.V. delivery	Enterovirus 68	GLP Pharm/ Tox	N/A

administered 3, 4, or 5 days postinfection (Qiu et al. 2014). This cocktail was named ZMapp and provided to patients through expanded access/compassionate use in August of 2014, just 8 months after the antibody sequences were provided to KBio. ZMapp was tested under investigational new drug (IND) clinical trials beginning in February 2015, only 12 months from the first purified protein product at R&D scale.

7.5 The Plant Advantage: Speed to IND

Speed to the clinic is a significant advantage offered by plant-based manufacturing. Aside from the time-critical needs of pandemic responses, rapid assessment of candidate therapeutics in early-stage clinical trials improves the speed of decision-making and provides a competitive advantage for biopharmaceutical companies (Bolisetty et al. 2020; Sun et al. 2022).

The timelines in Fig. 7.6 show typical durations of an ideal manufacturing process from identification of a therapeutic protein sequence to IND submission for both plant- and mammalian-based production systems. The more rapid production of GMP materials using plant-based manufacturing not only supports more rapid entry into the clinic, but also reduces the timeline and costs compared to

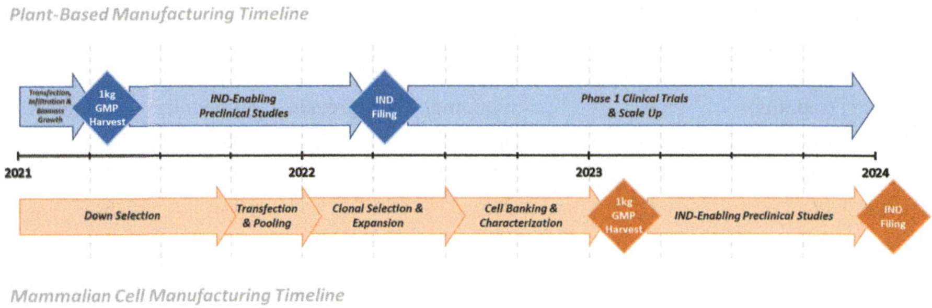

Fig. 7.6 Timeline from antibody sequence identification to IND

mammalian cell-based production. For example, mammalian manufacturing processes require roughly 18 months for cell expression selection and an additional 6 months of time for cell banking, whereas plant-based systems require as little as 3 months of time from initial screening to production of material for nonclinical studies. Furthermore, financial modeling suggests that plants may be superior to other platforms when comparing manufacturing costs as a function of regulatory process and and product overall costs (Farid et al. 2020; Scott 2020).

7.6 Conclusions

Production of biologics using plants has evolved over the past decade, resulting in commercialization of therapeutics for different indications in different countries (Tables 7.1 and 7.9). Ongoing research with plant-based manufacturing platforms is demonstrating improvements in yield, mammalian-like posttranslational modifications, and efficacy. As discussed here, plant-based GMP manufacturing of antibodies has a significant speed advantage, with lower capital investment, compared to traditional mammalian cell culture-based platforms, creating products with equivalent physical and functional characteristics compared with those from mammalian cells. Multiple plant-made therapeutics have been studied in human clinical trials and been shown to be safe and efficacious. Recent WHO guidance includes considerations for plant-produced therapeutics, indicating the growing acceptance of plant-based manufacturing by the global regulatory community. As more clinical data is generated and additional products are approved for the market, plant production systems will likely play a significant role in the future of biotherapeutics.

Addendum

Pogue, G.P., Swope, K. Rininger, J., Schoukroun-Barnes, L. Morton, J. Hume, S, Hamorsky, K., Fuqua, J., Royal, J.M., Pauly, M.H., Brennan, M., Zeitlin, L., Whaley, K., Stevens, S., Bratcher, B. 2023. *Plant-Based Antibody Manufacturing*. In: *Tools*

& Techniques of Plant Molecular Farming. Springer Nature Publishing. Kole, C., Chaurasia, A., Hefferon, K.L., Panigrahi, J. (eds). 209-233; https://doi.org/10.1007/978-981-99-4859-8_7.

Disclosure of Interest with Commercial Company

All authors acknowledge financial interest in KBio, Ltd. KBio produces recombinant therapeutic proteins using plant systems and thus may potentially benefit financially from the research findings presented here.

Addendum

Today, biologics are typically produced in prokaryotic and/or eukaryotic cell-based systems through recombinant DNA technologies[1]. The cell bank expansion process coupled with long culture cycles allowing multiple harvest of particularly secreted products puts the cultures of cells at risk from adventitious agent contamination. Principal threats are viruses, bacteria, and mycoplasma. The impact of contaminants includes poor production yields due to premature plateauing in growth density, health risks from toxin impurities, and risk of human infection. What is the risk of adventitious agents or contaminating pathogens in plant-based biomanufacturing?

The authors can only speak with authority concerning the system described in this article with regard to risk and mitigation methods. The primary distinguishing concerns in our plant-based production system are plant pathogens and culture contaminants. Human viral and bacterial pathogens have limited ability to be harbored in plants, and no studies have demonstrated propagation of human pathogenic agents using plants as an intermediate host. Producing recombinant proteins in intact plants employs an open system during the plant growth phase that can risk exposure to environmental microbes. As with all protein expression systems, there are risks that certain contaminates may produce metabolites, for example, toxins, that could be problematic to humans.

[1]Barone, P.W., Wiebe, M.E., Leung, J.C. *et al.* Viral contamination in biologic manufacture and implications for emerging therapies. *Nat Biotechnol* **38**, 563–572 (2020). https://doi.org/10.1038/s41587-020-0507-2; A. Zahn, J., & C. Halter, M. (2020). Surveillance and Elimination of Bacteriophage Contamination in an Industrial Fermentation Process. IntechOpen. https://doi.org/10.5772/intechopen.81151. Malik, P., Mukherjee, S., Mukherjee, T.K. (2023). Microbial Contamination of Mammalian Cell Culture. In: Mukherjee, T.K., Malik, P., Mukherjee, S. (eds) Practical Approach to Mammalian Cell and Organ Culture. Springer, Singapore. https://doi.org/10.1007/978-981-19-1731-8_5-1.

In this manuscript, we describe the consistency of mAbs, in this case the three mAb ZMapp cocktail, produced by transient expression in *Nicotiana* plants. The exigencies of the Ebola epidemic of 2014 necessitated the KBio research and development (R&D) facility to operate beyond the intended design by supporting sustained cGMP production intensity at near commercial capacity for a period of over 18 months, yielding 70+ ZMapp lots. Such usage over an extended period allowed all aspects of the R&D manufacturing system, facility, and utilities to be stress tested. Throughout the over year and a half ZMapp campaign, two contamination events were noted in the upstream plant growth area. The contamination events were identified through visual inspection as the R&D facility did not have routine environmental or water monitoring of plant growth systems at that time. Root cause analysis was performed, and additional mitigation strategies were implemented to improve the plant growth environmental systems, including enhanced water monitoring programs, redesign of irrigation and fertigation systems, HVAC updates, and move to an abiotic plant growth media.

The ZMapp product purified from plants grown during the environmental contamination was subjected to extensive testing for presence of possible microbial excretion products. Using validated assays, no evidence of microbial products or toxins were noted in associated DS or DP. Further, comparing data with other ZMapp lots indicated no significant deviation in values obtained from relevant identity, purity, potency, and safety release assays[2]. These data, coupled with results presented in this chapter demonstrating highly consistent glycosylation, protein charge, and other highly derivative product properties, support the contention that the mAb production system described in this chapter has exhibited excellent separation capabilities, producing products free from impurities, even when challenged by environmental contamination. The U.S. FDA vetted the updated plant growth and monitoring systems and product derived under these conditions in recent Drug Master File submissions. Further, the U.S. FDA has supported subsequent clinical investigation of KBio produced mAb lots.

References

Alam A, Jiang L, Kittleson GA, Steadman KD, Nandi S, Fuqua JL et al (2018) Technoeconomic modeling of plant-based Griffithsin manufacturing. Front Bioeng Biotechnol 6:1–13

Antibody Society (2022) Antibody therapeutics approved or in regulatory review in the EU or US - The Antibody Society. https://www.antibodysociety.org/resources/approved-antibodies/. Accessed 8 Apr 2023

Bakker HAC, Jan HB (2010) Mammalian-type glycosylation in plants. Glycobiology 1:1–22

[2] Swope K, Morton J, Pogue GP, Burden L, Partain N, Hume S, Shepherd J, Simpson CA, Brennan MB, Furman TC, Kingrey-Gebe S, Martinez T, McDonough J, Pauly MH, Whaley KJ, Zeitlin L, Bratcher B, Haydon H. Reproducibility and flexibility of monoclonal antibody production with Nicotiana benthamiana. MAbs. 2022 Jan-Dec;14(1):2013594. https://doi.org/10.1080/19420862.2021.2013594. and Kbio, unpublished data.

Bardor M, Faveeuw C, Fitchette AC, Gilbert D, Galas L, Trottein F et al (2003) Immunoreactivity in mammals of two typical plant glyco-epitopes, core α(1,3)-fucose and core xylose. Glycobiology 13:427–434

Barone PW, Wiebe ME, Leung JC, Hussein ITM, Keumurian FJ, Bouressa J et al (2020) Viral contamination in biologic manufacture and implications for emerging therapies. Nat Biotechnol 38:563–572

Bio (2014) 4 benefits of plant-made pharmaceuticals more on this topic. Biotechnology Innovation Organization, pp 1–2

Bolisetty P, Tremml G, Xu S, Khetan A (2020) Enabling speed to clinic for monoclonal antibody programs using a pool of clones for IND-enabling toxicity studies. mAbs 12:1–8

Buyel JF, Fischer R (2012) Predictive models for transient protein expression in tobacco (*Nicotiana tabacum L.*) can optimize process time, yield, and downstream costs. Biotechnol Bioeng 109:2–4

Buyel JF, Twyman RM, Fischer R (2017) Very-large-scale production of antibodies in plants: the biologization of manufacturing. Biotechnol Adv 35:458–465

Centers for Disease Control and Prevention (2019) Years of Ebola virus disease outbreaks - cases and outbreaks of EVD by year. Cdc.Gov, pp 1–33

Chen Q, Davis KR (2016) The potential of plants as a system for the development and production of human biologics. F1000Res 5:1–8

Diamos AG, Hunter JGL, Pardhe MD, Rosenthal SH, Sun H, Foster BC et al (2020) High level production of monoclonal antibodies using an optimized plant expression system. Front Bioeng Biotechnol 7:1–15

Dubald M, Cropscience B, Peters J (2009) A case study for plant-made pharmaceuticals comparing different plant expression and production systems. Methods Mol Biol 483:209–221

Ecker DM, Jones SD, Levine HL (2015) The therapeutic monoclonal antibody market. mAbs 7:9–14

Farid SS, Baron M, Stamatis C, Nie W, Coffman J (2020) Benchmarking biopharmaceutical process development and manufacturing cost contributions to R&D. mAbs 12:1754999

Gomord V, Fitchette AC, Menu-Bouaouiche L, Saint-Jore-Dupas C, Plasson C, Michaud D et al (2010) Plant-specific glycosylation patterns in the context of therapeutic protein production. Plant Biotechnol J 8:564–587

Gregersen JP (2008) A risk-assessment model to rate the occurrence and relevance of adventitious agents in the production of influenza vaccines. Vaccine 26:3297–3304

Hiatt A, Pauly M (2006) Monoclonal antibodies from plants: a new speed record. Proc Natl Acad Sci U S A 103:14645–14646

KBP-201 COVID-19 Vaccine Trial in Healthy Volunteers (2020)

Larsen JS, Karlsson RTG, Tian W, Schulz MA, Matthes A, Clausen H et al (2020) Engineering mammalian cells to produce plant-specific N-glycosylation on proteins. Glycobiology 30:528–538

Liu L (2015) Antibody glycosylation and its impact on the pharmacokinetics and pharmacodynamics of monoclonal antibodies and Fc-fusion proteins. J Pharm Sci 104:1866–1884

Liu J, Eris T, Li C, Cao S, Kuhns S (2016) Assessing analytical similarity of proposed Amgen biosimilar ABP 501 to Adalimumab. BioDrugs 30:321–338

LONZA (2020) Mammalian protein expression

Lu RM, Hwang YC, Liu IJ, Lee CC, Tsai HZ, Li HJ et al (2020) Development of therapeutic antibodies for the treatment of diseases. J Biomed Sci 27:1–30

Ma JK-C, Christou P, Chikwamba R, Haydon H, Paul M, Ferrer MP et al (2013) Realising the value of plant molecular pharming to benefit the poor in developing countries and emerging economics. Plant Biotechnol J 11:1029–1033

Margolin E, Allen JD, Verbeek M, van Diepen M, Ximba P, Chapman R et al (2021) Site-specific glycosylation of recombinant viral glycoproteins produced in *Nicotiana benthamiana*. Front Plant Sci 12:1–12

Medicago (2022) Plant-based technology

Merlin M, Gecchele E, Capaldi S, Pezzotti M, Avesani L (2014) Comparative evaluation of recombinant protein production in different biofactories: the green perspective. Biomed Res Int 2014:136419

Mir-Artigues P, Twyman RM, Alvarez D, Cerda Bennasser P, Balcells M, Christou P et al (2019) A simplified techno-economic model for the molecular pharming of antibodies. Biotechnol Bioeng 116:2526–2539

Moussavou G, Ko K, Lee JH, Choo YK (2015) Production of monoclonal antibodies in plants for cancer immunotherapy. Biomed Res Int 2015:306164

Nandi S, Kwong AT, Holtz BR, Erwin RL, Marcel S, McDonald KA (2016) Techno-economic analysis of a transient plant-based platform for monoclonal antibody production. mAbs 8:1456–1466

Politch JA, Cu-Uvin S, Moench TR, Tashima KT, Marathe JG, Guthrie KM et al (2021) Safety, acceptability, and pharmacokinetics of a monoclonal antibody-based vaginal multipurpose prevention film (MB66): a phase I randomized trial. PLoS Med 18:1–24

Protalix Biotherapeutics (2022) Protalix Pipeline

Qiu X, Wong G, Audet J, Bello A, Fernando L, Alimonti JB et al (2014) Reversion of advanced Ebola virus disease in nonhuman primates with ZMapp. Nature 514:47–53

Ramírez-Alanis IA, Renaud JB, García-Lara S, Menassa R, Cardineau GA (2018) Transient co-expression with three O-glycosylation enzymes allows production of GalNAc-O-glycosylated granulocyte-colony stimulating factor in N. benthamiana. Plant Methods 14:1–14

ReportLinker (2021) Monoclonal antibodies (mAbs) global market report 2021. The Business Research Company, pp 1–10

Rup B, Alon S, Amit-Cohen BC, Almon EB, Chertkoff R, Tekoah Y et al (2017) Immunogenicity of glycans on biotherapeutic drugs produced in plant expression systems - the taliglucerase alfa story. PLoS One 12:1–18

Sack M, Rademacher T, Spiegel H, Boes A, Hellwig S, Drossard J et al (2015) From gene to harvest: insights into upstream process development for the GMP production of a monoclonal antibody in transgenic tobacco plants. Plant Biotechnol J 13:1094–1105

Schillberg S, Raven N, Spiegel H, Rasche S, Buntru M (2019) Critical analysis of the commercial potential of plants for the production of recombinant proteins. Front Plant Sci 10:1–10

Scott C (2020) Making MAbs: bioprocess advancements challenge platform assumptions. Bioprocess Int:1–18

Sifniotis V, Cruz E, Eroglu B, Kayser V (2019) Current advancements in addressing key challenges of therapeutic antibody design, manufacture, and formulation. Antibodies 8:36

Song Y, Zhang M, Yin L, Wang K, Zhou Y et al (2020) COVID-19 treatment: close to a cure? A rapid review of pharmacotherapies for the novel coronavirus (SARS-CoV-2). Int J Antimicrob Agents 56:1–9

Strasser R, Stadlmann J, Schähs M, Stiegler G, Quendler H, Mach L et al (2008) Generation of glyco-engineered *Nicotiana benthamiana* for the production of monoclonal antibodies with a homogeneous human-like N-glycan structure. Plant Biotechnol J 6:392–402

Sun D, Gao W, Hu H, Zhou S (2022) Why 90% of clinical drug development fails and how to improve it? Acta Pharm Sin B 12:3049

Swope K, Morton J, Pogue GP, Hume S, Pauly MH, Shepherd J et al (2021) Manufacturing plant-made monoclonal antibodies for research or therapeutic applications. Methods Enzymol 660:239–263

Swope K, Morton J, Pogue GP, Burden L, Partain N, Hume S et al (2022) Reproducibility and flexibility of monoclonal antibody production with *Nicotiana benthamiana*. mAbs 14:1–13

Tusé D (2011) Safety of plant-made pharmaceuticals: product development and regulatory considerations based on case studies of two autologous human cancer vaccines. Hum Vaccin 7:322–330

Tusé D, McDonald KA (2014) Manufacturing economics of plant-made biologics: case studies in therapeutic and industrial enzymes. Plant Made Biologics 7:1–16

Walwyn DR, Huddy SM, Rybicki EP (2015) Introduction recombinant protein expression in plants. Biotechnol Lett 37:265–279

Chapter 8
Turnip Mosaic Virus Nanoparticles: A Versatile Tool in Biotechnology

Daniel A. Truchado, Sara Rincón, Lucía Zurita, and Fernando Ponz

Abstract Within plant molecular farming (PMF), the use of viruses and virus-like particles (VLPs) is increasingly gaining momentum due to the vast array of possibilities they offer. In addition to the wide application of viruses as vectors of genes for their transient expression in plants, viral particles are being exploited as natural nanoparticles amenable to production in plants and functionalization with very different purposes. One important group of plant viruses exploited in this context is formed by viruses with flexuous elongated virions of a high aspect ratio. One of these viruses is turnip mosaic virus (TuMV), a potyvirus. TuMV virions and VLPs have been produced in plants in different functionalized manners for an ample range of applications. They have also been chemically functionalized "in vitro" after purification of their natural unmodified forms. The chapter describes and discusses the work carried out so far for the development and applications of TuMV in PMF nanobiotechnology.

Keywords Turnip mosaic virus · Viral nanoparticles · Plant nanobiotechnology · Virus functionalization

Abbreviations

CAL-B	Lipase B from *Candida antarctica*
CP	Capsid protein
EGCG	Epigallocatechin gallate
EGF	Epithelial growth factor
GelMA	Gelatin methacryloyl

D. A. Truchado · S. Rincón · L. Zurita · F. Ponz (✉)
Centro de Biotecnología y Genómica de Plantas (CBGP), Instituto Nacional de Investigación y Tecnología Agraria y Alimentaria (INIA-CSIC), Universidad Politécnica de Madrid (UPM), Madrid, Spain
e-mail: daniel.truchado@upm.es; sara.rincon@upm.es; lucia.zurita@inia.csic.es; fponz@inia.csic.es

C. Kole et al. (eds.), *Tools & Techniques of Plant Molecular Farming*, Concepts and Strategies in Plant Sciences, https://doi.org/10.1007/978-981-99-4859-8_8

GFP	Green fluorescent protein
LTP	Lipid transfer protein
TuMV	Turnip mosaic virus
VEGFR-3	Vascular endothelial growth factor receptor 3
VLPs	Virus-like particles
VNPs	Viral nanoparticles

8.1 Viral Nanoparticles in Molecular Farming

During the last three decades, plant molecular farming has become a useful technology for the production of recombinant proteins (Fischer and Buyel 2020; Horn et al. 2004; Schillberg and Finnern 2021). The main reasons why plant molecular farming has become so extended are its low cost, its greater scalability, and its greater biosafety for humans compared to other platforms such as mammalian cell cultures since plants do not allow the replication of mammalian viruses (Fischer and Buyel 2020). One of the products of molecular farming gaining popularity over the last few years is viral nanoparticles (VNPs), which are subsequently used as biotechnological tools (Chung et al. 2020; Rybicki 2020; Shukla et al. 2014, 2018; Wu et al. 2022). Viruses, as naturally occurring nanoparticles, present many interesting characteristics for their use as biotechnological tools, such as biocompatibility, self-assembly, different shapes (icosahedral, flexuous, rod-shaped, etc.), their easily modifiable composition, or their selective permeability to small compounds (Jeevanandam et al. 2018). Among all VNPs with biotechnological interest, plant viruses offer an especially safe alternative because they lack tropism towards mammalian tissue. Thus, there are currently several plant VNPs being developed as biotechnological devices with interest in nanomedicine (Chung et al. 2020; Rybicki 2020). One of these plant viruses is turnip mosaic virus (TuMV), whose versatility as a platform for the development of different biotechnological tools will be discussed along this chapter (see also chapters "Plant Viral Vectors: Important Tools for Biologics Production" and "Medical Applications of Plant Virus Nanoparticles").

8.2 Turnip Mosaic Virus Nanoparticles

TuMV is a well-known plant pathogen of the family *Potyviridae*. Since it was first reported in 1921 (Gardner and Kendrick 1921; Schultz 1921), TuMV has been found to be infecting more than 300 species distributed globally, including mono- and dicotyledonous plants (Nellist et al. 2022). Due to the economic impact of its infections on agriculture, TuMV was classified as one of the most important pathogens of field-grown vegetable crops (Tomlinson 1987).

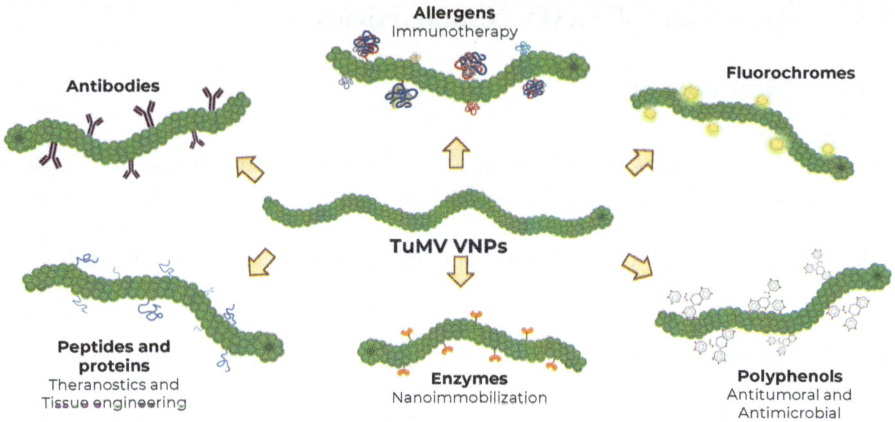

Fig. 8.1 Schematic representation of the different TuMV-based nanoparticles developed so far and their potential application in different fields of biotechnology

However, what makes TuMV attractive to nanobiotechnology is not its epidemiology but its structure. Like other potyviruses, TuMV is a non-enveloped, elongated, and flexuous virus of around 720 nm long and 12 nm wide. TuMV virions consist of approximately 2000 copies of the capsid protein surrounding the genomic RNA in a helical fashion, creating a flexuous tube. The capsid protein molecule exposes its N-terminal region to the exterior of the virion, while the C-terminal region is projected towards the inner part of the tube (Cuesta et al. 2019; Nellist et al. 2022). Despite not having the genomic RNA, TuMV virus-like particles (VLPs) keep this structure, although they seem to be more variable in length than virions (Cuesta et al. 2019). This structure of TuMV VNPs provides two main characteristics that make them appealing to be exploited as biotechnological tools. First, the great number of copies of capsomeres per VNP and its known spatial arrangement facilitate their functionalization by chemical conjugation or genetic engineering (Yuste-Calvo et al. 2019). Second, the high aspect ratio of TuMV VNPs is a major advantage in processes such as drug delivery in tumors, where elongated viral nanoparticles perform better than their spherical counterparts (Chung et al. 2020). This was explained by the fact that elongated nanoparticles present enhanced tumor homing, tumor retention, and permeability through tissues and membranes when compared with spherical viral nanoparticles (Lee et al. 2013). Therefore, the functionalization of TuMV VNPs provides a wide variety of tools with potential interest in different fields of nanoscience (Fig. 8.1).

8.3 Applications of TuMV Nanoparticles

8.3.1 Theranostics

During the last decades, the importance of diagnostics, therapy, and relationship between them has been a central point in medicine. Research on TuMV nanoparticles has contributed to improving the diagnostics and the therapy in three different fields of theranostics: immunotherapy, antibody detection, and cancer therapy.

8.3.1.1 Immunotherapy

For immunotherapy, TuMV VLPs were functionalized by gene fusion to create an allergen-coated nanoparticle for the treatment of food allergy. Food allergy is an adverse immune response whose prevalence is estimated to be increasing in the last years, especially in developed countries, where 5–10% of children and 3–4% of adults are allergic to some type of food (Sampath et al. 2021; Sicherer and Sampson 2010). The molecules responsible for allergy triggering are called allergens and are frequently lipoproteins (Breiteneder and Mills 2005; Scheurer et al. 2021). One of such allergens is Pru p 3, a small lipid transfer protein (LTP) present in peaches which represents one of the most important allergens in the Mediterranean region (Pazos-Castro et al. 2022). Since Pru p 3 has also become a good model for studying LTP allergy, TuMV VLPs were genetically functionalized with Pru p 3 via gene fusion to develop a nanotool for allergen-specific immunotherapy. Gene fusion has long been used in biotechnology for the obtention of recombinant protein products (Uhl et al. 1992). In the context of VNP functionalization, gene fusion allows the production of capsid proteins (CPs) containing a recombinant amino acid sequence of interest in a desired region. In the case of TuMV, this region is the N-terminal as it is exposed to the solvent. Gene fusion has the advantage of ensuring one copy of the protein of interest per CP. However, sometimes, it seems to affect VNP assembly, and genetic constructions turn out to be unsuccessful (Yuste-Calvo et al. 2019). This was not the case of TuMV-Pru p 3, whose production by molecular farming took place successfully. TuMV-Pru p 3 VLP production started with a gene construct containing the CP of TuMV fused to Pru p 3 through a flexible linker. This gene construct was cloned in pEAQ as an expression vector and was subsequently agroinfiltrated in *Nicotiana benthamiana* plants using *Agrobacterium tumefaciens.*

TuMV-Pru p 3 VLPs resulted to be nontoxic both in vitro and in vivo and were transported by Caco-2 cells without affecting epithelial integrity, showing the possibility of being delivered orally for immunotherapy. The administration of TuMV-Pru p 3 VLPs in previously sensitized mice significantly reduced sIgG2a levels in allergic specimens in almost a twofold ratio, and also, a downward tendency in sIgE levels was observed (Pazos-Castro et al. 2022). These results support the use of functionalized TuMV VLPs in allergen-specific immunotherapy as well as in other immune-based pathologies (see also chapter "Antigen-Specific Immunotherapy for Allergic and Autoimmune Diseases Using Plant-Made Antigens").

8.3.1.2 Immunization and Antibody Sensing

Gene fusion was first used in TuMV to functionalize VNPs with a small peptide derived from the human vascular endothelial growth factor receptor 3 (VEGFR-3) for immunization and peptide-directed antibody sensing. The use of peptides for immunization and antibody sensing has several advantages when compared to complex proteins. For immunization, peptides provide more simplicity, and the antibodies produced against them are practically monoclonal. However, they are poorly immunogenic, and its use implies several technical problems (Sánchez et al. 2013). One solution to get round these problems is the use of viruses as nanoscaffolds for peptide presentation, and, as aforementioned, TuMV provided an attractive structure for this purpose. In the case of antibody sensing, peptides have shown to successfully detect antibodies directed to a given epitope in complex serum samples (Andresen and Grötzinger 2009; Larman et al. 2011). Nevertheless, the use of peptide-functionalized VNPs for this purpose had not been addressed until TuMV-VEGFR-3 VNPs were deployed (Sánchez et al. 2013).

The production of TuMV-VEGFR-3 VLPs started with a synthetic DNA fragment containing the sequence encoding the first amino acid of the CP of TuMV, then the sequence of VEGFR-3, and, finally, the sequence of the rest of TuMV CP. This fragment was subsequently cloned in vectors p35Tunos-vec01 and p35Tunos-vec0-Nat 1 (Touriño et al. 2008), and plants of the species *Arabidopsis thaliana* and *Brassica juncea* were inoculated with them. The functionalized VNPs were purified through a protocol using a CsCl gradient (Sánchez et al. 2013).

To assess the immunization potential of the VNPs, three different groups of five 8-week-old male BALB/c mice were inoculated with 10 μg of wild-type or chimeric purified VNPs or with 0.7 μg of free VEGFR-3 (as this was the amount of peptide carried by the chimeric VNPs). Analyzed sera from immunized mice showed that those inoculated with TuMV-VEGFR-3 VNPs had significantly higher titers of anti-VEGFR-3 antibodies than those inoculated with the free peptide. Moreover, increased anti-VEGFR-3 sensing properties of the VNPs were shown as higher dilutions of sera resulted to be positive in ELISA assays when plates were coated with TuMV-VEGFR-3 VNPs as compared with those coated with the free peptide. Both results demonstrate the potential of these chimeric VNPs in the field of immunology, the first case reported in plant virus VNPs until then (Sánchez et al. 2013).

8.3.1.3 Anti-tumor Therapy

Plant-derived polyphenol interest has been increasing over the years due to their biological activities, such as their antimicrobial and antibiofilm properties against a vast array of microorganisms, including bacteria, fungi, and viral pathogens (Hui et al. 2017). In addition, epidemiological studies and associated meta-analyses have suggested that these compounds have a wide range of health-promoting traits in humans, like protection against cardiovascular diseases, cancer, or diabetes (Pandey and Rizvi 2009). One of the major sources of polyphenolic compounds for pharmaceutical and medical applications are the flavonoids present in green tea (*Camellia sinensis*). In fact, the purported health benefits of green tea are usually thought to derive from the chemopreventive activity of its high concentrations of catechins.

Epigallocatechin gallate (EGCG) is the most abundant and studied of them all, constituting about 50% of the catechin pool and accounting for between 100 and 200 mg in a cup of brewed green tea (Gopal et al. 2016). Its role in disease management is partly attributed to the number of hydroxyl groups and the presence of structural groups characteristic of catechins, which have a major impact on their antioxidant activity (Pradhan and Dubey 2021). Furthermore, this flavonoid has been reported to be responsible for most of the therapeutic benefits either in clinical, animal, or cell culture studies and to have the most potent antiproliferative effects targeting biochemical and genetic functions that are unique to cancer cells (Du et al. 2012; Hastak et al. 2003).

Although the extent of EGCG interaction mechanisms at the molecular level is still not well understood, in vitro studies in mouse models have shown that it induces a decreased risk of cancer development through binding to several key proteins and downregulation of the expression of other pathways (Luo et al. 2017; Rady et al. 2018; Wei et al. 2018), followed by growth inhibition due to apoptosis or suppression of angiogenesis and metastasis (Fujimura et al. 2012; Tachibana 2011), protection of DNA from damage and/or methylation in normal cells, and inhibition of oncogenic gene expression (Chen et al. 2011). Moreover, the chemoprotective activities of EGCG have been observed at different stages of the cell cycle, which is one of the main results found in preclinical studies.

Despite the numerous properties and promising outcomes of this compound, its applicability as an alternative to chemotherapy has proved to be limited due to inefficient systemic administration and bioavailability (Mereles and Hunstein 2011). The limiting factor of EGCG, and catechins in general, is their instability in their free form; being highly reactive molecules, they are susceptible to autoxidation and very sensitive to changes in temperature and pH, with the latter being the most critical factor for their stability under certain conditions. EGCG degrades in neutral solutions (pH 6.5–7.8), and other factors such as oxygen or protein concentration, antioxidant levels, or presence of metal ions affect its polymerization and decomposition (Zeng et al. 2017), making it easily degradable or metabolizable after injection or oral ingestion.

For this reason, designing a vehicle that maintains the effects attributed to EGCG is an attractive strategy to overcome its lack of stability in its free form. Nanoparticle-based delivery systems are believed to be plausible options to protect EGCG against adverse conditions, and consequently, they would open the way for the development of its numerous beneficial activities. In this context, protein-based nanoparticles are a very favorable platform for nanoscale drug delivery since they lack the toxic effects of metal-based nanoparticles (avoiding tissue accumulation and unspecific interactions with proteins and DNA) while presenting no environmental drawbacks.

As mentioned above, the structure of TuMV is highly modifiable and symmetric, and the outward arrangement of multiple copies of capsid protein subunits on the surface makes them versatile nanoparticles and very useful for functionalization with a wide range of molecules (Yuste-Calvo et al. 2019). Chemical conjugation of EGCG to the lysine residues of the capsid protein by the Mannich condensation reaction yields EGCG-TuMV nanoparticles with a high number of available molecules concentrated within a relatively small area (Velázquez-Lam et al. 2020). This functionalization also allows the EGCG molecule to be exposed to the external environment, which is important since negatively charged EGCG binds to the outer surface of positively charged cell membranes, thus damaging or fragmenting the lipid bilayer (Das et al. 2014). This makes TuMV-NPs rapidly and efficiently internalized by normal human and tumor cells in vitro.

Recent results have shown for the first time the potentiation of a drug effect by the use of functionalized TuMV nanoparticles for cancer therapy. The administration of this flavonoid via VNPs enhanced the antiproliferative effect and cytotoxic efficacy of EGCG in cell lines of different types of tumors, including lung, colorectal, breast, and head and neck cancer (Velázquez-Lam et al. 2022), which has laid the groundwork for future work in targeting chemotherapeutic drugs specifically to tumor cells by conjugating molecules that recognize tumor-specific components with TuMV-NP for early diagnosis or reduction of side effects of chemotherapeutic agents.

8.3.2 Tissue Engineering

The field of tissue engineering emerged more than 40 years ago as an alternative to organ transplantation, as the latter implied two important problems to overcome: shortage of organ donors and immune rejection. Tissue engineering offers a good alternative because it allows the regeneration of the patient's own tissues and organs by developing artificial organs that are highly biocompatible and functional. The regeneration of new tissues involves three basic components: cells, a scaffold, and growth factors (Ikada 2006). Alternatively, tissue engineering has recently been proposed as a tool to create the so-called clean meat, which consists of an animal-free option based on artificial skeletal muscle tissue (Ben-Arye and Levenberg 2019).

One of the most important aspects to take into account in tissue engineering is the creation of the proper environment for cells to grow successfully and promote tissue regeneration. Functionalized VNPs have been used to this end thanks to their biosafety, their spatial arrangement, and their amenability to be chemically conjugated to growth factors. Thus, scaffolds containing VNPs in tissue improved cell adhesion and cell orientation (Zhao et al. 2015).

In this context, TuMV VNPs functionalized with the epithelial growth factor (EGF) were developed as a new platform for muscle tissue regeneration. EGF was chemically conjugated to lysine residues present in the TuMV CPs via the Staudinger reaction using phosphine and azide linkers (González-Gamboa et al. 2022). Then, TuMV-EGF VNPs were used as an additive to engineer gelatin methacryloyl (GelMA) hydrogels for enhanced attachment, proliferation, and alignment of fibroblasts in artificial muscle fibers. Increased proliferation and adhesion rates were observed in fibroblast cultures on TuMV-EGF-GelMA compared to those cultures on TuMV-GelMA and EGF-GelMA (González-Gamboa et al. 2022). Fibroblasts on TuMV-EGF-GelMA continued to proliferate rapidly after adhesion to the gel, which indicated that TuMV-EGF worked both as a nano-carrier for EGF and as a physical-mechanical cue for cells. These results showed that TuMV-EGF can be used as an additive to GelMA-based bioinks for tissue engineering since it creates a microenvironment, which promotes the adhesion and proliferation of fibroblasts (González-Gamboa et al. 2022).

8.3.3 Industrial Biotechnology

Application of enzymes is continuously increasing in the manufacture of different products in many fields such as food, textile, and pharmaceutical industries or even in biosensor production. Unlike free enzymes in solution, immobilized enzymes provide several advantages, the main ones showing more stability and a higher resistance to environmental disturbances (Homaei et al. 2013). However, some types of enzyme immobilization can lead to a loss of enzyme-specific activity (Sheldon and van Pelt 2013).

One form of enzyme immobilization is attachment to nanoparticles. In particular, protein nanoparticles stand out because they share the same nature with enzymes and give a desirable chemical homogeneity to the complexes. Also, protein nanoparticles have an improved biocompatibility when compared to their inorganic counterparts and provide a wide variety of groups to which the enzymes of interest chemically conjugate (Cuenca et al. 2016). VNPs are a special type of these proteinaceous nanoparticles, as they combine some properties of inorganic and some others of organic nanoparticles. During the last years, VNPs, and in particular, plant VNPs, have been used as nano-carriers for enzymes (Cardinale et al. 2012; Eiben et al. 2019). In some cases, not only did enzyme immobilization using plant VNPs provide robustness but also increased catalytic activity compared to controls with the same amount of free enzyme (Koch et al. 2015).

The elongated and flexuous geometry of TuMV VNPs makes them good candidates for enzyme immobilization, so the enzyme lipase B from *Candida antarctica* (CAL-B) was selected to be immobilized in TuMV nanonets using glutaraldehyde as a two-headed conjugating agent. CAL-B is commonly used in the pharmaceutical and chemical industries because it is one of the best catalyzers in the production of nitrogen-containing compounds (Gotor-Fernández et al. 2006; Villar-Barro et al. 2017), so its immobilization was very interesting from the standpoint of industrial biotechnology (Idris and Bukhari 2012). TuMV-based nanonets as a platform for enzyme immobilization were obtained for the first time in this study (Cuenca et al. 2016). Large aggregates of CAL-B could be observed on those nanonets, and, more interestingly, CAL-B showed higher specific activity than the enzyme in its free form. These results showed that, although CAL-B had been previously immobilized on different supports (Cipolatti et al. 2014), TuMV-based nanonets were a good alternative that even increased the catalytic activity of the enzyme (Cuenca et al. 2016).

8.3.4 Agricultural Applications

In the agricultural sector, antimicrobial control is becoming a growing global problem due to a number of factors such as increasing food demand, climate change, or regulatory constraints on plant protection products, leading to a reduction in the effectiveness of current approaches to attack infections (see also chapter "Plant Molecular Farming of Antimicrobial Peptides for Plant Protection and Stress Tolerance").

Antimicrobial resistance happens naturally and is a dynamic threat. The development of this risk is accelerated by the emergence of multiresistant bacteria, the lack of new drugs, or the environmentally damaging effects of certain control methods. All these aspects lead to economic, ecological, and life losses, so there is a need for novel and powerful agents to replace or enhance existing antimicrobials (Coates et al. 2002). Biofilm formation is another critical area to consider when discussing antimicrobial resistance, as bacteria within these multicellular aggregates show increased resistance.

As previously mentioned, the chemical functionalization of EGCG to TuMV viral nanoparticles allows an enhancement of the intrinsic properties of this reactive flavonoid, including its reported antimicrobial and antibiofilm effects (Blanco et al. 2005). Studies over the past 20 years have shown that polyphenolic catechins from green tea can inhibit the growth of a wide range of Gram-positive and Gram-negative bacterial species with moderate potency (Jeon et al. 2014).

Conjugation of EGCG with TuMV was also shown to promote the ability of TuMV to act as an antimicrobial agent (Velázquez-Lam et al. 2020). EGCG works by affecting bacteria in different ways, such as inhibiting enzymes which affect DNA (Das et al. 2014), causing oxidative stress (Cui et al. 2012), or binding to proteins and phospholipids of the lipid bilayer, damaging the cell membrane and increasing its permeability (Papuc et al. 2017). Although it was shown that TuMV VNPs functionalized with this flavonoid by chemical conjugation to lysine residues not only maintained EGCG activity, but also showed a remarkable enhancement, further studies are needed to understand the specific mechanism to know whether they are able to cross the membrane and act internally.

8.4 Future Prospects

So far, many TuMV-based nanotools have been developed by two main functionalization techniques: chemical conjugation and genetic engineering. However, a third method for functionalization has been successfully applied to TuMV VNPs: the SpyTag/SpyCatcher technology, which combines a first gene fusion followed by a chemical conjugation. This technology is based on intramolecular isopeptide bonds occurring in cell surface adhesion proteins of Gram-positive bacteria (Hae et al. 2007). More specifically, SpyTag and SpyCatcher are peptides derived from the division of the CnaB2 domain of the *Streptococcus pyogenes* fibronectin-binding protein FbaB (Zakeri et al. 2012). Within this domain, a spontaneous isopeptide bond takes place between the amine of Lys^{31} and the carboxylic group of Asp^{117}. This reaction is resistant to a wide spectrum of pH, temperature, and buffer conditions. Taking this into account, a novel technology for protein coupling was developed by splitting the CnaB2 domain into two peptides that would eventually reconstitute via this spontaneous isopeptide bond: SpyTag, a small 13-amino acid sequence bearing the Asp^{117}, and SpyCatcher, with the rest of the protein (Zakeri et al. 2012). By genetically fusing SpyTag and SpyCatcher to either the N-terminus or the C-terminus of two proteins of interest, they will later bind to each other robustly and with a high yield.

The SpyTag/SpyCatcher technology was tested in TuMV using the green fluorescent protein (GFP) as a protein of interest for coating the VLPs. GFP had been previously selected to decorate VLPs of an icosahedral virus by using SpyTag/SpyCatcher (Peyret et al. 2020), but, to our knowledge, this was the first attempt to coat flexuous VNPs with this protein coupling technology.

Four different synthetic genes were designed, so all the possible combinations between the two peptides (SpyTag and SpyCatcher) with TuMV CP and GFP were represented. Transient expression of the constructs was carried out by agroinfiltration of transformed *Agrobacterium tumefaciens* strains in *Nicotiana benthamiana* plants. Fluorescence analysis revealed the production of GFP and its correct folding in the correspondent agroinfiltrated leaves. Western blot analyses of protein extracts showed that all constructs were produced in plants. Moreover, we observed that the interaction between SpyTag and SpyCatcher took place in vivo in the two combinations co-agroinfiltrated (SpyTag-CP + SpyCatcher-GFP and SpyCatcher-CP + SpyTag-GFP), with this being the first time a flexuous VNP is functionalized using this technique (Truchado et al. 2023).

Among all the uses SpyTag/SpyCatcher technology has been applied to, the functionalization of VNPs seems promising as it occurs in a very efficient manner compared to other alternatives. The successful results regarding the coating of TuMV VLPs with GFP open new doors to a large number of fields in which TuMV VNPs can be implemented, such as the creation of potential novel vaccines (Peyret et al. 2020; Tan et al. 2021), cancer therapy (Wang et al. 2019), or functionalization of VNPs with a wide variety of other peptides of interest (Hatlem et al. 2019; Reddington and Howarth 2015).

In summary, TuMV is a good platform for the creation of novel tools in biotechnology via molecular farming in *Nicotiana benthamiana* plants. TuMV VNPs offer an interesting architecture for their functionalization, and, once derivatized by chemical conjugation or genetic engineering, they have been reported to be applied successfully in multiple fields in nanoscience. Moreover, a novel protein coupling technique called SpyTag/SpyCatcher offers a promising method for a more robust and efficient functionalization of TuMV VNPs. This reveals that not only have TuMV-derived nanoplatforms proved their versatility, but also there is still a huge potential to be developed in the following years for these functionalized VNPs.

Acknowledgments Work on TuMV nanobiotechnological functionalization has benefited from different grants obtained along the years. These are P2018/BAA-4574, COV20/00114, and PanGreen-CM from the Comunidad de Madrid; RTA2015-00017-00-00 from INIA; and ARIMNet-2 618127, an ERANet project. Daniel A. Truchado is a postdoctoral associate funded by a contract of the "Margarita Salas" program of the Spanish Ministry of Universities. Sara Rincón is funded by the P2018/BAA-4574 grant.

References

Andresen H, Grötzinger C (2009) Deciphering the antibodyome-peptide arrays for serum antibody biomarker diagnostics. Curr Proteomics 6:1–12

Ben-Arye T, Levenberg S (2019) Tissue engineering for clean meat production. Front Sustain Food Syst 3. https://doi.org/10.3389/fsufs.2019.00046

Blanco AR, Sudano-Roccaro A, Spoto GC, Nostro A, Rusciano D (2005) Epigallocatechin gallate inhibits biofilm formation by ocular staphylococcal isolates. Antimicrob Agents Chemother 49:4339–4343. https://doi.org/10.1128/AAC.49.10.4339-4343.2005

Breiteneder H, Mills ENC (2005) Molecular properties of food allergens. J Allergy Clin Immunol 115:14–23. https://doi.org/10.1016/j.jaci.2004.10.022

Cardinale D, Carette N, Michon T (2012) Virus scaffolds as enzyme nano-carriers. Trends Biotechnol 30:369. https://doi.org/10.1016/j.tibtech.2012.04.001

Chen D, Biao Wan S, Yang H, Yuan J, Hang Chan T, Ping Dou Q (2011) EGCG, green tea polyphenols and their synthetic analogs and prodrugs for human cancer prevention and treatment. Adv Clin Chem 53:155–177

Chung YH, Cai H, Steinmetz NF (2020) Viral nanoparticles for drug delivery, imaging, immunotherapy, and theranostic applications. Adv Drug Deliv Rev 156:214–235. https://doi.org/10.1016/j.addr.2020.06.024

Cipolatti EP, Silva MJA, Klein M, Feddern V, Feltes MMC, Oliveira JV, Ninow JL, De Oliveira D (2014) Current status and trends in enzymatic nanoimmobilization. J Mol Catal B Enzym 99:56–67. https://doi.org/10.1016/j.molcatb.2013.10.019

Coates A, Hu Y, Bax R, Page C (2002) The future challenges facing the development of new antimicrobial drugs. Nat Rev Drug Discov 1:895. https://doi.org/10.1038/nrd940

Cuenca S, Mansilla C, Aguado M, Yuste-Calvo C, Sánchez F, Sánchez-Montero JM, Ponz F (2016) Nanonets derived from turnip mosaic virus as scaffolds for increased enzymatic activity of immobilized *Candida antarctica* lipase B. Front Plant Sci 7. https://doi.org/10.3389/fpls.2016.00464

Cuesta R, Yuste-Calvo C, Gil-Cartón D, Sánchez F, Ponz F, Valle M (2019) Structure of Turnip mosaic virus and its viral-like particles. Sci Rep 9:15396. https://doi.org/10.1038/s41598-019-51823-4

Cui Y, Oh YJ, Lim J, Youn M, Lee I, Pak HK, Park W, Jo W, Park S (2012) AFM study of the differential inhibitory effects of the green tea polyphenol (−)-epigallocatechin-3-gallate (EGCG) against Gram-positive and Gram-negative bacteria. Food Microbiol 29:80–87. https://doi.org/10.1016/j.fm.2011.08.019

Das S, Tanwar J, Hameed S, Fatima Z (2014) Antimicrobial potential of epigallocatechin-3-gallate (EGCG): a green tea polyphenol. J Biochem Pharmacol Res 2:167–174

Du GJ, Zhang Z, Wen XD, Yu C, Calway T, Yuan CS, Wang CZ (2012) Epigallocatechin gallate (EGCG) is the most effective cancer chemopreventive polyphenol in green tea. Nutrients 4:1679–1691. https://doi.org/10.3390/nu4111679

Eiben S, Koch C, Altintoprak K, Southan A, Tovar G, Laschat S, Weiss IM, Wege C (2019) Plant virus-based materials for biomedical applications: trends and prospects. Adv Drug Deliv Rev 145:96–118. https://doi.org/10.1016/j.addr.2018.08.011

Fischer R, Buyel JF (2020) Molecular farming—the slope of enlightenment. Biotechnol Adv 40:107519. https://doi.org/10.1016/j.biotechadv.2020.107519

Fujimura Y, Sumida M, Sugihara K, Tsukamoto S, Yamada K, Tachibana H (2012) Green tea polyphenol EGCG sensing motif on the 67-kDa laminin receptor. PLoS One 7:e37942. https://doi.org/10.1371/journal.pone.0037942

Gardner MW, Kendrick JB (1921) Turnip mosaic. J Agric Res 22:123

González-Gamboa I, Velázquez-Lam E, Lobo-Zegers MJ, Frías-Sánchez AI, Tavares-Negrete JA, Monroy-Borrego A, Menchaca-Arrendondo JL, Williams L, Lunello P, Ponz F, Alvarez MM, Trujillo-de Santiago G (2022) Gelatin-methacryloyl hydrogels containing turnip mosaic virus for fabrication of nanostructured materials for tissue engineering. Front Bioeng Biotechnol 10. https://doi.org/10.3389/fbioe.2022.907601

Gopal J, Muthu M, Paul D, Kim DH, Chun S (2016) Bactericidal activity of green tea extracts: the importance of catechin containing nano particles. Sci Rep 6:1–14. https://doi.org/10.1038/srep19710

Gotor-Fernández V, Busto E, Gotor V (2006) *Candida antarctica* lipase B: an ideal biocatalyst for the preparation of nitrogenated organic compounds. Adv Synth Catal 348:797. https://doi.org/10.1002/adsc.200606057

Hae JK, Coulibaly F, Clow F, Proft T, Baker EN (2007) Stabilizing isopeptide bonds revealed in gram-positive bacterial pilus structure. Science 318:1625–1628. https://doi.org/10.1126/science.1145806

Hastak K, Gupta S, Ahmad N, Agarwal MK, Agarwal ML, Mukhtar H (2003) Role of p53 and NF-κB in epigallocatechin-3-gallate-induced apoptosis of LNCaP cells. Oncogene 22:4851–4859. https://doi.org/10.1038/sj.onc.1206708

Hatlem D, Trunk T, Linke D, Leo JC (2019) Catching a SPY: using the SpyCatcher-SpyTag and related systems for labeling and localizing bacterial proteins. Int J Mol Sci 20. https://doi.org/10.3390/ijms20092129

Homaei AA, Sariri R, Vianello F, Stevanato R (2013) Enzyme immobilization: an update. J Chem Biol 6:185–205. https://doi.org/10.1007/s12154-013-0102-9

Horn ME, Woodard SL, Howard JA (2004) Plant molecular farming: systems and products. Plant Cell Rep 22:711–720. https://doi.org/10.1007/s00299-004-0767-1

Hui X, Hua SH, Wu QQ, Li H, Gao WY (2017) Antimicrobial mechanism of epigallocatechin gallate and gallocatechin gallate: they target 1-deoxy-D-xylulose 5-phosphate reductoisomerase, the key enzyme of the MEP terpenoid biosynthetic pathway. Arch Biochem Biophys 622:1–8. https://doi.org/10.1016/j.abb.2017.04.007

Idris A, Bukhari A (2012) Immobilized *Candida antarctica* lipase B: hydration, stripping off and application in ring opening polyester synthesis. Biotechnol Adv 30:550–563. https://doi.org/10.1016/j.biotechadv.2011.10.002

Ikada Y (2006) Challenges in tissue engineering. J R Soc Interface 3:589. https://doi.org/10.1098/rsif.2006.0124

Jeevanandam J, Barhoum A, Chan YS, Dufresne A, Danquah MK (2018) Review on nanoparticles and nanostructured materials: history, sources, toxicity and regulations. Beilstein J Nanotechnol 9:1050–1074. https://doi.org/10.3762/bjnano.9.98

Jeon J, Kim JH, Lee CK, Oh CH, Song HJ (2014) The antimicrobial activity of (−)-epigallocatehin-3-gallate and green tea extracts against *Pseudomonas aeruginosa* and *Escherichia coli* isolated from skin wounds. Ann Dermatol 26:564–569. https://doi.org/10.5021/ad.2014.26.5.564

Koch C, Wabbel K, Eber FJ, Krolla-Sidenstein P, Azucena C, Gliemann H, Eiben S, Geiger F, Wege C (2015) Modified TMV particles as beneficial scaffolds to present sensor enzymes. Front Plant Sci 6. https://doi.org/10.3389/fpls.2015.01137

Larman HB, Zhao Z, Laserson U, Li MZ, Ciccia A, Gakidis MAM, Church GM, Kesari S, Leproust EM, Solimini NL, Elledge SJ (2011) Autoantigen discovery with a synthetic human peptidome. Nat Biotechnol 29:535–541. https://doi.org/10.1038/nbt.1856

Lee KL, Hubbard LC, Hern S, Yildiz I, Gratzl M, Steinmetz NF, Author BS (2013) Shape matters: the diffusion rates of TMV rods and CPMV icosahedrons in a spheroid model of extracellular matrix are distinct. Biomater Sci 1:581. https://doi.org/10.1039/b000000x/NIH

Luo KW, Chen W, Lung WY, Wei XY, Cheng BH, Cai ZM, Huang WR (2017) EGCG inhibited bladder cancer SW780 cell proliferation and migration both in vitro and in vivo via down-regulation of NF-κB and MMP-9. J Nutr Biochem 41:56–64. https://doi.org/10.1016/j.jnutbio.2016.12.004

Mereles D, Hunstein W (2011) Epigallocatechin-3-gallate (EGCG) for clinical trials: more pitfalls than promises? Int J Mol Sci 12:5592. https://doi.org/10.3390/ijms12095592

Nellist CF, Ohshima K, Ponz F, Walsh JA (2022) Turnip mosaic virus, a virus for all seasons. Ann Appl Biol 180:312–327. https://doi.org/10.1111/aab.12755

Pandey KB, Rizvi SI (2009) Plant polyphenols as dietary antioxidants in human health and disease. Oxidative Med Cell Longev 2:270–278. https://doi.org/10.4161/oxim.2.5.9498

Papuc C, Goran GV, Predescu CN, Nicorescu V, Stefan G (2017) Plant polyphenols as antioxidant and antibacterial agents for shelf-life extension of meat and meat products: classification, structures, sources, and action mechanisms. Compr Rev Food Sci Food Saf 16:1243–1268. https://doi.org/10.1111/1541-4337.12298

Pazos-Castro D, Margain C, Gonzalez-Klein Z, Yuste-Calvo C, Garrido-Arandia M, Zurita L, Esteban V, Tome-Amat J, Diaz-Perales A, Ponz F (2022) Suitability of potyviral recombinant virus-like particles bearing a complete food allergen for immunotherapy vaccines. Front Immunol 13. https://doi.org/10.3389/fimmu.2022.986823

Peyret H, Ponndorf D, Meshcheriakova Y, Richardson J, Lomonossoff GP (2020) Covalent protein display on Hepatitis B core-like particles in plants through the in vivo use of the SpyTag/SpyCatcher system. Sci Rep 10:17095. https://doi.org/10.1038/s41598-020-74105-w

Pradhan S, Dubey RC (2021) Beneficial properties of green tea. In: Antioxidant properties and health benefits of green tea. Nova Science Publishers, New York, pp 27–56

Rady I, Mohamed H, Rady M, Siddiqui IA, Mukhtar H (2018) Cancer preventive and therapeutic effects of EGCG, the major polyphenol in green tea. Egypt J Basic Appl Sci 5:1–23. https://doi.org/10.1016/j.ejbas.2017.12.001

Reddington SC, Howarth M (2015) Secrets of a covalent interaction for biomaterials and biotechnology: SpyTag and SpyCatcher. Curr Opin Chem Biol 29:94–99. https://doi.org/10.1016/j.cbpa.2015.10.002

Rybicki EP (2020) Plant molecular farming of virus-like nanoparticles as vaccines and reagents. Wiley Interdiscip Rev Nanomed Nanobiotechnol 12. https://doi.org/10.1002/wnan.1587

Sampath V, Abrams EM, Adlou B, Akdis C, Akdis M, Brough HA, Chan S, Chatchatee P, Chinthrajah RS, Cocco RR, Deschildre A, Eigenmann P, Galvan C, Gupta R, Hossny E, Koplin JJ, Lack G, Levin M, Shek LP, Makela M, Mendoza-Hernandez D, Muraro A, Papadopoulous NG, Pawankar R, Perrett KP, Roberts G, Sackesen C, Sampson H, Tang MLK, Togias A, Venter C, Warren CM, Wheatley LM, Wong GWK, Beyer K, Nadeau KC, Renz H (2021) Food allergy across the globe. J Allergy Clin Immunol 148:1347–1364. https://doi.org/10.1016/j.jaci.2021.10.018

Sánchez F, Sáez M, Lunello P, Ponz F (2013) Plant viral elongated nanoparticles modified for log-increases of foreign peptide immunogenicity and specific antibody detection. J Biotechnol 168:409–415. https://doi.org/10.1016/j.jbiotec.2013.09.002

Scheurer S, Van Ree R, Vieths S (2021) The role of lipid transfer proteins as food and pollen allergens outside the Mediterranean area. Curr Allergy Asthma Rep 21:1–13. https://doi.org/10.1007/s11882-020-00982-w/Published

Schillberg S, Finnern R (2021) Plant molecular farming for the production of valuable proteins—critical evaluation of achievements and future challenges. J Plant Physiol 258–259:153359. https://doi.org/10.1016/j.jplph.2020.153359

Schultz ES (1921) A transmissible mosaic disease of Chinese cabbage, mustard, and turnip. J Agric Res 22:173

Sheldon RA, van Pelt S (2013) Enzyme immobilisation in biocatalysis: why, what and how. Chem Soc Rev 42:6223–6235. https://doi.org/10.1039/c3cs60075k

Shukla S, Dickmeis C, Nagarajan AS, Fischer R, Commandeur U, Steinmetz NF (2014) Molecular farming of fluorescent virus-based nanoparticles for optical imaging in plants, human cells and mouse models. Biomater Sci 2:784–797. https://doi.org/10.1039/c3bm60277j

Shukla S, Dickmeis C, Fischer R, Commandeur U, Steinmetz NF (2018) In planta production of fluorescent filamentous plant virus-based nanoparticles. In: Methods in molecular biology. Humana Press Inc., pp 61–84. https://doi.org/10.1007/978-1-4939-7808-3_5

Sicherer SH, Sampson HA (2010) Food allergy. J Allergy Clin Immunol 125:S116. https://doi.org/10.1016/j.jaci.2009.08.028

Tachibana H (2011) Green tea polyphenol sensing. Proc Japan Acad B Phys Biol Sci 87:66–80. https://doi.org/10.2183/pjab.87.66

Tan TK, Rijal P, Rahikainen R, Keeble AH, Schimanski L, Hussain S, Harvey R, Hayes JWP, Edwards JC, McLean RK, Martini V, Pedrera M, Thakur N, Conceicao C, Dietrich I, Shelton H, Ludi A, Wilsden G, Browning C, Zagrajek AK, Bialy D, Bhat S, Stevenson-Leggett P, Hollinghurst P, Tully M, Moffat K, Chiu C, Waters R, Gray A, Azhar M, Mioulet V, Newman J, Asfor AS, Burman A, Crossley S, Hammond JA, Tchilian E, Charleston B, Bailey D, Tuthill TJ, Graham SP, Duyvesteyn HME, Malinauskas T, Huo J, Tree JA, Buttigieg KR, Owens RJ, Carroll MW, Daniels RS, McCauley JW, Stuart DI, Huang KYA, Howarth M, Townsend AR (2021) A COVID-19 vaccine candidate using SpyCatcher multimerization of the SARS-CoV-2 spike protein receptor-binding domain induces potent neutralising antibody responses. Nat Commun 12:542. https://doi.org/10.1038/s41467-020-20654-7

Tomlinson JA (1987) Epidemiology and control of virus diseases of vegetables. Ann Appl Biol 110:661–681

Touriño A, Sánchez F, Fereres A, Ponz F (2008) High expression of foreign proteins from a biosafe viral vector derived from Turnip mosaic virus. Spanish J Agric Res 6:48–58. https://doi.org/10.5424/sjar/200806s1-373

Truchado DA, Rincón S, Zurita L, Sánchez F, Ponz F (2023) Isopeptide bonding *in planta* allows functionalization of elongated proteinaceous viral nanoparticles, including non-viable constructs by other means. Viruses 15:375. https://doi.org/10.3390/v15020375

Uhl M, Forsberg G, Moks T, Hartmanis M, Nilsson B (1992) Fusion proteins in biotechnology. Curr Opin Biotechnol 3:363–369

Velázquez-Lam E, Imperial J, Ponz F (2020) Polyphenol-functionalized plant viral-derived nanoparticles exhibit strong antimicrobial and antibiofilm formation activities. ACS Appl Bio Mater 3:2040–2047. https://doi.org/10.1021/acsabm.9b01161

Velázquez-Lam E, Tome-Amat J, Segrelles C, Yuste-Calvo C, Asensio S, Peral J, Ponz F, Lorz C (2022) Antitumor applications of polyphenol-conjugated turnip mosaic virus-derived nanoparticles. Nanomedicine 17:999. https://doi.org/10.2217/nnm-2022-0067

Villar-Barro Á, Gotor V, Brieva R (2017) Enantioselective chemoenzymatic synthesis of a key segment of neuronal nitric oxide synthase inhibitors and several related 3-aminopyridinylmethyl-4-hydroxypyrrolidines. Green Chem 19:436–446. https://doi.org/10.1039/c6gc01965j

Wang W, Liu Z, Zhou X, Guo Z, Zhang J, Zhu P, Yao S, Zhu M (2019) Ferritin nanoparticle-based SpyTag/SpyCatcher-enabled click vaccine for tumor immunotherapy. Nanomed Nanotechnol Biol Med 16:69–78. https://doi.org/10.1016/j.nano.2018.11.009

Wei R, Mao L, Xu P, Zheng X, Hackman RM, MacKenzie GG, Wang Y (2018) Suppressing glucose metabolism with epigallocatechin-3-gallate (EGCG) reduces breast cancer cell growth in preclinical models. Food Funct 9:5682–5696. https://doi.org/10.1039/c8fo01397g

Wu Z, Zhou J, Nkanga CI, Jin Z, He T, Borum RM, Yim W, Zhou J, Cheng Y, Xu M, Steinmetz NF, Jokerst JV (2022) One-step supramolecular multifunctional coating on plant virus nanoparticles for bioimaging and therapeutic applications. ACS Appl Mater Interfaces 14:13692–13702. https://doi.org/10.1021/acsami.1c22690

Yuste-Calvo C, González-Gamboa I, Pacios LF, Sánchez F, Ponz F (2019) Structure-based multifunctionalization of flexuous elongated viral nanoparticles. ACS Omega 4:5019–5028. https://doi.org/10.1021/acsomega.8b02760

Zakeri B, Fierer JO, Celik E, Chittock EC, Schwarz-Linek U, Moy VT, Howarth M (2012) Peptide tag forming a rapid covalent bond to a protein, through engineering a bacterial adhesin. Proc Natl Acad Sci 109:E690–E697. https://doi.org/10.1073/pnas.1115485109/-/DCSupplemental

Zeng L, Ma M, Li C, Luo L (2017) Stability of tea polyphenols solution with different pH at different temperatures. Int J Food Prop 20:1–18. https://doi.org/10.1080/10942912.2014.983605

Zhao X, Lin Y, Wang Q (2015) Virus-based scaffolds for tissue engineering applications. Wiley Interdiscip Rev Nanomed Nanobiotechnol. https://doi.org/10.1002/wnan.1327

Chapter 9
Targeting Chloroplasts for Plant Molecular Farming

Kiran Saba, Fatima Ijaz, Muhammad Suleman Malik, Neelam Batool, Andreas Gunter Lössl, and Mohammad Tahir Waheed

Abstract Chloroplast transformation has emerged as a promising platform for the development of modified transplastomic plants expressing many useful products including pharmaceutical drugs, enzymes, biomaterials, and products related to the agriculture industry. Due to high copy number of chloroplasts in green plants, it is an attractive technology for the high mass production of expressed foreign proteins. Moreover, multiple genes can be expressed into plants in a single transformation event without any epigenetic effects because of operon system of chloroplasts like prokaryotes. Plastid transformations solve the problem of expressed transgene containment, gene silencing, and minimum pleiotropic effects, which are predominant in nuclear transformation. In this chapter, we have discussed the benefits of chloroplast transformation in plant molecular farming, different methods used for plastid transformation, application of chloroplast bioengineering in difference fields of science, and advantages and limitations of chloroplast transformation for plant molecular farming.

Keywords Plant molecular pharming · Chloroplast transformation · Total soluble protein (TSP)

K. Saba
Department of Biochemistry, Shaheed Benazir Bhutto Women University, Peshawar, Pakistan

Department of Biochemistry, Faculty of Biological Sciences, Quaid-i-Azam University, Islamabad, Pakistan
e-mail: kiransaba@sbbwu.edu.pk

F. Ijaz · M. S. Malik · N. Batool · M. T. Waheed (✉)
Department of Biochemistry, Faculty of Biological Sciences, Quaid-i-Azam University, Islamabad, Pakistan
e-mail: tahirwaheed@qau.edu.pk

A. G. Lössl
Department of Applied Plant Science and Plant Biotechnology, University of Natural Resources and Life Sciences (BOKU), Vienna, Austria
e-mail: andreas@lossl.de

C. Kole et al. (eds.), *Tools & Techniques of Plant Molecular Farming*, Concepts and Strategies in Plant Sciences, https://doi.org/10.1007/978-981-99-4859-8_9

Abbreviations

CNT	Carbon nanotube
DW	Dry weight
GFP	Green fluorescent protein
HIV	Human immunodeficiency virus
HPV	Human papillomavirus
LEEP	Lipid exchange envelope penetration
ND	Not determined
PCR	Polymerase chain reaction
PEG	Polyethylene glycol
PHAs	Polyhydroxyalkanoates
PHB	Polyhydroxybutyrate
TCP	Total cellular protein
TSP	Total soluble protein
UV	Ultraviolet
YFP	Yellow fluorescent protein

9.1 Introduction

9.1.1 Chloroplast Transformation

Commercial production of recombinant products through manipulation of plastome of algae and plants is described as "transplastomics," which is an increasingly active field (Maliga 2004; Rascón-Cruz et al. 2021). A number of current studies have shown that by utilizing tobacco chloroplasts, significant amounts of recombinant protein can be obtained; in some cases, the proteins have been accumulated, achieving 5–40% of total soluble protein (TSP) (Ruhlman et al. 2010) and up to 70% of total soluble protein in leaves of tobacco plants (Wang et al. 2018; Corigliano et al. 2019; Morgenfeld et al. 2020). The chloroplast was genetically transformed about two decades ago (Boynton et al. 1988; Svab et al. 1990). Apart from having the potential of a high-level production of foreign proteins, other charms of the chloroplast transformation are its effectiveness as a highly precise genetic engineering technique (due to integration of transgene through homologous recombination), nonexistence of gene silencing mechanisms and epigenetic effects in plastids, ease of stacking multiple transgenes under a single operon system, and containment of transgene due to the maternal mode of inheritance in chloroplast, which eliminates chloroplasts (and thus plastid transgenes) from transmission through pollens (Saba et al. 2019; Latif et al. 2022).

Chloroplasts are involved in various ceullular functions such as: biosynthesis of fatty acids, Vitamins, purines, pyrimidines, isoprenoids, starch and pigments. Chloroplasts are also implicated in the metabolism of nitrogen and amino acids, and phytohormones such as cytokinins, abscisic acid, and gibberellins. Thus, as a result, any interruption of its normal metabolism can be lethal in plants (Farquhar et al. 2011).

9.1.2 Chloroplast Transformation and Molecular Farming

Development of transgenic plants as expression factories for the production of biopharmaceuticals is an area generally considered as molecular farming and is recently getting great attention (Raskin et al. 2002; Rascón-Cruz et al. 2021). Researchers are interested in the development and synthesis of plant-based antibodies, edible vaccines, industrial enzymes, and human therapeutic proteins. Reduced costs of production and delivery, ease of scale-up, and high safety standards (referring to extremely reduced risk of product contamination by endotoxins and human pathogens) are the main attractions of utilizing plants as platforms for producing pharmaceuticals (Scotti et al. 2012; Ahmad and Mukhtar 2013; Saba et al. 2020; Latif et al. 2022). Due to the greater ability of the chloroplast for expression and accumulation of foreign proteins, it seems particularly attractive to exploit plastids for the production of proteinaceous pharmaceuticals, such as antimicrobials, antigens, and antibodies. Researchers (Daniell et al. 2009; Lössl and Waheed 2011; Maliga and Bock 2011; Scotti et al. 2012; Ahmad and Mukhtar 2013) have reported several advantages and limitations regarding chloroplast-based plant molecular farming. Currently, there are many available drugs which have been derived from the plants. Plants are also manipulated to increase the production of food and different compounds. Until the nineteenth century, all improvements in crops were brought by plant breeding techniques (Chemat et al. 2019).

The first chloroplast-based product that proved to be immunologically active in experimental animals was a candidate subunit vaccine against *Clostridium tetani*, the agent causing tetanus (Tregoning et al. 2003). Recently, numerous promising steps have been taken in this direction. So far, majority of efforts have been made for the increased production of antigens to be used as vaccines and their tests for immunological efficacy in animal studies. Encouraging progress has been made with the development of a chloroplast-based vaccine against many infectious diseases (Saba et al. 2020).

9.2 Methods of Chloroplast Transformation

Introduction of required and new foreign genetic material into the host or targeted cells is known as the transformation. Generally, transformation of plants can be performed via direct or indirect gene transfer, both of which have different working principles. Physical or chemical reactions are involved in the direct gene transfer approach, while biological vectors are utilized in the indirect gene transfer method to introduce genes into the targeted cell/tissue (Low et al. 2018). The indirect method in chloroplast transformation is a novel technique, whereby alteration must be performed on the VirD2 protein of *Agrobacterium* that plays a major role in *Agrobacterium*-mediated transformation (Matsuoka 2015). The following sections discuss the commonly used approaches in chloroplast transformation.

9.2.1 Biolistic Method

The most reliable and efficient method for the introduction of DNA into the chloroplast is microparticle bombardment. The biolistic method is not something completely new in the world of genetic engineering. It was first recognized and applied to the field in the late 1980s, where it was primarily tested on plants for transformation studies. The term "biolistic" originated from the term biology, and ballistics refers to its mode of action that is similar to a gun. The entire concept of biolistic method is dependent on high pressure, whereas the desired DNA is projected into the host via gene gun at high speed with the help of pressurized helium gas (Jinturkar et al. 2011; Matsumoto and Gonsalves 2012; Bhatia et al. 2015; Carter and Shieh 2015; Yu et al. 2020).

The microparticles used in the bombardment process are made of either tungsten or gold; despite gold being more expensive, it is the preferred material. Gold is preferred because (1) the gold particles have more uniform size; (2) they are biologically inert in nature as compared to tungsten; and (3) tungsten has a tendency to catalyze the slow degradation of DNA bound to it. The biolistic process initiates with the purification of gold or tungsten particles that are used as the delivery vector. The selected particles are purified through the treatment of isopropanol and glycerol to obtain the purest form of the particle. This step ensures that the particles are free of contaminants to prevent the DNA construct from being affected during the process. The DNA construct is coated on metal particles, using the precipitation principle. This step uses calcium chloride and isopropyl to form DNA-tungsten/gold complex. The complex formed is kept on the macro-carrier known as a plastic bullet. The micro-carrier is loaded into the chamber, where it is located around 7 in. away from the rupture disk. After the macro-carrier is loaded into the biolistic chamber, the petri dish containing the desired chloroplast to be transformed is placed at the bottom of the biolistic chamber and the pressure within the chamber is reduced to the desired level. The pressure applied varies depending on the type of cells used and the distance between the rupture disk and the petri dish (Ramesh et al. 2011; Lacroix and Citovsky 2020).

After this setup, the bombardment process can take place. Generally, two methods are used for the bombardment: one involves charged electricity, and the other involves pressurized gas. It is common to use helium gas, where pressurized helium gas is applied on the macro-carrier to transfer the DNA complex into the chloroplast at high speed. After the bombardment, the DNA construct is incorporated into the desired cells followed by 2 weeks of incubation for the healing of the cells, and further process is carried out once the callus is formed (Jinturkar et al. 2011; Carter

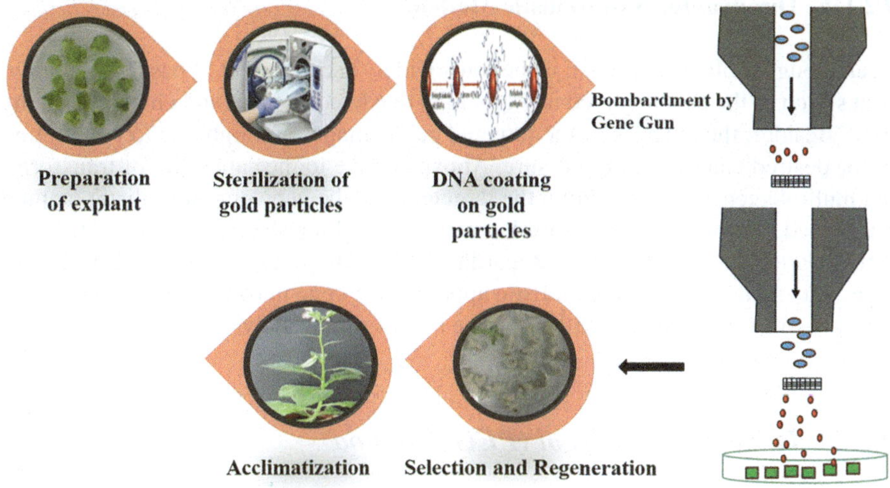

Fig. 9.1 Schematic diagram showing major steps involved in the generation of transgenic plants using biolistic bombardment method

and Shieh 2015; Matsuoka 2015; Lacroix and Citovsky 2020). The major steps involved in biolistic gene method are shown in Fig. 9.1.

The biolistic approach has been applied on *Artemisia annua* to obtain a higher yield of artemisinin for a drug against malaria infection (Raskin et al. 2002). In another study, the chloroplast genome of potatoes has been transformed through the biolistic DNA delivery approach. This study was performed to analyze the efficiency of transformation of macro-chloroplast, and the amount of the heterologous protein being produced was higher as compared to the conventional approach (Occhialini et al. 2020).

9.2.1.1 Advantages of Biolistic Method

The biolistic method is largely applied in plant transformation because it is not limited to a certain species or plant cells. It is the most favorable approach for embryogenic callus. Another factor that the biolistic method is frequently used is because of its capability to successfully transform stable desired plastid gene, other than only focused on nuclear transformation. Other than that, in the biolistic method, the introduction of the desired genes to the target cells is carried out by bypassing the cell wall barrier, and the gene transferred is not attacked or interrupted by any other cellular components on the cell surface, as the gene is directly introduced to a specific location of the desired cell (Jinturkar et al. 2011; Bhatia et al. 2015; Carter and Shieh 2015).

9.2.1.2 Disadvantages of Biolistic Method

Even though biolistic approach is the most widely used method, it has certain drawbacks such as the cost involved is pretty high due to equipment and chemicals being used. Besides, the chances of the targeted cell damages are high due to penetration of the desired gene via high pressure. Above all, due to the capability of transforming multiple genes at once within the targeted cell, the gene gun or biolistic method might lead to gene silencing (Jinturkar et al. 2011; Ramesh et al. 2011; Bhatia et al. 2015; Carter and Shieh 2015). Regardless of all these advantages and disadvantages, the choice of using the biolistic approach for gene introduction in desired cell target is still very much dependent on the results of the study.

9.2.2 Polyethylene Glycol (PEG) Method

Polyethylene glycol (PEG)-mediated transformation is the second most commonly used plant transformation approach. PEG is normally utilized when the target cell for transformation is protoplast. It is one of the renowned protoplast or chloroplast transformation techniques due to its straightforward utilization of equipment and minimal cost (Liu and Friesen 2012). The method involves simple and easy steps that can even be performed in any laboratory setting, with the presence of a biosafety cabinet. Due to its efficacy and ability to give an expected outcome, this technique has become one of the preferred protoplast transformation techniques. PEG-mediated transformation introduces the gene of interest to the targeted cell through the disruption of the cell membrane's dynamic by raising the permeability of the cell membrane. In this process, the chloroplast is co-cultured with PEG, where the desired DNA construct is passed through the cell membrane in the form of vesicles to transform the chloroplast (Mathur and Koncz 1998; Liu and Vidali 2011; Low et al. 2018; Yu et al. 2020).

Recent studies showed that PEG has not shown considerable evidence in promoting the synergy between the DNA construct and the cell membrane; therefore, the function and process of PEG-mediated transformation remain unclear. However, it has been hypothesized that the osmotic condition of protoplast can be controlled by PEG, which assists in the uptake of the DNA construct. The whole process of PEG-mediated chloroplast transformation involves three basic steps, i.e., (a) isolation of protoplast, (b) introduction of DNA construct into the chloroplast, and (c) regeneration of protoplast. The PEG-mediated transformation starts with the acquisition of protoplast suspension through various enzymatic treatments of the plant cells. The protoplast suspension is then allowed to undergo a centrifugation process (Liu and Vidali 2011; Liu and Friesen 2012). The protoplast pellet obtained from the centrifugation of suspension is mixed with the desired DNA construct that is incorporated with a selectable marker such as GFP. This is followed by the introduction of 60% PEG solution to the mixture and further incubated for a specific time. The concentration of PEG solution and the incubation period depend on the quantity of chloroplast and DNA construct used. After 3 days of introducing PEG to the mixture, the protoplast is grown in selectable conditions such as in the presence of antibiotics and antifungals, to avoid the growth of bacteria and fungi. The transformants are monitored by the expression of the selectable marker incorporated in the DNA construct, and the transformation efficiency is also identified. Further confirmation on the identification of the transformants is performed by using PCR. After the screening of the transformants, the transformed chloroplast is treated with osmotic buffer and sorbitol followed by the culturing of transformant on regeneration medium for recovery process (Liu and Vidali 2011; Liu and Friesen 2012; Masani et al. 2014; Nanjareddy et al. 2016; Rehman et al. 2016; Díaz et al. 2019; Zienkiewicz et al. 2019).

9.2.2.1 Advantages and Disadvantages of PEG

The main advantage of PEG-mediated transformation is that a large-sized DNA construct can be taken up by the protoplast without causing any physical damages to the cell membrane or the protoplast. Studies have also shown that PEG-mediated transformation has higher transformation efficiency in protoplast (Matsumoto and Gonsalves 2012). However, the main drawback of PEG-mediated transformation is the production of a large amount of transient transformants that normally lead to retrieval of a huge amount of active protoplast (Liu and Friesen 2012).

9.2.3 Carbon Nanotube Carriers

Carbon nanotube carriers (CNTs) are documented as a delivery system for transport of biomolecules and drug components to their targeted location. The physical properties and unique structure of CNT enable them to be a successful transporter (Son et al. 2016). Nanoparticles of 10–20 nm in size and having a positive charge are used for transport of plasmid DNA (pDNA) into the chloroplasts with higher efficiency as the charged nanoparticles have been proven to move across the chloroplast envelope. This uptake was considered as the lipid exchange envelope penetration (LEEP), and the charge on nanoparticles is the major contributor to this hypothesis (Newkirk et al. 2021). But it has also been stated that the nanoparticles designed to target the chloroplast using peptide did not stick to the LEEP mechanism. However, this is only seen in leaf mesophyll cells of Arabidopsis with greater than 75% success rate (Newkirk et al. 2021). It is important to consider that CNT should be designed with the ability to take up and subsequently deliver the pDNA to the chloroplast selectively. The delivery process using CNT as demonstrated by Kwak et al. is relatively straightforward, in which designed CNT was incubated together with mesophyll protoplast of *Arabidopsis thaliana* allowing the uptake of pDNA. The transformed protoplast was then identified via the expression of the yellow fluorescence protein (YFP) in the chloroplast. This delivery approach utilizing CNT is economical that does not require any specialized complex equipment (Kwak et al. 2019).

9.2.4 *UV Laser Microbeam*

UV laser microbeam-mediated gene delivery is an attractive delivery system due to its huge spatial control over the laser beam produced by the optical fiber. There are four types of laser-assisted gene delivery methods which are (a) optoinjection, (b) photochemical internalization, (c) transfection via laser-induced waves, and (d) selective cell targeting with light-absorbing particles. The UV laser beam is being directed to the cell causing the cell perforation and formation of a hole on the cell membrane (Yao et al. 2008). The hole formed is approximately 0.5 μm in size, which can self-heal and recover within 5 s. During this process of membrane opening, the external DNA or the plasmid DNA is taken up by the cells resulting in the transformation event to happen. Weber et al. used UV laser microbeam approach to introduce DNA into the chloroplast of *Brassica napus* protoplast. Fluorescence microscopy was used to verify the uptake via this approach (Weber et al. 1988). Nevertheless, this approach is not broadly used due to the high cost of the equipment that can generate a laser beam having the dimensions of 100 nm. Above all, there is also risk of damage to the cell and chloroplast by the UV laser radiation (Rivera et al. 2012). Till now, there are only few chloroplast transformation studies that have been utilizing this approach. Therefore, more studies are needed for understanding the efficacy of this approach.

9.2.5 **Agrobacterium**-*Mediated Transformation*

Agrobacterium-mediated transformation is based on one of the *Agrobacterium* species, i.e., *A. tumefaciens*. This is one of the most recommended gene delivery systems, which is noninvasive as *A. tumefaciens* is capable of simulating the natural plant transformation process. In this technique, the pathogen infects and subsequently transfers the gene of interest into the chloroplast of the host plant. This approach involves five steps where it begins with (a) signal recognition, (b) T-DNA processing, (c) T-DNA movement to host cell, (d) T-DNA integrating with the host genome, and (e) expression of T-DNA (Pratiwi and Surya 2020). *A. tumefaciens* would attach to the plant cells and the gene products within the pathogen, which would be transferred to the T-DNA found in the tumor-inducing (Ti) plasmid. The T-DNA would integrate with the chloroplast genome via nonhomologous recombination. After the integration of the gene of interest, the chloroplast further regenerates and proliferates (Pratiwi and Surya 2020). De Block et al. have successfully transformed the chloramphenicol-resistant genes into the chloroplast genome using the *Agrobacterium*-mediated transformation technique (De Block et al. 1985).

9.3 Plants Transformed by Chloroplast Transformation

Although plastid transformation is >20 years old, it has been successfully achieved in relatively few species. Plastid transformation is a tissue culture-dependent process because exposure of plastids (cells) to the selective agent and gradual replacement of plastid genome copies can be best accomplished in the cell culture environment. Sustained plant regeneration capability is an important characteristic because it allows selective elimination of wild-type plastid genome copies before plant regeneration. The main crop species used for plastid transformation are in the Solanaceae, including tobacco, tomato (*S. lycopersicum*), potato (*S. tuberosum*), pepper (*Capsicum annuum*), and eggplant (*S. melongena*). Plastid transformation has also been reported in other important crop species, including lettuce (*Lactuca sativa*, Asteraceae), soybean (*Glycine max*, Fabaceae), cotton (*Gossypium* spp., Malvaceae), cauliflower (*Brassica oleracea* var. botrytis, Brassicaceae), carrot (*Daucus carota*, Umbelliferae), rice (*Oryza sativa*, Poaceae), and bitter squash (*Momordica charantia*, Cucurbitaceae). Until now, the plastids of over 20 flowering plants have been transformed. In addition to the crops mentioned above, recent successes in plastid transformation have been reported in the plant species bitter melon (Narra et al. 2018), and the medicinal plant sweet wormwood (*Artemisia annua*) (Kaushal et al. 2020) and licorice weed (*Scoparia dulcis*) (Muralikrishna et al. 2016; Kota et al. 2019a). The list of species in which plastid transformation has been achieved is given in Table 9.1.

9.4 Applications of Plastid Transformation

For the production of plant-based recombinant proteins and the improvement of crop yield, chloroplast engineering has its promising significance. Due to specific integration of transgene into chloroplast genome of the plants, it shows high copy number, and expression level of foreign protein will be high as compared to nuclear transformation (Meyers et al. 2010). There are various applications of this technology including better understanding of structure and function of plastids, metabolism, and evolutionary relationship with ancient ancestors; enhancement of the quality and performance of the plants to survive in the extreme conditions; and expression of the beneficial and economical agriculture traits and metabolic pathways to improve the quality of applied research in the field of biotechnology (Bock 2007). Following are some significant applications of plastid transformation.

Table 9.1 Species in which plastid transformation has been demonstrated

	Family	Species	Name	References
Plants				
Dicots	Solanaceae	*Nicotiana tabacum* var. Petit Havana, *Nicotiana benthamiana*, and *Nicotiana sylvestris*	Tobacco	Svab et al. (1990), Svab and Maliga (1993), Davarpanah et al. (2009), Maliga and Svab (2011)
		Solanum lycopersicum	Tomato	Ruf et al. (2001)
		Solanum tuberosum cv. Desirée and line 1607	Potato	Sidorov et al. (1999), Steinfeld et al. (2015)
		Petunia hybrida var. Pink Wave	Petunia	Zubko et al. (2004)
		Solanum melongena	Eggplant	Singh et al. (2010)
		Capsicum annuum var. G4	Pepper	Kota et al. (2019b)
	Brassicaceae	*Arabidopsis thaliana*	Arabidopsis	Sikdar et al. (1998)
		Brassica napus	Oilseed rape	Hou et al. (2003)
		Lesquerella fendleri	Lesquerella	Skarjinskaia et al. (2003)
		Brassica oleracea var. botrytis	Cauliflower	Nugent et al. (2006)
		Brassica oleracea var. capitata	Cabbage	Liu et al. (2007)
		Brassica napus cv. FY-4	Rapeseed	Cheng et al. (2010)
	Malvaceae	*Gossypium hirsutum* cv. Coker310FR	Cotton	Kumar et al. (2004b)
	Apiaceae	*Daucus carota* cv. Half long	Carrot	Kumar et al. (2004a)
	Fabaceae	*Glycine max*	Soybean	Dufourmantel et al. (2004)
		Medicago sativa cv. Longmu 803	Alfalfa	Wei et al. (2011)
	Asteraceae	*Lactuca sativa* cv. Verônica, cv Flora and cv. Cisco	Lettuce	Lelivelt et al. (2005), Kanamoto et al. (2006)
		Artemisia annua	Artemisa	Kaushal et al. (2020)
	Salicaceae	*Populus alba*	Poplar	Okumura et al. (2006)
	Amaranthaceae	*Beta vulgaris*	Sugar beet	De Marchis et al. (2009)
	Plantaginaceae	*Scoparia dulcis*	Sweet broom	Muralikrishna et al. (2016)
	Cucurbitaceae	*Momordica charantia*	Bitter melon	Narra et al. (2018)

	Family	Species	Name	References
Monocots	Poaceae	*Oryza sativa* var. Japonica line 19 and Hwa-Chung	Rice	Wang et al. (2018)
Algae				
Green algae	Chlamydomonadaceae	*Chlamydomonas reinhardtii*	_	Goldschmidt-Clermont (1991)
	Euglenaceae	*Euglena gracilis*	_	Doetsch et al. (2001)
	Haematococcaceae	*Haematococcus pluvialis*	_	Gutierrez et al. (2012)
	Dunaliellaceae	*Dunaliella tertiolecta*	_	Georgianna et al. (2013)
Red algae	Porphyridiophyceae	*Porphyridium* sp. UTEX 637	_	Lapidot et al. (2002)
	Bangiaceae	*Pyropia yezoensis*	_	Kong et al. (2017)
	Cyanidiaceae	*Cyanidioschyzon merolae*	_	Zienkiewicz et al. (2017)
Microalgae	Phaeodactylaceae	*Phaeodactylum tricornutum*	_	Xie et al. (2014)
	Monodopsidaceae	*Nannochloropsis oceanica*	_	Gan et al. (2018)
	Isochrysidaceae	*Tisochrysis lutea*	_	Bo et al. (2020)
Moss	Funariaceae	*Physcomitrella patens*	Moss	Sugiura and Sugita (2004)
Liverworts	Marchantiaceae	*Marchantia polymorpha*	Umbrella liverwort	Chiyoda et al. (2007)

9.4.1 Applications in Basic Science

Chloroplast transformation was first reported in 1988 in *Chlamydomonas reinhardtii* (Boynton et al. 1988), and the normal function of plant was restored by replacing the mutant gene of chloroplast with the normal wild-type gene. Since then, this technique has been used significantly to study and better understand different metabolic pathways and function of plastid genes, which play an important role in functional genomics of plants (Bock 2001; Daniell 2002; Maliga 2004). Different studies have been carried out including site-specific mutagenesis; replacement, deletion, and insertion of gene; high expression of foreign and native proteins of the plants; study of the plastid metabolic pathways; and structural and functional genomics and proteomics (Maliga 2004; Bock 2007; Koop et al. 2007).

9.4.2 Antigen Vaccines and Protein-Based Drugs

Protein-based drugs produced in transplastomic plants may solve many of the associated issues without compromising the drug efficacy and rise in cost (Adem et al. 2017). Many vaccine antigens and biopharmaceuticals have been successfully produced from the chloroplasts of flowering plants. Unlike microorganisms, plant chloroplasts can perform posttranslational modifications of protein-based drugs and promote their proper folding: phosphorylation and disulfide bond formation (Řepková 2010). Human coagulation factors made from plants have also been shown to improve immune tolerance in hemophilia murine and canine models (Herzog et al. 2017; Kwon et al. 2018). In addition, high-level expression of vaccine antigens and therapeutic proteins has been achieved in plant chloroplasts (leaves and roots) or chromoplasts (fruits) for antigens associated with the plague, tetanus, human immunodeficiency virus (HIV), cholera, malaria, Alzheimer's disease, and hemophilia (Tregoning et al. 2003; Herzog et al. 2017). Table 9.2 provides a partial list of vaccine antigens and drug proteins expressed in the plant's chloroplast.

9.4.3 Industrial Enzymes and Biomaterials

Several industrial enzymes and biomaterials have been expressed in plastid genomes of different plants. The most important biodegradable polyester biopolymer is polyhydroxyalkanoate (PHA), which is naturally synthesized by microorganisms that can be used as an alternative to petroleum-based plastics (Dobrogojski et al. 2018). The most well-known and studied PHA is polyhydroxybutyrate (PHB). To date, however, the highest level of PHB accumulation was achieved in tobacco plastids, with levels of 18.8% of dry weight (DW). The tobacco system was based on an

Table 9.2 Vaccine antigens and biopharmaceuticals engineered via chloroplast genome of higher plants

Traits	Expression	Host plant	References
Insulin	14.3% TSP	Tobacco	Kwon et al. (2013)
Hemophilia B	3.8% TSP in tobacco	Tobacco	Verma et al. (2010)
HIV	7–8% TSP	Tobacco	Scotti et al. (2009)
HPV	3–8% TSP	Tobacco	Morgenfeld et al. (2014)
Cholera	7.3% TSP 13.2% TSP	Tobacco Lettuce	Davoodi-Semiromi et al. (2010)
Tuberculosis	1.2–7.5% TSP	Tobacco	Lakshmi et al. (2013)
Tuberculosis	>0.035% TSP	Carrot	Permyakova et al. (2015)
Dengue virus	0.8–1.6% TSP	Tobacco	Gottschamel et al. (2016)
Polio virus	4–5% TSP	Tobacco	Lakshmi et al. (2013)
Bacterial phage lytic protein	>70% TSP	Tobacco	
Interferon-α2b (IFN-α2b)	21% TSP	Tobacco	Wang et al. (2015)
Basic fibroblast growth factor (bFGF)	0.1% TSP	Tobacco	Morgenfeld et al. (2009)

TSP total soluble protein

operon extension strategy to synthesize high PHB levels by introducing a bacterial operon, consisting of three genes encoding enzymes necessary for PHB biosynthesis, into the tobacco chloroplast genome (Bohmert-Tatarev et al. 2011). The high amounts of PHB produced in this system stem from the high flux of the PHB biosynthetic precursor acetyl-CoA released during fatty acid biosynthesis (Snell and Peoples 2009). Typical examples of industrial enzymes and biomaterials obtained through plastid transformation are given in Table 9.3.

9.5 Limitations of Chloroplast Transformation in Plant Molecular Farming

Undoubtedly, plastid genetic engineering holds great promise for plant biotechnology; however, certain challenges should be considered before the technology can reach its full potential. Of utmost significance is the need to extend the range of crops for plastid transformation. Although progress has been made in developing plastid transformation for some important crops (Ruf et al. 2001; Kumar et al. 2004b), still there is a lack of workable protocols for cereal species, including the

Table 9.3 Industrial enzymes and biomaterials expressed in chloroplast genome of tobacco

Traits	Gene	Expression	Host plant	References
Cellulases	*bgl1C, cel6B, cel9A, xeg74*	5–40% TSP	Tobacco	Petersen and Bock (2011)
	CelA, CelB	22–23 mg/g TSP	Tobacco	Espinoza-Sánchez et al. (2016)
Elastin-derived polymer	*eg121*	ND	Tobacco	Guda et al. (2000)
Endo-1,4-beta-glucanase	*celA*	10.7% TSP	Tobacco	Gray et al. (2009)
Exo-cellobiohydrolase	*celB*	3% TSP	Tobacco	Yu et al. (2007)
Fibronectin extra domain A	*EDA*	2% TCP	Tobacco	Farran et al. (2008)
Monellin	*monellin*	2.5% TSP	Tobacco	Roh et al. (2006)
p-Hydroxybenzoic acid	*ubi*C	13–18% TSP	Tobacco	Viitanen et al. (2004)
Polyhydroxybutyrate	*phb* operon	18.8% DW	Tobacco	Bohmert-Tatarev et al. (2011)
Xylanase	xynA	6% TSP	Tobacco	Leelavathi et al. (2003)
	xyn	35% TSP	Tobacco	Castiglia et al. (2016)
β-Glucosidase	Bgl1	20 mg/g TSP	Tobacco	Espinoza-Sánchez et al. (2016)
	celB	60–70% TSP	Tobacco	Castiglia et al. (2016)
Endo-glucanase	Endo	≤2% TSP	Tobacco	Castiglia et al. (2016)
Superoxide dismutase	Cu/Zn SOD	9% TSP	Tobacco	Madanala et al. (2015)

TCP total cellular proteins, *TSP* total soluble proteins, *ND* not determined

main staple foods of the world. Most likely, it demands major investments in optimization of the currently available protocols of tissue culture, regeneration, and selection, before plastid transformation could become a reality (Dufourmantel et al. 2004). So far, majority of studies utilizing transplastomic technology have been carried out in tobacco, mostly limiting the analysis of transgene expression to leaf chloroplasts. If the technology is to be broadly applied in food crops, a better understanding is needed for gene expression and its control in nongreen plastid types found in most fruits, tubers, and seeds.

Chloroplast transformation is normally significant in dicotyledonous plants and especially the members belonging to *Solanaceae* family. The main reasons of limitation of plastid transformation in monocotyledonous plant species is the recalcitrant nature of these plants and the fact that it could not regenerate in artificial conditions of tissue culturing (Lee et al. 2006). These plants are also resistant to antibiotics used for chloroplast transformation, e.g., spectinomycin, kanamycin, chloramphenicol, and ampicillin (Li et al. 2011). It has been reported that plastids have the potential to move among the cells, and hence it has the ability of grafting donor plant tissue containing transformed plastids on to a recipient untransformed plant tissue; thus, it has the ability to transfer transformed plastids into untransformed plants (Stegemann et al. 2012).

Transformation of plastid compartment of plants offers many advantages like transgene containment and high protein levels that have increased the interest of scientists (Maliga 2004). Mostly, plants are transformed by constitutive expression, but many times, constitutive expression has deleterious effects on plants after the expression of transgene product either due to the interference with the metabolic pathways of plants or sometimes due to their toxic effects on the plants (Daniell et al. 2001; Lössl and Waheed 2011). Hence, constitutive promoters cannot be used for the transgene product whose low or high production has harmful effects on plants. Transgene products could interact with metabolism at different stages and thus could lead to phenotypic alterations in the transformed plants. The solution of this problem is the use of inducible expression systems. With the use of this system, it becomes possible to regulate gene expression in transgenic plants. Moreover, the use of an appropriate promoter-targeted transgene expression could be restricted to a particular plant organ (Gatz and Lenk 1998). There are several inducible systems, and the main requirement of activating a promoter is achieved by different chemicals like ethanol, steroids, pesticides, antibiotics, phytohormones, and metallic ions (Mühlbauer and Koop 2005; Tungsuchat et al. 2006; Verhounig et al. 2010; Emadpour et al. 2015). Some other factors are also used like light, pathogen infection, and stress (Johnson et al. 2003).

References

Adem M, Beyene D, Feyissa T (2017) Recent achievements obtained by chloroplast transformation. Plant Methods 13(1):1–11

Ahmad N, Mukhtar Z (2013) Green factories: plastids for the production of foreign proteins at high levels. Gene Ther Mol Biol 15:14–29

Bhatia S, Bera T, Dahiya R, Bera T, Bhatia S, Bera T (2015) Classical and nonclassical techniques for secondary metabolite production in plant cell culture. In: Modern applications of plant biotechnology in pharmaceutical sciences. Academic Press, London, pp 231–291

Bo Y, Wang K, Wu Y, Cao H, Cui Y, Wang L (2020) Establishment of a chloroplast transformation system in Tisochrysis lutea. J Appl Phycol 32(5):2959–2965

Bock R (2001) Transgenic plastids in basic research and plant biotechnology. J Mol Biol 312(3):425–438

Bock R (2007) Structure, function, and inheritance of plastid genomes. In: Cell and molecular biology of plastids. Springer, Berlin, pp 29–63

Bohmert-Tatarev K, McAvoy S, Daughtry S, Peoples OP, Snell KD (2011) High levels of bioplastic are produced in fertile transplastomic tobacco plants engineered with a synthetic operon for the production of polyhydroxybutyrate. Plant Physiol 155(4):1690–1708

Boynton JE, Gillham NW, Harris EH, Hosler JP, Johnson AM, Jones AR et al (1988) Chloroplast transformation in Chlamydomonas with high velocity microprojectiles. Science 240(4858):1534–1538

Carter M, Shieh JC (2015) Guide to research techniques in neuroscience. Academic Press

Castiglia D, Sannino L, Marcolongo L, Ionata E, Tamburino R, De Stradis A, Scotti N (2016) High-level expression of thermostable cellulolytic enzymes in tobacco transplastomic plants and their use in hydrolysis of an industrially pretreated Arundo donax L. biomass. Biotechnol Biofuels 9(1):1–16

Chemat F, Abert-Vian M, Fabiano-Tixier AS, Strube J, Uhlenbrock L, Gunjevic V, Cravotto G (2019) Green extraction of natural products. Origins, current status, and future challenges. TrAC Trends Anal Chem 118:248–263

Cheng L, Li HP, Qu B, Huang T, Tu JX, Fu TD, Liao YC (2010) Chloroplast transformation of rapeseed (Brassica napus) by particle bombardment of cotyledons. Plant Cell Rep 29(4):371–381

Chiyoda S, Linley PJ, Yamato KT, Fukuzawa H, Yokota A, Kohchi T (2007) Simple and efficient plastid transformation system for the liverwort Marchantia polymorpha L. suspension-culture cells. Transgenic Res 16(1):41–49

Corigliano MG, Albarracín RM, Vilas JM, López EFS, Luoni SAB, Deng B, Clemente M (2019) Heat treatment alleviates the growth and photosynthetic impairment of transplastomic plants expressing Leishmania infantum Hsp83-Toxoplasma gondii SAG1 fusion protein. Plant Sci 284:117–126

Daniell H (2002) Gene flow from genetically modified crops: current and future technologies for transgene containment. Nat Biotechnol 20:581–586

Daniell H, Lee SB, Panchal T, Wiebe PO (2001) Expression of the native cholera toxin B subunit gene and assembly as functional oligomers in transgenic tobacco chloroplasts. J Mol Biol 311(5):1001–1009

Daniell H, Singh ND, Mason H, Streatfield SJ (2009) Plant-made vaccine antigens and biopharmaceuticals. Trends Plant Sci 14(12):669–679

Davarpanah SJ, Jung SH, Kim YJ, Park YI, Min SR, Liu JR, Jeong WJ (2009) Stable plastid transformation in Nicotiana benthamiana. J Plant Biol 52(3):244–250

Davoodi-Semiromi A, Schreiber M, Nalapalli S, Verma D, Singh ND, Banks RK, Daniell H (2010) Chloroplast-derived vaccine antigens confer dual immunity against cholera and malaria by oral or injectable delivery. Plant Biotechnol J 8(2):223–242

De Block M, Schell J, Van Montagu M (1985) Chloroplast transformation by *agrobacterium* tumefaciens. EMBO J 4(6):1367–1372

De Marchis F, Wang Y, Stevanato P, Arcioni S, Bellucci M (2009) Genetic transformation of the sugar beet plastome. Transgenic Res 18(1):17–30

Díaz A, Villanueva P, Oliva V, Gil-Durán C, Fierro F, Chávez R, Vaca I (2019) Genetic transformation of the filamentous fungus Pseudogymnoascus verrucosus of Antarctic origin. Front Microbiol 10:2675

Dobrogojski J, Spychalski M, Luciński R, Borek S (2018) Transgenic plants as a source of polyhydroxyalkanoates. Acta Physiol Plant 40(9):1–17

Doetsch NA, Favreau MR, Kuscuoglu N, Thompson MD, Hallick RB (2001) Chloroplast transformation in Euglena gracilis: splicing of a group III twintron transcribed from a transgenic psbK operon. Curr Genet 39(1):49–60

Dufourmantel N, Pelissier B, Garcon F, Peltier G, Ferullo JM, Tissot G (2004) Generation of fertile transplastomic soybean. Plant Mol Biol 55(4):479–489

Emadpour M, Karcher D, Bock R (2015) Boosting riboswitch efficiency by RNA amplification. Nucleic Acids Res 43(10):e66

Espinoza-Sánchez EA, Torres-Castillo JA, Rascón-Cruz Q, Zavala-García F, Sinagawa-García SR (2016) Production and characterization of fungal β-glucosidase and bacterial cellulases by tobacco chloroplast transformation. Plant Biotechnol Rep 10(2):61–73

Farquhar J, Zerkle AL, Bekker A (2011) Geological constraints on the origin of oxygenic photosynthesis. Photosynth Res 107(1):11–36

Farran I, Río-Manterola F, Íñiguez M, Gárate S, Prieto J, Mingo-Castel AM (2008) High-density seedling expression system for the production of bioactive human cardiotrophin-1, a potential therapeutic cytokine, in transgenic tobacco chloroplasts. Plant Biotechnol J 6(5):516–527

Gan Q, Jiang J, Han X, Wang S, Lu Y (2018) Engineering the chloroplast genome of oleaginous marine microalga Nannochloropsis oceanica. Front Plant Sci 9:439

Gatz C, Lenk I (1998) Promoters that respond to chemical inducers. Trends Plant Sci 3(9):352–358

Georgianna DR, Hannon MJ, Marcuschi M, Wu S, Botsch K, Lewis AJ, Mayfield SP (2013) Production of recombinant enzymes in the marine alga Dunaliella tertiolecta. Algal Res 2(1):2–9

Goldschmidt-Clermont M (1991) Transgenic expression of aminoglycoside adenine transferase in the chloroplast: a selectable marker for site-directed transformation of Chlamydomonas. Nucleic Acids Res 19(15):4083–4089

Gottschamel J, Lössl A, Ruf S, Wang Y, Skaugen M, Bock R, Clarke JL (2016) Production of dengue virus envelope protein domain III-based antigens in tobacco chloroplasts using inducible and constitutive expression systems. Plant Mol Biol 91(4):497–512

Gray BN, Ahner BA, Hanson MR (2009) High-level bacterial cellulase accumulation in chloroplast-transformed tobacco mediated by downstream box fusions. Biotechnol Bioeng 102(4):1045–1054

Guda C, Lee SB, Daniell H (2000) Stable expression of a biodegradable protein-based polymer in tobacco chloroplasts. Plant Cell Rep 19(3):257–262

Gutierrez CL, Gimpel J, Escobar C, Marshall SH, Henríquez V (2012) Chloroplast genetic tool for the green microalgae Haematococcus pluvialis (Chlorophyceae, Volvocales) 1. J Phycol 48(4):976–983

Herzog RW, Nichols TC, Su J, Zhang B, Sherman A, Merricks EP, Daniell H (2017) Oral tolerance induction in hemophilia B dogs fed with transplastomic lettuce. Mol Ther 25(2):512–522

Hou BK, Zhou YH, Wan LH, Zhang ZL, Shen GF, Chen ZH, Hu ZM (2003) Chloroplast transformation in oilseed rape. Transgenic Res 12(1):111–114

Jinturkar KA, Rathi MN, Misra A (2011) Gene delivery using physical methods. In: Challenges in delivery of therapeutic genomics and proteomics. Elsevier, pp 83–126

Johnson C, Boden E, Arias J (2003) Salicylic acid and NPR1 induce the recruitment of trans-activating TGA factors to a defense gene promoter in Arabidopsis. Plant Cell 15(8):1846–1858

Kanamoto H, Yamashita A, Asao H, Okumura S, Takase H, Hattori M, Tomizawa KI (2006) Efficient and stable transformation of Lactuca sativa L. cv. Cisco (lettuce) plastids. Transgenic Res 15(2):205–217

Kaushal C, Abdin MZ, Kumar S (2020) Chloroplast genome transformation of medicinal plant Artemisia annua. Plant Biotechnol J 18(11):2155

Kong F, Zhao H, Liu W, Li N, Mao Y (2017) Construction of plastid expression vector and development of genetic transformation system for the seaweed Pyropia yezoensis. Mar Biotechnol 19(2):147–156

Koop HU, Herz S, Golds TJ, Nickelsen J (2007) The genetic transformation of plastids. In: Cell and molecular biology of plastids. Springer, Berlin, pp 457–510

Kota S, Hao Q, Narra M, Anumula V, Rao AV, Hu Z, Abbagani S (2019a) Improved plastid transformation efficiency in Scoparia dulcis L. J Plant Biotechnol 46(4):323–330

Kota S, Lakkam R, Kasula K, Narra M, Qiang H, Rao Allini V, Abbagani S (2019b) Construction of a species-specific vector for improved plastid transformation efficiency in Capsicum annuum L. 3 Biotech 9(6):1–11

Kumar S, Dhingra A, Daniell H (2004a) Plastid-expressed betaine aldehyde dehydrogenase gene in carrot cultured cells, roots, and leaves confers enhanced salt tolerance. Plant Physiol 136(1):2843–2854

Kumar S, Dhingra A, Daniell H (2004b) Stable transformation of the cotton plastid genome and maternal inheritance of transgenes. Plant Mol Biol 56(2):203–216

Kwak SY, Lew TTS, Sweeney CJ, Koman VB, Wong MH, Bohmert-Tatarev K et al (2019) Chloroplast-selective gene delivery and expression in planta using chitosan-complexed single-walled carbon nanotube carriers. Nat Nanotechnol 14(5):447–455

Kwon KC, Nityanandam R, New JS, Daniell H (2013) Oral delivery of bioencapsulated exendin-4 expressed in chloroplasts lowers blood glucose level in mice and stimulates insulin secretion in beta-TC 6 cells. Plant Biotechnol J 11(1):77–86

Kwon KC, Sherman A, Chang WJ, Kamesh A, Biswas M, Herzog RW, Daniell H (2018) Expression and assembly of largest foreign protein in chloroplasts: oral delivery of human FVIII made

in lettuce chloroplasts robustly suppresses inhibitor formation in haemophilia A mice. Plant Biotechnol J 16(6):1148–1160
Lacroix B, Citovsky V (2020) Biolistic approach for transient gene expression studies in plants. In: Biolistic DNA delivery in plants. Humana Press, New York, pp 125–139
Lakshmi PS, Verma D, Yang X, Lloyd B, Daniell H (2013) Low cost tuberculosis vaccine antigens in capsules: expression in chloroplasts, bio-encapsulation, stability and functional evaluation in vitro. PLoS One 8(1):e54708
Lapidot M, Raveh D, Sivan A, Arad SM, Shapira M (2002) Stable chloroplast transformation of the unicellular red alga Porphyridium species. Plant Physiol 129(1):7–12
Latif S, Gottschamel J, Syed T, Younus I, Gull K, Sameeullah M, Waheed MT (2022) Inducible expression of human papillomavirus-16 L1 capsomeres in the plastomes of Nicotiana tabacum: transplastomic plants develop normal flowers and pollen. Biotechnol Appl Biochem 69(2):596–611
Lee SM, Kang K, Chung H, Yoo SH, Xu XM, Lee SB, Kim M (2006) Plastid transformation in the monocotyledonous cereal crop, rice (Oryza sativa) and transmission of transgenes to their progeny. Mol Cells 21(3):401
Leelavathi S, Gupta N, Maiti S, Ghosh A, Siva Reddy V (2003) Overproduction of an alkali- and thermo-stable xylanase in tobacco chloroplasts and efficient recovery of the enzyme. Mol Breed 11(1):59–67
Lelivelt CL, McCabe MS, Newell CA, Desnoo CB, Van Dun KM, Birch-Machin I et al (2005) Stable plastid transformation in lettuce (Lactuca sativa L.). Plant Mol Biol 58(6):763–774
Li W, Ruf S, Bock R (2011) Chloramphenicol acetyltransferase as selectable marker for plastid transformation. Plant Mol Biol 76(3):443–451
Liu Z, Friesen TL (2012) Polyethylene glycol (PEG)-mediated transformation in filamentous fungal pathogens. In: Plant fungal pathogens. Humana Press, pp 365–375
Liu YC, Vidali L (2011) Efficient polyethylene glycol (PEG) mediated transformation of the moss Physcomitrella patens. J Vis Exp (JoVE) 50:e2560
Liu CW, Lin CC, Chen JJ, Tseng MJ (2007) Stable chloroplast transformation in cabbage (Brassica oleracea L. var. capitata L.) by particle bombardment. Plant Cell Rep 26(10):1733–1744
Lössl AG, Waheed MT (2011) Chloroplast-derived vaccines against human diseases: achievements, challenges and scopes. Plant Biotechnol J 9(5):527–539
Low LY, Yang SK, Kok DXA, Ong-Abdullah J, Tan NP, Lai KS (2018) Transgenic plants: gene constructs, vector and transformation method. In: New visions in plant science. IntechOpen, London, pp 41–61
Madanala R, Gupta V, Pandey AK, Srivastava S, Pandey V, Singh PK, Tuli R (2015) Tobacco chloroplasts as bioreactors for the production of recombinant superoxide dismutase in plants, an industrially useful enzyme. Plant Mol Biol Report 33(4):1107–1115
Maliga P (2004) Plastid transformation in higher plants. Annu Rev Plant Biol 55:289–313
Maliga P, Bock R (2011) Plastid biotechnology: food, fuel, and medicine for the 21st century. Plant Physiol 155(4):1501–1510
Maliga P, Svab Z (2011) Engineering the plastid genome of Nicotiana sylvestris, a diploid model species for plastid genetics. In: Plant chromosome engineering. Humana Press, Totowa, NJ, pp 37–50
Masani MYA, Noll GA, Parveez GKA, Sambanthamurthi R, Prüfer D (2014) Efficient transformation of oil palm protoplasts by PEG-mediated transfection and DNA microinjection. PLoS One 9(5):e96831
Mathur J, Koncz C (1998) PEG-mediated protoplast transformation with naked DNA. In: Arabidopsis protocols. Humana Press, pp 267–276
Matsumoto TK, Gonsalves D (2012) Biolistic and other non-*Agrobacterium* technologies of plant transformation. In: Plant biotechnology and agriculture. Academic Press, pp 117–129
Matsuoka A (2015) Genetic engineering of Agrobacterium tumefaciens to target chloroplasts and identification of a novel nuclear localization signal of VIRD2 protein. University of Lethbridge, Lethbridge

Meyers B, Zaltsman A, Lacroix B, Kozlovsky SV, Krichevsky A (2010) Nuclear and plastid genetic engineering of plants: comparison of opportunities and challenges. Biotechnol Adv 28(6):747–756

Morgenfeld M, Segretin ME, Wirth S, Lentz E, Zelada A, Mentaberry A, Bravo-Almonacid F (2009) Potato virus X coat protein fusion to human papillomavirus 16 E7 oncoprotein enhance antigen stability and accumulation in tobacco chloroplast. Mol Biotechnol 43(3):243–249

Morgenfeld M, Lentz E, Segretin ME, Alfano EF, Bravo-Almonacid F (2014) Translational fusion and redirection to thylakoid lumen as strategies to enhance accumulation of human papillomavirus E7 antigen in tobacco chloroplasts. Mol Biotechnol 56(11):1021–1031

Morgenfeld MM, Vater CF, Alfano EF, Boccardo NA, Bravo-Almonacid FF (2020) Translocation from the chloroplast stroma into the thylakoid lumen allows expression of recombinant epidermal growth factor in transplastomic tobacco plants. Transgenic Res 29(3):295–305

Mühlbauer SK, Koop HU (2005) External control of transgene expression in tobacco plastids using the bacterial lac repressor. Plant J 43(6):941–946

Muralikrishna N, Srinivas K, Kumar KB, Sadanandam A (2016) Stable plastid transformation in Scoparia dulcis L. Physiol Mol Biol Plants 22(4):575–581

Nanjareddy K, Arthikala MK, Blanco L, Arellano ES, Lara M (2016) Protoplast isolation, transient transformation of leaf mesophyll protoplasts and improved *Agrobacterium*-mediated leaf disc infiltration of Phaseolus vulgaris: tools for rapid gene expression analysis. BMC Biotechnol 16(1):1–14

Narra M, Kota S, Velivela Y, Ellendula R, Allini VR, Abbagani S (2018) Construction of chloroplast transformation vector and its functional evaluation in Momordica charantia L. 3 Biotech 8(3):1–11

Newkirk GM, Allende P, Jinkerson RE, Giraldo JP (2021) Nanotechnology approaches for chloroplast biotechnology advancements. Front Plant Sci 12:691295

Nugent GD, Coyne S, Nguyen TT, Kavanagh TA, Dix PJ (2006) Nuclear and plastid transformation of Brassica oleracea var. botrytis (cauliflower) using PEG-mediated uptake of DNA into protoplasts. Plant Sci 170(1):135–142

Occhialini A, Pfotenhauer AC, Frazier TP, Li L, Harbison SA, Lail AJ et al (2020) Generation, analysis, and transformation of macro-chloroplast potato (Solanum tuberosum) lines for chloroplast biotechnology. Sci Rep 10(1):1–13

Okumura S, Sawada M, Park YW, Hayashi T, Shimamura M, Takase H, Tomizawa KI (2006) Transformation of poplar (Populus alba) plastids and expression of foreign proteins in tree chloroplasts. Transgenic Res 15(5):637–646

Permyakova NV, Zagorskaya AA, Belavin PA, Uvarova EA, Nosareva OV, Nesterov AE, Deineko EV (2015) Transgenic carrot expressing fusion protein comprising M. tuberculosis antigens induces immune response in mice. BioMed Res Int 2015:417565

Petersen K, Bock R (2011) High-level expression of a suite of thermostable cell wall-degrading enzymes from the chloroplast genome. Plant Mol Biol 76(3):311–321

Pratiwi RA, Surya MI (2020) *Agrobacterium*-mediated transformation. In: Genetic transformation in crops. IntechOpen

Ramesh VM, Bingham SE, Webber AN (2011) A simple method for chloroplast transformation in Chlamydomonas reinhardtii. In: Photosynthesis research protocols. Humana Press, Totowa, NJ, pp 313–320

Rascón-Cruz Q, González-Barriga CD, Iglesias-Figueroa BF, Trejo-Muñoz JC, Siqueiros-Cendón T, Sinagawa-García SR, Espinoza-Sánchez EA (2021) Plastid transformation: advances and challenges for its implementation in agricultural crops. Electron J Biotechnol 51:95–109

Raskin I, Ribnicky DM, Komarnytsky S, Ilic N, Poulev A, Borisjuk N, Fridlender B (2002) Plants and human health in the twenty-first century. Trends Biotechnol 20(12):522–531

Rehman L, Su X, Guo H, Qi X, Cheng H (2016) Protoplast transformation as a potential platform for exploring gene function in Verticillium dahliae. BMC Biotechnol 16(1):1–9

Řepková J (2010) Potential of chloroplast genome in plant breeding. Czech J Genet Plant Breed 46(3):103–113

Rivera AL, Gómez-Lim M, Fernández F, Loske AM (2012) Physical methods for genetic plant transformation. Phys Life Rev 9(3):308–345

Roh KH, Shin KS, Lee YH, Seo SC, Park HG, Daniell H, Lee SB (2006) Accumulation of sweet protein monellin is regulated by the psbA 5′ UTR in tobacco chloroplasts. J Plant Biol 49(1):34–43

Ruf S, Hermann M, Berger IJ, Carrer H, Bock R (2001) Stable genetic transformation of tomato plastids and expression of a foreign protein in fruit. Nat Biotechnol 19(9):870–875

Ruhlman T, Verma D, Samson N, Daniell H (2010) The role of heterologous chloroplast sequence elements in transgene integration and expression. Plant Physiol 152(4):2088–2104

Saba K, Gottschamel J, Younus I, Syed T, Gull K, Lössl AG, Waheed MT (2019) Chloroplast-based inducible expression of ESAT-6 antigen for development of a plant-based vaccine against tuberculosis. J Biotechnol 305:1–10

Saba K, Sameeullah M, Asghar A, Gottschamel J, Latif S, Lössl AG, Waheed MT (2020) Expression of ESAT-6 antigen from Mycobacterium tuberculosis in broccoli: an edible plant. Biotechnol Appl Biochem 67(1):148–157

Scotti N, Alagna F, Ferraiolo E, Formisano G, Sannino L, Buonaguro L, Cardi T (2009) High-level expression of the HIV-1 Pr55gag polyprotein in transgenic tobacco chloroplasts. Planta 229(5):1109–1122

Scotti N, Rigano MM, Cardi T (2012) Production of foreign proteins using plastid transformation. Biotechnol Adv 30(2):387–397

Sidorov VA, Kasten D, Pang SZ, Hajdukiewicz PT, Staub JM, Nehra NS (1999) Stable chloroplast transformation in potato: use of green fluorescent protein as a plastid marker. Plant J 19(2):209–216

Sikdar SR, Serino G, Chaudhuri S, Maliga P (1998) Plastid transformation in Arabidopsis thaliana. Plant Cell Rep 18(1):20–24

Singh AK, Verma SS, Bansal KC (2010) Plastid transformation in eggplant (Solanum melongena L.). Transgenic Res 19(1):113–119

Skarjinskaia M, Svab Z, Maliga P (2003) Plastid transformation in Lesquerella fendleri, an oilseed Brassicaceae. Transgenic Res 12(1):115–122

Snell KD, Peoples OP (2009) PHA bioplastic: a value-added coproduct for biomass biorefineries. Biofuels Bioprod Biorefin 3(4):456–467

Son KH, Hong JH, Lee JW (2016) Carbon nanotubes as cancer therapeutic carriers and mediators. Int J Nanomedicine 11:5163

Stegemann S, Keuthe M, Greiner S, Bock R (2012) Horizontal transfer of chloroplast genomes between plant species. Proc Natl Acad Sci 109(7):2434–2438

Steinfeld B, Scott J, Vilander G, Marx L, Quirk M, Lindberg J, Koerner K (2015) The role of lean process improvement in implementation of evidence-based practices in behavioral health care. J Behav Health Serv Res 42(4):504–518

Sugiura C, Sugita M (2004) Plastid transformation reveals that moss tRNAArg-CCG is not essential for plastid function. Plant J 40(2):314–321

Svab Z, Maliga PAL (1993) High-frequency plastid transformation in tobacco by selection for a chimeric aadA gene. Proc Natl Acad Sci 90(3):913–917

Svab Z, Hajdukiewicz P, Maliga P (1990) Stable transformation of plastids in higher plants. Proc Natl Acad Sci 87(21):8526–8530

Tregoning JS, Nixon P, Kuroda H, Svab Z, Clare S, Bowe F, Maliga P (2003) Expression of tetanus toxin fragment C in tobacco chloroplasts. Nucleic Acids Res 31(4):1174–1179

Tungsuchat T, Kuroda H, Narangajavana J, Maliga P (2006) Gene activation in plastids by the CRE site-specific recombinase. Plant Mol Biol 61(4):711–718

Verhounig A, Karcher D, Bock R (2010) Inducible gene expression from the plastid genome by a synthetic riboswitch. Proc Natl Acad Sci 107(14):6204–6209

Verma D, Moghimi B, LoDuca PA, Singh HD, Hoffman BE, Herzog RW, Daniell H (2010) Oral delivery of bioencapsulated coagulation factor IX prevents inhibitor formation and fatal anaphylaxis in hemophilia B mice. Proc Natl Acad Sci 107(15):7101–7106

Viitanen PV, Devine AL, Khan MS, Deuel DL, Van Dyk DE, Daniell H (2004) Metabolic engineering of the chloroplast genome using the Escherichia coli ubi C gene reveals that chorismate is a readily abundant plant precursor for p-hydroxybenzoic acid biosynthesis. Plant Physiol 136(4):4048–4060

Wang YP, Wei ZY, Zhong XF, Lin CJ, Cai YH, Ma J, Xing SC (2015) Stable expression of basic fibroblast growth factor in chloroplasts of tobacco. Int J Mol Sci 17(1):19

Wang Y, Wei Z, Xing S (2018) Stable plastid transformation of rice, a monocot cereal crop. Biochem Biophys Res Commun 503(4):2376–2379

Weber G, Monajembashi S, Greulich KO, Wolfrum J (1988) Genetic manipulation of plant cells and organelles with a laser microbeam. In: Progress in plant protoplast research. Springer, Dordrecht, pp 319–322

Wei Z, Liu Y, Lin C, Wang Y, Cai QA, Dong Y, Xing S (2011) Transformation of alfalfa chloroplasts and expression of green fluorescent protein in a forage crop. Biotechnol Lett 33(12):2487–2494

Xie WH, Zhu CC, Zhang NS, Li DW, Yang WD, Liu JS et al (2014) Construction of novel chloroplast expression vector and development of an efficient transformation system for the diatom Phaeodactylum tricornutum. Mar Biotechnol 16(5):538–546

Yao CP, Zhang ZX, Rahmanzadeh R, Huettmann G (2008) Laser-based gene transfection and gene therapy. IEEE Trans Nanobioscience 7(2):111–119

Yu LX, Gray BN, Rutzke CJ, Walker LP, Wilson DB, Hanson MR (2007) Expression of thermostable microbial cellulases in the chloroplasts of nicotine-free tobacco. J Biotechnol 131(3):362–369

Yu Y, Yu PC, Chang WJ, Yu K, Lin CS (2020) Plastid transformation: how does it work? Can it be applied to crops? What can it offer? Int J Mol Sci 21(14):4854

Zienkiewicz M, Krupnik T, Drożak A, Golke A, Romanowska E (2017) Transformation of the Cyanidioschyzon merolae chloroplast genome: prospects for understanding chloroplast function in extreme environments. Plant Mol Biol 93(1):171–183

Zienkiewicz M, Krupnik T, Drożak A, Kania K (2019) PEG-mediated, stable, nuclear and chloroplast transformation of Cyanidioschizon merolae. Bio-Protocol 9:e3355

Zubko MK, Zubko EI, Zuilen KV, Meyer P, Day A (2004) Stable transformation of petunia plastids. Transgenic Res 13(6):523–530

Chapter 10
Plant Molecular Farming for Developing Countries: Current Status and Future Perspectives

Muhammad Suleman Malik, Neelam Batool, Fatima Ijaz, Kiran Saba, Andreas Gunter Lössl, Muhammad Sameeullah, and Mohammad Tahir Waheed

Abstract Plant molecular farming is a nascent but promising biotechnology-based industry. It is an alternate system for the development of pharmaceutical and non-pharmaceutical products cost-effectively in a bulk amount and in short timescale as compared to already established expression systems. This chapter covers the different strategies used for plant molecular farming (PMF) such as stable or transient transformation methods and what advantages they offer, how PMF can help the developing countries to eradicate the different health-related problems they face, what problems are encountered during the establishment of plant molecular pharming as local industry, and what steps should be taken to solve them. We also discuss how PMF helped and is helping the world in different sectors of life globally. Further, the current status of the products that are made via PMF is also discussed, including antibodies, enzymes, growth factors, other pharmaceutical products, and

M. S. Malik · N. Batool · F. Ijaz · M. T. Waheed (✉)
Department of Biochemistry, Faculty of Biological Sciences, Quaid-i-Azam University, Islamabad, Pakistan
e-mail: tahirwaheed@qau.edu.pk

K. Saba
Department of Biochemistry, Faculty of Biological Sciences, Quaid-i-Azam University, Islamabad, Pakistan

Department of Biochemistry, Shaheed Benazir Bhutto Women University, Peshawar, Pakistan
e-mail: kiransab@sbbwu.edu.pk

A. G. Lössl
Department of Applied Plant Science and Plant Biotechnology, University of Natural Resources and Life Sciences (BOKU), Vienna, Austria
e-mail: andreas@lossl.de

M. Sameeullah
Center for Innovative Food Technologies Development, Application and Research, Bolu Abant Izzet Baysal University, Bolu, Turkey

C. Kole et al. (eds.), *Tools & Techniques of Plant Molecular Farming*, Concepts and Strategies in Plant Sciences, https://doi.org/10.1007/978-981-99-4859-8_10

non-pharmaceutical or industrial products, which have the greatest impact on health, economy, poverty, and industry of developed and developing countries.

Keywords Plant molecular farming · Stable transformation · Transient transformation · Pharmaceutical products · Cost-effective

Abbreviations

Bt	*Bacillus thuringiensis*
CAS	CRISPR associated
CRISPR	Clustered regularly interspaced short palindromic repeats
EPI	Exocrine pancreatic insufficiency
ETEC	Enterotoxigenic *Escherichia coli*
FAO	Food and Agriculture Organization
Gb3	Globotriaosylceramide
GDP	Gross domestic product
GM	Genetically modified
GMOs	Genetically modified organisms
GOI	Gene of interest
GRAS	Generally recognized as safe
IGg	Immunoglobulin
IP	Intellectual property
mAB	Monoclonal antibody
PEG	Polyethylene glycol
PHA	Polyhydroxyalkanoate
PHB	3-Hydroxybutyrate
PMF	Plant molecular farming
Ri	Root inducing
SARS-CoV-2	Severe acute respiratory syndrome coronavirus 2
ScFv	Single-chain variable fragment
SDGs	Sustainable Development Goals
SRL	Socially responsible licensing
Ti	Tumor inducing
VLPs	Viruslike particles

10.1 Plant Molecular Farming

Plant molecular farming (PMF), the latest branch of plant biotechnology, involves the genetic engineering of plants to make and produce large quantities of recombinant pharmaceuticals and industrial proteins. This technology rests on the genetic transformation of plants, which can be accomplished by the methods like stable gene transfer (gene transfer to nucleus and chloroplasts) and unstable gene transfer methods (viral vector) (Gupta et al. 2017). The increase in search for biomedicines has been associated with factors such as high costs and inefficient production systems (insect cells, bacterial, microbial eukaryotes, mammalian cells, and transgenic animals). Thus, a much safer and cost-effective production system is required (Bhatia 2018). Transgenic plants have the potential to produce safe recombinant proteins (vaccines, enzymes, growth factors, antibodies, etc.) and also attribute to large-scale, low-cost production system. Hence, they have been the subject of considerable attention (Alireza and Nader 2015).

Currently, there are many available drugs, which have been derived from the plants. Plants are also manipulated to increase the production of food and different compounds. Until the nineteenth century, all improvements in crops were brought by plant breeding techniques (Chemat et al. 2019). After the nineteenth century, advancements in genetically modified plants were made by introduction of different methods. One of the two main such methods includes plant's organs, tissues, cells, and protoplast in the tissue culture, and the second is the recombinant DNA technology involving manipulation of genes. These two fields became an important part of biotechnology in the past few decades. Recombinant DNA technology has great contribution to the development of plant biotechnology and has been used for a wide variety of plants with precise manipulation of genetic material (Chawla 2011).

10.2 Plant Transformation Strategies

Genetic transformation is an important development in plant organ and tissue culture, which permits the transfer of foreign genes with desirable traits into the host plant, and as a result, only transformed plants are selected (Hinchee et al. 1988). Prevention from biotic and abiotic stresses, improved quality, higher yield, resistance to diseases and pests, and other horticulture traits are the significant characteristics which can be engineered into plants via genetic transformation (Parmar et al. 2017). This method is extremely helpful in the transgene integration from not only the unrelated plants but also genes from bacterial, viral, fungal, and animal genomes (Ow et al. 1986; Spörlein and Koop 1991; Onouchi et al. 1991; Huang et al. 2003; Wang et al. 2009), thus improving the quality and yield for the future world.

Techniques of plant improvement are revolutionized upon the integration of classical breeding methods of plants with recent discoveries in the area of protoplast and recombinant DNA technology. In plant cells, DNA is present in nucleus, plastids, and mitochondria (Oldenburg and Bendich 2015). Plant cells can also be transformed either by transfer of naked DNA into protoplast (vector-independent methods) or via vector containing pieces of DNA (vector-dependent methods). Vector-independent methods are microinjections, liposome fusion, silicon carbide, and electroporation (Potrykus et al. 1985; Deshayes et al. 1985; Klein et al. 1987). Vector-dependent methods are *Agrobacterium*-mediated and high-velocity microprojectile bombardment method. Many different approaches have been tried for transformation, and among them, three are widely used, i.e., *Agrobacterium*-mediated transformation, biolistic gene gun, and direct DNA transfer method via

PEG-mediated transformation, and are of great importance in plant transformations (Dale et al. 1993). Transformation results in the expression of foreign gene in host organism, which can be either stable or transient.

10.2.1 Transient Expression

To develop the improved genetic transformation methods and to study different metabolic processes, transient expression method is much appropriate because in this method expression of transferred gene can be studied in a very short period of time (Gheysen et al. 1998). In transient expression system, there are no position effects because foreign genes cannot be integrated into the host genome as well as cannot be transferred to the next progeny. Extrachromosomal transgene expression can be observed within 3 h after DNA incorporation, and expression reaches to maximum within 2 days and remains for 10 days (Whitham et al. 2015).

10.2.2 Stable Transformation

In stable transformation, foreign genes are stably integrated into the host plant genome and are heritable. It is used to transform both nuclear and chloroplast DNA of different plants (Daniell et al. 2016). Though stable transformation is time consuming, there are different advantages of stable transformation as it gives high expression of transferred genes. It provides a stable and continuous genetic source in the form of transformed seeds. Stable transformation is not much costly as it requires investment for the first time for developing transgenic plants. After that, with the availability of transformed seeds, it becomes quite easy to grow transformed plants (Waheed et al. 2015). According to Keshavareddy (Keshavareddy et al. 2018), *Agrobacterium*-mediated transformation and particle bombardment method are the two most important and frequently used methods of plant transformation. Genes responsible for the production of vaccines can be expressed in transgenic plants via different stable transformation strategies like nuclear transformation (Hammond and Nemchinov 2009) and chloroplast transformation (Daniell et al. 2009).

10.2.2.1 *Agrobacterium*-Mediated Nuclear Transformation

Transformation through *Agrobacterium* has become a popular method in plants. This technique has been widely used for the introduction of desired genes in plant genome and successful regeneration of transgenic plants. This is the most effective way of nuclear transformation in plants under laboratory conditions. Naturally, *Agrobacterium tumefaciens* is a plant pathogen that infects wound sites in dicotyledons and causes tumor in plants (Binns and Thomashow 1988). Virulent strain has a tumor-inducing ability due to the presence of Ti-plasmid (tumor inducing). There are many strains of *Agrobacterium* containing different mega plasmids, e.g., in *A. tumefaciens* and *A. rhizogenes*, and their Ti-plasmid and Ri-plasmid cause crown gall and hairy root disease, respectively (Klee et al. 1987). For the purpose of stable transformation, the *Agrobacterium*-mediated transformation is considered better over the other methods of direct gene transfer because it results in the reduction of copy number of integrated gene in transgenic species, thereby decreasing the problems caused by transgene's co-suppression behavior, therefore resulting in a more stable expression of foreign gene by using a simplified technology with low cost (Hansen and Wright 1999).

Agrobacterium has the ability to transfer and integrate specific DNA segment (T-DNA) of Ti-plasmid in the genome of host cells and take over host cell's machin ery for its beneficial processes. In infected cells, T-DNA is transcribed by using host machinery, and its protein causes tumor production (Bandurska et al. 2016). In T-DNA, there are two types of genes: genes for opine synthesis and oncogenic genes. Opines are synthesized by condensation of amino acids and carbohydrates and consumed by *Agrobacterium* as nitrogen and carbon source. Oncogenic genes are involved in the production of cytokines and auxin leading to tumor formation (Hooykaas and Schilperoort 1992). In Ti-plasmid, genes involved in T-DNA transfer and opine catabolism are located outside the T-DNA segment (Zupan et al. 1996).

10.2.2.2 Chloroplast Transformation

The chloroplast, also known as plastid, is one of the organelles in plant cells and eukaryotic algae. Due to the presence of chlorophyll, chloroplasts are the site of photosynthesis and a source of world's food (Verma and Daniell 2007). It is a self-replicating organelle like DNA and mitochondria with genome size of ~120–150 kb. There are more than 100–300 plastids per cell in most plant species, accounting for 1000–10,000 copies of plastomes (Boffey and Leech 1982). In the field of biotechnology, plastid transformation has great potential as compared to nuclear transformation because of many advantages (Meyers et al. 2010). One of the most important advantages of chloroplast transformation is that the plastid transgene expression can be very high and desired protein product may account for up to 70–72% of the total leaf or soluble protein (Ruhlman et al. 2010). The reason behind this fact is the polyploidy of the plastid genome with up to 10,000 copies of the chloroplast genome in each plant cell, so a high amount of transgenic protein is produced in each cell. In higher plants, maternal inheritance of plastid genes takes place like in most angiosperm species (Hagemann 2004), so the transgenes are not dispersed into the environment by pollen. Transgene containment thus makes plastid transformation a valuable technique due to the lower risk of environment contamination for the production of genetically modified plants (Svab and Maliga 2007).

10.2.2.2.1 Biolistic or Gene Gun

Chloroplast transformation is mainly obtained by using biolistic approach, and the first successful chloroplast transformation was reported in *Chlamydomonas reinhardtii* via bombardment method by Boynton (Boynton et al. 1988). This technique was first described in plants as a gene transfer method (Klein et al. 1987). This direct method of gene transfer is suitable for many biological systems. This method has the potential to surpass physical barriers for genetic transformation in many cells, e.g., cell wall in plant cells. This technique can be applied for transient expression as well as for generation of stable transformants (Christou 1992).

In the biolistic method, expression cassette of transgene is inserted into the chloroplast genome through homologous recombination of flanking sequences (Maliga 2003). It is a physical technique of transformation in which subcellular sized, high-density small gold or tungsten particles (microprojectiles) are coated with the DNA plasmid vector containing GOI and antibiotic resistance gene. Then microprojectiles are fired against retaining mesh by a strong pressure wave of helium gas under vacuum gaining speed of several hundred meters per second. Then microprojectile decelerates and particles are thrown off from the surface microprojectile due to their momentum and small size. Eventually, particles enter the target tissue. This technique allows the transformation of cells within a tissue because particles can penetrate several layers of cells (Zhang et al. 2014). In plastid transformation, transgene expression cassette integrates via homologous recombination between the plastid genome and the transformation vector.

This method has several advantages over others. Once this technique is established in the laboratory, it is easy to handle. A single shot can provide many hits. This method does not require protoplasts; rather, it has the ability to act directly on tissues or cells. It is applicable in studying transient gene expression in differentiated tissues (Klein et al. 1987). This technique works on not only dicots but also monocots, i.e., wheat or maize. However, this technique has certain drawbacks, e.g., in embryonic tissue, bombardment affects regeneration capacity and causes sterility or inactivation of transgene (Mubeen et al. 2016; see also Chap. 9).

10.3 Developing Countries and Their Problems

Developed and developing countries are present around the globe (shown in Fig. 10.1). The independent and prosperous countries are called developed countries, while the countries at the dawn of industrialization are called developing countries. Further, developed countries have a much higher GDP compared to developing countries. Some of the problems associated with developing countries include population growth, soaring debt, poverty, export marginalization, climate vulnerability, and poor healthcare systems. All these major problems need immense attention.

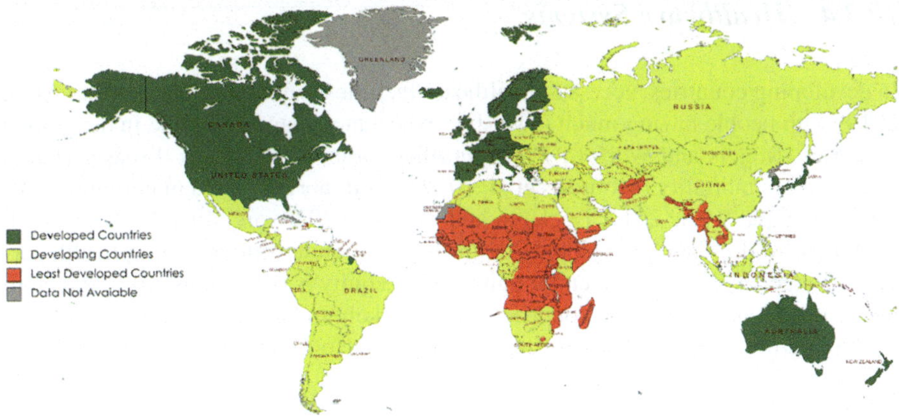

Fig. 10.1 World map showing the developed, developing, and least developed countries

10.3.1 Population Growth

Overpopulation has been linked to poverty. In the last few decades, increase in global population has been associated with developing countries. Higher birth rates have been observed in such countries. Also, with an increase in population, the demand of food increases (Nabi et al. 2020).

10.3.2 Poverty

Poverty, as described by Sen (1982), is "a matter of deprivation." It has often been associated with overpopulation.

10.3.3 Climate Vulnerability

Developing countries have not been associated with climate change (Kirtman et al. 2013). 79% of the carbon emissions are due to developed countries (CGD 2015). Owing to changes in climate, these countries may face challenges in adapting to it. In 2012, a report by Climate Vulnerability Monitor showed that in such countries, change in climate, on an average, causes 400,000 deaths each year because of hunger and contagious diseases (Climate vulnerability monitor report 2012). These effects have been quite harsh for the poorest countries of the world.

10.3.4 Healthcare Systems

In developing countries, access to healthcare facilities is quite low (Alhaji and Alam 2019) with people having much lower life expectancy than the people in developed countries (Rogers and Wofford 1989). Further, cases of infectious diseases (Fauci 2001), infant mortality (Molitoris et al. 2019), child mortality, and maternal mortality (Girum and Wasie 2017; Declercq and Zephyrin 2020) are much higher in such countries. Additionally, developing countries have quite little access to health services (Peters et al. 2008). Even though vaccine equity is of immense importance to tackle pandemics, still these countries are less likely to have the required resources for buying/producing and administering vaccines (Hotez and Bottazzi 2021).

10.4 Developing Countries and Plant Molecular Farming

Plant molecular farming has the potential to change the life standards of poor, especially in developing countries. It was recognized that it would have major impact on the health around the world by producing the cost-effective modern medicine (Twyman et al. 2005; Ma et al. 2013). Improving global health and decreasing the poverty have been discussed since the establishment of this field by making the global access to modern medicines at a lower cost. Developed countries have been helping the developing countries through donations, necessities, and manufacturing pharmaceuticals in established pharmaceutical industries to resolve the global health needs. But there is a need for alternative approach to address the health and financial problems of developing and low-income countries (Tschofen et al. 2016). Plant molecular farming offers various solutions in this respect. Cost-effectiveness of PMF attracts the governments of developing and commercial entities to invest in this technology for the production of potential targets associated with health and industry. This will help in improving the standards of living in developing countries (Ma et al. 2013; Rappuoli et al. 2002; Cohen et al. 2016; Murad et al. 2020).

PMF is still a nascent technology but has the potential to transform the pharmaceutical industry. PMF offers some benefits such as cost-effectiveness, scalability, and highly skilled labor, which are attractive points for developing countries to invest in the establishment of plant molecular industries. It is the engrained fact that the plant molecular farming results in the reduction of manufacturing costs as it involves growing of plants. The downstream processing and purification costs are still a major concern associated with plant molecular farming, and it offers limited savings (Fischer and Buyel 2020). However, plant molecular farming still has significant financial advantage as it saves the huge investment associated with the establishment of infrastructure and early stages of product development through cell fermentation systems (Ma et al. 2013). Another important benefit that PMF offers is production scalability. Developing countries need any industrial and pharmaceutical products in large quantities due to overpopulation. In many developing countries, the production facilities cannot meet such huge demands, which is a main reason why the developing countries face health issues and poverty. There is need of new technology, which answers this issue. PMF could be the potent alternative to solve the product scalability problem by production of desired product at unlimited scale (Shanmugaraj et al. 2020). The third most attractive benefit associated with the PMF is that it does not require highly skilled labor as it involves plant cultivation. Around the globe, the basic agriculture skills are widely available, which makes the transfer of the technology to developing countries easy. These benefits could be attractive points to stimulate the developing countries and commercial entities to establish a local industry (Ma et al. 2013; Murad et al. 2020; see also Chap. 11).

10.5 Issues Faced by Developing Countries for Adopting PMF as Local Industries

In developed countries, plant molecular farming is advancing, but the approach of developing countries towards this new technology is different. The risk/benefit analysis is different in developing countries as compared to developed countries. Developing countries have high debts, slow-growing economy, and limited budget related to health (Ma et al. 2013). By keeping in view these things, the governments of developing countries raise some questions before investing in new ventures. These questions are (Murad et al. 2020) the following:

1. Why a country with limited industrial infrastructure should invest in PMF, rather than upgrade the already present system, which decreases the risk of loss?
2. Which products to be prioritized as they have limited financial sources to invest and avoid risk of failures?
3. How could they get extra finances to invest in this new venture? Who would be the partner if needed?
4. How the skillful labor will be developed to work in a new industry?
5. How new regulatory methods and bodies will work efficiently in a developed country?

Poor and developing countries face many barriers for investing in new ventures like PMF such as lack of local government to prioritize the health of its people over other needs and limited budgets for each sector especially for the development of new industries. Therefore, developing countries need the financial and other technical support of developed countries to develop molecular farming technologies by addressing the abovementioned questions. Also, developed innovative technologies of developing countries provide opportunities to invest for the development of PMF and increase their accessibility to local and international markets. The above problems associated with the development of plant molecular farming can be solved by the below-mentioned steps (Ma et al. 2013; see also Chap. 13).

10.5.1 Selection of the Potential Targets for PMF

Vigilant assortment of products to be produced via PMF is very perilous in developing countries. Many factors affect the decision of making the product that is commercially beneficial, but global needs are different which are not mostly addressed during such decision-makings, especially for developing countries (Rybicki 2010). For molecular farming to have a huge impact, the product that has a high demand should be focused, especially in the field of medicine and which should provide motivation for commercial investors. Cost and scalability are benefits of PMF, but nowadays, other advantages are also associated with it. Transient expression is the fastest method to develop a desired product (Landry et al. 2010). Plants have the ability to produce complex proteins with proper posttranslational modifications, which are not achievable via other established methods. Moreover, the use of edible plant as the expression system enables the oral delivery of vaccines and other pharmaceutical products, which may suppress the need of post-purification processes, delivery systems, skilled labor, transport, risk of contaminations, disposal of used items, and storage needs, ultimately affecting the cost (Arntzen and Ahoney 2004; see also Chap. 1).

The collaboration of both developed and developing countries is needed to select a candidate, by mutual integration of scientists from both sides, to be produced via PMF, which should benefit the developing countries in many ways. Also, there is need to include the opinion of commercial partners by keeping in mind the financial benefits. This will help in selecting a handful number of products to be targets and decreasing the risk of failure. These benefits and steps taken can be possible answers to the first two questions of why the developing countries should invest in PMF, and the chances of risk of failures are limited and less costly (Ma et al. 2013; see also Chap. 2).

10.5.2 Support to Developing Countries in Technology Transfer

Young scientists from developing countries learn molecular farming in the research institutes of developed countries. But upon their return to native countries, due to lack of infrastructure, they cannot benefit their country with the new technology learnt. Especially, the field of molecular pharming, as other fields of plant biotechnology associated with the development of resistant crops and production of biofuels, can easily get grants at both national and international levels (Ma et al. 2013). Producing skilled labor, increasing capacity, and transfer of technology are logical steps that would help the developing countries establish PMF industries. For this to happen, the developing countries have to introduce flagship projects and facilities for new technology to be established; for this purpose, the developing countries require funding from developed countries for the betterment of their people. Also, developing countries need to produce a resolution at national level on the need of molecular farming technology to lay out a clear strategy for the investors and collaborators at both national and international levels.

10.5.3 Need of Regulatory Bodies and Framework

PMF technology is matured now in developed countries, and they have national regulatory bodies. To ensure the early effect of any development in PMF technology on developing countries, the coordination between the regulatory bodies of both developing countries and developed countries is necessary. A number of low-income countries such as Argentina, Brazil, and South Africa have gained extensive exper tise in this technology and management of risks associated with technology, such as control of genetically modified crops and their environmental risks and assessments, as compared to other developing countries. However, these countries too lack the regulatory framework for plant pharming (Rybicki et al. 2012). The regulations present for GM crops are not always suitable for plant molecular pharming, which is an emerging industry that can play an important role in the healthcare system. Therefore, extensive workout and collaboration are required by both developed and developing countries to regulate the products produced by plant molecular farming and plant molecular pharming (Sparrow et al. 2007; see also Chap. 14).

10.5.4 Encourage Suitable Intellectual Property Management and Socially Responsible Licensing

Handling universal health and poverty outcomes for developing countries is linked with policies of intellectual property resources management. Filing a patent strategy is both bad and good by creating delay to access or likely for promotion of certain products. For developing countries, this may not be appropriate because filing a patent may not be of value in certain regions where markets are nonprofitable. But there may be chance that if the patent is not filed, nearby countries or other developed countries may acquire the technology or innovation for its benefits or import the product to its country as it is cost effective. Another risk associated with not filing a patent is that the industries will not take up the technology at the risk of financial losses, especially in developing countries, where there is need to incentivize the local industries to invest in this new technology. Issuing patent and IP rights to local companies enables the licensor to implement the conditions for technology availability in other potential markets (Ma et al. 2013).

Another strategy that has to be implemented by developing countries for the success of PMF is socially responsible licensing (SRL). In SRL, the licensing of intellectual assets is discussed and managed in a conductive way by providing access to necessary drugs and other life-improving products globally. It is also social responsibility to not affect the economic objectives especially in developing countries as the developed countries are affected much by such implementations (Mimura 2007). The main objectives of SRL should be the following (Busang et al. 2011):

- To ensure the benefits of society due to the product that has to be produced by publicly funded projects
- To make sure that the product market price is appropriate, economic, and accessible to community by keeping in view all the investment done on the project
- To implement the statutory requirements and help in the commercial social investment
- To enhance the reputation of product by making it safe and accessible
- To upsurge the approval of technology universally and to generate substitute models of commercialization

An intensive and synchronized struggle is now required to ensure that the vision of establishment of PMF is taken into practice. This may need to be led from public funded organizations and by the collaborations of industrial associates with the governments of developing countries especially (see also Chap. 15).

10.6 Benefits Achieved So Far Via PMF

The different benefits so for achieved globally via plant molecular farming are numerous. These benefits mostly help developing countries to get rid of problems associated with them such as food security, hunger, low-quality diet, poverty, health, and economy (Shanmugaraj et al. 2020). Plant molecular farming may have provided solutions to these questions (Fig. 10.2). Some benefits achieved via molecular faming are briefly described below:

10.6.1 Poverty, Food Shortage, and Hunger

Extreme poverty can be eliminated, and world without hunger is possible only if enough food is produced and evenly distributed. Most of undernourished and poor people live in developing countries. Different studies have shown that extreme poverty can be reduced only by strengthening the smallholder farmers that implies that improvement in the agricultural field is necessary for a strong economy of developing countries (Cai et al. 2017). Around the world, according to the FAO report published in 2019, more than 800 million are chronically hungry and around 2 billion people are micronutrient deficient. For most of the various infectious diseases, other mental and physical impairments, and far too high figure of premature deaths, malnutrition is one of the causes. Huge health problems can be caused by low dietary quality and food insecurity. Therefore, according to SDGs, to reduce the hunger

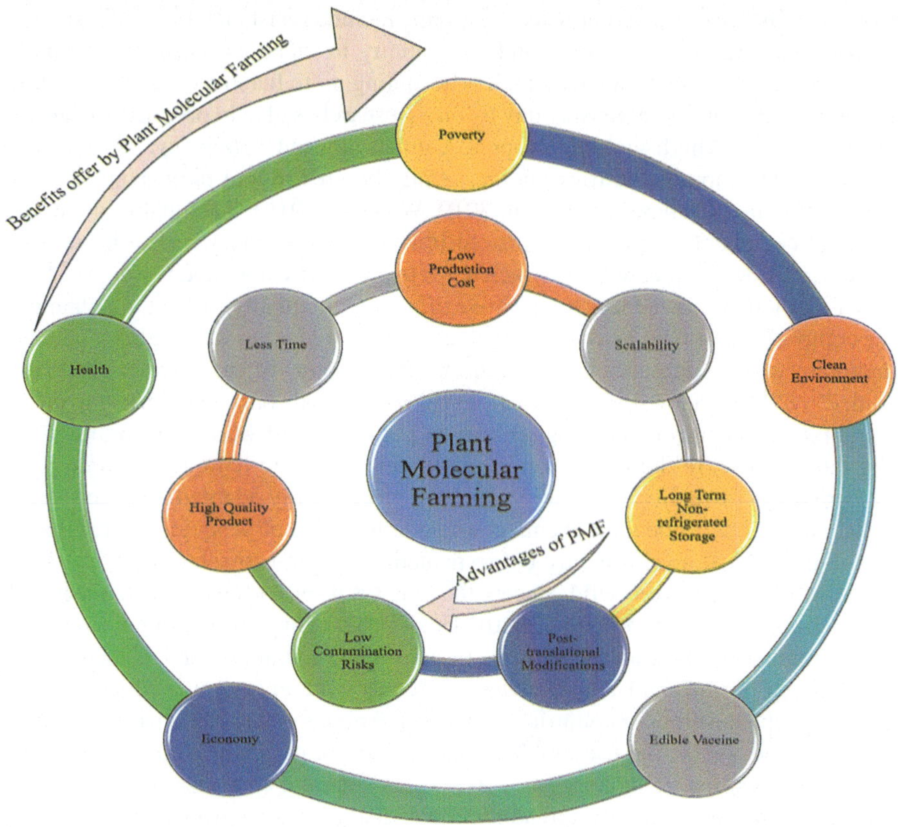

Fig. 10.2 Advantages of plant molecular farming and benefits offered by it to developing countries

around the world, the goal of "zero hunger and improved nutrition" has been put on second after "no poverty". Since the beginning of agriculture, the growing of enough food has been a challenge which has not yet overwhelmed. For this goal to be achieved, the major step is transformation in global food systems. Import roles can be played by agriculture technologies and strategies (Qaim 2016; Meemken and Qaim 2018; Springmann et al. 2018; Zaidi et al. 2019).

A new horizon has been opened for the agriculture field with GMOs. Recombinant DNA technology is helpful in the introduction of individual desired genes in the plants without any undesired effects on the genetic makeup of plant (Qaim 2016). The traits introduced so far in the plants are to make plant tolerant to herbicides, diseases, and pesticides. The first GM insect resistance crop was Bt-cotton having foreign gene from *Bacillus thuringiensis*. Other traits introduced in different species of crops are fungal, bacterial, and viral resistances via gene editing (Oliva et al. 2019). Introduction of such traits in plants reduced the use of chemical pesticides and prevented their drastic effect of plant itself and environment (Qaim and Zilberman 2003; Bailey-Serres et al. 2019). Other agronomic traits on which the work is going on by different research groups include resistance to abiotic stresses such as drought, heat, flooding, and soil salinity to make crops more tolerant to continuously changing climate, which is becoming more harsh due to global warming and water scarcity increasing day by day. Research is also in progress to develop nutrient-rich and high-yielding crops to combat nutrient deficiency, especially in infants in developing countries via increasing the growth and photosynthetic efficiency of plants (Bailey-Serres et al. 2019; Wu et al. 2019). The quality of several fruits and vegetables has been improved in North and South America through the use of CRISPR/Cas technology, which helps to prevent the loss and wastage of fruit. Other methods followed to get a positive effect on the quality of fruit are changing the composition of fatty acid in oil-producing crops, reducing wheat gluten content, and/or increasing important micronutrient contents in different fruits (De Steur et al. 2012; Modrzejewski et al. 2019). Different methods have been in use from the beginning of advancement in agriculture to get that breed which has higher yield. For this purpose, re-domestication is an efficient option in helping to get crops of higher genetic diversity and making the crop more locally adapted, resistant to climate, and less dependent on chemical usage in the form of pesticides or fertilizers. For re-domestication of already domesticated plants and to reintroduce the local genes of resistance from wild species that has been lost or never fully integrated, gene editing can be used (Qaim 2020). This introduction of traits to get resistance crops, improving the quality of crops through gene mutations and reducing their dependency on different chemical inputs, reduces the energy use which benefits especially the developing countries to stabilize their economy (Fischer et al. 2015).

In the mid-1990s, the use of GMOs had begun for commercial purposes. From that time, boost for technology has started. In developing countries, since then, this technology has been rapidly spreading. Since 2011, the area grown with GMOs in developing countries has been larger than the area in industrialized countries. 14% of the total worldwide cropland had been planted with GM plants in 2018. These 192 million ha of lands were cultivated by 17 million farmers in 26 countries. Majority of these countries belong to South and North America followed by Asian countries. A very few countries have adopted GMOs in Europe due to their non-acceptability by their community. The countries with the shares of the total GMO area in 2018 were Pakistan (1%), South Africa (1%), Paraguay (2%), China (2%), India (6%), Canada (7%), Argentina (12%), Brazil (27%), the USA (39%), and a number of other countries (Qaim 2016, 2020).

10.6.2 Medicine and Health

In developing countries, the mortality rate is more than 45% due to infectious diseases and one-third of deaths are caused by the infectious agents (Kumar et al. 2013). Vaccines have great importance in the prevention of infectious diseases. Vaccines provide immune protection by producing antibodies in animals and humans. Unavailability of vaccines for many life-threatening diseases is a dilemma that needs the development of safer, cheaper, and effective ways of vaccine production (Doherty et al. 2016). Several reasons count for the unavailability of vaccines against several diseases. The production of some vaccines is costly, and some pathogens cannot grow in exogenous media. Another reason could be conversion of pathogens from attenuated form to virulent form in vaccines, so that proper storage and distribution are required, which affects the cost of vaccines (Glick and Patten 2017). Higher costs of vaccines are mostly related to the costly fermenter-based production systems needing costly media, technical handling, stringent purification due to potential endotoxin contaminations, and low scalability (Lössl and Waheed 2011). For the development of next-generation vaccines against many infectious diseases, cost-effectiveness of the produced vaccines needs to be taken into account, considering the fact that the disease burden largely lies in developing countries.

Dr. Arntzen and colleagues gave idea to produce and deliver subunit vaccine in transgenic plants to overcome limitations in conventional procedures of vaccine production and delivery (Mason and Arntzen 1995; Haq et al. 1995). Plants can be used as alternate production platforms for valuable and effective products of pharmaceutical importance (Bock 2015; Sack et al. 2015). Plant molecular farming has been very well established over the past two decades. Induction of immune response through the administration of plant-derived antigens orally, subcutaneously, or intramuscularly is reported in the literature (Permyakova et al. 2015). Many plant-based vaccines have been developed, and some are in clinical trials (Naderi and Fakheri 2015; see also chapter "Plant Molecular Farming for Vaccine Development"). Plants offer several advantages directly related to cost-effective vaccine production. These include high scalability, low cost of production, ease of manipulation, high biomass, high protein yield, co-expression of two or more coupled antigens together as fusion proteins, and less stringent purification requirements because plants are not host to human pathogens, contrary to bacterial fermenter-based production systems (Lössl and Waheed 2011; Hoelscher et al. 2018; Saba et al. 2019). In addition to that, plants can be grown at sites where the production is needed, which minimizes the need of cooling chain and results in less or no transportation-related costs (Buyel et al. 2015; Hoelscher et al. 2018; Yang et al. 2018). In the form of seeds, vaccine source can be stored for a longer period at room temperature and transported worldwide easily (Sack et al. 2015). Due to eukaryotic nature of plants, post-translational modifications of foreign proteins can be carried out, which is essential for their proper folding and function (Lössl and Waheed 2011).

Today, the biopharmaceutical products made through plants have become a reality. The pandemic caused by SARS-CoV-2 virus (COVID-19) has a huge impact on the health of humans and on the social, economic, and global relations. By August 2020, it affected the lives of more than 24 million people and 800,000 deaths throughout the world (Sharma et al. 2020). Such scenarios demand the productions of potent antibodies, drugs, and vaccine at high speed. These kinds of situations have a drastic effect on low-income and developing countries. The most efficient way to fight against the emerging pandemics or bioterrorism is by manufacturing of vaccines in a short time that can be achieved through PMF (see also Chap. 12). PMF has helped in producing pharmaceutical products against infectious and chronic diseases that was not possible through other technologies before. So far, through plant molecular farming, the highly medicinal valuable proteins such as antibodies, enzymes, hormones, and vaccines have been produced (Stoger et al. 2014; Rosales-Mendoza 2020). Different pharmaceutically important proteins have been produced through PMF that can help to combat infectious diseases. For production of proteins of medicinal importance or potent vaccines, edible plants are preferred because they were given the status of GRAS (generally recognized as safe) such as banana, lettuce, potato, and tomato (Schillberg and Finnern 2021).

Many studies have been done on the recombinant proteins produced via PMF for their functionality through different biological assays, but proteins of medicinal importance need additional testing, that is, animal trials. For example, E1E2 heterodimer protein of hepatitis C virus produced in lettuce has shown immunogenic properties in mice model (Clarke et al. 2017). Similarly, domain III of envelope protein of West Nile virus expressed in tobacco plant has also shown immune response in mice when challenged with lethal infection of West Nile virus (Sun and Chen 2018). A strong antibody response was induced in mice when they were given tobacco co-expressing dengue structural proteins and truncated nonstructural RNA-dependent RNA polymerase in the form of viruslike particles (VLPs) (Lai et al. 2018; Ponndorf et al. 2021). One plant that produced medicinal important product, taliglucerase alfa, has completed human clinical trials and reached the market for treatment of Gaucher's disease (Hollak et al. 2010; Zimran et al. 2018). Other plants have produced products that are in later stages of clinical trials including pegunigalsidase alfa, a recombinant form of human globotriaosylceramide (Gb3), for the therapy of Fabry disease, a plant-derived vaccine for seasonal influenza (van der Veen et al. 2020; Ward et al. 2020). ZMapp is a monoclonal antibody that has been transiently expressed in tobacco plant used as treatment for Ebola patients in West Africa in 2014 (Park and Wi 2016; Pettit et al. 2016). Two publicly funded projects are in progress for the production of HIV-neutralizing antibodies in tobacco for clinical trials, and also in the seeds of maize for production of potent antimicrobial products for the developing countries (Hefferon 2013; see also chapter "Plant Molecular Pharming to Overcome the Global Impact of Neglected Tropical Diseases").

Other proteins produced in plants that are under clinical trials are gastric lipase for cystic fibrosis, insulin for diabetes, interleukin for Crohn's disease, and glucocerebrosidase for Gaucher's diseases (Einsiedel and Medlock 2005; Yao et al. 2015; Spiegel et al. 2018). Oral vaccine has also been tested on humans. Potato expressing VLP proteins of Norwalk virus was given to humans who volunteered, and they developed immune response to antigen. As the acidic environment destroys most of the plant tissue before reaching lymphoid tissues, high level of protein-expressing plants are required with strong encapsulations to counter the acidic environment of stomach to develop strong immune responses (Tacket et al. 2000; Arevalo-Villalobos et al. 2020; Kurup and Thomas 2020).

Another important most advanced category that attracts the researchers is the direct production of monoclonal antibodies via plant machinery. Monoclonal antibodies are large multimeric glycoproteins that bind to respective antigens and help the immune system to fight against the disease. The high increase of demand for antibodies exerts pressure for their production in bulk amount at reasonable costs (Frigerio et al. 2000; Chargelegue et al. 2004; Wieland et al. 2006). Different plant-made monoclonal antibodies are in different phases of clinical trials. Various ScFv antibody fragments against non-Hodgkin's disease, Rhino Rx for treating respiratory syncytial diseases, and IgG antibody for prevention of common cold are in Phase I of clinical trials, whereas CaroRx for blocking the adherence of bacteria that causes cavities and IgG, an antibody against cancer, are in Phase II of clinical trials. Recently, a plant-based ScFv mAB product got approval to be produced as vaccine at a large scale in Cuba (Pujol et al. 2005).

Animal vaccines also have crucial importance for better health of animals, humans, and economy of developing countries. Plants are suitable for production of veterinary vaccines because animals feed on plant or crude extract of the plant that can easily be given to them. By this, the downstream purification of protein from transgenic plants is not needed, which decreases the cost of vaccine productions, which in turn benefits the developing countries by directly feeding the animals with transgenic vaccine-producing plants (Phan et al. 2020). Piglets were immunized against enterotoxigenic *E. coli* (ETEC) by feeding on transgenic *Arabidopsis thaliana* seeds expressing anti-ETEC antibodies. Also, immunogenic response was reported in chickens when they were given crude extract of tobacco expressing H5 trimer of avian influenza virus (H5N1) (Virdi et al. 2013; Topp et al. 2016). These research findings show that plants are good factories for production of vaccines and other biopharmaceutical products (see also chapter "Plant-Based Veterinary Vaccines").

10.6.3 Industrial Products and Economy

As explained earlier, the plants are commercially economic system for production of pharmaceutical and non-pharmaceutical products. The production of non-pharmaceutical products via PMF has outnumbered the pharmaceutical product production showing the growth of such products due to less development time and regulatory burden as compared to production of pharmaceutical products. This helps in cost reductions and higher scalability due to requirement of less costly downstream processes in comparison to pharmaceutical products (Tschofen et al. 2016).

A number of non-pharmaceutical products have been produced via PMF. The enzymes which are involved in the conversion of biomass are required in higher quantities, so they must be produced very expensively. This group includes hydrolases, glycosidases, laccase, and proteases. So, these enzymes are potent candidates to be produced via PMF (Hood et al. 2003; Bailey et al. 2004). Different plant-derived proteins such as avidin, aprotinin, β-glucuronidase, lactoferrin, lysozyme, thyroid-stimulating hormone receptor, trypsin, peroxidase, laccase, and cellulase have reached the market (see also chapters "Molecular Farming of Industrial

Enzymes: Products and Applications" and "Plant Molecular Farming for the Bulk Production of Industrial Enzymes"). Avidin is primarily used as diagnostic reagent and was first time produced in corn via genetic transformation (Basaran and Rodríguez-Cerezo 2008; Obembe et al. 2011). GUS β-glucuronidase, β-D-glucuronide glucuronosohydrolase, is a homotetrameric hydrolase, first produced in corn at a commercial level having same properties as the original source of protein possess. Trypsin is used in different commercial applications and has significant market value. It has been genetically produced in maize for commercial uses. Trypsin in name TrypZean reached the market produced in transgenic maize (Horn et al. 2004). An inhibitor of chymotrypsin, kallikrein, trypsin, and pepsin known as aprotinin has also been produced in maize with different transgene technologies. It has high demand in medical site as wound healer. Collagen is another commercially important non-pharmaceutical protein used in the form of gelatin as biomaterials in the medical, pharmaceutical, and cosmetic industries. It has been expressed in transgenic tobacco as achieved via stable transformation (Ruggiero et al. 2000; Horn et al. 2004). Human gastric lipase is a protein that is important for the digestion of food lipids whose deficiency leads to special condition known as exocrine pancreatic insufficiency (EPI). A research is in progress to develop maize-produced mammalian lipase by a French company and to bring it to clinical trials. In another research, the gastric lipase from *canine* source has been expressed in tobacco (Gruber et al. 2001; Horn et al. 2004). Another product produced via plant machinery through gene modifications is human lactoferrin produced in tobacco, rice, and maize (Samyn-Petit et al. 2001; Nandi et al. 2002). Fungal enzymes such as peroxidases and laccases have important usage in the paper industry, so they are needed in high quantities. They have been expressed in maize (Basaran and Rodríguez-Cerezo 2008). The demand for fuel is increasing day by day, and its price is increasing continuously as well. So, the production of biofuel from plants is getting enormous importance. The enzymes cellulases, hemicellulases, ligninases, and xylanases have important value in the field of biofuel. The important enzyme cellulase is required to break down the long chains of cellulose to produce cellulosic ethanol. The cellulase has been expressed in maize to be produced in larger quantities for industrial use (Sticklen 2008; Mei et al. 2009; Chatterjee et al. 2010). Plastic usage has drastically affected the ecosystem of the world. Biodegradable plastics are environmentally friendly alternative to petrochemical polymers. Polyhydroxybutyrate (PHB) is the first plastic-like product produced via PMF in 1992. Other biodegradable plastic-like compounds such as cyanophycin, poly (3-hydroxybutyrate) (PHB), and polyhydroxyalkanoate (PHA) copolymer have been optimized for production via PMF. Spider silk proteins and elastins have also been expressed in transgenic plants (Christou et al. 2004; Scheller and Conrad 2004; Scheller et al. 2004; Conrad 2005; Matsumoto et al. 2009).

10.7 Conclusions

In view of all the above reports and data, it can be concluded that the plant molecular farming has a great potential for playing a significant role in the economic upliftment of the developing countries. The decision-making bodies at government level, and also the public-private industries and/or entrepreneurs, could opt for the novel platforms and technologies offered by modern biotechnology for the production of industrial products that are cost effective and affordable for people.

References

Alhaji MM, Alam S (2019) Health policy and system research in global south: bridging the gap through citizen participation. JPMI J Postgrad Med Inst 33(1):3

Alireza T, Nader RE (2015) Molecular farming in plants. In: Plants for the future. Intech Open, London, pp 25–41

Arevalo-Villalobos JI, Alonso DOG, Rosales-Mendoza S (2020) Using carrot cells as biofactories and oral delivery vehicles of ltb-syn: a low-cost vaccine candidate against synucleinopathies. J Biotechnol 309:75–80

Arntzen CJ, Ahoney RTM (2004) Plant-derived vaccines: a new approach to international public health. J Food Sci 69(1):CRH8–CRH10

Bailey MR, Woodard SL, Callaway E, Beifuss K, Magallanes-Lundback M, Lane JR et al (2004) Improved recovery of active recombinant laccase from maize seed. Appl Microbiol Biotechnol 63(4):390–397

Bailey-Serres J, Parker JE, Ainsworth EA, Oldroyd GED, Schroeder JI (2019) Genetic strategies for improving crop yields. Nature 575(7781):109–118

Bandurska K, Berdowska A, Krol M (2016) Transformation of medicinal plants using agrobacterium tumefaciens. Postepy Hig Med Dosw (Online) 70:1220–1228

Basaran P, Rodríguez-Cerezo E (2008) Plant molecular farming: opportunities and challenges. Crit Rev Biotechnol 28(3):153–172

Bhatia S (2018) Introduction to genetic engineering. In: Introduction to pharmaceutical biotechnology, basic techniques and concepts, vol 1. IOP Publishing Ltd. https://doi.org/10.1088/978-0-7503-1299-8

Binns AN, Thomashow MF (1988) Cell biology of agrobacterium infection and transformation of plants. Annu Rev Microbiol 42(1):575–606

Bock R (2015) Engineering plastid genomes: methods, tools, and applications in basic research and biotechnology. Annu Rev Plant Biol 66:211–241

Boffey SA, Leech RM (1982) Chloroplast DNA levels and the control of chloroplast division in light-grown wheat leaves. Plant Physiol 69(6):1387–1391

Boynton JE, Gillham NW, Harris EH, Hosler JP, Johnson AM, Jones AR et al (1988) Chloroplast transformation in chlamydomonas with high velocity microprojectiles. Science 240(4858):1534–1538

Busang R, Boisjoly-Letourneau K, Fourie B, Mulder M, Thangaraj H (2011) Health technologies: South Africa socially responsible licensing of health technologies: policy and practice in South Africa

Buyel JF, Twyman RM, Fischer R (2015) Extraction and downstream processing of plant-derived recombinant proteins. Biotechnol Adv 33(6):902–913. https://doi.org/10.1016/J.BIOTECHADV.2015.04.010

Cai Y, Golub AA, Hertel TW (2017) Agricultural research spending must increase in light of future uncertainties. Food Policy 70:71–83

Centre for Global Development (2015) Commitment to development index. http://www.cgdev.org/cdi-2015. Accessed 12 Dec 2022

Chargelegue D, Drake PMW, Obregon P, Ma JKC (2004) Production of secretory IgA in transgenic plants. In: Molecular farming: plant-made pharmaceuticals and technical proteins. Wiley, Weinheim, pp 159–169

Chatterjee A, Das NC, Raha S, Babbit R, Huang Q, Zaitlin D et al (2010) Production of xylanase in transgenic tobacco for industrial use in bioenergy and biofuel applications. In Vitro Cell Dev Biol 46(2):198–209

Chawla HS (2011) Introduction to plant biotechnology, 3rd edn. CRC Press, Boca Raton. https://doi.org/10.1201/9781315275369

Chemat F, Abert-Vian M, Fabiano-Tixier AS, Strube J, Uhlenbrock L, Gunjevic V et al (2019) Green extraction of natural products. Origins, current status, and future challenges. TrAC Trends Anal Chem 118:248–263

Christou P (1992) Genetic transformation of crop plants using microprojectile bombardment. Plant J 2(3):275–281

Christou P, Stoger E, Twyman RM (2004) Monocot expression systems for molecular farming. In: Molecular farming: plant-made pharmaceuticals and technical proteins. Wiley, Weinheim, pp 55–67

Clarke JL, Paruch L, Dobrica MO, Caras I, Tucureanu C, Onu A et al (2017) Lettuce-produced hepatitis c virus e1e2 heterodimer triggers immune responses in mice and antibody production after oral vaccination. Plant Biotechnol J 15(12):1611–1621. https://doi.org/10.1111/pbi.12743

Cohen JP, Silva L, Cohen A, Awatin J, Sturgeon R (2016) Progress report on neglected tropical disease drug donation programs. Clin Ther 38(5):1193–1204

Conrad U (2005) Polymers from plants to develop biodegradable plastics. Trends Plant Sci 10(11):511–512

Dale PJ, Irwin JA, Scheffler JA (1993) The experimental and commercial release of transgenic crop plants. Plant Breed 111(1):1–22

Daniell H, Singh ND, Mason H, Streatfield SJ (2009) Plant-made vaccine antigens and biopharmaceuticals. Trends Plant Sci 14(12):669–679

Daniell H, Lin C-S, Yu M, Chang W-J (2016) Chloroplast genomes: diversity, evolution, and applications in genetic engineering. Genome Biol 17:1–29

De Steur H, Gellynck X, Blancquaert D, Lambert W, Van Der Straeten D, Qaim M (2012) Potential impact and cost-effectiveness of multi-biofortified rice in China. New Biotechnol 29(3):432–442

Declercq E, Zephyrin L (2020) Maternal mortality in the United States: a primer. Commonwealth Fund

Deshayes A, Herrera-Estrella L, Caboche M (1985) Liposome-mediated transformation of tobacco mesophyll protoplasts by an Escherichia coli plasmid. EMBO J 4(11):2731–2737

Doherty M, Buchy P, Standaert B, Giaquinto C, Prado-Cohrs D (2016) Vaccine impact: benefits for human health. Vaccine 34(52):6707–6714

Einsiedel EF, Medlock J (2005) A public consultation on plant molecular farming. AgBioForum 8(1):26–32

Fauci AS (2001) Infectious diseases: considerations for the 21st century. Clin Infect Dis 32(5):675–685

Fischer R, Buyel JF (2020) Molecular farming—the slope of enlightenment. Biotechnol Adv 40:107519

Fischer K, Ekener-Petersen E, Rydhmer L, Edvardsson Björnberg K (2015) Social impacts of gm crops in agriculture: a systematic literature review. Sustainability 7(7):8598–8620

Frigerio L, Vine ND, Pedrazzini E, Hein MB, Wang F, Ma JKC et al (2000) Assembly, secretion, and vacuolar delivery of a hybrid immunoglobulin in plants. Plant Physiol 123(4):1483–1494

Gheysen G, Angenon G, Van Montagu M (1998) *Agrobacterium*-mediated plant transformation: a scientifically intriguing story with significant applications. In: Transgenic plant research. Harwood Academic Publishers, Amsterdam, pp 1–33

Girum T, Wasie A (2017) Correlates of maternal mortality in developing countries: an ecological study in 82 countries. Matern Health Neonatol Perinatol 3(1):1–6

Glick BR, Patten CL (2017) Molecular biotechnology: principles and applications of recombinant DNA. John Wiley & Sons

Gruber V, Berna PP, Arnaud T, Bournat P, Clément C, Mison D et al (2001) Large-scale production of a therapeutic protein in transgenic tobacco plants: effect of subcellular targeting on quality of a recombinant dog gastric lipase. Mol Breed 7(4):329–340

Gupta V, Sengupta M, Prakash J, Tripathy BC, Gupta V, Sengupta M et al (2017) Transgenic animals and plants. In: Basic and applied aspects of biotechnology. Springer, Singapore, pp 103–123. https://doi.org/10.1007/978-981-10-0875-7

Hagemann R (2004) The sexual inheritance of plant organelles. In: Molecular biology and biotechnology of plant organelles chloroplasts and mitochondria. Springer, Dordrecht, pp 93–113. https://doi.org/10.1007/978-1-4020-3166-3

Hammond RW, Nemchinov LG (2009) Plant production of veterinary vaccines and therapeutics. In: Plant-produced microbial vaccines. Springer, Berlin, pp 79–102

Hansen G, Wright MS (1999) Recent advances in the transformation of plants. Trends Plant Sci 4(6):226–231

Haq TA, Mason HS, Clements JD, Arntzen CJ (1995) Oral immunization with a recombinant bacterial antigen produced in transgenic plants. Science 268(5211):714–716

Hefferon K (2013) Plant-derived pharmaceuticals for the developing world. Biotechnol J 8(10):1193–1202

Hinchee MAW, Connor-Ward DV, Newell CA, McDonnell RE, Sato SJ, Gasser CS et al (1988) Production of transgenic soybean plants using agrobacterium-mediated DNA transfer. Bio/Technology 6(8):915–922

Hoelscher M, Tiller N, Teh AYH, Wu G-Z, Ma JKC, Bock R (2018) High-level expression of the HIV entry inhibitor griffithsin from the plastid genome and retention of biological activity in dried tobacco leaves. Plant Mol Biol 97(4–5):357–370. https://doi.org/10.1007/s11103-018-0744-7

Hollak CEM, vom Dahl S, Aerts JMFG, Belmatoug N, Bembi B, Cohen Y et al (2010) Force majeure: therapeutic measures in response to restricted supply of imiglucerase (Cerezyme) for patients with Gaucher disease. Blood Cells Mol Dis 44(1):41–47

Hood EE, Bailey MR, Beifuss K, Magallanes-Lundback M, Horn ME, Callaway E et al (2003) Criteria for high-level expression of a fungal laccase gene in transgenic maize. Plant Biotechnol J 1(2):129–140

Hooykaas PJJ, Schilperoort RA (1992) *Agrobacterium* and plant genetic engineering. Plant Mol Biol 19:15–38

Horn ME, Woodard SL, Howard JA (2004) Plant molecular farming: systems and products. Plant Cell Rep 22(10):711–720

Hotez PJ, Bottazzi ME (2021) Whole inactivated virus and protein-based COVID-19 vaccines. Annu Rev Med 73:55

Huang Y, Liang W, Pan A, Zhou Z, Huang C, Chen J et al (2003) Production of FaeG, the major subunit of k88 fimbriae, in transgenic tobacco plants and its immunogenicity in mice. Infect Immun 71(9):5436–5439

Keshavareddy G, Kumar ARV, Ramu VS (2018) Methods of plant transformation—a review. Int J Curr Microbiol Appl Sci 7(7):2656–2668

Matsumoto K, Murata T, Nagao R, Nomura CT, Arai S, Arai Y et al (2009) Production of short-chain-length/medium-chain-length polyhydroxyalkanoate (pha) copolymer in the plastid of Arabidopsis thaliana using an engineered 3-ketoacyl-acyl carrier protein synthase III. Biomacromolecules 10(4):686–690

Kirtman B, Power SB, Adedoyin AJ, Boer GJ, Bojariu R, Camilloni I et al (2013) Near-term climate change: projections and predictability. In: Climate change 2013—the physical science basis: working group I contribution to the fifth assessment report of the Intergovernmental Panel on Climate Change. Cambridge University Press, Cambridge, pp 953–1028. https://doi.org/10.1017/CBO9781107415324

Klee H, Horsch R, Rogers S (1987) *Agrobacterium*-mediated plant transformation and its further applications to plant biology. Annu Rev Plant Physiol 38(1):467–486

Klein TM, Wolf ED, Wu R, Sanford JC (1987) High-velocity microprojectiles for delivering nucleic acids into living cells. Nature 327(6117):70–73

Kumar BV, Raja TK, Wani MR, Sheikh SA, Lone MA, Nabi G et al (2013) Transgenic plants as green factories for vaccine production. Afr J Biotechnol 12(43):6147–6158

Kurup VM, Thomas J (2020) Edible vaccines: promises and challenges. Mol Biotechnol 62(2):79–90

Lai H, Paul AM, Sun H, He J, Yang M, Bai F et al (2018) A plant-produced vaccine protects mice against lethal West Nile virus infection without enhancing zika or dengue virus infectivity. Vaccine 36(14):1846–1852

Landry N, Ward BJ, Trépanier S, Montomoli E, Dargis M, Lapini G et al (2010) Preclinical and clinical development of plant-made virus-like particle vaccine against avian h5n1 influenza. PLoS One 5(12):e15559

Lössl AG, Waheed MT (2011) Chloroplast-derived vaccines against human diseases: achievements, challenges and scopes. Plant Biotechnol J 9(5):527–539

Ma JKC, Christou P, Chikwamba R, Haydon H, Paul M, Ferrer MP et al (2013) Realising the value of plant molecular pharming to benefit the poor in developing countries and emerging economies. Plant Biotechnol J 11(9):1029–1033

Maliga P (2003) Progress towards commercialization of plastid transformation technology. Trends Biotechnol 21(1):20–28

Mason HS, Arntzen CJ (1995) Transgenic plants as vaccine production systems. Trends Biotechnol 13(9):388–392

Meemken E-M, Qaim M (2018) Organic agriculture, food security, and the environment. Annu Rev Resour Econ 10:39–63

Mei C, Park SH, Sabzikar R, Callista Ransom CQ, Sticklen M (2009) Green tissue-specific production of a microbial endo-cellulase in maize (Zea mays L.) endoplasmic-reticulum and mitochondria converts cellulose into fermentable sugars. J Chem Technol Biotechnol 84(5):689–695

Meyers B, Zaltsman A, Lacroix B, Kozlovsky SV, Krichevsky A (2010) Nuclear and plastid genetic engineering of plants: comparison of opportunities and challenges. Biotechnol Adv 28(6):747–756

Mimura C (2007) Technology licensing for the benefit of the developing world: UC Berkeley's socially responsible licensing program. Ind High Educ 21(4):295–301

Modrzejewski D, Hartung F, Sprink T, Krause D, Kohl C, Wilhelm R (2019) What is the available evidence for the range of applications of genome-editing as a new tool for plant trait modification and the potential occurrence of associated off-target effects: a systematic map. Environ Evid 8(1):1–33

Molitoris J, Barclay K, Kolk M (2019) When and where birth spacing matters for child survival: an international comparison using the dhs. Demography 56(4):1349–1370

Mubeen H, Naqvi RZ, Masood A, Shoaib MW, Raza S (2016) Gene transformation: methods, uses and applications. J Pharm Biol Sci 4(2):54

Murad S, Fuller S, Menary J, Moore C, Pinneh E, Szeto T et al (2020) Molecular pharming for low and middle income countries. Curr Opin Biotechnol 61:53–59

Nabi AA, Shahid ZA, Mubashir KA, Ali A, Iqbal A, Zaman K (2020) Relationship between population growth, price level, poverty incidence, and carbon emissions in a panel of 98 countries. Environ Sci Pollut Res 27:31778–31792

Naderi S, Fakheri B (2015) Overview of plant-based vaccines. Res J Fish Hydrobiol 10(10):275–289

Nandi S, Suzuki YA, Huang J, Yalda D, Pham P, Wu L et al (2002) Expression of human lactoferrin in transgenic rice grains for the application in infant formula. Plant Sci 163(4):713–722

Obembe OO, Popoola JO, Leelavathi S, Reddy SV (2011) Advances in plant molecular farming. Biotechnol Adv 29(2):210–222

Oldenburg DJ, Bendich AJ (2015) DNA maintenance in plastids and mitochondria of plants. Front Plant Sci 6:883

Oliva R, Ji C, Atienza-Grande G, Huguet-Tapia JC, Perez-Quintero A, Li T et al (2019) Broad-spectrum resistance to bacterial blight in rice using genome editing. Nat Biotechnol 37(11):1344–1350

Onouchi H, Yokoi K, Machida C, Matsuzaki H, Oshima Y, Matsuoka K et al (1991) Operation of an efficient site-specific recombination system of Zygosaccharomyces rouxii in tobacco cells. Nucleic Acids Res 19(23):6373–6378

Ow DW, Wood KV, DeLuca M, De Wet JR, Helinski DR, Howell SH (1986) Transient and stable expression of the firefly luciferase gene in plant cells and transgenic plants. Science 234(4778):856–859

Park KY, Wi SJ (2016) Potential of plants to produce recombinant protein products. J Plant Biol 59(6):559–568

Parmar N, Singh KH, Sharma D, Singh L, Kumar P, Nanjundan J et al (2017) Genetic engineering strategies for biotic and abiotic stress tolerance and quality enhancement in horticultural crops: a comprehensive review. 3 Biotech 7(4):1–35

Permyakova NV, Zagorskaya AA, Belavin PA, Uvarova EA, Nosareva OV, Nesterov AE et al (2015) Transgenic carrot expressing fusion protein comprising *M. tuberculosis* antigens induces immune response in mice. Biomed Res Int 2015:1. https://doi.org/10.1155/2015/417565

Peters DH, Garg A, Bloom G, Walker DG, Brieger WR, Hafizur Rahman M (2008) Poverty and access to health care in developing countries. Ann N Y Acad Sci 1136(1):161–171

Pettit DK, Rogers RS, Arthur K, Brodsky Y, Clark RH, Crowell C et al (2016) Cho cell production and sequence improvement in the 13c6fr1 anti-Ebola antibody. MAbs 8(2):347–357. https://doi.org/10.1080/19420862.2015.1127492

Phan HT, Pham VT, Ho TT, Pham NB, Chu HH, Vu TH et al (2020) Immunization with plant-derived multimeric h5 hemagglutinins protect chicken against highly pathogenic avian influenza virus h5n1. Vaccine 8(4):593

Ponndorf D, Meshcheriakova Y, Thuenemann EC, Dobon Alonso A, Overman R, Holton N et al (2021) Plant-made dengue virus-like particles produced by co-expression of structural and non-structural proteins induce a humoral immune response in mice. Plant Biotechnol J 19(4):745–756

Potrykus I, Shillito RD, Saul MW, Paszkowski J (1985) Direct gene transfer state of the art and future potential. Plant Mol Biol Report 3(3):117–128

Pujol M, Ramírez NI, Ayala M, Gavilondo JV, Valdés R, Rodríguez M et al (2005) An integral approach towards a practical application for a plant-made monoclonal antibody in vaccine purification. Vaccine 23(15):1833–1837

Qaim M (2016) Genetically modified crops and agricultural development, 1st edn. Palgrave Macmillan, New York. https://doi.org/10.1057/9781137405722

Qaim M (2020) Role of new plant breeding technologies for food security and sustainable agricultural development. Appl Econ Perspect Policy 42(2):129–150

Qaim M, Zilberman D (2003) Yield effects of genetically modified crops in developing countries. Science 299(5608):900–902

Rappuoli R, Miller HI, Falkow S (2002) The intangible value of vaccination. Science 297:937–939

Rogers RG, Wofford S (1989) Life expectancy in less developed countries: socioeconomic development or public health? J Biosoc Sci 21(2):245–252

Rosales-Mendoza S (2020) Will plant-made biopharmaceuticals play a role in the fight against COVID-19? Expert Opin Biol Ther 20(6):545–548

Ruggiero F, Exposito JY, Bournat P, Gruber V, Perret S, Comte J et al (2000) Triple helix assembly and processing of human collagen produced in transgenic tobacco plants. FEBS Lett 469(1):132–136

Ruhlman T, Verma D, Samson N, Daniell H (2010) The role of heterologous chloroplast sequence elements in transgene integration and expression. Plant Physiol 152(4):2088–2104

Rybicki EP (2010) Plant-made vaccines for humans and animals. Plant Biotechnol J 8(5):620–637

Rybicki EP, Chikwamba R, Koch M, Rhodes JI, Groenewald J-H (2012) Plant-made therapeutics: an emerging platform in South Africa. Biotechnol Adv 30(2):449–459

Saba K, Gottschamel J, Younus I, Syed T, Gull K, Lössl AG et al (2019) Chloroplast-based inducible expression of ESAT-6 antigen for development of a plant-based vaccine against tuberculosis. J Biotechnol 305:1–10

Sack M, Hofbauer A, Fischer R, Stoger E (2015) The increasing value of plant-made proteins. Curr Opin Biotechnol 32:163–170

Samyn-Petit B, Gruber V, Flahaut C, Wajda-Dubos JP, Farrer S, Pons A et al (2001) N-glycosylation potential of maize: the human lactoferrin used as a model. Glycoconj J 18(7):519–527

Scheller J, Conrad U (2004) Production of spider silk proteins in transgenic tobacco and potato. In: Molecular farming: plant-made pharmaceuticals and technical proteins. Wiley, Weinheim, pp 171–181

Scheller J, Henggeler D, Viviani A, Conrad U (2004) Purification of spider silk-elastin from transgenic plants and application for human chondrocyte proliferation. Transgenic Res 13(1):51–57

Schillberg S, Finnern R (2021) Plant molecular farming for the production of valuable proteins–critical evaluation of achievements and future challenges. J Plant Physiol 258:153359

Sen A (1982) Poverty and famines: an essay on entitlement and deprivation. Oxford University Press, London

Shanmugaraj BI, Bulaon CJ, Phoolcharoen W (2020) Plant molecular farming: a viable platform for recombinant biopharmaceutical production. Plants 9(7):842

Sharma O, Sultan AA, Ding H, Triggle CR (2020) A review of the progress and challenges of developing a vaccine for COVID-19. Front Immunol 11:585354

Sparrow PAC, Irwin JA, Dale PJ, Twyman RM, Ma JKC (2007) Pharma-planta: road testing the developing regulatory guidelines for plant-made pharmaceuticals. Transgenic Res 16:147–161

Spiegel H, Stöger E, Twyman RM, Buyel JF (2018) Current status and perspectives of the molecular farming landscape. In: Molecular pharming: applications, challenges, and emerging areas. Wiley, Hoboken, NJ, pp 3–23

Spörlein B, Koop HU (1991) Lipofectin: direct gene transfer to higher plants using cationic liposomes. Theor Appl Genet 83(1):1–5

Springmann M, Clark M, Mason-D'Croz D, Wiebe K, Bodirsky BL, Lassaletta L et al (2018) Options for keeping the food system within environmental limits. Nature 562(7728):519–525

Sticklen MB (2008) Plant genetic engineering for biofuel production: towards affordable cellulosic ethanol. Nat Rev Genet 9(6):433–443

Stoger E, Fischer R, Moloney M, Ma JKC (2014) Plant molecular pharming for the treatment of chronic and infectious diseases. Annu Rev Plant Biol 65:743–768

Sun H, Chen Q (2018) Vaccines against West Nile virus. In: Prospects of plant-based vaccines in veterinary medicine. Springer, Cham, pp 121–146

Svab Z, Maliga P (2007) Exceptional transmission of plastids and mitochondria from the transplastomic pollen parent and its impact on transgene containment. Proc Natl Acad Sci 104(17):7003–7008

Tacket CO, Mason HS, Losonsky G, Estes MK, Levine MM, Arntzen CJ (2000) Human immune responses to a novel Norwalk virus vaccine delivered in transgenic potatoes. J Infect Dis 182(1):302–305

Topp E, Irwin R, McAllister T, Lessard M, Joensuu JJ, Kolotilin I et al (2016) The case for plant-made veterinary immunotherapeutics. Biotechnol Adv 34(5):597–604

Tschofen M, Knopp D, Hood E, Stöger E (2016) Plant molecular farming: much more than medicines. Annu Rev Anal Chem (Palo Alto Calif) 9(1):271–294

Twyman RM, Schillberg S, Fischer R (2005) Transgenic plants in the biopharmaceutical market. Expert Opin Emerg Drugs 10(1):185–218

van der Veen SJ, Hollak CEM, van Kuilenburg ABP, Langeveld M (2020) Developments in the treatment of fabry disease. J Inherit Metab Dis 43(5):908–921

Verma D, Daniell H (2007) Chloroplast vector systems for biotechnology applications. Plant Physiol 145(4):1129–1143

Virdi V, Coddens A, De Buck S, Millet S, Goddeeris BM, Cox E et al (2013) Orally fed seeds producing designer IgAs protect weaned piglets against enterotoxigenic Escherichia coli infection. Proc Natl Acad Sci 110(29):11809–11814

Waheed MT, Ismail H, Gottschamel J, Mirza B, Lössl AG (2015) Plastids: the green frontiers for vaccine production. Front Plant Sci 6:1005

Wang H-H, Yin W-B, Hu Z-M (2009) Advances in chloroplast engineering. J Genet Genomics 36(7):387–398

Ward BJ, Makarkov A, Séguin A, Pillet S, Trépanier S, Dhaliwall J et al (2020) Efficacy, immunogenicity, and safety of a plant-derived, quadrivalent, virus-like particle influenza vaccine in adults (18–64 years) and older adults (≥65 years): two multicentre, randomised phase 3 trials. Lancet 396(10261):1491–1503

Whitham SA, Eggenberger AL, Zhang C, Chowda-Reddy RV, Martin KM, Hill JH (2015) Recent advances in in planta transient expression and silencing systems for soybean using viral vectors. In: Recent advancements in gene expression and enabling technologies in crop plants. Springer, pp 423–451

Wieland WH, Lammers A, Schots A, Orzáez DV (2006) Plant expression of chicken secretory antibodies derived from combinatorial libraries. J Biotechnol 122(3):382–391

Wu J, Lawit SJ, Weers B, Sun J, Mongar N, Van Hemert J et al (2019) Overexpression of zmm28 increases maize grain yield in the field. Proc Natl Acad Sci 116(47):23850–23858

Yang M, Sun H, Lai H, Hurtado J, Chen Q (2018) Plant-produced zika virus envelope protein elicits neutralizing immune responses that correlate with protective immunity against zika virus in mice. Plant Biotechnol J 16(2):572–580. https://doi.org/10.1111/pbi.12796

Yao J, Weng Y, Dickey A, Wang KY (2015) Plants as factories for human pharmaceuticals: applications and challenges. Int J Mol Sci 16(12):28549–28565

Zaidi SS-E-A, Vanderschuren H, Qaim M, Mahfouz MM, Kohli A, Mansoor S et al (2019) New plant breeding technologies for food security. Science 363(6434):1390–1391

Zhang D, Das DB, Rielly CD (2014) Potential of microneedle-assisted micro-particle delivery by gene guns: a review. Drug Deliv 21(8):571–587

Zimran A, Gonzalez-Rodriguez DE, Abrahamov A, Cooper PA, Varughese S, Giraldo P et al (2018) Long-term safety and efficacy of taliglucerase alfa in pediatric Gaucher disease patients who were treatment-naive or previously treated with imiglucerase. Blood Cells Mol Dis 68:163–172

Zupan JR, Citovsky V, Zambryski P (1996) *Agrobacterium* vire2 protein mediates nuclear uptake of single-stranded DNA in plant cells. Proc Natl Acad Sci 93(6):2392–2397

Chapter 11
Plant Molecular Farming: A Boon for Developing Countries

Tamlyn Shaw, Sandra Jordaan, Tarin Ramsaroop, Francisco Pera, and Maribanyana Lebeko

Abstract Developing countries are often at the back of the queue to receive much-needed vaccines and diagnostics. This chapter discusses the diseases prevalent to a handful of these countries and how plant molecular farming can be an effective, feasible, and timely solution to these challenges. The chapter has a strong focus on health care, particularly diagnostic and vaccines/therapeutic proteins as the main contribution. The chapter suggests that PMF could be a tool to uplift, empower, and build capability within developing countries. Instead of relying on foreign "aid" from the developing world, these countries can take matters into their own hands, with homegrown solutions. A case study of the author's own molecular farming company in Cape Town, South Africa, is presented as proof of concept. From edible vaccines to the One Health concept, the chapter explores how the plant-based platform is a boon for developing nations.

Keywords Plant molecular farming · Developing country · Recombinant proteins · Africa · LMIC · Developed country · Diagnostic · Health · Therapeutics · Vaccines · *Nicotiana benthamiana*

Abbreviations

API	Active pharmaceutical ingredient
BRU	Biopharming Research Unit

T. Shaw (✉) · F. Pera
Cape Bio Pharms Pty Ltd., Unit 3, The Powder Mill, Ndabeni, Cape Town, South Africa

Cape Biologix Technologies Pty Ltd., Unit 3, The Powder Mill, Ndabeni, Cape Town, South Africa
e-mail: tamlyn@capebiopharms.com; tamlyn@capebiologix.com; francisco@capebiologix.com

S. Jordaan · T. Ramsaroop · M. Lebeko
Cape Biologix Technologies Pty Ltd., Unit 3, The Powder Mill, Ndabeni, Cape Town, South Africa
e-mail: sandra@capebiologix.com; tarin@capebiologix.com; mariba@capebiologix.com

C. Kole et al. (eds.), *Tools & Techniques of Plant Molecular Farming*, Concepts and Strategies in Plant Sciences, https://doi.org/10.1007/978-981-99-4859-8_11

BVDV	Bovine viral diarrhea virus
CBP	Cape Bio Pharms
CBT	Cape Biologix Technologies
CDC	Centers for Disease Control
cGMP	Current good manufacturing practice
COPD	Chronic obstructive pulmonary disease
COVID-19	Coronavirus disease 2019
CSIR	Council for Scientific and Industrial Research
CVD	Cardiovascular disease
FDA	Food and Drug Administration
FIND	Foundation for Innovative New Diagnostics
FMD	Foot and mouth disease
GCD	β-Glucocerebrosidase
GD	Gaucher disease
GLP	Good laboratory practice
HDI	Human Development Index
HIV	Human immunodeficiency virus
INTA	National Institute of Agricultural Technology
ISPMF	International Society for Plant Molecular Farming
KBP	Kentucky BioProcessing
LMIC	Low- and middle-income countries
NAb	Neutralizing antibody
NCD	Non-communicable diseases
NVD	Newcastle disease virus
OECD	Organization for Economic Cooperation and Development
PALM	Pathology and Laboratory Medicine
PAVM	Partnerships for African Vaccine Manufacturing
PBFD	Psittacine beak and feather disease
PMF	Plant molecular farming
RHD	Rabbit hemorrhagic disease
RNA	Ribonucleic acid
SAHPRA	South African Health Products Regulatory Authority
TB	Tuberculosis
UN	United Nations
VLP	Virus-like particle
WHO	World Health Organization

11.1 Introduction

Plant molecular farming (PMF) is the production of recombinant proteins in plants for a variety of uses, including as research reagents, diagnostic proteins, therapeutics, and vaccines. PMF is on the rise and is well aligned to assist with meeting many of the UN Sustainable Development Goals accepted and published in 2015

(Cookson and Stirk 2019) (United Nations: Department of Economic and Social Affairs, accessed in 2022 through https://sdgs.un.org/2030agenda).

This chapter will be a literature consolidation of the field (if the reader would pardon the pun), as well as including examples and anecdotes from our very own proof-of-concept PMF company. At the time of writing, the group consists of the primary enterprise, Cape Bio Pharms, and a manufacturing subsidiary, Cape Biologix Technologies, South Africa (Fig. 11.1), as well as the future cGMP R&D and manufacturing facility in Mauritius, Cape Biologix Technologies, Mauritious.

A "boon" refers to something that is helpful or beneficial, also described as a timely blessing. The aim of this chapter is to showcase how plant molecular farming represents a boon for developing countries, particularly in light of the numerous health crises of an epidemic or pandemic nature faced by such countries over the last century and which they were comparatively poorly equipped to contend with on a technological level.

There is an African proverb that states: "If you want to go fast, go alone. If you want to go far, go together." It is this spirit of Ubuntu, another African concept, meaning "I am because *we* are," that emphasizes the value of sharing ideals, common initiatives, and collaborating to bring ideas into fruition.

Among many other health challenges, the recent COVID-19 pandemic has further highlighted that countries operating in isolation are not protected from global threats, but that environmental, economic, and health-related phenomena affecting one nation will impact many, if not all, in one way or another. There, therefore, needs to be a drastic shift in our thinking to combat some of the challenges we now face as a global collective.

The worldwide COVID-19 pandemic response initiatives further demonstrated how crucial and impactful collaboration between stakeholders is in working towards

Fig. 11.1 Cape Biologix's multi-level indoor hydroponic grow room growing *Nicotiana benthamiana* plants in Cape Town, South Africa

a common goal. Scientists working together and focused on one shared objective resulted in the fastest vaccine rollout in history. Developing countries, in particular, would benefit from collaborating with each other to change the current narrative of suffering and victimhood that has been attached to them by the developed world.

It has become all too much of a cliché to assume that developing countries need help or aid from developed countries. Historically, it has been accepted as the duty of developed countries to "rescue" these nations, in part due to their role in colonizing them (Nations 2020). This chapter does not serve to point out the failings of more advanced nations in coming to the aid of developing countries, but rather, through a case study of our own proof of concept in South Africa, to suggest that plant molecular farming could be a way for developing countries to not only become self-sufficient in producing and supplying recombinant proteins, but also consequently thrive as experts in such technologies and contribute towards the global biopharmaceutical industry. Utilisation of inherent environmental attributes and human capacity is one tool whereby developing countries can help themselves to overcome these challenges.

A developing country is defined as a sovereign state with a less developed industrial base and a relatively low Human Development Index (HDI) (O'Sullivan and Sheffrin 2003). This is still not a universally agreed-upon term, and there is debate on precisely which countries fit this definition. Low- and middle-income countries (LMICs) is a term that is used interchangeably with "developing countries," while the latter description is based purely on the economies of these countries. For the purposes of this chapter, the term developing countries will be used and the focus will be on the agreed-upon countries in this category. A list of LMIC countries can be found at the World Bank website, https://data.worldbank.org/income-level/Low-and-middle-income, and the list of developing countries we will be sampling from is found here: https://www.isi-web.org/resources/developing-countries.

11.2 A General Look at Developing Countries

So-called developing countries tend to have the following socio-economic characteristics in common:

- They generally have high levels of unemployment, poor sanitation, and high pollution levels (Ferronato and Torretta 2019).
- Access to safe drinking water, sanitation, and hygiene services is limited (Johnston et al. 2011).
- There are high levels of infectious and tropical diseases that are compounded by vulnerable living conditions (De Rycker et al. 2018).
- Widespread government corruption hinders progress to address these issues (Damoah et al. 2018).

Developing countries worldwide are often the last in the queue to receive much-needed medicines, diagnostics, and healthcare advances, in general, which are largely manufactured and distributed by developed nations (Chamas et al. 2022). This has been shown in real time as the COVID-19 vaccines were stockpiled in developed countries, while those in most developing regions were left without access to them or had them in short supply (Padma 2021). This highlights the crucial need for developing countries to generate their own manufacturing capabilities and not rely on the "charity" of developed countries to protect and treat their own populations. It is time for developing countries to, where possible, no longer rely on "aid" and instead build their local capacity to uplift themselves and thrive (Amon and Torreele 2021).

One of the areas to start with is the local production of recombinant proteins. Such proteins play a pivotal role in scientific research to find treatments and vaccines against local diseases. They are also used as raw materials for producing diagnostic tests to detect illnesses and can be used both as vaccines to prevent and therapeutics to treat diseases. An economy cannot thrive if the population is continuously fighting against and frequently succumbing to preventable/treatable diseases (Huber et al. 2018; Smith et al. 2019).

In addition, molecular farming also has untapped potential to contribute to advances in building materials, biofuels, and multiple other industries (Tschofen et al. 2016). These other uses will also be discussed further in this chapter.

Developing countries face many healthcare challenges, which are often overshadowed by other higher priority governmental crises. Corruption is also common in developing countries; therefore, budgets that are well intentioned to address the above challenges can also be mismanaged (Damoah et al. 2018; Onwujekwe et al. 2018). In addition, the cost of medicines can be outside of LMIC budgets if those medicines are patented and imported (Islam et al. 2019; Bhatt et al. 2022). Local production of key treatments and vaccines is only recently becoming a reality for many LMICs.

The findings of the 2018 Lancet Series on Pathology and Laboratory Medicine (PALM) in low-income and middle-income countries highlighted that although diagnostics is central to health care, access to diagnostic testing in PALM countries is poor and inequitable in many parts of the world (Fleming et al. 2021). This is exacerbated in the cities of developing countries, as the movement of people from rural to urban centers causes rapid expansion of these urban centers. As a result, slums develop on the outskirts of these cities and become disease hotspots due to the lack of structured sanitation and overcrowding (Wang'ombe 1995).

Global access to health is now identified as a critical international goal (Cookson and Stirk 2019). In 2022, the WHO presented their new open-access resource database aimed at helping all countries provide healthcare services to all populations. In their words "the COVID-19 pandemic has revealed urgent gaps in countries' current ability to locate health facilities, impeding progress to provide equitable access to therapeutics, diagnostics, and vaccinations through the ACT-Accelerator and other initiatives." The recent/ongoing COVID-19 pandemic highlighted the disparities that were already existing and motivated more urgency in addressing them (WHO Global Health Facilities Database, accessed at https://www.who.int/news/

item/10-03-2022-who-global-health-facilities-database-ensuring-access-to-primary-healthcare-and-uhc).

11.3 Common Diseases of Developing Countries

As the list of developing countries is extensive, this section of this chapter aimed at finding overarching commonalities, as well as sampling certain countries from the list and taking a quick snapshot of their highest-ranking health issues. This is certainly not comprehensive or prioritized, and it does highlight that each country should focus its effort on targeting the country's own unique disease profile. There is no "one-size-fits-all" solution for developing countries; however, PMF could be a boon for a large majority of this list, including both communicable and noncommunicable diseases (see also Chap. 10).

People living in developing countries are more likely to die of a communicable disease than a noncommunicable disease. Diarrheal diseases remain a significant health challenge, remaining in the top ten causes of death in LMICs and the top five in low-income countries. Deaths and illness due to diarrheal disease are, however, decreasing (WHO: the Top Ten Causes of Death, access at https://www.who.int/news-room/fact-sheets/detail/the-top-10-causes-of-death). Malaria, tuberculosis, and HIV/AIDS are consistently found in the top ten; however, all are decreasing significantly. The so-called lifestyle diseases are on the opposite trend and are increasing in developing countries, particularly in the cities (Arokiasamy 2018; Bigna and Noubiap 2019). These include diabetes, cancers, and ischemic heart disease (Katzmarzyk et al. 2022).

There are a handful of diseases that are responsible for 90% of deaths worldwide and predominantly affect developing countries. These include malaria, dengue, HIV/AIDS, tuberculosis, hepatitis, pneumonia, and liver and cervical cancer (Waheed et al. 2016). Considering the vast number of people affected by these core diseases, it is obvious that, to date, the PMF products selected for manufacture are not focused on "need," but rather on economics. One exception to this is the Pharma Planta Initiative, a publicly funded European consortium focusing on producing an HIV microbicide. For this statistic to change, developing countries need to produce proteins themselves, which are focused on their own populations' needs.

The human population in Africa is expected to double from 1.1 to 2.3 billion by 2050 (Kaneda et al. 2021). This will amplify the current healthcare issues and requires forward planning both to meet basic healthcare needs and to mitigate healthcare challenges. Communicable diseases account for two-thirds of all deaths in Africa, with the other third being non-communicable and injuries (Gide 2014).

Africa, like other developing regions, is at the beginning of the wave of non-communicable diseases (NCDs) and needs to know how to treat/prevent cancers, diabetes, heart disease, cardiovascular diseases (CVDs), and chronic obstructive pulmonary diseases (COPDs). In 2014, the WHO estimated that by 2020, NCDs will account for 80% of the global burden of disease, causing seven out of every ten deaths in developing countries, where half of these will be premature, i.e., under the age of 70. In the year 2016, more than three-quarters of NCD deaths happened in LMICs with almost 46% of deaths occurring in those below the age of 70 years. Tobacco use, poor diet, physical inactivity, and alcohol are the four most common modifiable risk factors for NCDs. Mental health conditions have only recently been categorized by the WHO as a NCD (Islam et al. 2014). In India, the percentage of total deaths from NCDs is rising; in 1990, the deaths made up 53.6%, whereas these increased to 61.8% in 2016 (Thakur et al. 2020). Therefore, if we are to prepare for what is coming, molecular farming efforts will need to include the production of vaccines and treatments for cancers and diabetes, and other NCDs if possible, in its arsenal.

Malaria is another major cause of death in many countries and has the highest disease burden in Africa, where 90% of all malaria-related deaths occur. The mortality rate among children is very high in this region—approximately one child dies of malaria every minute (Davoodi-Semiromi et al. 2010). There is drug resistance reported for *Plasmodium*, against chloroquine and other chemotherapeutic drugs. Multiple antimalarial drugs are therefore needed to effectively treat this disease—a costly exercise for those in developing countries.

India has focused huge efforts on developing manufacturing capabilities; however, many people still do not have access to these medicines. The main communicable diseases affecting the population also include malaria (as the eighth highest) as well as pneumonia, diarrhea, tuberculosis, COVID-19, hepatitis (B and C), and HIV/AIDS among the top causes of death (Chandranand 2021).

During the 1960s/1970s, the rural population in Brazil experienced high levels of malnutrition, lack of sanitation, and proper hygiene. There were resulting high levels of parasitic infectious diseases, mainly mansonic schistosomiasis, Chagas disease, and malaria. Vaccine-preventable diseases were also prevalent during this time, such as measles and poliomyelitis. In 1989, the country eliminated polio, and measles since 2001, thanks to a country-wide campaign. Recently, some diseases have been re-emerging, such as dengue fever and yellow fever in 2010 and chikungunya and Zika since 2014 (Waldman and Sato 2016).

According to a World Bank Report entitled "The Global Burden of Disease: Generating Evidence, Guiding Policy: Middle East and North Africa Regional Edition," non-communicable diseases such as heart disease (up by 44%), stroke (up by 35%), and diabetes (up by 87%) in the Middle East and North Africa are causing more premature death and disability than they did in the past (Human Development Network, The World Bank, Institute for Health Metrics and Evaluation 2016). Novel emerging and re-emerging infectious diseases have been a huge health concern recently. In Asia alone, Zika virus and Nipah virus in Southeast Asia have occurred on an epidemic scale, not to mention SARS-CoV-2, which originated in Asia (Hu 2020).

In Russia, as in many other countries, emerging natural focal infectious diseases (EIDs) are on the increase, including tick-borne encephalitis, ixodid tick-borne borreliosis, hemorrhagic fever with renal syndrome, Crimean–Congo hemorrhagic fever, West Nile fever, Astrakhan spotted fever, leptospirosis, and tularemia (Malkhazova et al. 2020).

Over 95% of new cases of TB and deaths occur in the developing world. Treatment of multidrug-resistant TB is becoming increasingly difficult, and new strategies are therefore crucial. The only current approved vaccine is the BCG vaccine, which is effective in childhood but has questionable efficacy in adults.

Neglected tropical diseases (NTDs) predominantly affect the world's poorest and receive very little attention from the EU and other high-income countries. The big killers such as HIV/AIDS often overshadow NTDs. However, the death rates of NTDs, and the effects they have on the quality of life and the economies of developing countries, are alarming. Blindness, mental retardation, and diverse disability conditions that result as a consequence of having NTDs affect millions in Africa and Latin America (Boutayeb 2007). Sadly, these diseases could be treated if enough focus was directed towards their research. The profits and motivation for it, however, are lacking. Public spending on drugs is less than US $6 in sub-Saharan Africa compared to an estimated US $240 spent in countries of the Organisation for Economic Co-operation and Development (OECD) (Trouiller et al. 2002).

Access to diagnostics has a long way to go in many developing countries. Diagnostic tools make the treatment of disease and management of outbreaks easier, as they allow appropriate direction of resources. According to John Nkengasong, the director of the Africa Centres for Disease Control and Prevention (CDC) in Addis Ababa, Ethiopia, "Lack of access to diagnostics is Africa's Achilles' heel" (Nkengasong 2020). It is clear from the list of neglected diseases and diseases relevant to developing countries that attention and focus are required to tackle these global health challenges that may seem not to hold economic value to developed nations and big pharmaceutical companies. A "for the people, by the people" approach is needed, as well as capacity building within developing countries to empower them to solve their own challenges—as "aid" is not coming.

11.4 The Plant Molecular Farming Platform

After identifying this need for self-sufficiency in the supply of diagnostics and therapeutics, the next step is to consider what platforms and technologies would best serve the unique requirements of these countries. Modern medicine underwent a revolution in the 1980s with the unveiling (and subsequent FDA approval) of synthetic insulin, expressed in bacteria, for the treatment of diabetes. As gene sequencing technologies matured, synthetic (recombinant) protein technology has been catapulted to the forefront of disease therapeutics, generating a myriad of synthetic hormones, enzymes, signaling peptides such as growth factors, immune system components such as antibodies, and, at the heart of vaccine development, antigens. Recombinant proteins are designed to mimic the properties of a protein occurring naturally in one species (typically the species they are intended to treat); however, their DNA is modified to possess genetic elements which allow their expression in host organisms, which can be prolifically cultivated for protein production.

Many different types of recombinant proteins are created to be used as reagents in basic scientific research, often to study disease and contribute to the development of future treatments and vaccines. These "reagents" require a low compliance level, such as good laboratory practice (GLP), to produce as they are not intended for ingestion/injection in human subjects, nor for diagnostic testing of patient samples. Another rung up the compliance ladder are diagnostic proteins used in medical devices: these should ideally be produced under ISO 13485 compliance due to their potential, but indirect, impact on human health. Finally, at the top rung of the ladder are therapeutic proteins and vaccines, which are intended for direct administration to human patients and therefore require current good manufacturing practice (cGMP), including stringent and controlled quality processes. In addition, these proteins must enter and pass preclinical (nonhuman) and clinical (human) trials in order to win health authority approval, allowing their injection into or consumption by humans.

Host organisms frequently employed for recombinant protein expression include bacteria, yeast, insect cells, live animals, or mammalian cell culture. Each of these expression platforms is associated with unique biosafety and bioethical implications and considerations. The capital expenditure in terms of infrastructure, as well as the operational running costs associated with the production of protein reagents, diagnostics, or therapeutics, is notoriously high. This is largely due to the specialized equipment, building infrastructure, and sterile working environment required for the cultivation of microorganisms and mammalian cells. The overhead expenses associated with producing recombinant proteins under sterile conditions are further inflated by the necessity for specialist skills required both in upstream cultivation processes and in downstream processing for recombinant protein recovery. Furthermore, the susceptibility of microbial and mammalian cell cultivation to contamination or infection, which frequently results in potentially catastrophic work-in-progress losses, poses a significant financial risk at an industrial level.

Taken together, the above-mentioned challenges have, for the most part, held the biopharma industry mostly out of reach for developing economies, save for a select few government-funded and academic institutions, and limited such recombinant protein technology primarily to research scale.

As mentioned, there are multiple platforms that can be used to produce recombinant proteins, such as mammalian cell culture, insect cells, bacteria, yeast, and even live animals.

Plants represent the "new kid on the block," although the technology has been in development for over 30 years. Molecular farming is the production of recombinant proteins in a plant host.

The alternate spelling is "molecular pharming," in reference to the pharmaceutical applications of recombinant proteins. The production of recombinant mammalian proteins in plants was first demonstrated in 1989, with the successful expression of functional full-sized antibodies in transgenic tobacco (Hiatt et al. 1989; see also chapter "Tobacco Plants as a Versatile Host for the Expression of Glycoproteins"), followed quickly in the 1990s with human serum albumin (HSA) produced in transgenic (stable) tobacco and potato plants (Sijmons et al. 1990). The field gained traction and established itself by the early 2000s with the successful production of therapeutics, recombinant enzymes, and human and veterinary vaccines (Twyman et al. 2003).

Slowly, leading biotech companies have evolved from this early work and have produced promising results for several clinical conditions such as cystic fibrosis and non-Hodgkin's lymphoma (European et al. 2005). Plant molecular farming (PMF) is picking up steam worldwide (Fischer and Buyel 2020), with the big players—who are cGMP compliant—mainly focusing on producing therapeutic proteins and vaccines. This growth is evident based on the growing attendance of the international society of plant molecular farming (ISPMF) conferences. The 2021 conference, which was held online in lieu of the global COVID-19 pandemic, reached its record high attendance to date (http://www.ispmf.org/events). The recent conference, from 26 to 28 September 2022 in Rome, was limited to 100 people that allowed to register for safety reasons, and this trend could therefore unfortunately not be confirmed for this year.

In the beginning, the focus of PMF was mostly on expressing proteins in transgenic or stable plants. With the advent of the transient expression system using the *A. tumefaciens* vacuum infiltration and deconstructed viral vectors in the early 2000s (see also chapter "Plant Viral Vectors: Important Tools for Biologics Production"), PMF has really taken off as a viable protein production platform. Transgenic plants can take months to develop and are suitable for the production of proteins needed in high volumes. Transgenic plants containing the genes of interest for proteins that are needed on a large scale in developing countries could be grown in the field, and farming skills can be repurposed for their cultivation provided that the necessary controls are put in place to ensure that these GMO crops are contained (Rhodes and Mandivenyi 2020).

This chapter is not just theoretical envisioning; in fact, PMF is in the late stages of developing and producing products in developing countries from continents as distant as Brazil, South Africa, Argentina, and Thailand. All these countries have a high level of expertise in the area, connections with local research and veterinary centers, and focus on region-specific ailments (Rybicki et al. 2013). There are many lab-scale proofs of concept of protein production available that are relevant to developing countries. A great example is the laboratory that Cape Bio Pharms and Cape Biologix Technologies spun out of, namely the University of Cape Town's Bio Pharming Research Unit (BRU). This unit, led by Professor Ed Rybicki, is responsible for producing an impressive range of proteins, largely focused on developing country diseases.

Argentina, alongside South Africa, has also been one of the top developing countries in molecular farming advancement. The National Institute of Agricultural Technology (INTA) in Argentina has gained expertise in producing plant-based proteins as vaccine candidates. Both countries have had a large focus on veterinary vaccines and diagnostics. This is due to the high need for these products, and their essential role in animal production, coupled with the lower regulatory requirements (Rybicki et al. 2013). The focus of both countries' choices of products includes rabbit hemorrhagic disease virus (RHD), foot and mouth disease virus (FMD), bovine viral diarrhea virus (BVDV), bovine rotaviruses, Newcastle disease virus (NVD), rabbit and human papillomaviruses, bluetongue virus, and psittacine beak and feather disease virus (PBFD) (Rybicki et al. 2013; see also chapter "Plant-Based Veterinary Vaccines").

Molecular farming efforts are also blossoming in Brazil. The country has identified PMF as an investment area, and a US $180 million investment has been made into a facility, which started building in 2017 and was expected to open its doors in 2022. Part of the investment into the facility was a collaboration with the Israeli company, Protalix Biotherapeutics, with an agreement to establish their plant cell culture platform in Brazil to produce taliglucerase alfa as a product (Murad et al. 2020; see also chapters "Scaling Up the Plant Molecular Farming via Bioprocessing of Plant Cell Suspension Culture" and "Production of Recombinant Proteins Using Plant Cell Suspension Cultures and Bioreactor Engineering: A Short Review").

In Thailand, the government has recognized biopharmaceuticals as a growth engine to focus on. Baiya Phytopharm is a success story in the country and is progressing towards vaccines and therapeutics by producing cosmetic and diagnostic proteins. In fact, similarly to Cape Biologix in South Africa, Baiya Phytopharm produced diagnostic antigens and antibodies for the COVID-19 pandemic, and their plant-made proteins were incorporated into approved diagnostic test kits (Rattanapisit et al. 2021a, b; see also Chap. 12).

PMF technology allows for research and development (R&D), as well as production hubs to be built quickly and affordably in developing countries, so that these nations may produce their own supply of proteins for diagnostics, reagents, as well as vaccines and therapeutics. To this purpose, Cape Biologix has secured a European Investment Bank Facility to allow for future construction of large-scale diagnostic and cGMP facilities in Mauritius.

In some instances, the exact same protein can be used for all the above applications, provided that the required compliance level is met, for example GLP for protein reagents, ISO 13485 (medical devices) for diagnostic protein materials to comply with test kit manufacturer requirements, and finally, cGMP to produce therapeutic proteins and vaccines.

For this PMF technology to prosper and fulfill its potential of leveling out the healthcare needs for developing nations, regulatory agency swiftness will need to parallel these advancements. It would be a travesty for regulatory red tape to hold back the development of such a crucial and versatile technology.

Plant molecular farming has a range of untapped potential, including producing biomaterials such as spider silk protein (Weichert et al. 2014; Scheller and Conrad 2005), biofuels, cosmetic proteins, and food proteins. Plant matter has long been recognized as a source of sugars for fermentation to biofuels and other biomaterials. However, the tough plant cell wall is resistant to microbial and enzymatic breakdown and blocks access to the sugars within the plant fibers. This resistance has been termed "biomass recalcitrance." This is a barrier to the cost-effectiveness of using plants for fuel; however, plant molecular farming may just be the solution. Expression of key proteins by transient expression within the plants that allow ease of breakthrough into the cell wall or ease of extraction of sugars could open a huge potential for developing countries to grow their own, cost-effective fuel.

Fig. 11.2 Images from Cape Biologix's grow room at their Cape Town facility

One particular protein of relevance is leghemoglobin, the active ingredient in the Impossible™ Burger and other meat replacement products (Impossible Foods 2016). As one of the large contributors to global warming, the meat industry requires alternatives. In 2019, Cape Bio Pharms proved that *Nicotiana benthamiana* plants can express this protein; however, functionality tests were paused due to the shift to focus on proteins relevant to the pandemic (Cape Bio Pharms internal reports). And finally, PMF could provide affordable building materials to replace the makeshift "shacks" of the slums, thereby providing dignity to those living in informal settlements, as well as small dwellings for the homeless.

The setup of molecular farming platforms in developing countries will be very different from those in the developed countries. For instance, one of the aims of building these facilities is to offer employment to unskilled and semiskilled workers. The ideal place for this is in the plant growth rooms and in the infiltration process. At Cape Biologix Technologies, a molecular pharming start-up based in Cape Town, South Africa, plant cultivation was intentionally not automated, such that all seeding, transplanting, watering, and infiltration are performed by hand by a team of "grow room keepers" (Fig. 11.2).

11.5 Advantages of the PMF Platform

In developing countries, one of the greatest challenges in terms of biopharmaceutical manufacturing capacity is the shortage of specialized skills among the working population. Traditional protein production platforms involve the need for advanced technician and specialist skills throughout both upstream and downstream processes. In contrast, the technical expertise required for upstream plant cultivation is not as niche as that of mammalian or bacterial cell culturing. Many developing countries have extensive agricultural sectors, and the skills related to this industry may be effectively transferred to molecular farming. The job creation alone for unskilled and semiskilled workers in the large grow rooms of molecular farming facilities would contribute to addressing high unemployment rates often found in developing countries. Cape Biologix Technologies may be used as a case study for job creation by skill transference, in that 15–20 unskilled workers were employed and successfully trained in plant cultivation techniques such as seeding, watering, transplanting, and infiltration of hydroponic plants.

Production scalability is another advantage of the plant-based platform, which greatly impacts the profitability of manufacturing to meet market demands. In a plant-based facility, upscaling source biomass requires an expansion of grow room space to accommodate more plants, as well as a relative increase in infiltration capacity. Increased capital investment, in terms of equipment and infrastructure, for scaling up production comes into play primarily in downstream processing. This is particularly beneficial for developing countries, where site costs are relatively low, and expansion of building infrastructure is less costly than upscaling sophisticated upstream technologies. In comparison to other bioprocessing platforms, PMF upstream processes require less stringent manufacturing practices, and cGMP for plant-produced biopharmaceuticals only becomes crucial in the downstream processes (Twyman et al. 2003; see also chapters "Molecular Farming of Pharmaceutical Proteins in Different Crop Systems: A Way Forward" and "Molecular Farming for the Production of Recombinant Pharmaceutical Proteins in Plants").

In terms of profitability and scaling, investment in large batch processing equipment would further benefit the profit margin for biopharmaceutical production in a developing economy, as the input costs for batches would not be significantly higher, and, in turn, larger batches would generate greater revenue without costing much more to produce. Epidemic-level diseases such as HIV, TB, malaria, and Ebola are wreaking havoc in developing countries, and addressing this medical need would translate into generating 10^3–10^6 g of diagnostic and therapeutic proteins per year (Ma et al. 2013).

For therapeutics, the plant-based system is ideally suited to producing plant-derived proteins that are toxic to other organisms or species. Plant-derived protein toxins, such as ricin (found naturally in castor beans), can be combined with disease-targeting antibodies and applied for the treatment of biomarker-characterized diseases such as cancer (Sehnke et al. 1994).

Another significant benefit of the PMF transient expression platform is speed to market. The time between protein molecular design and protein delivery ranges from several weeks to 1–2 months, depending on the accessibility of molecular services for protein design phases. This is particularly important for new vaccines and for producing proteins swiftly for novel diseases.

Transgenic plants ultimately offer the most cost-effective strategy to address the issue of global access to healthcare. This is due to the economy of scale and low technical and scientific expertise required; the medicines could be homegrown by local farmers (Paul et al. 2013). The cultivation of plants in greenhouses or open field is indeed significantly less expensive and more scalable than fermenter systems (Twyman et al. 2003).

11.6 Boosting Molecular Farming Efforts

Now is the perfect time for change, as the system is ripe to produce proteins at the scale that has been identified as critical for LMICs. The commercial aspect must focus not only on what is profitable, but, more importantly, on what is needed by the world—and perhaps governments and regulatory agencies need to play a stronger role incentivizing this shift.

Ma et al. (2013) suggest a few actions to aid this reprioritization. Cape Biologix Technologies has already followed a similar line of thinking, i.e., (1) prioritize important targets for molecular farming in LMICs and (2) support LMIC partners to develop expertise, assist in technology transfer, and build capacity, exemplified by the Bio Pharming Research Unit (BRU) at the University of Cape Town, who were taught in and exposed to molecular farming by scientists at the Fraunhofer Institute in Germany. The BRU then developed this technology, and it was eventually commercialized with the inception of Cape Bio Pharms, in Cape Town, in 2018. This was made possible by local government grant funding, and subsequent international grant and investor funding. (3) Increased collaboration between international regulatory bodies and developing of national regulatory frameworks in LMICs. The authors also mention that South Africa, Brazil, and Argentina are in a good position for regulatory approvals of molecular farming products due to their existing experience with GMO crops. (4) Appropriate intellectual property management and socially responsible licensing are also promoted.

Many of those in molecular farming transitioned from an academic background and had developing country solutions in mind when doing so. The platform has always been viewed as beneficial for poor socio-economic environments. The above-mentioned authors also did a case study on the partnership between a local not-for-profit entity, the Council for Scientific and Industrial Research (CSIR) in Pretoria, South Africa, and Kentucky BioProcessing in Texas, USA.

This was deemed a successful model of technology transfer and collaboration between entities residing in a developed and a developing country.

11.7 Cape Bio Pharms/Cape Biologix Technologies: An African Story

Cape Biologix Technologies (CBT) is a subsidiary of Cape Bio Pharms (CBP). CBT was formed in response to the COVID-19 pandemic as the manufacturing arm of CBP, focused exclusively on SARS-CoV-2 proteins, and has evolved to also manufacturing a variety of other reagents and diagnostic proteins. CBP opened its doors in 2018 and remains the primary research and intellectual property holding company.

Since inception, this start-up has moved swiftly from successfully producing and selling reagents to the local South African research market to the development and production of diagnostic proteins—one of which has already been incorporated into a SAHPRA-approved antibody lateral flow device. SAHPRA (South African Health Products Regulatory Authority) is the regulator for medical devices and medicines in the country. When the COVID-19 pandemic hit, CBP was able to quickly pivot from producing reagents to SARS-CoV-2 spike proteins and antibodies for diagnostic test kits. Samples of these proteins were sent to local diagnostic test kit manufacturers, and cooperative R&D between the kit manufacturers and the protein producer has resulted in the successful production of antibody and neutralizing antibody test kits, with antigen kits currently in development, at the time of writing. As the market shifted away from COVID test kits, these kits were not pursued in order to focus on more relevant reagents. A range of other diagnostic proteins are also in the pipeline, including those for Ebola, malaria, dengue, HIV, and yellow fever.

During late 2020, CBT was fortunate enough to receive grant funding from the Foundation for Innovative New Diagnostics (FIND) to scale up their facility (some images of the facility after scale-up can be seen in Fig. 11.3). This included expanding the facility from one small single-layer plant grow room to a 24-layered plant grow room and 25-layered incubation, a lab expansion with large-scale equipment, and a team growing from 7 staff to over 40 employees. The facility was able to increase its protein production capacity from 100 mg to 20,000 mg/month in the space of a few months. This exercise really demonstrated the "walk the talk" as it was seen in real time how rapidly the plant-based system could respond to a challenge by being agile and flexible in shifting products and scaling up as needed.

However, it has not always been smooth sailing for the molecular farming start-up in South Africa. Long lead times for other imported reagents and materials needed for production can cause huge delays, as well as regulatory hurdles and backlogs delaying entry to the market. In addition, energy supply shortages in the country lead to scheduled electricity blackouts (termed loadshedding) that need to be combatted with the installation and use of large, expensive generators and UPS systems to prevent protein stocks from thawing.

Getting a start-up off the ground during a global pandemic is a challenge in itself. The Department of Trade and Industry (DTI)'s Technology and Human Resources for Industry Programme (THRIP) can be seen as the catalyst that allowed the business to begin, giving a much-needed grant to allow commercialization of this technology from the University of Cape Town's Biopharming Research Unit (BRU), led by Prof. Ed Rybicki and managed by Prof. Inga Hitzeroth and Dr. Ann Meyers. The start-up honors these incredible IP founders. BRU trained scientists, Scott de Beer and Francisco Pera were part of the founding team of the start-up and have done an amazing job of bringing this technology from the lab to commercial scale, and building the scientific side of the business. CBP was also seeded by UCT's own investment group, the Evergreen Fund; however, once seed funding allows the

Fig. 11.3 Some images from Cape Biologix's Cape Town R&D and manufacturing facility after scaling up

building of the business, it takes time to create a product-market fit and begin making sustainable sales. This "valley of death" of most start-ups was brutal, and investor funding, as well as the continued support of the local university, the University of Cape Town, helped CBT to cross this valley. They now have a rapidly growing pipeline of products in R&D for various customers. One such customer, Mologic Ltd., has just transitioned from a for-profit company into a social enterprise model now trading as Global Access Diagnostics (GADx). Their goal is to target neglected diseases for developing countries. Led by George Soros' Economic Development Fund with support from the Bill & Melinda Gates Foundation, the venture is aimed at expanding affordable access to medical technology in LMICs. It is collaborations with organizations such as these that will allow molecular farming to be used as a tool to effectively contribute to solving the many health issues in developing countries that have historically been ignored.

11.8 Products of Molecular Farming

The products that have been commercially produced to date in the plant-based system, however, have largely not been geared towards developing countries or LMIC needs. The first commercial product to come out of the plant-based system was for the rare Gaucher's disease. There are currently 6000 people living with Gaucher's in the United States. The treatment consists of a recombinant form of the enzyme human β-glucocerebrosidase (GCD), termed ELELYSO, which became the first plant-derived biologic that was FDA approved for use in humans (Maharjan and

Choe 2021). This pharmaceutical therapeutic protein was developed by Protalix Biotherapeutics in Israel.

ZMapp, a plant-produced monoclonal antibody cocktail, was approved for emergency use for Ebola in the 2014–2016 Ebola outbreak in West Africa (Arntzen 2015; Davey et al. 2017). According to the Director-General of the WHO, Margaret Chan, "… Ebola has historically been confined to poor African nations … A profit-driven industry does not invest in products for markets that cannot pay" (Murad et al. 2020).

The other products produced to date have been utilizing the speed of the platform to produce seasonal vaccines and treatments for orphan diseases in the American and Canadian markets (see also chapter "Plant Molecular Farming for Vaccine Development").

Medicago Inc., for instance, is a Canadian molecular farming company focused on producing seasonal flu, norovirus, rotavirus, and pandemic flu vaccines. It has also produced a COVID-19 VLP vaccine. iBio Inc., based in California (R&D) and Texas (Manufacturing), has a pipeline focused mainly on oncology and fibrotic diseases. Since the time of writing, this company has since ceased operations, however the authors wish to recognise the contribution to the field of PMP made by the company.

11.9 Plant-Made Proteins Already Combatting Developing Country Diseases

A range of vaccine candidates, diagnostic proteins, and research reagents have been produced worldwide, in developing countries and developed countries, targeting diseases mainly affecting developing countries.

The BRU at UCT has produced an impressive range of proteins in plants, aimed at developing country diseases. These include human papillomavirus virus-like particles (HPV VLPs) (Naupu et al. 2020), rift valley fever virus N proteins (Mbewana et al. 2018), a host of animal proteins for African horse sickness (Dennis et al. 2018), porcine circovirus (Gunter et al. 2019), and various other diagnostic reagents and vaccine candidates (Meyers et al. 2008; Rybicki et al. 2012; Huddy et al. 2018; Dennis et al. 2018).

In addition, BRU has developed a variety of subunit vaccine candidates for human immunodeficiency virus (HIV). HIV-1 subtype C is the most prevalent variant of the virus and has, therefore, been the focus of vaccine development (Meyers et al. 2008). HIV has infected more than 40 million people worldwide. Most of these infections have occurred in sub-Saharan Africa. The BRU and similar institutes could be tapped for licensing agreements to manufacture and commercialize these reagents in bulk for LMICs.

Cape Bio Pharms and Cape Biologix Technologies have licensed and are in the process of licensing products from the BRU pipeline for manufacture. The Cape

Town facility received their ISO 13485 certification in 2022, and is already producing research and diagnostic reagents. The list of over 19 proof-of-concept proteins produced at CBP to date includes antibodies and antigens to HIV, SARS-CoV-2, cytokines, malaria, Ebola, serum proteins, and many more the company has also since released over 35 new antibodies in 2023 alone.

To list and discuss only a few proteins produced in plants targeting diseases mainly affecting developing countries will highlight the flexibility and agility of the platform.

Dengue fever, a neglected tropical disease affecting various developing countries, has also been the focus of plant-made VLPs. Co-expression of dengue virus structural proteins and truncated nonstructural RNA-dependent RNA polymerase in *N. benthamiana* resulted in the formation of virus-like particles (VLPs), which induced a strong antibody response in mice (Ponndorf et al. 2021; see also chapter "Plant Molecular Pharming to Overcome the Global Impact of Neglected Tropical Diseases").

Hepatitis C virus E1E2 heterodimer produced transiently in lettuce triggered immune responses in mice (Clarke et al. 2017). Hepatitis B VLPs were produced in transgenic plants as early as the 1990s. The hepatitis B vaccine was initially produced in yeast, and Arntzen and Mason were able to successfully produce it in plants (Mason et al. 1992).

Domain III of the West Nile virus envelope protein was produced in *N. benthamiana* and protected mice against lethal West Nile virus infections (Stander et al. 2021).

Various TB antigens as vaccine candidates have been expressed successfully in plants. These include Ag85B, ESAT-6, Mtb72F, MPT83, and MPT64 (Rosales-Mendoza et al. 2015). An effective subunit vaccine would consist of multiple antigens, in a cocktail, in order to activate multiple aspects of the immune system.

11.10 Edible Vaccines

The unrealized boon of molecular farming for developing countries is "edible" vaccines. These would remove the need for trained personnel for the administration of injections, cold chain for transport which is often a huge challenge in Africa, and storage of the proteins substantially reducing the overall costs of these medicines. Molecular farming is ideally suited to producing animal vaccines and diagnostics, as the cost of these made in plants can be very competitive. Animals can be fed or injected with crude plant extracts, saving production costs substantially.

Edible or oral vaccines have also been tested in humans (Kurup and Thomas 2020). Plants such as tomato, lettuce, and banana have been selected because they have been granted GRAS (generally recognized as safe) status. Bananas offer an additional advantage because their tough skin protects the vaccine-containing fleshy fruit from damage. Edible vaccines also produce high levels of IgA, or mucosal antibodies, as the epithelial lining of the bronchial system and the gut are the sources of these antibodies. These represent the body's first line of defense against pathogens. The highly acidic pH of the stomach, however, can easily denature and degrade proteins—as is its function. Contrary to this, plant tissues are able to protect plant-made proteins/vaccines from degradation in the gut. Edible vaccines have been proven to cause antibody responses. Volunteers who ate raw potato tubers peeled and cut into bite-sized chunks, in all cases, were seen to have antibody responses indicating mucosal immunization (Arntzen et al. 2005). In this study by Arntzen et al., 14 volunteers ate bite-sized pieces of raw potato that had been genetically engineered to produce a non-virulent version of the toxin secreted by *E. coli* causing diarrhea. Ten out of 11 volunteers (91%) who consumed the transgenic potatoes had a fourfold increase in serum antibodies after immunization, and 6 out of 11 (55%) also showed fourfold increase in intestinal antibodies (Ravi et al. 2013).

This edible vaccine technology could be useful to produce animal vaccines; however, the cGMP requirements for human vaccines call for uniform plant-by-plant dosages, and the controls needed to ensure exact dosages are expensive and are a huge impediment to the actualization of this technology. In addition, Arntzen (2015) points out that high levels of antigen are needed to accumulate in the plant tissue to equate to a reasonable amount of plant matter to ingest to attain immunity, and this is not possible with existing systems. Purified vaccine candidates from plants are far simpler for regulatory approvals and quality control as the end product is then comparable to proteins produced in other, more established production systems (Arntzen 2015).

11.11 Regulatory and Government Involvement

As mentioned earlier in this chapter, it would be a pity for regulations to lag behind the implementation of this technology in developing countries. The South African Government has made a concerted effort to promote molecular farming through government grants. Notably, the THRIP program from the Department of Trade and Industry was what allowed Cape Bio Pharms to begin, to enable commercialization of BRU's IP portfolio. As the company has focused initially on research reagents and diagnostic proteins, regulatory approval has not been an issue thus far. The only regulatory issues have been for the final products (lateral flow diagnostic test kits) that contain the plant-made proteins. Regulatory red tape may become an obstacle when producing therapeutic proteins and vaccines in plants. An example, albeit not from a developing country, is the Medicago Inc. SARS-CoV-2 VLP vaccine produced in tobacco plants. The WHO rejected Medicago Inc.'s application for the emergency use of their plant-based vaccine, Covifenz, due to the company being owned, in part, by the cigarette manufacturer, Philip Morris International. This, coupled with the negative connotations attached to GMO crops, can give PMF a bad image, which can slow progress in approvals and in governmental buy-in to embrace the technology for the boon that it can be.

11.12 One Health

The world is currently facing a health crisis due to drug-resistant pathogens. Currently, many vaccines are made from live attenuated viruses. This impacts the cold chain of the vaccines, which is an added expense to maintain, particularly in developing countries that tend to have warm climates like Africa, India, Brazil, and Asia. In addition, this poses a risk of infection for immunocompromised patients as well as those who are malnourished. There is a need for next-generation vaccine development, as well as prophylactic treatment (see also chapters "Delivery of Drugs and Vaccines Through Plant Molecular Farming" and "Development of Oral Prophylactic and Therapeutic Vaccines Against HPV on the Basis of Plant Expression System").

The One Health concept, promoted by Prof. Rybicki and his lab, is that the health of humans is interconnected with that of the animals, plants, and environment in which we live. It is crucial to focus on all three aspects for a holistic picture of health. This can be clearly seen with zoonotic diseases, which "jump" from one species to another, such as HIV, SARS-CoV-2, and Ebola viruses to name a few. These viruses are also transmitted from wild animals to domestic animals and livestock. According to the One Health website (accessed at https://onehealthinitiative.com/about/), the One Health concept is:

> *A worldwide strategy for expanding interdisciplinary collaborations and communications in all aspects of health care for humans, animals and the environment. The synergism achieved will advance health care for the 21st century and beyond by accelerating biomedical research discoveries, enhancing public health efficacy, expeditiously expanding the scientific knowledge base, and improving medical education and clinical care. When properly implemented, it will help protect and save untold millions of lives in our present and future generations.*

Indeed, the One Health concept is already assisting Brazil to improve the surveillance of venomous animal injury and vector-borne and zoonotic diseases (de Souza Leandro et al. 2021).

The plant-based system is well poised to achieve the One Health vision. The drawback of some of the other systems is that they are expensive to set up and maintain, as well as being not easily scalable (Rybicki 2017). Plant systems can be set up at a fraction of the cost—our facility, Cape Bio Pharms, established a pilot facility for less than R10m and scaled up for less than R50m. To scale up, one simply adds more plants, or another greenhouse/grow room. CBP's scale-up to CBT comprised of the addition of another indoor hydroponic grow room with the capacity of 20,000 plants and was completed in under 2 months.

11.13 Conclusions

In conclusion, molecular farming is a tool that allows for affordable, rapid, versatile, and large-scale production of recombinant proteins for developing countries. These proteins can be utilized as diagnostic reagents, scientific research reagents, vaccines, therapeutic proteins, as well as in industry as enzymes, biofuel, and a host of other applications.

However great the tool or platform, it will be hampered by a lack of integration into regulatory language and continent-wide harmonization/connection. Each R&D and manufacturing facility needs to be located within the region it serves and the product base focused on the needs of the people surrounding it. This will require political will or grant funding to initiate mechanisms to ensure economic sustainability for these platforms. As an old Chinese proverb states, "The best time to plant a tree was 20 years ago. The second-best time is now."

Acknowledgments The Biopharming Research Unit at the University of Cape Town.

Research Contracts and Innovation (RC&I) office at the University of Cape Town.

DTI, THRIP (South African Government Department of Trade and Industry).

University of Cape Town.

The Evergreen Fund.

University Technology Fund (UTF).

Markus Sack, consultant to Cape Biologix and original designer of SARS-CoV-2 spike sequences.

References

Amon JJ, Torreele E (2021) Virtual roundtable—equitable COVID-19 vaccine access. Health Hum Rights J 23:273–288

Arntzen C (2015) Plant-made pharmaceuticals: from "edible vaccines" to Ebola therapeutics. Plant Biotechnol J 13:1013–1016

Arntzen C, Plotkin S, Dodet B (2005) Plant-derived vaccines and antibodies: potential and limitations. Vaccine 23:1753–1756

Arokiasamy P (2018) India's escalating burden of non-communicable diseases. Lancet Glob Health 6:e1262–e1263

Bhatt H, Santosh K, Thapa P (2022) Drug manufacturing and access to medicine in Nepal—A literature review of challenges and proposed remediation. Dhaka Univ J Pharm Sci 20:373–379

Bigna JJ, Noubiap JJ (2019) The rising burden of non-communicable diseases in sub-Saharan Africa. Lancet Glob Health 7:e1295–e1296

Boutayeb A (2007) Developing countries and neglected diseases: challenges and perspectives. Int J Equity Health 6:4–7

Chamas C, Barbeitas MM, Correa M, Kameda K, de Oliveira ACD, Villarinho L (2022) Innovation in diagnostics: addressing gaps in low- and middle-income countries. Bull World Health Organ 100:467–467A

Chandranand P (2021) Review on infectious diseases in India and their respective diagnostic platforms. Int J Res Appl Sci Eng Technol 9:1862–1871

Clarke JL, Paruch L, Dobrica MO, Caras I, Tucureanu C, Onu A, Ciulean S, Stavaru C, Eerde A, Wang Y et al (2017) Lettuce-produced hepatitis C virus E1E2 heterodimer triggers

immune responses in mice and antibody production after oral vaccination. Plant Biotechnol J 15:1611–1621
Cookson MD, Stirk PMR (2019) UN sustainable development goals
Damoah IS, Akwei CA, Amoako IO, Botchie D (2018) Corruption as a source of government project failure in developing countries: evidence from Ghana. Proj Manag J 49:17–33
Davey RT, Dodd L, Proschan MA, Neaton J (2017) A randomized, controlled trial of ZMapp for Ebola virus infection. N Engl J Med 176:139–148
Davoodi-Semiromi A, Schreiber M, Nalapalli S, Verma D, Singh ND, Banks RK, Chakrabarti D, Daniell H (2010) Chloroplast-derived vaccine antigens confer dual immunity against cholera and malaria by oral or injectable delivery. Plant Biotechnol J 8:223–242
De Rycker M, Baragaña B, Duce SL, Gilbert IH (2018) Challenges and recent progress in drug discovery for tropical diseases. Nature 559:498–506
de Souza Leandro A, Lopes RD, Martins CA, Rivas AV, da Silva I, Galvão SR, Maciel-de-Freitas R (2021) The adoption of the one health approach to improve surveillance of venomous animal injury, vector-borne and zoonotic diseases in Foz do Iguaçu, Brazil. PLoS Negl Trop Dis 15:1–8
Dennis SJ, O'Kennedy MM, Rutkowska D, Tsekoa T, Lourens CW, Hitzeroth II, Meyers AE, Rybicki EP (2018) Safety and immunogenicity of plant-produced African horse sickness virus-like particles in horses. Vet Res 49:105
European T, Framework U, Consortium P (2005) Molecular farming for new drugs and vaccines. EMBO Rep 6:593
Ferronato N, Torretta V (2019) Waste mismanagement in developing countries: a review of global issues. Int J Environ Res Public Health 16:1060
Fischer R, Buyel JF (2020) Molecular farming—the slope of enlightenment. Biotechnol Adv 40:107519
Fleming KA, Horton S, Wilson ML, Atun R, DeStigter K, Flanigan J, Sayed S, Adam P, Aguilar B, Andronikou S et al (2021) The lancet commission on diagnostics: transforming access to diagnostics. Lancet 6736:1997
Gide A (2014) The health of the people-what works. The African Regional Health Report
Gunter CJ, Regnard GL, Rybicki EP, Hitzeroth II (2019) Immunogenicity of plant-produced porcine circovirus-like particles in mice. Plant Biotechnol J 17:1751–1759
Hiatt A, Cafferkey R, Bowdish K (1989) Production of antibodies in transgenic plants. Nature 342:76–78
Hu Y (2020) Emerging and reemerging infectious diseases: global overview. Res Soc Strat Mobil 68:1–8
Huber C, Finelli L, Stevens W (2018) The economic and social burden of the 2014 Ebola outbreak in West Africa. J Infect Dis 218:S698–S704
Huddy SM, Hitzeroth II, Meyers AE, Weber B, Rybicki EP (2018) Transient expression and purification of horseradish peroxidase C in Nicotiana benthamiana. Int J Mol Sci 19:1–13
Human Development Network, The World Bank, Institute for Health Metrics and Evaluation (2016) The global burden of disease generating evidence, guiding policy. Middle East and North Africa Regional Edition
Impossible Foods (2016) The Impossible Burger. Agric Food E-Newsletter, pp 1–4
Islam SMS, Purnat TD, Phuong NTA, Mwingira U, Schacht K, Fröschl G (2014) Non communicable diseases (NCDs) in developing countries: a symposium report. Glob Health 10:1–7
Islam MD, Kaplan WA, Trachtenberg D, Thrasher R, Gallagher KP, Wirtz VJ (2019) Impacts of intellectual property provisions in trade treaties on access to medicine in low and middle income countries: a systematic review. Glob Health 15:1–14
Johnston RB, Berg M, Johnson CA, Tilley E, Hering JG (2011) Water and sanitation in developing countries: geochemical aspects of quality and treatment. Elements 7:163–168
Kaneda T, Greenbaum C, Haub C (2021) World Population Data Sheet 2021
Katzmarzyk PT, Friedenreich C, Shiroma EJ, Lee IM (2022) Physical inactivity and non-communicable disease burden in low-income, middle-income and high-income countries. Br J Sports Med 56:101–106

Kurup VM, Thomas J (2020) Edible vaccines: promises and challenges. Mol Biotechnol 62:79–90
Ma JKC, Christou P, Chikwamba R, Haydon H, Paul M, Ferrer MP, Ramalingam S, Rech E, Rybicki E, Wigdorowitz A et al (2013) Realising the value of plant molecular pharming to benefit the poor in developing countries and emerging economies. Plant Biotechnol J 11:1029–1033
Maharjan PM, Choe S (2021) Plant-based COVID-19 vaccines: current status, design, and development strategies of candidate vaccines. Vaccine 9:1–21
Malkhazova S, Pestina P, Prasolova A, Orlov D (2020) Emerging natural focal infectious diseases in Russia: a medical–geographical study. Int J Environ Res Public Health 17:1–12
Mason HS, Lam DMK, Arntzen CJ (1992) Expression of hepatitis B surface antigen in transgenic plants. Proc Natl Acad Sci U S A 89:11745–11749
Mbewana S, Meyers AE, Weber B, Mareledwane V, Ferreira ML, Majiwa PAO, Rybicki EP (2018) Expression of Rift Valley fever virus N-protein in Nicotiana benthamiana for use as a diagnostic antigen. BMC Biotechnol 18:1–10
Meyers A, Chakauya E, Shephard E, Tanzer FL, Maclean J, Lynch A, Williamson A, Rybicki EP (2008) Expression of HIV-1 antigens in plants as potential subunit vaccines. BMC Biotechnol 15:1–15
Murad S, Fuller S, Menary J, Moore C, Pinneh E, Szeto T, Hitzeroth I, Freire M, Taychakhoonavudh S, Phoolcharoen W et al (2020) Molecular pharming for low and middle income countries. Curr Opin Biotechnol 61:53–59
Nations U (2020) Statistical yearbook 2020 edition
Naupu PN, van Zyl AR, Rybicki EP, Hitzeroth II (2020) Immunogenicity of plant-produced human papillomavirus (HPV) virus-like particles (VLPS). Vaccine 8:1–15
Nkengasong J (2020) Let Africa into the market for COVID-19 diagnostics. Nature 580:565
O'Sullivan A, Sheffrin SM (2003) Economics: principles in action. Prentice Hall, Needham, MA
Onwujekwe O, Agwu P, Orjiakor C, Mbachu C, Hutchinson E, Odii A, Obi U, Ogbozor A, Ichoku H, Mckee M et al (2018) Corruption in the health sector in Anglophone West Africa: common forms of corruption and mitigation strategies. In: Anti-corruption evidence, pp 1–32
Padma TV (2021) Will Covid become a disease of the young? Nature 595:342–343
Paul M, Teh A, Twyman R, Ma J (2013) Target product selection—where can molecular pharming make the difference? Curr Pharm Des 19:5478–5485
Ponndorf D, Meshcheriakova Y, Thuenemann EC, Dobon Alonso A, Overman R, Holton N, Dowall S, Kennedy E, Stocks M, Lomonossoff GP et al (2021) Plant-made dengue virus-like particles produced by co-expression of structural and non-structural proteins induce a humoral immune response in mice. Plant Biotechnol J 19:745–756
Rattanapisit K, Bulaon CJI, Khorattanakulchai N, Shanmugaraj B, Wangkanont K, Phoolcharoen W (2021a) Plant-produced SARS-CoV-2 receptor binding domain (RBD) variants showed differential binding efficiency with anti-spike specific monoclonal antibodies. PLoS One 16:1–15
Rattanapisit K, Yusakul G, Shanmugaraj B, Kittirotruji K, Suwatsrisakul P, Prompetchara E, Taychakhoonavud S, Phoolcharoen W (2021b) Plant-produced recombinant SARS-CoV-2 receptor-binding domain; an economical, scalable biomaterial source for COVID-19 diagnosis. Biomater Transl 2:43–49
Ravi I, Baunthiyal M, Saxena J (2013) Edible vaccines. Adv Biotechnol:1–259. https://doi.org/10.1007/978-81-322-1554-7_12
Rhodes JI, Mandivenyi W (2020) South Africa—synthetic biology regulatory considerations and biodiversity—a legal perspective for South Africa. In: Chaurasia A, Hawksworth DL, de Miranda M (eds) GMOs: implications for biodiversity conservation and ecological processes. Springer International Publishing, Cham, pp 495–499
Rosales-Mendoza S, Ríos-Huerta R, Angulo C (2015) An overview of tuberculosis plant-derived vaccines. Expert Rev Vaccines 14:877–889
Rybicki EP (2017) Plant-made vaccines and reagents for the one health initiative. Hum Vaccines Immunother 13:2912–2917
Rybicki EP, Chikwamba R, Koch M, Rhodes JI, Groenewald JH (2012) Plant-made therapeutics: an emerging platform in South Africa. Biotechnol Adv 30:449–459

Rybicki E, Hitzeroth I, Meyers A, Santos M, Wigdorovitz A (2013) Developing country applications of molecular farming: case studies in South Africa and Argentina. Curr Pharm Des 19:5612–5621

Scheller J, Conrad U (2005) Production of spider silk proteins in transgenic tobacco and potato. In: Molecular farming: plant-made pharmaceuticals and technical proteins, pp 171–181

Sehnke PC, Pedrosa L, Paul AL, Frankel AE, Ferl RJ (1994) Expression of active, processed ricin in transgenic tobacco. J Biol Chem 269:22473–22476

Sijmons PC, Dekker BMM, Schrammeijer B, Verwoerd TC, Van Den Elzen PJM, Hoekema A (1990) Production of correctly processed human serum albumin in transgenic plants. Bio/Technology 8:217–221

Smith KM, Machalaba CC, Seifman R, Feferholtz Y, Karesh WB (2019) Infectious disease and economics: the case for considering multi-sectoral impacts. One Health 7:100080

Stander J, Chabeda A, Rybicki EP, Meyers AE (2021) A plant-produced virus-like particle displaying envelope protein domain III elicits an immune response against West Nile virus in mice. Front Plant Sci 12:1–12

Thakur JS, Paika R, Singh S (2020) Burden of noncommunicable diseases and implementation challenges of national NCD Programmes in India. Med J Armed Forces India 76:261–267

Trouiller P, Olliaro P, Torreele E, Orbinski J, Laing R, Ford N (2002) Drug development for neglected diseases: a deficient market and a public-health policy failure. Lancet 359:2188–2194

Tschofen M, Knopp D, Hood E, Stöger E (2016) Plant molecular farming: much more than medicines. Annu Rev Anal Chem 9:271–294

Twyman RM, Stoger E, Schillberg S, Christou P, Fischer R (2003) Molecular farming in plants: host systems and expression technology. Trends Biotechnol 21:570–578

Waheed MT, Sameeullah M, Khan FA, Syed T, Ilahi M, Gottschamel J, Lössl AG (2016) Need of cost-effective vaccines in developing countries: what plant biotechnology can offer? Springerplus 5:1–9

Waldman EA, Sato APS (2016) Path of infectious diseases in Brazil in the last 50 years: an ongoing challenge. Rev Saude Publica 50:1–18

Wang'ombe JK (1995) Public health crises of cities in developing countries. Soc Sci Med 41:857–862

Weichert N, Hauptmann V, Menzel M, Schallau K, Gunkel P, Hertel TC, Pietzsch M, Spohn U, Conrad U (2014) Transglutamination allows production and characterization of native-sized ELPylated spider silk proteins from transgenic plants. Plant Biotechnol J 12:265–275

Chapter 12
Plant Molecular Pharming: A Promising Solution for COVID-19

Maribanyana Lebeko, Tamlyn Shaw, Sandra Jordaan, Tarin Ramsaroop, and Francisco Pera

Abstract Plant-based systems, particularly plant molecular pharming of therapeutically relevant proteins, that could be used as subunit vaccines, antigens (in rapid antigen test kits), as well as antibodies, especially therapeutic and/or diagnostic monoclonal antibodies and antibody cocktails, offer a promising solution for coronavirus disease 2019 (COVID-19). To date, out of almost 40 vaccines, there is only one FDA-approved vaccine against COVID-19 produced in plants. Similarly, there are only a few monoclonal antibody therapies derived from plant-based expression systems in preclinical development, with none approved even for emergency use. The lack of a solid footprint of plant molecular pharming in combating the debilitating effects of SARS-CoV-2 implores the feasibility of using a plant expression system as a suitable platform to produce effective, safe, and affordable SARS-CoV-2 vaccines and therapeutics. This chapter seeks to unpack significant efforts placed in exploring the possibility of utilizing such systems as alternative quick, adaptable, and low-cost strategies for the production of therapeutics against SARS-CoV-2.

Keywords Plant molecular pharming · Plant-derived · SARS-CoV-2 · Vaccines · Antibodies · Antigens · Therapeutics

---

M. Lebeko · S. Jordaan
Cape Biologix Technologies Pty Ltd., Unit 3, Cape Town, Western Cape, South Africa
e-mail: mariba@capebiologix.com; sandra@capebiologix.com

T. Shaw · F. Pera (✉)
Cape Biologix Technologies Pty Ltd., Unit 3, Cape Town, Western Cape, South Africa

Cape Bio Pharms Pty Ltd, Unit 3, Cape Town, Western Cape, South Africa
e-mail: tamlyn@capebiologix.com; francisco@capebiopharms.com; francisco@capebiologix.com

T. Ramsaroop
Cape Bio Pharms Pty Ltd, Unit 3, Cape Town, Western Cape, South Africa
e-mail: tarin@capebiologix.com

C. Kole et al. (eds.), *Tools & Techniques of Plant Molecular Farming*, Concepts and Strategies in Plant Sciences, https://doi.org/10.1007/978-981-99-4859-8_12

Abbreviations

ACE-2	Angiotensin-converting enzyme 2
API	Active pharmaceutical ingredient
BAT	British American Tobacco
bFGF	Basic fibroblast growth factor
BY-2	Tobacco bright yellow-2
CDC	Centers for Disease Control
COVID-19	Coronavirus disease 2019
CpG	Cytidine-phospho-guanosine
FDA	Food and Drug Administration
FIND	Foundation for Innovative New Diagnostics
GCD	β-Glucocerebrosidase
GD	Gaucher disease
GM-CSF	Granulocyte-macrophage colony-stimulating factor
HC	Heavy constant
HIV	Human immunodeficiency virus
IL-1	Interleukin-1
IL-6	Interleukin-6
KBP	Kentucky BioProcessing
LC	Light constant
LMIC	Low- and middle-income countries
mAb	Monoclonal antibody
NAb	Neutralizing antibody
PAVM	Partnerships for African Vaccine Manufacturing
RBD	Receptor-binding domain
RNA	Ribonucleic acid
SAHPRA	South African Health Products Regulatory Authority
TB	Tuberculosis
UN	United Nations
VLP	Viruslike particle
WHO	World Health Organization

12.1 Introduction

Since the first reported case in Wuhan, China, in December 2019 (Chen et al. 2020), the novel coronavirus that causes severe acute respiratory syndrome (SARS-CoV-2) has become a pandemic with over half a billion confirmed positive cases, and a resultant 6.5 million deaths due to coronavirus disease 2019 (COVID-19)-related complications (Worldometer 2022).

The first confirmed case of COVID-19 in Africa was from Egypt in February 2020 (Africa CDC 2020; African Union and Africa CDC 2021). Africa now accounts for approximately 2.11% of the cumulative global infections, with over 257,000 deaths to date (Worldometer 2022). South Africa alone contributed to 33% of the cumulative cases in Africa (Worldometer 2022).

12.1.1 Structure of SARS-CoV-2

Coronaviruses (Fig. 12.1) are single-stranded (ss), positive-sense, enveloped RNA viruses with a diameter of 80–120 nm and genome length of approximately 27–34 kb (Lu et al. 2020; Sipulwa et al. 2016). Four strains of coronaviruses have been identified (α-CoV, β-CoV, γ-CoV, and δ-CoV), which mainly affect vertebrates. SARS-CoV-2 is a member of the beta genera of coronaviruses. This novel coronavirus contains four structural proteins, namely spike (S), envelope (E), membrane (M), and nucleocapsid (N) proteins. The name "corona," meaning crown in Spanish, is derived from the spike proteins that protrude outward from the virus' surface, which resemble a "crown" when viewed under the microscope. Aside from these structural proteins, the genome codes for many nonstructural proteins whose functions include replication and viral assembly processes (Naqvi et al. 2020).

The viral life cycle begins when the S1 portion of the spike protein binds to host ACE-2 receptors, allowing entry of the virions into host cells. Once through the cell membrane, the SARS-CoV-2 ssRNA genome attaches to host ribosomes, resulting in the production of a large polyprotein that is then proteolytically cleaved by enzymes into smaller peptides for the folding and assembly of new virions (Yan et al. 2020).

Most patients recover from the viral infection within 1–3 weeks; however, a small proportion (~5%) develops severe illness that can progress into acute respiratory distress syndrome (ARDS), which can lead to death. An overactive, hyperinflammatory immune response with excess release of cytokines is the main driver of disease development and of tissue damage in these patients (Vanderbeke et al. 2021).

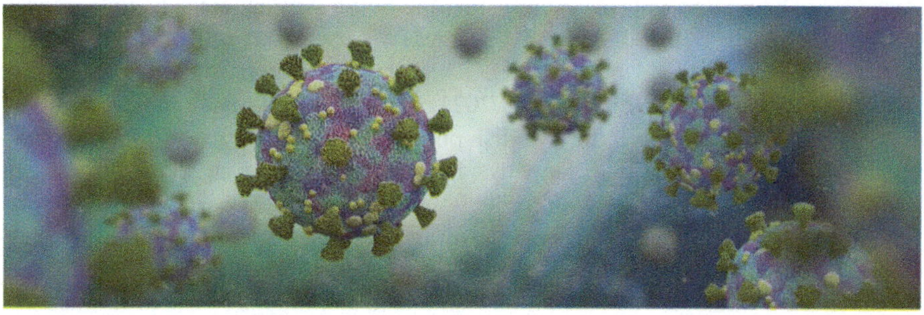

Fig. 12.1 Illustration of SARS-CoV-2 virus. (Reference: Adobe bought Stock image)

12.1.2 The Global Impact of SARS-CoV-2

The pandemic has significantly impacted the global economy, affecting more than 200 countries worldwide, thereby prompting leaders to reassess existing crisis response and management strategies. Likewise, vaccination has proven to be a promising approach to curbing the COVID-19 pandemic at individual and population levels. To date, 175 vaccine candidates are in the clinical trial stages, with 33 vaccines already approved for emergency use in at least one country (Basta and Moodie 2020). The World Health Organization (WHO) has fully approved 10 vaccines for global use (Basta and Moodie 2020; WHO 2022). As of January 2022, over 9.6 billion vaccine doses were administered globally, with almost 60% of the world's population having received at least one dose of the vaccine (Our World in Data 2020).

12.1.3 The Impact of SARS-CoV-2 Vaccine Equity on the African Continent

Despite the significant strides in vaccine development, there still exist problems associated with lack of vaccine equity, and active pharmaceutical ingredient (API) manufacturing capacity in low–middle-income countries. For instance, Africa has, to date, fully vaccinated only 21% of its population and 6% have received at least a single dose (Our World in Data 2020). This will have a direct impact on Africa's return to normal; therefore, appropriate strategies must be put in place to mitigate these shortcomings in pandemic readiness. Taking this into account, the Africa Centres for Disease Control (CDC) has proposed an initiative aimed at accelerating access to vaccines in Africa. This can only be achieved if we produce our own vaccines. The Partnerships for African Vaccine Manufacturing's (PAVM) vision is to manufacture 60% of its vaccines in the continent by 2040 (African Union and Africa CDC 2021). The PAVM has made considerable progress in leveraging pan-African and global partnerships to scale up vaccine manufacturing in several African countries. Their efforts include (1) an accord to finance vaccine production at Institut Pasteur de Dakar, Senegal; (2) establishing the South African consortium comprising Biovac, Afrigen Biologics and Vaccines, a network of universities, and the Africa CDC to establish the first COVID-19 mRNA vaccine technology hub in Africa (SAMRC News 2021; Sawahel 2021); (3) partnering several pharmaceutical companies in Morocco with the Swedish pharmaceutical contract development and manufacturing organization company, Recipharm; (4) an agreement between Holding Company for Biological Products and Vaccines (VACSERA) in Egypt and Sinovac; and (5) partnering with Algeria and Russia to produce the Sputnik V COVID-19 vaccine (Africa CDC 2021).

Given these strategies by PAVM, its vision is to manufacture 60% of its vaccines in the continent by 2040. We strongly believe that alternative vaccine and biologics manufacturing not be limited to already existing platforms but be expanded to those platforms that are more agile and adaptive to the rapid changes inherent to pandemics. Considering these, the chapter introduces plant-molecular pharming as a suitable alternative, complementary platform to address not only vaccine equity challenges in Africa, but also production of biologics more rapidly, at a large scale, and at more affordable rates.

12.2 Why Plant-Based Platforms?

Since it was discovered in the 1980s that plants can be genetically manipulated to produce proteins from other organisms (Barta et al. 1986; Barton et al. 1983), plants have been gaining traction as an alternative protein manufacturing platform, a technology that is today called "molecular pharming/farming." In "molecular pharming," conventional bioreactors are replaced with plants. In effect, each plant is a mini bioreactor. This simple, yet powerful, concept gives molecular pharming certain advantages that have yet to be rivaled by more traditional systems (see also Chap. 1).

However, it is important to understand that molecular pharming should not replace other expression systems, but rather complement them. For several years, the biotechnology industry focused on using a small number of production platforms. Bacteria (mostly *E. coli*) was first used for producing simpler proteins (Johnson 1983). Yeast species and, eventually, a selection of insect and mammalian cells were used to achieve efficient production of more complex molecules (Brondyk 2009; Cereghino and Cregg 1999; Sanchez and Demain 2012). Each of these systems has their own advantages and disadvantages; therefore, selection of expression system/s must be done on a case-by-case basis. The demand for increasingly diverse and complex proteins is on the rise, but no single expression system is a one-size-fits-all. It is natural to see the benefit of having a diverse arsenal of possible expression systems to choose from. Taking this into account, we can ask the question: What advantage does molecular farming bring to the protein manufacturing industry?

Cultivation of plants is one of humanity's oldest practices. Compared to bioreactor fermentation, growing plants is a relatively simple process that has and continues to be extensively studied throughout many different fields. Plants do not require complex and expensive growth media that is often required in cell cultures. Plants are also inherently safe to work with as they do not support the growth of human pathogens, resulting in a low risk of process-related contamination. Furthermore, since plants can rely on their innate immune system, there is also no need for creating a sterile environment, allowing the setup of plant protein manufacturing facilities to be simpler, and ultimately more cost effective than bioreactor-based platforms (Buyel 2019).

It is important to note that these advantages relate mostly to the upstream activities of molecular pharming and that downstream processing can be comparable to other expression systems. Nevertheless, a simpler and more cost-effective system lowers the entry barrier for those who want to venture into the recombinant protein manufacturing space. This is particularly relevant for developing countries who want to produce their own vaccines and therapeutics and reduce their dependence on developed countries (see also Chap. 11).

In the context of preparedness for events such as epidemics and pandemics, where protein manufacturers must rapidly shift and ramp up their routine production schedules to accommodate the large amounts of specific recombinant proteins that are sporadically required for research, diagnostic, and therapeutic use, plants may again offer unique advantages in terms of safety, speed, flexibility, and scale-up potential (Kelly-Cirino et al. 2019).

Using transient expression in plants, it is possible to achieve recombinant protein production from 6 to 8 weeks after receiving the physical DNA gene sequence. Furthermore, since each plant can be considered as a "mini bioreactor," producing different proteins in parallel becomes relatively easy, which allows for the high-throughput screening of numerous construct variants and simple upstream linear scalability potential from research-scale (e.g., 100–500 g biomass) to large-scale manufacturing (e.g., >500 kg biomass). Having a platform that has short development timelines and that is flexible enough to easily allocate some of its production capacity towards new products in a rapid and cost-effective way is crucial for preventing outbreaks like the recent SARS-CoV-2 pandemic (see also Chap. 2).

12.2.1 A Brief Account of Our Story as Cape Bio Pharms

What follows is a brief story of our first-hand experience at Cape Bio Pharms working with the plant-based platform. Cape Bio Pharms is a private small South African company dedicated to producing plant-based proteins to be used as reagents and diagnostics to benefit global health. Before SARS-CoV-2, our focus at Cape Bio Pharms was on producing different antibodies and antigens to supply the African continent with alternative sources of these research reagents. As SARS-CoV-2 began to spread in China and gain momentum, we decided to order the DNA coding for the spike protein, anticipating that it could be useful as a diagnostic reagent for research and positive serum screening. We received the DNA constructs in the beginning of March 2020, before the WHO declared it as a pandemic. Because we had no idea of how the protein would fold or express, we designed the protein in such a way that it could be truncated and fused at several key regions that were extrapolated from previously reported SARS and MERS spike proteins.

By the end of April, we had identified two main conformations of the spike protein that could be successfully expressed in our *Nicotiana benthamiana* plants (Fig. 12.2) and could be detected by positive serum (Makatsa et al. 2021). From these two proteins, one received immense interest from test kit manufacturers to create a lateral flow device that could detect positive serum.

While the test kits were being formulated and validated, we worked on scaling up our production capacity to cope with the possible demand for the spike protein. By the end of 2020, we had increased our plant biomass capacity from about 1 kg a week to 100 kg. However, we faced a bottleneck in purification because we could not source large enough chromatography columns due to the lockdown restrictions. We believe that both the scale-up time and purification challenges could have been addressed faster if it was not for the lockdown constraints. Nevertheless, by the end of 2020, we could produce enough recombinant protein to supply test kit manufacturers with enough reagent to produce test kits in the order of millions.

By that time, antibody sequences against SARS had been discovered, and research was shifting towards developing second-generation diagnostics that could detect the presence of antigen instead of positive serum. To respond to this, we also produced several antibodies against both the spike and the nucleocapsid protein, as well as the ACE2 receptor protein fused to an fc region, which are currently selling to test kit manufacturers. Eventually, as the COVID market became saturated with test kits and reagents from all over the world, we decided to start producing reagents against other diseases.

Overall, in little under 2 years, we have radically shifted production strategies to address the market demand. We were able to do this gradually and almost effortlessly thanks to the plant system. Similarly, in their publication, Diego-Martin et al. (2020) explain how it would be possible to scale up the production of recombinant antibodies from milligrams to gram amounts in little more than 6 weeks using little infrastructure requirements and *N. benthamiana* as a production platform.

Fig. 12.2 *Nicotiana benthamiana* plant close-up picture. (Source: Cape Bio Pharms Photo library)

We hope that our story was able to demonstrate the flexibility of molecular pharming and its place in fighting outbreaks like the recent SARS-CoV-2 pandemic. The next section therefore describes the available plant-based expression systems for addressing COVID-19.

12.3 Available Plant-Based Platforms to Produce Recombinant Vaccines and Biologics

The plant-based platform for recombinant vaccines and biologics production began with the use of a multitude of different plant species, including tobacco, algae, tomatoes, maize, and many others (see also chapter "Molecular Farming of Pharmaceutical Proteins in Different Crop Systems: A Way Forward"). After some time, the platforms have converged on a few key species. One of the most popular of these, *Nicotiana benthamiana*, is used by Medicago Inc., Kentucky BioProcessing, Icon Genetics, iBio, and UniBio, as well as others (Nosaki et al. 2021). During its evolution in the Australian desert, *N. benthamiana* adapted to its environment by sacrificing its pathogen defense for an accelerated reproductive cycle (Bally et al. 2018; see also chapter "Tobacco Plants as a Versatile Host for the Expression of Glycoproteins").

Transgenic plants were the pioneers of plant-based platforms in recombinant protein production. The process involves stably transforming the gene of interest into the genome of the plant at the nuclear, chloroplast, or plastid levels. Stable or transgenic plants require longer development times (approximately 4–6 months, or longer) compared to transient systems; however, the major advantage is that its upstream management only involves growing the plants that contain the gene of interest (Ghag et al. 2021). This method is useful for producing large volumes of a single vaccine or therapeutic and is ideal for most large-scale production platforms. However, the yields of product from stable transgenics can also be lower than transient systems.

There are two types of transient expression systems, those based on viral plant vectors (see also chapter "Plant Viral Vectors: Important Tools for Biologics Production") and those based on *Agrobacterium tumefaciens* infiltration/agroinfiltration. Transient expression has the advantage of being fast and flexible and can conveniently be used to confirm that the gene product is translated appropriately, and is functional, before upscaling transient production (Fischer et al. 1999).

12.3.1 Plant Cell Suspension Cultures

The first licensed recombinant pharmaceutical protein made in plants, taliglucerase alfa (Elelyso®), was produced in carrot cell suspension cultures by Protalix Biotherapeutics (Grabowski et al. 2014). The production system used consists of novel bioreactors that were designed in-house. Large polyethylene bags filled with media and supplied with sterile air are used to produce recombinant human β-glucocerebrosidase (Tekoah et al. 2015). The setup and its running costs are more affordable than those of mammalian systems (Nosaki et al. 2021). The regulatory requirements, however, are comparable to mammalian cell cultures, which makes for a smoother approval process. Cell suspension cultures are based on classical fermentation technology and are therefore relatively straightforward to scale up (Santos et al. 2016).

The technology of producing proteins from plant cell suspensions is now more than 25 years old (Sijmons et al. 1990). Recent significant yield increases in plant cells could create a boom for this technology again. Currently, whole plants and transient expression have taken center stage.

The plant cell types typically used for therapeutics, vaccines, and pharmaceutical proteins are carrot cells and tobacco BY-2 cells. The plant cell culture media consists of simple plant nutrient media, salts, and no added proteins. Therefore, those proteins excreted into the media become relatively easy to purify with no contaminating proteins from the media (Ghag et al. 2021). This platform is ideal for proteins required in large quantities (see also chapters "Scaling Up the Plant Molecular Farming via Bioprocessing of Plant Cell Suspension Culture" and "Production of Recombinant Proteins Using Plant Cell Suspension Cultures and Bioreactor Engineering: A Short Review").

12.3.2 Transient Expression Systems

Plant molecular pharming does offer a promising solution for COVID-19. This expression platform, with its variations, is ideal for a fast response to pandemics and for producing novel proteins. Potentially, a combination of transient expression systems can determine if proteins can be expressed, and if they are conformationally and posttranslationally suited to combat a certain virus, and subsequently if transgenic cells or whole plants can produce a given therapeutic or vaccine *en masse* at a fraction of the cost. A prime example of this is Medicago's recently approved VLP vaccine against SARS-CoV-2, that was a success despite the Food and Drug Administration's (FDA) initial apprehension (Medicago Vaccine 2022; see also chapter "Medical Applications of Plant Virus Nanoparticles").

One of the major challenges of plant-produced therapeutic protein and vaccines is the presence of plant-specific glycans. This can be overcome by genetic engineering of plant host glycosylation machinery as well as other methods. The co-expression of a *Leishmania major* oligosaccharyltransferase can also be used to improve the glycan occupancy of plant-derived proteins (Castilho et al. 2018).

12.3.3 Vaccines

In 2012, the first plant-derived biopharmaceutical was approved by the US Food and Drug Administration (FDA). Protalix Biotherapeutics, based in Karmiel, Israel, utilized the transgenic carrot cell suspension system to produce *taliglucerase alfa* (ELELYSO), a recombinant human β-glucocerebrosidase (GCD) for commercial use in enzyme replacement therapy of Gaucher disease (GD) (Mor 2015). However, for the past 30 years, plant-based systems have been used for therapeutics and vaccine candidates targeting diseases such as hepatitis B (Joung et al. 2016), influenza (Hodgins et al. 2019), HIV (Scotti et al. 2010), cancers (Massa et al. 2007; Sarkar et al. 2015), malaria, and cholera (Davoodi-Semiromi et al. 2010) amongst many others (see also chapter "Plant Molecular Farming for Vaccine Development").

Although many plant-based systems have been investigated for the production of biologics, *Nicotiana benthamiana* has widely been used as the mainstream production host choice (LeBlanc et al. 2020). Several companies have made strides in the development of biopharmaceuticals utilizing this platform.

Since the COVID-19 pandemic, there have been efforts to identify vaccine or therapeutic production systems that are agile and adaptable for "pandemic preparedness." Bearing this in mind, Diego-Martin et al. (2020) undertook a proof-of-concept pilot study aimed at rapid *Agrobacterium*-mediated transient expression of SARS-CoV-2-related proteins. The group investigated whether this system had the ability to produce milligram amounts of six different recombinant monoclonal antibodies against SARS-CoV-2 in *N. benthamiana*, within a few weeks. The data shows that gram amounts of antibodies against SARS-CoV-2 can potentially be achieved in just over 6 weeks using repurposed greenhouses as biomanufacturing facilities (Diego-Martin et al. 2020). This exercise was pivotal in evaluating the potential of plant-based systems in addressing the shortage in local supply (mostly in Africa) of biopharmaceuticals during the pandemic.

The next sections of this chapter deal with the existing preclinical as well as clinical vaccine candidates against COVID-19 infections. Lastly, we will give a brief account of one approved vaccine. To date, globally, there are 186 vaccine candidates in various stages of clinical trials. Forty-eight vaccine candidates are currently registered in Phase 1; 67 in Phase 2; and 71 candidates in Phase 3. Of these, plant-derived platforms account for only 2% of the other vaccine platforms such as mammalian culture systems (Fig. 12.3).

12.3.3.1 Plant-Derived Vaccine Candidates in Preclinical Trials

To date, a variety of plant-derived vaccine candidates against severe COVID-19 cases are in their early stages of development (preclinical studies). Many of these candidates target the structural proteins of SARS-CoV-2 such as spike (S) and nucleocapsid (N) and are therefore either subunit or VLP-based vaccine candidates. Institutions such as IBio, Akdeniz University, and G+FLAS Life Sciences have a variety of these candidates in their vaccine manufacturing pipeline (Fig. 12.4).

For instance, G+FLAS Life Sciences in South Korea used the glycoengineered tobacco plant, *Nicotiana benthamiana*, to express the receptor-binding domain (RBD) of the SARS-CoV-2 spike. This candidate subunit vaccine elicited potent humoral responses in BALB/cAnHsd mice via the induction of highly neutralizing antibodies (Maharjan et al. 2021). Similarly, researchers from the Akdeniz University in Turkey engineered the nucleocapsid (N) protein and co-expressed it with the RBD of SARS-CoV-2 in *N. benthamiana* plant to produce an antigen cocktail vaccine candidate. Following immunization of Balb/c mice with 5 μg of antigen cocktail at days 0 and 21, high-titer neutralizing antibodies were elicited with the cocktail compared to RBD or N proteins individually (Mamedov et al. 2021).

Lastly, iBio in the USA uses its FastPharming® System, which combines an automated hydroponic system, vertical farming, and glycan engineering technology (Maharjan and Choe 2021) to address the challenges associated with the ever-evolving SARS-CoV-2, by designing and developing second-generation vaccines that address the current durability, access, and variant-inclusion [DAVi] challenges we still face globally. Their strategy is to design "variant-inclusive" vaccine

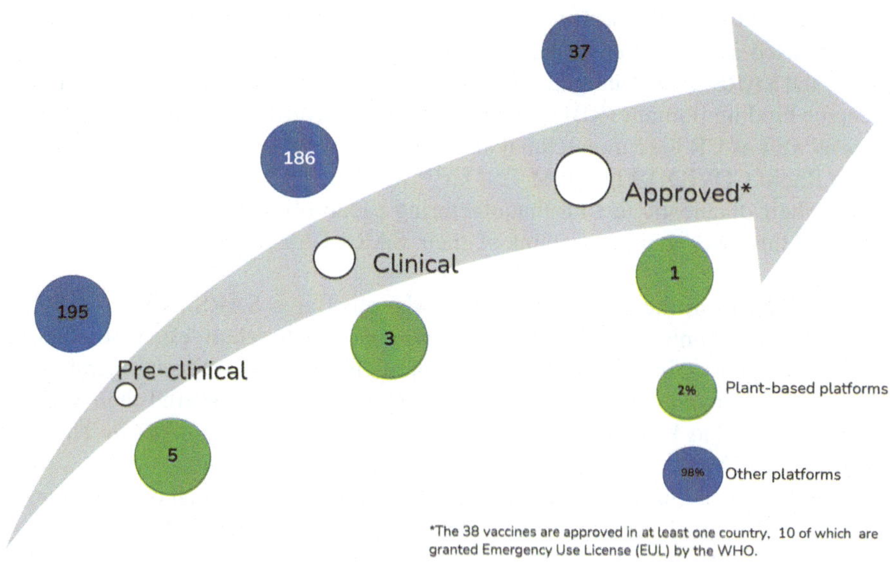

Fig. 12.3 The number of SARS-CoV-2 vaccine candidates and approved vaccines manufactured from various platforms, in comparison to plant-based expression

candidates that potentially confer protection against a broad spectrum of variants, i.e., a pan-coronavirus vaccine. To this end, they have designed a variety of both VLP and subunit-based variants such as IBIO-200, IBIO-201, and IBIO-202 (which are believed to address challenges by conferring long-lasting broad-spectrum protection and offering alternative vaccine administration routes) (Oliveira et al. 2020; IBioVaccine 2022).

12.3.3.2 Plant-Derived Vaccine Candidates in Clinical Trials

This section aims to highlight a few of the vaccine candidates that have progressed from preclinical to clinical trials (Phases 1–3). In each phase, an example of a plant-derived vaccine candidate will be given, describing the nature of the trial, country or countries where the trials are/were conducted, number of participants, and expected outcomes. We also give a brief history of the company and technology employed.

12.3.3.2.1 Phase 1: Baiya SARS-CoV-2 VAX 1

Founded in 2018 as a spin-off company from the faculty of pharmaceutical sciences in Chulalongkorn University, Thailand, Baiya Phytopharm™ aims to deliver cutting-edge, plant-based technologies that are quick, simple, and scalable (Baiyaphytopharm.com 2022). In their pipeline lies production of actives for cosmetic reagents, SARS-CoV-2 antigens, and antibodies as well as diagnostic test kits. Baiya Phytopharm™ manufactures the basic fibroblast growth factor (bFGF), or Baiya Plant FGF™, and basic epidermal growth factor (bEGF), or Baiya Plant EGF™, used as actives in antiaging cosmetics (Baiyaphytopharm.com 2022). As part of the SARS-CoV-2 antigens, and antibodies, the company produces the spike's receptor-binding domain (RBD) and angiotensin-converting enzyme (ACE2) antigens as well as CR3022 monoclonal antibody (Siriwattananon et al. 2021a, b). The Baiya Rapid COVID-19 IgG/IgM Test kit™ was launched in March 2020, forming part of their diagnostic test kit manufacturing capabilities (Baiyaphytopharm.com 2022). Below we give an account of their SARS-CoV-2 vaccine manufacturing journey.

Stemming from their preclinical data, wherein Baiya SARS-CoV-2 VAX 1, a recombinant subunit vaccine candidate, was shown to elicit efficient humoral immune responses in both mice and cynomolgus monkeys (Shanmugaraj et al. 2021; Siriwattananon et al. 2021c), Baiya Phytopharm™ registered their vaccine candidate into the Phase 1 clinical trials (Trial ID: NCT04953078) in Thailand (September 2021).

This study is a Phase 1, open-label, randomized, first-in-human dose escalation clinical trial to evaluate the safety, tolerability, and reactogenicity of escalating doses of Baiya SARS-CoV-2 VAX1 vaccine in participants aged 18–60 for adult groups and 61–75 for elderly groups. Each group will consist of three cohorts to evaluate different doses (low, medium, high) of Baiya SARS-CoV-2 VAX vaccine.

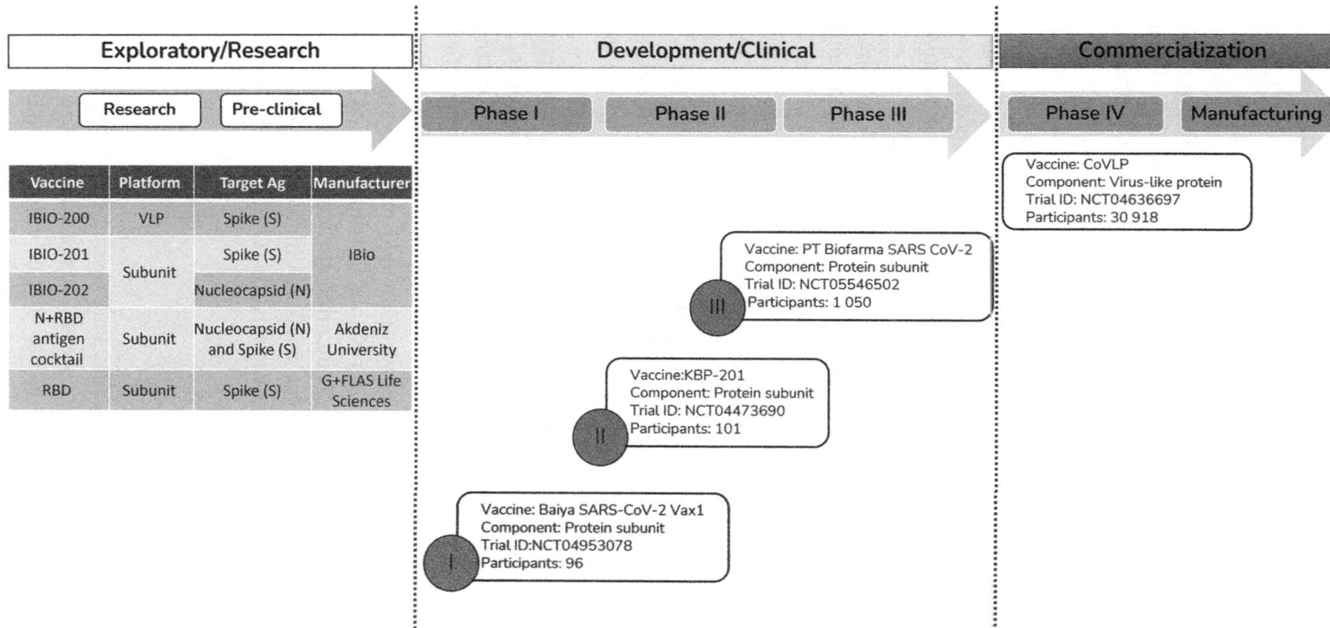

Fig. 12.4 Outlines of the global coverage of plant-derived SARS-CoV-2 vaccine candidates in preclinical and clinical stages of development, as well as one that is approved for use

Participants will be injected with two doses of the investigational product on day 1, and another on day 21 (Clinical Trials.Gov 2022).

12.3.3.2.2 Phase 2: KBP-201

Kentucky BioProcessing (KBP) is the US-based biotechnology subsidiary of Reynolds American Inc. (RAI), owned by the British American Tobacco (BAT) group (BAT News Release, April 2020 (BAT Vaccine 2020)). Compared to conventional vaccine manufacturing technology, BAT employs its proprietary, fast-growing tobacco plant technology, which has several advantages. These include fast (due to quicker accumulation of VLPs within the tobacco plants) and robust (production of 3 million vaccine-harboring plants within 6 weeks) manufacturing capabilities, as well as vaccine safety (since tobacco plants cannot host pathogens, which cause human disease), immunogenicity, and stability (vaccine will not require cold storage due to its stability at room temperature) (BAT Vaccine 2020).

Since its acquisition in 2014, KBP has made significant strides in the biotechnology industry. In 2014, they manufactured the anti-Ebola monoclonal antibody cocktail, ZMapp™, in partnership with Mapp Biopharmaceuticals, Inc. (Tran et al. 2016). In 2015, KBP produced an antibody (MB66) that protects against sexual transmission of herpes and HIV (Politch et al. 2021; see also chapter "Development of Oral Prophylactic and Therapeutic Vaccines Against HPV on the Basis of Plant Expression System"). In 2020, both their quadrivalent flu and SARS-CoV-2 vaccine candidates entered Phase 1 clinical trials (Kentucky Bio 2022). Below is an account of their contribution to curbing the COVID-19 pandemic within the vaccine manufacturing scope.

KBP is a plant-based vaccine developer that has secured the third position in the race to develop a plant-based COVID-19 vaccine. In December 2020, KBP embarked on the first-in-human (FIH), observer-blinded, randomized, placebo-controlled, parallel group study to evaluate the safety and immunogenicity of a protein subunit vaccine (KBP-COVID-19) (Trial ID: NCT04473690), coupled with a CPG adjuvant vaccine. Healthy adults were divided into two age groups, Part A (18–49 years) and Part B (50–85 years) (Clinical Trials.Gov 2022). The primary objective was to evaluate the side effects incurred at vaccination administration sites within 14 days. The secondary outcome measures included serious adverse effects in vaccination groups up to 1 year postvaccination (Clinical Trials.Gov 2022).

12.3.3.2.3 Phase 3: PT Bio Farma

PT Bio Farma (Persero) is an Indonesian state-owned enterprise (SOE) located in Bandung, West Java, and is the only local vaccine manufacturer in Indonesia. Since its inception in the 1980s, PT Bio Farma has built an impressive portfolio, including production of an array of vaccines and sera for both local and international markets. Its antiviral vaccine manufacturing pipeline includes vaccines against viral diseases such as poliomyelitis (both mono- and polyvalent oral vaccines), influenza, measles, as well as SARS-CoV-2. Other vaccines are designed for the fight against diphtheria, tetanus, TB, meningitis, and varicella.

PT Bio Farma registered (Trial ID: NCT05313035) a Phase 2/3, double-blind, randomized, placebo-controlled study to evaluate the efficacy, safety, and immunogenicity of SARS-CoV-2 protein subunit recombinant vaccine (Arthur and BioPharma Reporter 2021) adjuvanted with alum + cpg 1018 in healthy populations aged 18 years and above in Indonesia. Phase 2 was a dose-ranging study which recruited 360 subjects to compare two vaccine formulations and to evaluate their subsequent safety and immunogenicity. One vaccine group and placebo group proceeded into a Phase 3 (1050 individuals) trial aimed at evaluating the vaccine efficacy until 6 months post-primary outcome measures (Trial ID: NCT05546502). The primary outcomes included immunogenicity and efficacy (prevention of severe COVID-19 cases) of the vaccine 14 days to 6 months post the last vaccine dose.

12.3.3.3 Are There Any Approved Plant-Derived Vaccines Against SARS-CoV-2?

Plant-based platforms account for only 2% of the vaccine candidates at various stages of development. To date, there is only one vaccine approved in at least one country for commercial use. Medicago Inc.'s coronavirus-like particle vaccine (CoVLP) has recently been approved for use in Canada (Arthur and BioPharma Reporter 2022; Medicago Vaccine 2022). This is based on the transient expression of full-length S glycoprotein of the SARS-CoV-2 strain in non-transgenic *Nicotiana benthamiana* plants, wherein *Agrobacterium tumefaciens* is used as a transfer vector for transfection into plant cells (Pillet et al. 2022). Here, we provide a brief case on how the CoVLP vaccine was developed, and its preclinical, clinical, and approval journey.

12.3.4 The "Scientific Journey" of CoVLP

Pillet et al. (2022) reported the results of their preclinical study conducted in nonhuman primates. Here, they evaluated the immunogenicity and protection induced in rhesus macaques by intramuscular injections of a CoVLP vaccine candidate formulated with or without Adjuvant System 03 (AS03, from GSK) or cytidine-phosphoguanosine (CpG) 1018. The S protein trimers self-assemble (thus forming VLPs), subsequently bud off the plant cell surface, and accumulate in the interstitial spaces between the cell membrane and cell wall. These were isolated from the plant matrix and subsequently purified. Male Indian rhesus macaques from 3.5 to 8 years old were used as the animal model for their preclinical study. The study design was such that animals received 15 μg of CoVLPs by intramuscular injection with/without adjuvant at day 0, followed by another dose at day 28.

The data shows that the AS03-adjuvated vaccine induced a polyfunctional interleukin-2 (IL-2)-driven response and IL-4 expression in CD4 T cells. Furthermore, lower viral replication following nasal swabs was evident. Lung observations showed fewer infected cells as well as reduced immune cell infiltration. Reduced pro-inflammatory cytokine levels were evident as per observation of the bronchoalveolar lavage fluid (BALF). Lastly, vaccinated animals showed no signs of clinical, pathologic, or virologic evidence of vaccine-associated enhanced disease, demonstrating the safety of the vaccine. The adjuvated CoVLP was therefore selected for vaccine development and clinical trials (Pillet et al. 2022).

Ward et al. (2021) embarked a Phase 2 observer-blinded, dose escalation, randomized controlled study aimed at evaluating the safety and immunogenicity of the CoVLP vaccine candidate. In their study, neutralizing antibody (NAb) and cellular responses were used to assess both short-term as well as long-term (12 months post-vaccination) tolerability/safety and immunogenicity of CoVLP. The enrolled 180 adults (18–55 years) received either 3.75, 7.5, or 15 μg of vaccine intramuscularly (IM). The dosage interval was 21 days apart, with day 42 as the endpoint. The study shows that all formulations were well tolerated, with mild-to-moderate adverse effects in adjuvanted groups. Furthermore, the study shows that there was a tenfold increase in NAb titers with AS03 adjuvant. These responses were comparable to those seen in convalescent serum/plasma, as well as those seen in hospitalized patients (Ward et al. 2021). Unadjuvanted doses resulted in biased Th-1 responses, which were augmented by co-administration with CpG 1018. On the other hand, the AS03 adjuvant elicited the faster and more balanced Th1/Th2 response. In general, CoVLP elicited both humoral and T-cell-mediated immunity and was therefore set to progress to Phase 3 clinical trials.

To ensure greater vaccine coverage and diversity, the Phase 2/3 efficacy trial of CoVLP with AS03 adjuvant was conducted in up to 11 countries in North, Central, and South America, and Europe (Gobeil et al. 2021; Ward et al. 2021). Over 30,000 participants were enrolled in this event-driven, randomized, observer-blinded, placebo-controlled study aimed at evaluating the safety, efficacy, and immunogenicity of the CoVLP formulation in adults aged 18 years and older (Newton Clinical Trial News 2021). The primary endpoint was laboratory-confirmed symptomatic SARS-CoV-2 infection. The CoVLP+AS03 vaccine was effective in preventing COVID-19 caused by a spectrum of variants, excluding Omicron that was not in circulation during the study (Newton Clinical Trial News 2021). In the New England Journal of Medicine (NEJM), Hager et al. (2022) reported that the vaccine efficacy was shown to range from 69.5% (against symptomatic infection) to 78.8% (against moderate-to-severe disease) across all tested variants. However, in the Clinical Trials interview, Ward stated that they were busy designing the booster study based on the modified vaccine, which will share many mutations with the Omicron (B.1.1.529) variant. Over 1000 participants who previously received CoVLP will be enrolled in this study.

This is the first plant-based vaccine to be authorized for prevention of severe COVID-19. It has since acquired the commercial name Covifenz (Medicago Vaccine 2022).

12.3.5 Therapeutic Interventions

12.3.5.1 Antiviral Therapies

Recombinant human ACE-2 (rhACE-2) is currently under development for the treatment of acute lung injury and pulmonary arterial hypertension. The idea is that ACE-2 can also act as a decoy for SARS-CoV-2 by binding to the spike proteins and inhibiting cellular entry (Kiplin Guy et al. 2020). Indeed, Cape Biologix Technologies is partnering with Boku University in Vienna, Austria, and Prof Zatloukal at the University of Graz, Austria, to produce an ACE-2 decoy protein in plants. This type of partnership between a low- and middle-income country (LMIC) institution and a high-income nation meets one of the United Nations' Sustainability Goals (UN Sustainability Goals 2022). Cape Biologix Technologies has already been producing ACE-2 that is fused to a human IgG Fc tag, which has been successfully used as a reagent and diagnostic protein, demonstrating proof of concept for this protein. A research group in Thailand has also produced an ACE-2 fusion in plants and showed that treatment with this fusion protein post-viral infection dramatically inhibited SARS-CoV-2 infectivity in Vero cells with an IC50 value of 0.84 μg/mL (Siriwattananon et al. 2021a).

12.3.5.2 Monoclonal Antibody (mAb) Therapies

The last decade of plant molecular farming development has been dubbed "a decade of plantibodies" (Schillberg et al. 2002; Xu et al. 2011). Over 200 polyclonal, monoclonal, and cocktail antibodies are actively being evaluated in preclinical and clinical trials as therapeutic antibody therapies for COVID-19 (Yang et al. 2020). In their review, Ning et al. (2021a) detail the progress made in therapeutic COVID-19 antibody development and application. They further elaborate on the challenges faced by these therapies and suggest new strategies and solutions for them (Ning et al. 2021a). As of August 2021, there were over 120 antibodies in discovery and preclinical phases, while Phase 1, 2, and 3 of clinical studies recorded 11, 5, and 7 antibody therapy studies, respectively.

Monoclonal antibodies account for approximately 82% of candidate therapeutic antibodies (Yang et al. 2020). To date, only 7 mAbs against SARS-CoV-2 have received emergency use authorization (EUA), namely bamlanivimab (LY-CoV555), etesevimab (LY-CoV016), casirivimab (REGN10933), imdevimab (REGN10987), sotrovimab (S309/VIR-7831), cilgavimab (COV2-2130), and tixagevimab (COV2-2196) (Hwang et al. 2022).

12.3.5.2.1 The Future of Plant-Derived mAbs as Therapeutics and/or Diagnostics

Rattanapisit et al. (2020) undertook a proof-of-concept study using plants as an expression system to rapidly produce SARS-CoV-2 antigens and antibodies, as well as with a potential for diagnostic reagents. The rapid production is ideal for response to pandemics. In their study, the group expressed and purified the well-known monoclonal antibody (CR3022) isolated by Tian et al. (2020) from convalescent plasma of a COVID-19 patient. The yield of the mAb was estimated to be 130 μg/g of fresh leaf weight. In vitro binding studies reflected that plant-derived CR3022 binds to SARS-CoV-2 RBD, in a similar manner to the positive control serum. It was however showed that the mAb exhibited no neutralizing activity against the virus (Rattanapisit et al. 2020).

One of the major clinical presentations of COVID-19 is atypical immune activation, depicted by an increased cytokine expression, termed a cytokine storm (Chen et al. 2020; Vanderbeke et al. 2021). Herein, cell-mediated immunity is highly activated, thereby resulting in increased expression of cytokines such as tumor necrosis factor (TNF), interleukin-1 beta (IL-1β), interleukin-6 (IL-6), and granulocyte-macrophage colony-stimulating factor (GM-CSF), which contribute to disease severity (Hwang et al. 2022; To et al. 2020). The management of the resultant cytokine storm entails the use of immune modulators (both immunosuppressant and immunostimulatory). Monoclonal antibodies have therefore been widely used as such, not only for COVID-19, but also for many other immune-related conditions. Below we give a brief account of potential plant-derived immune modulators.

Potential Role of Plant-Derived mAbs in Combating SARS-CoV-2 Induced Cytokine Storm

Jugler et al. (2021) offered the first report of a glycoengineered plant-derived anti-interleukin-6 receptor monoclonal antibody (IL-6RmAb) and its activity against SARS-CoV-2-related cytokine signaling. Herein, the geminivirus bean yellow dwarf virus-based vector system was used to transiently express both the HC and LC of the IL-6RmAb in 5–7-week-old *N. benthamiana* plants and harvested at 5 days post-infiltration, resulting in an expression level of 55.95 μg/g fresh leaf weight. Binding specificity and affinity to IL-6R were evaluated via ELISA, wherein the binding affinity was comparable to the KD value reported for the mammalian cell-produced anti-IL6R mAb. Furthermore, in vitro cell-based luciferase reporter assays for IL-6 signaling depicted that the plant-derived anti-IL6R effectively inhibited the IL-6 signaling (Jugler et al. 2021).

Given the above studies, we believe that the use of plant expression systems has the potential to curb the devastating effects of cytokine storms in COVID-19 patients.

Compared with highly variable and heterogeneous therapeutic polyclonal antibodies such as convalescent plasma, mAb therapies comprise totally identical antibodies. They usually target a single epitope and are drug specific. These can be precisely engineered and optimized to serve specific treatment purposes, which render such therapies safer and more effective in comparison to polyclonal antibody therapies. Even though therapeutic polyclonal antibodies are more robust and resistant to SARS-CoV-2 mutations and variants (due to multiple drug targets and increased epitope-binding capacity), polyclonal antibodies vary vastly from donor to donor and are therefore not reproducible. On the other hand, monoclonal antibodies are easier to control; they are scalable and reproducible (due to their independence of donors) (Ning et al. 2021b). To increase epitope coverage, thereby increasing efficacy, mAb cocktail therapies have been explored.

Therapeutic Cocktail mAb Therapies

Therapeutic monoclonal antibody cocktails are the best of both worlds. They contain well-characterized components as they are monoclonal antibodies—and therefore contain homogenous components—yet can target multiple epitopes or have synergy from the different components' effects. One cocktail that put the plant-based platform in the spotlight was a treatment for Ebola ZMapp. A combination of bamlanivimab and etesevimab has been approved for emergency use for COVID-19, after it was found that at day 11 a statistically significant reduction in viral load occurred (Ning et al. 2021a). Various other monoclonal antibody cocktails are in clinical trials to treat COVID-19. Such include, but are not limited to, mAbs such as BRII (BRII-196 and BRII-198), REGN-COV2, ADM03820, and AZD7442. Many of these antibodies target the SARS-CoV-2 spike protein, and antibodies targeting the same antigen that do not overlap often have complementary, enhancing effects. To our knowledge, there are no plant-derived mAbs in use thus far; however, there are a few candidates that have been and are currently in discovery stages. Below we give an account of a few of such.

Candidate Plant-Derived Cocktail mAb Therapies

Shanmugaraj et al. (2020) offered the first report of functional anti-SARS-CoV-2 mAbs produced in plants. Their aim was to evaluate the ability of their plant expression system to rapidly produce human anti-SARS-CoV-2 mAbs (B38 and H4). These mAbs were part of the four neutralizing antibodies isolated from a convalescent COVID-19 patient by Wu et al. (2020). B38 and H4 were shown to able to block Ace-2 from binding to the RBD of the SARS-CoV-2 spike protein. Because these two antibodies occupy varying epitopes of RBD, they are said to be two non-competing, thus making them potential therapeutic mAb cocktails.

Shanmugaraj et al. (2020) showed for the first time that co-expression of the heavy-chain and light-chain sequences of both B38 and H4 mAbs in the geminiviral vector plant expression system resulted in their rapid accumulation in *Nicotiana benthamiana* leaves within 4 days post-infiltration. In vitro studies further showed that the antibody cocktail exhibited increased receptor-binding kinetics as well as potent neutralization activity in SARS-CoV-2-infected Vero-E6 cell lines.

Similarly (Jugler et al. 2022) recently undertook a study where they used the similar plant expression system to transiently express both CA1 and CB6 mAbs (isolated from a convalescent COVID-19 patient) (Duan et al. 2020), wherein 6-week-old *Nicotiana benthamiana* plants were agroinfiltrated and leaves were harvested at 7 days post-infiltration. Receptor binding studies showed that both mAbs were able to specifically bind to RBD and that their dissociation constants (K_D) were comparable to those produced in mammalian systems. Furthermore, in vitro neutralization studies showed that the two antibodies were individually able to neutralize SARS-CoV-2 in Vero-E6 cell line.

However, the group showed that the two mAbs cannot be used as a cocktail, as they competitively bind to RBD. It was therefore for this reason that they paired the plant-derived CA1 with the other two hybridoma-made 3C4 and 11D7 mAbs. Their neutralization studies indicate that the three-mAb cocktail synergistically neutralizes SARS-CoV-2 in vitro (Jugler et al. 2022).

Taken together, these studies demonstrate the feasibility of using a plant expression system as a suitable platform to produce effective, safe, and affordable SARS-CoV-2 mAbs. These studies allude to the possibility of exploring plant expression systems as alternative quick, adaptable, and low-cost strategies for production of therapeutics against SARS-CoV-2.

Diagnostic Monoclonal Antibodies

Cape Biologix Technologies began producing SARS-CoV-2 antigens and antibodies in early 2020 in response to the early stages of the pandemic (Makatsa et al. 2021). Through close relationships with diagnostic test kit manufacturers in South Africa, a serology-based lateral flow device, that detects IgM and IgG antibodies against SARS-CoV-2, was approved by the local regulatory authority, South African Health Products Regulatory Authority (SAHPRA). The company was able to scale up production and sell hundreds of mg of histidine-tagged S1 spike protein produced in *N. benthamiana* plants. In addition, a collaboration with researchers at Fraunhofer Institute in Germany utilized the same plant-made S1-His protein antigen in a novel diagnostic assay (Pietschmann et al. 2021). The company is currently developing antibodies for use in antigen test kits to detect viral infection, in collaboration with a local diagnostic test kit manufacturer.

12.3.5.3 Repurposing of Existing Therapies

Due to the maladaptive, hyper-inflammatory immune response associated with diseases, drugs that already target inflammatory pathways can be used as possible treatments. Repurposing of biologic agents that target cytokines would be a logical strategy. Interleukins 1α and 1β (IL-1α and IL-1β) are both involved in the inflammatory pathway, whose overactivation creates much of the damaging disease state and life-threatening effects associated with SARS-CoV-2 infection. IL-1α is released from damaged lung tissue, and IL-1β is produced by myeloid cells that enter the affected area. Anakinra is a recombinant form of an IL-1 receptor antagonist that is used to treat diseases associated with excess cytokine production, such as rheumatoid arthritis (Cavalli and Dinarello 2015). Anakinra has been tested for treating COVID-19, with promising results (Cavalli et al. 2020; Huet et al. 2020). It has also been recommended for approval by the European Medicines Agency (EMA) (EMA 2022). This protein could theoretically be produced in plants, although no studies have been found so far in our literature search. Ideally, any protein currently showing promise against COVID-19 could also be produced in plants and tested and may very well be a bio-better to those currently on the market. Examples include LCB1, an ultrapotent mini-peptide showing greater efficacy than some of the best antibodies at blocking SARS-CoV-2 (Case et al. 2021).

Vascular endothelial growth factor-D (VEGF-D) has been identified as the most important indicator related to the severity of disease in SARS-CoV-2-infected individuals, thereby implying its potential use as a biomarker for disease progression (Kong et al. 2020). Bevacizumab, a humanized mAb, has widely been used as an anti-VEGF in the treatment of cancers (Achen et al. 2000; Stacker et al. 2001), and therefore has the potential to be used as an immunomodulator in COVID-19 patients. Using the transgenic rice callus as an alternative gene expression system, Chen et al. (2016) in their earlier studies reported the first plant-derived anti-VEGF (bevacizumab).

12.4 Conclusions

In conclusion, the plant-based system and particularly plant molecular pharming of therapeutically relevant proteins offer a promising solution for COVID-19. The approved and effective monoclonal antibodies and antibody cocktails can be produced in plants. Biosimilars or bio-betters could then be tested and produced *en masse* in plants at a large scale, and potentially more cost-effectively. The transient expression platform allows rapid production of novel proteins and can progress from novel sequence to protein for testing within weeks. This is ideal for responding to fast-mutating viruses such as SARS-CoV-2. Vaccines that are protein based could also be produced in plants, and perhaps the enhanced economics could assist with increasing access to the countries left behind due to stockpiling by richer countries.

Acknowledgments Our sincere gratitude goes to our funders: the South African Government's Department of Trade and Industry (DTI), Foundation for Innovative New Diagnostics (FIND), and the University Technology Fund (UTF). Thanks to the BioPharming Research Unit at the University of Cape Town, the University of Cape Town at large, as well as the Research Contracts and Innovation office at the University of Cape Town. Lastly, thanks Markus Sack for your consultations to Cape Biologix and for assisting with the original design of the SARS-CoV-2 spike sequences.

References

Achen MG, Roufail S, Domagala T, Catimel B, Nice EC, Geleick DM, Murphy R, Scott AM, Caesar C, Makinen T et al (2000) Monoclonal antibodies to vascular endothelial growth factor-D block its interactions with both VEGF receptor-2 and VEGF receptor-3. Eur J Biochem 267:2505–2515

Africa CDC (2020) Africa identifies first case of coronavirus disease: statement by the director of Africa CDC. Available at: https://africacdc.org/news-item/africa-identifies-first-case-of-coronavirus-disease-statement-by-the-director-of-africa-cdc/#:~:text=News%20%2F%20P

Africa CDC (2021) Statement on the remarkable progress made by several African countries as part of the partnerships for African vaccine manufacturing (PAVM), Jul 2021. Available at: https://africacdc.org/news-item/statement-on-the-remarkable-progress-made-by-several-afr

African Union and Africa CDC Post-Summit Communique (2021) African Union Press Release

Arthur R, BioPharma Reporter (2021) Plant-based COVID-19 vaccine candidate reports positive phase 3 results, 7 Dec 2021. Available at: https://www.biopharma-reporter.com/Article/2021/12/07/Plant-based-COVID-19-vaccine-reports-positive-Phase-3-results

Arthur R, BioPharma Reporter (2022) Medicago plant-based vaccine approved for use in Canada. Available at: https://www.biopharma-reporter.com/Article/2022/02/28/Plant-based-COVID-19-vaccine-approved-in-Canada. Accessed on 4 Apr 2022

Baiyaphytopharm.com (2022) Baiyaphytopharm website. Available at: https://baiyaphytopharm.com/. Accessed on 16 Aug 2022

Bally J, Jung H, Mortimer C, Naim F, Philips JG, Hellens R, Bombarely A, Goodin MM, Waterhouse PM (2018) The rise and rise of Nicotiana benthamiana: a plant for all reasons. Annu Rev Phytopathol 56:405–426

Barta A, Karin S, Diana T, Klaus H, Marjori A, M., and Antionius J. M., M. (1986) The expression of a nopaline synthase—human growth hormone chimaeric gene in transformed tobacco and sunflower callus tissue. Plant Mol Biol 6:347–357

Barton KA, Binns AN, Matzke AJM, Chilton MD (1983) Regeneration of intact tobacco plants containing full length copies of genetically engineered T-DNA, and transmission of T-DNA to R1 progeny. Cell 32:1033–1043

Basta NE, Moodie EMM, on behalf of the VIPER (Vaccines, Infectious Disease Prevention, and E.R.G.C.-19 V.D. and A.T.T.) (2020) COVID-19 vaccine development and approvals tracker. Available at: covid19.trackvaccines.org. Accessed on 18 Jan 2022

BAT Vaccine (2020) BAT working on potential COVID-19 vaccine through US bio-tech subsidiary, 1 Apr 2020. https://www.bat.com/group/sites/UK__9D9KCY.nsf/vwPagesWebLive/DOBN8QNL#. Accessed on 22 Jun 2022

Brondyk WH (2009) Chapter 11: Selecting an appropriate method for expressing a recombinant protein. Elsevier Inc.

Buyel JF (2019) Plant molecular farming—integration and exploitation of side streams to achieve sustainable biomanufacturing. Front Plant Sci 9:1–17

Case JB, Chen RE, Cao L, Ying B, Winkler ES, Johnson M, Goreshnik I, Pham MN, Shrihari S, Kafai NM et al (2021) Ultrapotent miniproteins targeting the SARS-CoV-2 receptor-binding domain protect against infection and disease. Cell Host Microbe 29:1151–1161.e5

Castilho A, Beihammer G, Pfeiffer C, Göritzer K, Montero-Morales L, Vavra U, Maresch D, Grünwald-Gruber C, Altmann F, Steinkellner H et al (2018) An oligosaccharyltransferase from Leishmania major increases the N-glycan occupancy on recombinant glycoproteins produced in Nicotiana benthamiana. Plant Biotechnol J 16:1700–1709

Cavalli G, Dinarello CA (2015) Treating rheumatological diseases and co-morbidities with interleukin-1 blocking therapies. Rheumatology (United Kingdom) 54:2134–2144

Cavalli G, Farina N, Campochiaro C, De Luca G, Della-Torre E, Tomelleri A, Dagna L (2020) Repurposing of biologic and targeted synthetic anti-rheumatic drugs in COVID-19 and hyper-inflammation: a comprehensive review of available and emerging evidence at the peak of the pandemic. Front Pharmacol 11:1–10

Cereghino GPL, Cregg JM (1999) Applications of yeast in biotechnology: protein production and genetic analysis. Curr Opin Biotechnol 10:422–427

Chen L, Yang X, Luo D, Yu W (2016) Efficient production of a bioactive bevacizumab monoclonal antibody using the 2A self-cleavage peptide in transgenic rice callus. Front Plant Sci 7:1–12

Chen N, Zhou M, Dong X, Qu J, Gong F, Han Y, Qiu Y, Wang J, Liu Y, Wei Y et al (2020) Epidemiological and clinical characteristics of 99 cases of 2019 novel coronavirus pneumonia in Wuhan, China: a descriptive study. Lancet 395:507–513

Clinical Trials.Gov (2022) Clinical Trails.gov. https://clinicaltrials.gov/ct2/show/NCT04636697. Accessed on 14 Apr 2022

Davoodi-Semiromi A, Schreiber M, Nalapalli S, Verma D, Singh ND, Banks RK, Chakrabarti D, Daniell H (2010) Chloroplast-derived vaccine antigens confer dual immunity against cholera and malaria by oral or injectable delivery. Plant Biotechnol J 8:223–242

Diego-Martin B, González B, Vazquez-Vilar M, Selma S, Mateos-Fernández R, Gianoglio S, Fernández-del-Carmen A, Orzáez D (2020) Pilot production of SARS-CoV-2 related proteins in plants: a proof of concept for rapid repurposing of indoor farms into biomanufacturing facilities. Front Plant Sci 11:612781

Duan K, Liu B, Li C, Zhang H, Yu T, Qu J, Zhou M (2020) The feasibility of convalescent plasma therapy in severe COVID- 19 patients: a pilot study. MedRxiv:1–22. https://doi.org/10.1101/2020.03.16.20036145

EMA (2022) European Medicines Agency. Available at: https://www.ema.europa.eu/en/news/ema-recommends-approval-use-kineret-adults-covid-19. Accessed on 15 Jul 2022

Fischer R, Vaquero-Martin C, Sack M, Drossard J, Emans N, Commandeur U (1999) Towards molecular farming in the future: transient protein expression in plants. Biotechnol Appl Biochem 30:113–116

Ghag SB, Adki VS, Ganapathi TR, Bapat VA (2021) Plant platforms for efficient heterologous protein production. Biotechnol Bioprocess Eng 26:546–567

Gobeil P, Pillet S, Boulay I, Séguin A, Makarkov A, Heizer G, Bhutada K, Mahmood A, Charland N, Trépanier S et al (2021) Phase 2 randomized trial of an AS03 adjuvanted plant-based virus-like particle vaccine for Covid-19 in healthy adults, older adults and adults with comorbidities. MedRxiv. https://doi.org/10.1101/2021.05.14.21257248

Grabowski G, Golembo M, Shaaltiel Y (2014) Taliglucerase alfa: an enzyme replacement therapy using plant cell expression technology. Mol Genet Metab 111:S49

Hager KJ, Pérez Marc G, Gobeil P, Diaz RS, Heizer G, Llapur C, Makarkov AI, Vasconcellos E, Pillet S, Riera F et al (2022) Efficacy and safety of a recombinant plant-based adjuvanted Covid-19 vaccine. N Engl J Med 386:2084–2096

Hodgins B, Pillet S, Landry N, Ward BJ (2019) Prime-pull vaccination with a plant-derived virus-like particle influenza vaccine elicits a broad immune response and protects aged mice from death and frailty after challenge. Immun Ageing 16:1–14

Huet T, Beaussier H, Voisin O, Jouveshomme S, Dauriat G, Lazareth I, Sacco E, Naccache JM, Bézie Y, Laplanche S et al (2020) Anakinra for severe forms of COVID-19: a cohort study. Lancet Rheumatol 2:e393–e400

Hwang YC, Lu RM, Su SC, Chiang PY, Ko SH, Ke FY, Liang KH, Hsieh TY, Wu HC (2022) Monoclonal antibodies for COVID-19 therapy and SARS-CoV-2 detection. J Biomed Sci 29:1–50

IBioVaccine (2022) Vaccines (IBIO-202). Available at: https://ibioinc.com/vaccines/ibio-202. Accessed on 20 Jan 2022

Johnson IS (1983) Human insulin from recombinant DNA technology author(s): Irving S. Johnson. Science 219:632–637

Joung YH, Park SH, Moon KB, Jeon JH, Cho HS, Kim HS (2016) The last ten years of advancements in plant-derived recombinant vaccines against hepatitis b. Int J Mol Sci 17:1–29

Jugler C, Sun H, Chen Q (2021) SARS-CoV-2 spike protein-induced interleukin 6 signaling is blocked by a plant-produced anti-interleukin 6 receptor monoclonal antibody. Vaccine 9:1–12

Jugler C, Sun H, Grill F, Kibler K, Esqueda A, Lai H, Li Y, Lake D, Chen Q (2022) Potential for a plant-made SARS-CoV-2 neutralizing monoclonal antibody as a synergetic cocktail component. Vaccine 10:772

Kelly-Cirino CD, Nkengasong J, Kettler H, Tongio I, Gay-Andrieu F, Escadafal C, Piot P, Peeling RW, Gadde R, Boehme C (2019) Importance of diagnostics in epidemic and pandemic preparedness. BMJ Glob Heal 4:1–8

Kentucky Bio (2022) Kentucky bioprocessing about us page. Available at: https://kentuckybioprocessing.com/about-us-2/. Accessed on 20 Jan 2022

Kiplin Guy R, DiPaola RS, Romanelli F, Dutch RE (2020) Rapid repurposing of drugs for COVID-19. Science 368:829–830

Kong Y, Han J, Wu X, Zeng H, Liu J, Zhang H (2020) VEGF-D: a novel biomarker for detection of COVID-19 progression. Crit Care 24:1–4

LeBlanc Z, Waterhouse P, Bally J (2020) Plant-based vaccines: the way ahead? Viruses 13:1–13

Lu R, Zhao X, Li J, Niu P, Yang B, Wu H, Wang W, Song H, Huang B, Zhu N et al (2020) Genomic characterisation and epidemiology of 2019 novel coronavirus: implications for virus origins and receptor binding. Lancet 6736:1–10

Maharjan PM, Choe S (2021) Plant-based COVID-19 vaccines: current status, design, and development strategies of candidate vaccines. Vaccine 9:1–21

Maharjan PM, Cheon J, Jung J, Kim H, Lee J, Song M, Jeong GU, Kwon Y, Shim B, Choe S (2021) Plant-expressed receptor binding domain of the SARS-CoV-2 spike protein elicits humoral immunity in mice. Vaccine 9:978

Makatsa MS, Tincho MB, Wendoh JM, Ismail SD, Nesamari R, Pera F, de Beer S, David A, Jugwanth S, Gededzha MP et al (2021) SARS-CoV-2 antigens expressed in plants detect antibody responses in COVID-19 patients. Front Plant Sci 12:589940

Mamedov T, Yuksel D, Ilgın M, Gürbüzaslan I, Gulec B, Mammadova G, Ozdarendeli A, Yetiskin H, Kaplan B, Islam Pavel ST et al (2021) Production and characterization of nucleocapsid and rbd cocktail antigens of SARS–CoV-2 in nicotiana benthamiana plant as a vaccine candidate against COVID-19. Vaccine 9:1–15

Massa S, Franconi R, Brandi R, Muller A, Mett V, Yusibov V, Venuti A (2007) Anti-cancer activity of plant-produced HPV16 E7 vaccine. Vaccine 25:3018–3021

Medicago Vaccine (2022) Medicago Covifenz COVID-19 vaccine (date modified: 2022-06-07). https://www.canada.ca/en/health-canada/services/drugs-health-products/covid19-industry/drugs-vaccines-treatments/vaccines/medicago.html

Mor T (2015) Molecular pharming's foot in the FDA's door: Protalix's trailblazing story. Biotechnol Lett 37:2147–2150

Naqvi AAT, Fatima K, Mohammad T, Fatima U, Singh IK, Singh A, Atif SM, Hariprasad G, Hasan GM, Hassan MI (2020) Insights into SARS-CoV-2: genome, structure, evolution, pathogenesis and therapies: structural genomics approach. BBA Mol Basis Dis 1866:1–17

Newton Clinical Trial News (2021) Newton. Clinical Trials Arena News, 8 Dec 2021. https://www.clinicaltrialsarena.com/news/company-news/medicago-previews-vlp-covid-19-vaccine-booster-trial-slated-for-january/. Accessed on 14 Apr 2022

Ning L, Abagna HB, Jiang Q, Liu S, Huang J (2021a) Development and application of therapeutic antibodies against covid-19. Int J Biol Sci 17:1486–1496

Ning S, Yu B, Wang Y, Wang F (2021b) SARS-CoV-2: origin, evolution, and targeting inhibition. Front Cell Infect Microbiol 11:1–19

Nosaki S, Hoshikawa K, Ezura H, Miura K (2021) Transient protein expression systems in plants and their applications. Plant Biotechnol 38:297–304

Oliveira SC, de Magalhães MTQ, Homan EJ (2020) Immunoinformatic analysis of SARS-CoV-2 nucleocapsid protein and identification of COVID-19 vaccine targets. Front Immunol 11:1–10

Our World in Data (2020) Coronavirus pandemic (COVID-19). Hannah R, Edouard M, Lucas R, Cameron A, Charlie G, Esteban O, Joe H, Bobbie M, Diana B, Max R. Available at: https://ourworldindata.org/

Pietschmann J, Voepel N, Voß L, Rasche S, Schubert M, Kleines M, Krause HJ, Shaw TM, Spiegel H, Schroeper F (2021) Development of fast and portable frequency magnetic mixing-based serological SARS-CoV-2-specific antibody detection assay. Front Microbiol 12:1–11

Pillet S, Arunachalam PS, Andreani G, Golden N, Fontenot J, Aye PP, Röltgen K, Lehmicke G, Gobeil P, Dubé C et al (2022) Safety, immunogenicity, and protection provided by unadjuvanted and adjuvanted formulations of a recombinant plant-derived virus-like particle vaccine candidate for COVID-19 in nonhuman primates. Cell Mol Immunol 19:222–233

Pogrebnyak N, Golovkin M, Andrianov V, Spitsin S, Smirnov Y, Egolf R, Koprowsky H (2005) Severe acute respiratory syndrome (SARS) S protein production in plants: development of recombinant vaccine. PNAS 102:9062–9067

Politch JA, Cu-Uvin S, Moench TR, Tashima KT, Marathe JG, Guthrie KM, Cabral H, Nyhuis T, Brennan M, Zeitlin L et al (2021) Safety, acceptability, and pharmacokinetics of a monoclonal antibody-based vaginal multipurpose prevention film (MB66): a phase I randomized trial. PLoS Med 18:1–24

Rattanapisit K, Shanmugaraj B, Manopwisedjaroen S, Purwono PB, Siriwattananon K, Khorattanakulchai N, Hanittinan O, Boonyayothin W, Thitithanyanont A, Smith DR et al (2020) Rapid production of SARS-CoV-2 receptor binding domain (RBD) and spike specific monoclonal antibody CR3022 in Nicotiana benthamiana. Sci Rep 10:1–11

SAMRC News (2021) South Africa establishes a world-class consortium to create the first technology transfer hub for Covid-19 vaccines in Africa, Jun 2021. Available at: https://www.samrc.ac.za/media-release/south-africa-establishes-world-class-consortium-creat

Sanchez S, Demain A (2012) Special issue on the production of recombinant proteins. Biotechnol Adv 30:1100–1101

Santos RB, Abranches R, Fischer R, Sack M, Holland T (2016) Putting the spotlight back on plant suspension cultures. Front Plant Sci 7:1–12

Sarkar S, Jain S, Rai V, Sahoo DK, Raha S, Suklabaidya S, Senapati S, Rangnekar VM, Maiti IB, Dey N (2015) Plant-derived SAC domain of PAR-4 (prostate apoptosis response 4) exhibits growth inhibitory effects in prostate cancer cells. Front Plant Sci 6:1–17

Sawahel W (2021) COVID-19 vaccine technology transfer hub for Africa. University World News: Africa Edition. Available at: https://www.universityworldnews.com/post.php?story=20210628104252566. Accessed on 18 Jan 2022

Schillberg S, Emans N, Fischer R (2002) Antibody molecular farming in plants and plant cells. Phytochem Rev 1:45–54

Scotti N, Buonaguro L, Tornesello ML, Cardi T, Buonaguro FM (2010) Plant-based anti-HIV-1 strategies: vaccine molecules and antiviral approaches. Expert Rev Vaccines 9:925–936

Shanmugaraj B, Rattanapisit K, Manopwisedjaroen S, Thitithanyanont A, Phoolcharoen W (2020) Monoclonal antibodies B38 and H4 produced in Nicotiana benthamiana neutralize SARS-CoV-2 in vitro. Front Plant Sci 11:1–11

Shanmugaraj B, Siriwattananon K, Malla A, Phoolcharoen W (2021) Potential for developing plant-derived candidate vaccines and biologics against emerging coronavirus infections. Pathogens 10:1–18

Sijmons PC, Dekker BMM, Schrammeijer B, Verwoerd TC, Van Den Elzen PJM, Hoekema A (1990) Production of correctly processed human serum albumin in transgenic plants. Bio/Technology 8:217–221

Sipulwa LA, Ongus JR, Coldren RL, Bulimo WD (2016) Molecular characterization of human coronaviruses and their circulation dynamics in Kenya, 2009-2012 coronaviruses: emerging and re-emerging pathogens in humans and animals Susanna Lau positive-strand RNA viruses. Virol J 13:18

Siriwattananon K, Manopwisedjaroen S, Kanjanasirirat P, Budi Purwono P, Rattanapisit K, Shanmugaraj B, Smith DR, Borwornpinyo S, Thitithanyanont A, Phoolcharoen W (2021a)

Development of plant-produced recombinant ACE2-fc fusion protein as a potential therapeutic agent against SARS-CoV-2. Front Plant Sci 11:1–12
Siriwattananon K, Manopwisedjaroen S, Shanmugaraj B, Prompetchara E, Ketloy C, Buranapraditkun S, Tharakhet K, Kaewpang P, Ruxrungtham K, Thitithanyanont A et al (2021b) Immunogenicity studies of plant-produced sars-cov-2 receptor binding domain-based subunit vaccine candidate with different adjuvant formulations. Vaccine 9:1–16
Siriwattananon K, Manopwisedjaroen S, Shanmugaraj B, Rattanapisit K, Phumiamorn S, Sapsutthipas S, Trisiriwanich S, Prompetchara E, Ketloy C, Buranapraditkun S et al (2021c) Plant-produced receptor-binding domain of SARS-CoV-2 elicits potent neutralizing responses in mice and non-human primates. Front Plant Sci 12:1–15
Stacker SA, Caesar C, Baldwin ME, Thornton GE, Williams RA, Prevo R, Jackson DG, Nishikawa SI, Kubo H, Achen MG (2001) VEGF-D promotes the metastatic spread of tumor cells via the lymphatics. Nat Med 7:186–191
Tekoah Y, Shulman A, Kizhner T, Ruderfer I, Fux L, Nataf Y, Bartfeld D, Ariel T, Gingis-Velitski S, Hanania U et al (2015) Large-scale production of pharmaceutical proteins in plant cell culture-the Protalix experience. Plant Biotechnol J 13:1199–1208
Tian X, Li C, Huang A, Xia S, Lu S, Shi Z, Lu L, Jiang S, Yang Z, Wu Y et al (2020) Potent binding of 2019 novel coronavirus spike protein by a SARS coronavirus-specific human monoclonal antibody. Emerg Microbes Infect 9:382–385
To KK-W, Tsang OT-Y, Leung W-S, Tam AR, Wu T-C, Lung DC, Yip CC-Y, Cai J-P, Chan JM-C, Chik TS-H et al (2020) Temporal profiles of viral load in posterior oropharyngeal saliva samples and serum antibody responses during infection by SARS-CoV-2: an observational cohort study. Lancet Infect Dis 3099:1–10
Tran EEH, Nelson EA, Bonagiri P, Simmons JA, Shoemaker CJ, Schmaljohn CS, Kobinger GP, Zeitlin L, Subramaniam S, White JM (2016) Mapping of ebolavirus neutralization by monoclonal antibodies in the ZMapp cocktail using cryo-electron tomography and studies of cellular entry. J Virol 90:7618–7627
United Nations Sustainability Goals (2022). Available at: https://www.globalgoals.org/goals/17-partnerships-for-the-goals/. Accessed on 10 Aug 2022
Vanderbeke L, Van Mol P, Van Herck Y, De Smet F, Humblet-Baron S, Martinod K, Antoranz A, Arijs I, Boeckx B, Bosisio FM et al (2021) Monocyte-driven atypical cytokine storm and aberrant neutrophil activation as key mediators of COVID-19 disease severity. Nat Commun 12:1–15
Ward BJ, Gobeil P, Séguin A, Atkins J, Boulay I, Charbonneau PY, Couture M, D'Aoust MA, Dhaliwall J, Finkle C et al (2021) Phase 1 randomized trial of a plant-derived virus-like particle vaccine for COVID-19. Nat Med 27:1071–1078
WHO (2022) WHO emergency use listed vaccines for COVID-19. Available at: https://extranet.who.int/pqweb/vaccines/vaccinescovid-19-vaccine-eul-issued. Accessed on 17 Aug 2022
Worldometer (2022) Worldometer Covid-19 coronavirus pandemic. Available at: https://www.worldometers.info/coronavirus/. Accessed on 18 Jan 2022
Wu Y, Wang F, Shen C, Peng W, Li D, Zhao C, Li Z, Li S, Bi Y, Yang Y et al (2020) A noncompeting pair of human neutralizing antibodies block COVID-19 virus binding to its receptor ACE2. Science 368:1274–1278
Xu J, Ge X, Dolan MC (2011) Towards high-yield production of pharmaceutical proteins with plant cell suspension cultures. Biotechnol Adv 29:278–299
Yan R, Zhang Y, Li Y, Xia L, Guo Y, Zhou Q (2020) Structural basis for the recognition of the SARS-CoV-2 by full-length human ACE2. Science 2762:1–10
Yang J, Petitjean SJL, Koehler M, Zhang Q, Dumitru AC, Chen W, Derclaye S, Vincent SP, Soumillion P, Alsteens D (2020) Molecular interaction and inhibition of SARS-CoV-2 binding to the ACE2 receptor. Nat Commun 11:4541

Chapter 13
Biopharming's Growing Pains

Kathleen L. Hefferon and Henry I. Miller

Abstract Obtaining medicines from plants is not new. Aspirin was first isolated from the bark of the willow tree in the eighteenth century. And many other common pharmaceuticals are purified from the world's flora. More recently, scientists have developed techniques that take this process a step further. "Biopharming," the use of molecular genetic engineering techniques to induce agricultural crops to synthesize high-value pharmaceuticals, has much to offer. If we are to realize its potential, however, we will need reasonable, science-based regulation. However, largely because of excessive risk aversion on the part of regulators, progress has been slow.

Keywords Biopharming · Regulation · Plant · Molecular farming

Abbreviations

BAT	British American Tobacco
CBP	Cape Bio Pharms
CPMV	Cowpea mosaic virus
eCPMV	Empty cowpea mosaic virus particles
FDA	Food and Drug Administration
GE	Genetically engineered
GMP	Good manufacturing practices
HIV	Human immunodeficiency virus
KBP	Kentucky Bio Pharm
LED	Light-emitting diode

K. L. Hefferon (✉)
Department of Microbiology, Cornell University, Ithaca, NY, USA
e-mail: Klh22@cornell.edu

H. I. Miller
American Council on Science and Health, New York, NY, USA; https://www.henrymillermd.org

C. Kole et al. (eds.), *Tools & Techniques of Plant Molecular Farming*, Concepts and Strategies in Plant Sciences, https://doi.org/10.1007/978-981-99-4859-8_13

MCB	Master Cell Bank
TCM	Traditional Chinese medicine
UK	United Kingdom
US	United States
USDA	United States Department of Agriculture
VLP	Viruslike particle

13.1 An Introduction to Biopharming

The COVID-19 pandemic has spurred profound changes in the scientific research community worldwide. Laboratories have reoriented their research to focus on various aspects of this scourge, and thousands of articles have already appeared on preprint servers and in journals. As scientific researchers race to find solutions, the production of high-value pharmaceuticals in plants, or "biopharming," a technology that has teetered on the brink of significant recognition for many years, is mushrooming. The pandemic could be an opportunity to prove its worth (Hefferon and Miller 2020).

Academics and biotech companies are using genetic engineering techniques to reprogram plants—which have included corn, potatoes, rice, and bananas, among others (discussed below)—to produce significant concentrations of pharmaceuticals, including vaccines (Drake et al. 2017). The concept is venerable. Many common medicines, such as morphine, codeine, the laxative Metamucil, and the anticancer drug Taxol, are all purified from plants. There are also a few examples of Chinese herbal treatments that have proved effective in clinical trials. One notable product that has emerged from traditional Chinese medicine, or TCM, is artemisinin. First isolated by Youyou Tu at the China Academy of Traditional Chinese Medicine in Beijing, the molecule is now a powerful treatment for malaria and led to Tu being awarded the Nobel Prize in Physiology or Medicine in 2015 (Su and Miller 2015).

But biopharming's great promise lies in using genetic engineering techniques to make old plants do radically new things.

There is also great potential for cost cutting in the process: The energy for product synthesis comes from the sun, and the primary raw materials are water and carbon dioxide. In addition, biopharming offers tremendous flexibility and economy when adjustments in production are necessary. Doubling the acreage of a crop requires far less capital than doubling the capacity of a bricks-and-mortar factory, making biopharmed drugs potentially much less expensive to produce than those made in conventional ways. As little as 2000 acres can provide the substrate for a year's supply of some products. Grain from a biopharmed crop can be stored safely for long periods with no loss of activity. The quality of the final drug can meet the same standards as current fermentation technology using microorganisms.

Biopharmed vaccines are especially promising (Fischer and Buyel 2020). They are inexpensive to produce, easy to upscale, and often do not require refrigeration, needles, or trained medical personnel, thus making them attractive for use in developing countries. Many research studies and clinical trials have shown that plant-made vaccines elicit a robust immune response in animals and humans and are safe and efficacious. Examples of plant-made vaccines and therapeutics produced by molecular pharming include vaccines to combat cholera, dengue fever virus, hepatitis B virus, Ebola virus, and monoclonal antibodies to HIV (Hefferon 2019; see also chapters "Plant Molecular Farming for Vaccine Development" and "Plant-Based Veterinary Vaccines").

Although such plant biologics have largely focused on the diseases of the poor in developing countries, they have found other niches as well (Tschofen et al. 2016). For example, several plant-made vaccines to combat pandemic influenza are currently completing clinical trials and may soon be on the market, and plant-based immunotherapies to treat a variety of cancers are in development. A plant-based therapeutic to provide the enzyme glucocerebrosidase in Gaucher's disease patients has also found a reliable market and is currently commercially available.

Several biopharming companies and academic research labs have taken up the challenge to combat COVID-19. Medicago, a Canadian biopharmaceutical company, successfully developed a biopharmed viruslike particle (VLP) of the coronavirus only 20 days after obtaining the SARS-CoV-2 genetic sequence. Instead of using egg-based methods to produce a vaccine, their technology inserts a genetic sequence that encodes the spike protein of COVID-19 into *Agrobacterium*, a common soil bacterium that is taken up by plants ((Krenek et al. 2015). The resulting plants produce a VLP that is composed of plant lipid membrane and COVID-19 spike protein, and which acts as the vaccine. The VLPs are similar in size and shape to the actual coronavirus but lack viral or plant nucleic acid and, therefore, are noninfectious.

Previously, Medicago made VLPs that contain influenza virus hemagglutinin and demonstrated their safety and efficacy in animal models as well as in human clinical trials (Ward et al. 2020, 2021). The cost of producing a plant-made vaccine based on VLPs is a small fraction compared to its conventional counterpart.

Also in Canada, the University of Western Ontario and Suncor are developing serological test kits for COVID-19 using algae as a production factory to make the viral spike proteins. Algae has long been considered a potential platform for generating pharmaceutical proteins as well as industrial proteins such as cellulases. Algae are a superior bio-factory alternative because it is easy to grow at scale and can be readily modified to produce the viral proteins (see also chapter "Microalgae as a Bioreactor for Molecular Farming of Oral Edible Vaccines Against Infectious Diseases of Humans and Animals").

British American Tobacco, through its biotech subsidiary in the USA, Kentucky BioProcessing (KBP), is developing a vaccine for COVID-19 that is in Phase 1–2 clinical trials. Researchers at KBP cloned a part of the genetic sequence of SARS-CoV-2, which they used to develop a potential antigen that was inserted into *Nicotiana benthamiana* plants for production. The vaccine elicited a positive immune response in preclinical testing. BAT could manufacture as much as 1–3 million COVID-19 vaccines per week. (They were able to make 10 million doses of flu vaccine and an Ebola vaccine in a month, using the same plant-based approach.)

South African company Cape Bio Pharms (CBP) is also responding to the COVID-19 pandemic with the production of plant-derived reagents that could be used in diagnostic kits. CBP is producing SARS-CoV-2 spike S1 reagents consisting of various regions of the glycoprotein attached to various fusion proteins. The company is also collaborating with antibody manufacturers to produce antibodies against these proteins (see also Chap. 12).

Another example of a biopharming solution to COVID-19 is being developed in Professor Nicole Steinmetz's lab at the University of California, San Diego, using cowpea mosaic virus viruslike nanoparticles (VLPs) based on the empty CPMV (eCPMV) virion (Shukla et al. 2020). Two key platforms are available: virus nanoparticles (VNPs) based on the complete CPMV virion, including the genomic RNA, and viruslike nanoparticles (VLPs) based on the empty CPMV (eCPMV) virion. It is unclear whether these platforms differ in terms of immunotherapeutic potential. They found that the formulations had similar effects on most secreted cytokines and immune cells, but the RNA-containing CPMV particles were "uniquely able to boost populations of potent antigen-presenting cells, such as tumor-infiltrating neutrophils and activated dendritic cells. Our results will facilitate the development of CPMV and eCPMV as immunotherapeutic vaccine platforms with tailored responses." These VLP vaccines can be administered via implants and microneedle technology, an advantage in low-income countries (see also chapter "Medical Applications of Plant Virus Nanoparticles").

A collaboration between research groups in Toronto, Canada, is working on a novel way to both prevent and treat COVID-19 using an antiviral protein that blocks virus replication. When loaded onto a plant virus nanoparticle, the protein can enter cells and block virus infection. It is possible that this biopharmed antiviral protein can be loaded into an inhaler and administered to the lungs of infected and uninfected patients. Similarly, a synthetic, plant-made antibody has been designed to prevent virus infection and block person-to-person transmission. It can be produced easily in plants engineered to synthesize antibodies that are as "humanized" as possible, reducing the likelihood that patients' immune system will reject them as "foreign" (see also Chap. 7).

13.2 Regulation of Plant "Biopharming"

Although plant molecular pharming has been under development in academic labs for over 20 years, the progress to commercialization has been slow and painful (Spök et al. 2008; Hundleby et al. 2018). A recent example is the research and development of Medicago, a Canadian company that estimates that biological proteins such as vaccines and monoclonal antibodies could be obtained from genetically engineered tobacco plants at 1/1000th the cost of current methods (Hefferon and Miller 2019).

These plant-derived pharmaceuticals are also easy to scale up and remain stable at room temperature for longer time periods.

The regulatory hurdles are virtually prohibitory. For example, in 2010, the biotech company Ventria Bioscience approached the FDA for recognition of two human proteins, lysozyme and lactoferrin, synthesized in genetically engineered rice, to be given "generally recognized as safe" status. These proteins were intended for use in oral rehydration solutions to treat diarrheal diseases. Research in Peru had shown that an oral rehydration solution with the proteins extracted from Ventria's rice substantially lessens the duration of diarrhea and reduces the rate of recurrence—a near-miraculous advance for people in the developing world. In this example, Ventria never received any response from the FDA, and the product was never marketed for use. This and other high-value proteins, such as the HIV/AIDS drug Truvada (which interferes with the replication of the virus, but costs $2000 a month), could make a difference if produced inexpensively in plants and made available to low- to middle-income countries (see also Chaps. 10 and 11).

Other HIV microbicides such as griffithsin, an antiviral protein derived from red algae, have been produced in plants and distributed in the form of active, crude extracts. This would cut costs by reducing the need for complicated production processes.

One potential regulatory success story in the near future may be found in the use of plant-made vaccines to prevent seasonal flu. The ability of influenza to infect multiple animal species (humans, birds, and pigs), as well as to rapidly alter its surface hemagglutinin protein, makes it difficult to develop an effective, "universal" vaccine, thus making the flu a continuing global public health challenge. Fortunately, flu vaccines can be produced rapidly and in large amounts in genetically engineered plants as "viruslike particles." They have exhibited safety and efficacy in clinical trials but have not yet been approved for use in humans.

13.3 Navigating Plant Molecular Pharming Regulation for Commercial Products

Without clear, predictable, and reasonable regulatory frameworks, it is not surprising that pharmaceutical companies, most of which have little experience of working with plants, are reluctant to make large up-front investments. Despite these hurdles, plant molecular pharming has expanded as a commercial business, and large-scale manufacturing facilities have been constructed in the USA, the UK, and elsewhere (Schillberg et al. 2019; Menzel et al. 2018). Companies such as Medicago, iBio, and Kentucky Bioprocessing can currently process thousands of kilograms of plant biomass grown in greenhouse settings into highly purified pharmaceutical proteins. None of these companies yet has an approved vaccine, however. Others, such as the UK's Leaf Expression Systems, have begun to sell a small number of diagnostic products and laboratory reagents, none of which is currently for human use.

It can be challenging to confine the entire plant-based protein expression systems within a clean room or greenhouse environment, in order to be compatible with good manufacturing practices (GMPs) (Fischer et al. 2012). Nevertheless, the use of contained systems for plant cultures has helped to lessen various regulatory and safety concerns that might pertain to open-field production systems.

For regulatory authorities, good manufacturing practice (GMP) regulations for biologic products make it easier to regulate plant cell lines, because they more closely resemble mammalian cell lines or bacterial cultures—with which drug regulators have vast experience—than whole plants. Thus, plant cell cultures represent the easiest, and open-field GE plants the most difficult, regulatory hurdles to overcome (EMA 2008; Health Canada 2008).

Similarly, there are varying levels of difficulty for the type of product regulated (FDA/USDA 2002; Hundleby et al. 2022). Regulation is more stringent for plant-derived drugs and biologics but less for proteins (such as collagen) to be used in cosmetics or for veterinary products (MacDonald et al. 2015). Finally, plant-derived research reagents or diagnostics generally have the lowest level of regulation of all and are thus often the first products commercialized by plant molecular farming companies.

In certain instances, products of molecular farming have undergone rapid approval. Examples include an orphan drug to treat rare diseases, such as taliglucerase alfa by the Israeli company Protalix to treat Gaucher's disease, and a tobacco-derived antibody cocktail called ZMapp that was cleared for emergency use to address an Ebola outbreak in Western Africa (Tekoah et al. 2015; Davey et al. 2016). More recently, plant-derived products to diagnose SARS-CoV-2 may have the opportunity to move more rapidly through the approval process by the U.S. COVID-19 Pandemic Emergency Task Force (https://www.fda.gov/drugs/coronavirus-covid-19-drugs/coronavirus-treatment-acceleration-program-ctap).

As mentioned above, regulations for products made using plant cell culture tend to be the most straightforward, as regulators generally view them as analogous to mammalian cell and bacterial culture systems. The fact that plant cells lack endotoxins like those in *E. coli* or mammalian pathogens such as hepatitis C virus makes them preferable production platforms from a safety and quality assurance standpoint (see also chapters "Production of Recombinant Proteins Using Plant Cell Suspension Cultures and Bioreactor Engineering: A Short Review" and "Scaling Up the Plant Molecular Farming via Bioprocessing of Plant Cell Suspension Culture").

However, the regulations quickly become more onerous when entire plants themselves are the bioreactors for protein production. For example, plant cell culture facilitates the creation of a "Master Cell Bank," or MCB, which is produced from the original product-synthesizing cell line. It is cryopreserved in multiple vials to prevent genetic variation and potential contamination by eliminating the total number of times a cell line is passaged or handled during the manufacturing process. The MCB is a hallmark of good manufacturing practice (GMP) conditions. An MCB is more difficult to obtain using transgenic intact plants, which may through the process of breeding differ genetically slightly from one plant to another (Buyel 2019; Hundleby et al. 2022).

The upstream manufacturing phase for molecular farming can include plant cultivation, infiltration, harvesting, and initial extraction steps. GMP rules usually begin at the point when a sterile extract is available as starting material for purification (Fischer et al. 2012). However, they can also be incorporated for upstream manufacturing under greenhouse conditions. For example, consistency improves for GMP manufacturing when LED lights, substrates such as rockwool (used for hydroponics, instead of soil which can differ substantially), and optimal spacing of plants are incorporated, properties that can control batch consistency.

Downstream processing begins when the product is separated from the production organism. For plant molecular farming, this is approached similarly to the way microbial fermentation or mammalian bioreactors would be handled, and no additional regulation should apply (Buyel et al. 2015). The result of either process should be a production system that can generate a batch-by-batch analysis of impurities, allergenicities, and other safety measures.

No plants for molecular farming have been approved for commercial field production, although some work is taking place in approved, confined research field trials (which require isolation, toxicity data, and oversight by inspectors to witness disposal of residual plant material). Rather, molecular pharming has up to now been restricted to laboratories and greenhouses, in order to prevent release into the environment. These restrictions are a response to concerns such as the possible movement of pollen and the unintentional introduction of plant material containing bioactive substances into food supply chains and accidental consumption by people, livestock, or wildlife.

The field testing of biopharmed plants has proved problematic. In 2003, the US Department of Agriculture announced onerous new rules for testing crops engineered to produce pharmaceuticals. The ostensible objective of the regulation is to avoid contaminating food supplies with drugs, especially when edible crops are used to produce them. But the food industry's worries that biopharmed plants could contaminate their products are overblown. And in any case, the risk can be mitigated in several ways, most obviously by using nonfood plants like tobacco. In fact, even if biopharmed plants were to contaminate food crops, the likelihood that consumers would end up with harmful amounts of drugs in their breakfast cereal, pasta, or tofu is very small.

One way to moderate the regulatory burden for molecular biopharming is to grow plants under containment conditions such as in a greenhouse, vertical farm, or specialized clean room. This would be the more straightforward pathway toward commercialization considering the current regulatory landscape. Companies that have deviated from this, such as Ventria Bioscience (which grows plots of transgenic rice expressing lactoferrin and lysozyme in both Colorado and the US Virgin Islands), have found the regulatory pathway excessively burdensome.

Under current oversight regimes for genetically engineered organisms that were discussed at the beginning of this chapter, the strictest oversight prevails for open-field genetically engineered crops of any sort, and unfortunately, using open fields for plant-derived pharmaceuticals adds an additional layer of complexity. (And that becomes even more exaggerated outside North America.)

There are significant disparities among geographical regions of the world such as North America and Europe with regard to genetically engineered crop regulation

(Masip et al. 2013). US and Canadian legislation lean more toward regulating the product, whereas the EU regulation tends to be triggered by the process used for the transformation event and applies a "guilty unless proven innocent" paradigm—the "Precautionary Principle." This disharmony in regulatory frameworks across national borders negatively impacts commercialization of plant-made pharmaceuticals grown using transgenic plants under open-field conditions (Sparrow et al. 2013; see also Chap. 14).

13.4 Conclusions

Protalix was the first company to commercialize a plant-based therapeutic agent, a product derived from a plant cell culture which very closely resembles conventional mammalian culture, making it easier to gain approval than full transgenic plants. Unfortunately, other vaccine products derived from plants lack a clear path to enable an easy navigation through the current regulatory landscape. In effect, this is a paradigm for biopharming.

Most plant-made vaccines and other biologics have been shown to be cheap, safe, and efficacious, but they have not yet entered the marketplace, largely due to regulatory constraints. The lack of an appropriate regulatory structure to guide plant-made vaccines through to commercial development has stalled efforts to provide lifesaving medicines to low- and middle-income countries. For example, plant biologics to treat emerging infectious diseases such as Ebola and Zika virus have been successfully produced in academic labs, but it is critical to get these and others into pharmacies.

Clearly, plant-made vaccines deserve a place in public health. Biologics such as these are inexpensive to produce, efficacious, and safe and do not require cold chain or sophisticated medical equipment to disseminate them. Most research and development for plant-made vaccines has originated at universities and publicly funded institutes, where freedom to operate is less restrictive than with other technologies. Plant-based commercial-scale manufacturing facilities can grow and process thousands of kilograms of plant biomass into purified forms of biologics, including vaccines and antibodies. Although mechanisms exist to provide safe, inexpensive plant-made vaccines to the world, regulatory hurdles have confounded their advancement to the marketplace. This deficit can have resounding effects on low- and middle-income countries that are already poorly equipped to protect their populations against emerging diseases (Tusé et al. 2020).

Similarly, poor countries do not possess the capacity to address the increasing number of chronic diseases such as cancer in their aging citizens, and steadily increasing populations will further strain healthcare systems. Medicines to treat these conditions could be generated in plants, if only regulatory requirements were risk based and scientifically defensible.

References

Buyel JF (2019) Plant molecular farming—integration and exploitation of side streams to achieve sustainable biomanufacturing. Front Plant Sci 9:1893. https://doi.org/10.3389/fpls.2018.01893

Buyel JF, Fischer R, Twyman RM (2015) Extraction and downstream processing of plant-derived recombinant proteins. Biotechnol Adv 33:902–913

Davey RT, Dodd L, Proschan MA et al (2016) A randomized, controlled trial of ZMapp for Ebola virus infection. N Eng J Med 375:1448–1456

Drake PM, Szeto TH, Paul MJ, Teh AY, Ma JK-C (2017) Recombinant biologic products versus nutraceuticals from plants—a regulatory choice? Br J Clin Pharmacol 83:82–87

EMA (now EMEA) Committee for Proprietary Medicinal Products (CPMP) (2008) Guideline on the quality of biological active substances produced by stable transgene expression in higher plants. EMEA/CHMP/BWP/48316/ 2006. EMEA, London

FDA/USDA (2002) Draft guidance: drugs, biologics and medical devices derived from bioengineered plants for use in humans and animals. FDA, Rockville

Fischer R, Buyel JF (2020) Molecular farming—the slope of enlightenment. Biotechnol Adv 40:107519. https://doi.org/10.1016/j.Biotechadv.2020.107519

Fischer R, Schillberg S, Hellwig S, Twyman RM, Drossard J (2012) GMP issues for recombinant plant-derived pharmaceutical proteins. Biotechnol Adv 30:434–439

Health Canada (2008) Plant molecular farming (PMF) applications: plant-derived biologic drugs for human use. Health Products and Food Branch, Ottawa

Hefferon KL (2019) The role of plant expression platforms in biopharmaceutical development: possibilities for the future. Expert Rev Vaccines 18(12):1301–1308. https://doi.org/10.1080/14760584.2019.1704264

Hefferon KL, Miller HI (2019) Cures for cancer could grow on trees poorly designed regulation blocks investment in 'pharming'—plant-based medications and vaccines. Wall Street Journal

Hefferon K, Miller H (2020) Could COVID-19 cause 'biopharming' to bloom? J Commercial Biotechnol 25(3). https://doi.org/10.5912/jcb941

Hundleby PAC, Sack M, Twyman RM (2018) Biosafety, risk assessment and regulation of molecular farming. In: Kermode A, Jiang L (eds) Molecular pharming: applications, challenges and emerging areas. Wiley, Hoboken, pp 329–351

Hundleby PAC, D'Aoust MA, Finkle C, Atkins J, Twyman RM (2022) Regulation of molecular farming products. Methods Mol Biol 2480:313–333. https://doi.org/10.1007/978-1-0716-2241-4_17

Krenek P, Samajova O, Luptovciak I, Doskocilova A, Komis G, Samaj J (2015) Transient plant transformation mediated by Agrobacterium tumefaciens: principles, methods and applications. Biotechnol Adv 33(6 Pt 2):1024–1042. https://doi.org/10.1016/j.biotechadv.2015.03.012. Epub 2015 Mar 25.

MacDonald J, Doshi K, Dussault M et al (2015) Bringing plant-based veterinary vaccines to market: managing regulatory and commercial hurdles. Biotechnol Adv 33:1572–1581

Masip G, Sabalza M, Pérez-Massot E, Banakar R, Cebrian D, Twyman RM, Capell T, Albajes R, Christou P (2013) Paradoxical EU agricultural policies on genetically engineered crops. Trends Plant Sci 18:312–324

Menzel S, Holland T, Boes A, Spiegel H, Fischer R, Buyel JF (2018) Downstream processing of a plant-derived malaria transmission-blocking vaccine candidate. Protein Expr Purif 152:122–130

Schillberg S, Raven N, Spiegel H, Rasche S, Buntru M (2019) Critical analysis of the commercial potential of plants for the production of recombinant proteins. Front Plant Sci 10:720. https://doi.org/10.3389/fpls.2019.00720

Shukla S, Wang C, Beiss V, Cai H, Washington T 2nd, Murray AA, Gong X, Zhao Z, Masarapu H, Zlotnick A, Fiering S, Steinmetz NF (2020) The unique potency of Cowpea mosaic virus (CPMV) *in situ* cancer vaccine. Biomater Sci 8(19):5489–5503. https://doi.org/10.1039/d0bm01219j

Sparrow PAC, Broer I, Hood EE, Eversole K, Hartung F, Schiemann J (2013) Risk assessment and regulation of molecular farming—a comparison between Europe and US. Curr Pharm Des 19:5513–5530

Spök A, Twyman RM, Fischer R, Ma JK-C, Sparrow PAC (2008) Evolution of a regulatory framework for pharmaceuticals derived from genetically modified plants. Trends Biotechnol 26:506–517

Su XZ, Miller LH (2015) The discovery of artemisinin and the Nobel Prize in Physiology or Medicine. Sci China Life Sci 58(11):1175–1179. https://doi.org/10.1007/s11427-015-4948-7

Tekoah Y, Shulman A, Kizhner T et al (2015) Large-scale production of pharmaceutical proteins in plant cell culture-the Protalix experience. Plant Biotechnol J 13:1199–1208

Tschofen M, Knopp D, Hood E, Stöger E (2016) Plant molecular farming: much more than medicines. Annu Rev Anal Chem 9:271–294

Tusé D, Nandi S, McDonald KA, Buyel JF (2020) The emergency response capacity of plant-based biopharmaceutical manufacturing—what it is and what it could be. Front Plant Sci 11:594019

Ward BJ, Makarkov A, Séguin A, Pillet S, Trépanier S, Dhaliwall J, Libman MD, Vesikari T, Landry N (2020) Efficacy, immunogenicity, and safety of a plant-derived, quadrivalent, virus-like particle influenza vaccine in adults (18-64 years) and older adults (65 years): two multicentre, randomised phase 3 trials. Lancet 396:1491–1503

Ward BJ, Séguin A, Couillard J, Trépanier S, Landry N (2021) Phase III: randomized observer-blind trial to evaluate lot-to-lot consistency of a new plant-derived quadrivalent virus like particle influenza vaccine in adults 18-49 years of age. Vaccine 39:1528–1533

Chapter 14
Biosafety, Risk Analysis, and Regulatory Framework for Molecular Farming in Europe

Lilya Kopertekh and Ralf Wilhelm

Abstract Over the last decade, molecular farming has matured as a commercially relevant production platform as shown by the approval of several plant-derived industrial and pharmaceutical proteins. Despite evident benefits of plants over the traditional mammalian and bacterial cell-based expression systems, the progress toward commercialization of plant-made recombinant proteins takes much effort and time. This reflects a combination of factors including technical issues (low expression level, high costs of downstream processing), limited information on process cost structure, and regulatory uncertainties. In the EU, plant-derived recombinant proteins are subjected to two or three regulatory frameworks, referring to the authorization of genetically modified organisms and other governing the produced protein as pharmaceutical or industrial product, resulting in slow and costly approval. Within this regulatory system, the specific guidelines for molecular farming are still evolving. In this review, we summarize the current state for the risk assessment and regulation of plant-made industrial and plant-made pharmaceutical products and discuss recent changes and the need for further development of specific molecular farming regulatory landscape.

Keywords Molecular farming · Process-specific regulation · Product-specific regulation · Risk assessment · Risk management

Abbreviations

CMS	Cytoplasmic male sterility
ECHA	European Chemicals Agency

L. Kopertekh (✉) · R. Wilhelm
Julius Kühn Institute (JKI), Federal Research Centre for Cultivated Plants, Institute for Biosafety in Plant Biotechnology, Quedlinburg, Germany
e-mail: lilya.kopertekh@julius-kuehn.de; ralf.wilhelm@julius-kuehn.de

C. Kole et al. (eds.), *Tools & Techniques of Plant Molecular Farming*, Concepts and Strategies in Plant Sciences, https://doi.org/10.1007/978-981-99-4859-8_14

EFSA	European Food Safety Authority
EMA	European Medicines Agency
FDA	Food and Drug Administration (USA)
GMM	Genetically modified microorganism
GMO	Genetically modified organism
GMP	Good manufacturing practice
GRAS	Generally recognized as safe
PMI	Plant-made industrial compound
PMP	Plant-made medicinal product
RAC	Committee for Risk Assessment
REACH	Registration, Evaluation, Authorisation and Restriction of Chemicals
SEAC	Committee for Socio-economic Analysis
USDA	United States Department of Agriculture

14.1 Introduction

Molecular farming refers to the production of recombinant proteins, peptides, and small molecules in plant cells. The use of plants as bioreactors began in 1989 by the expression of a functional full-size IgG1 antibody in transgenic tobacco (Hiatt et al. 1989). Since that time, different types of proteins for both pharmaceutical and industrial purposes have been produced in plants. Pharmaceutically relevant proteins include various antibody formats, vaccines, hormones, blood products, and enzymes (Ma et al. 2003; Paul and Ma 2011; Sabalza et al. 2014). The product profile of plant-made industrials (PMI) ranges from research-grade reagents and technical enzymes to cosmetic products (Tschofen et al. 2016; Fischer et al. 2013; Paul et al. 2013).

Plants offer several benefits over traditional bacterial and mammalian cell-based platforms for recombinant protein production (Schillberg et al. 2019; Schillberg and Finnern 2021). First, plants can fold and assemble complex proteins and carry out eukaryotic posttranslational modifications, which are required for optimal protein activity (Komori et al. 2009; Matsubayashi 2011; Strasser 2016). Moreover, the glycosylation engineering of host plants allowed the production of biobetters with humanlike or optimized glycan profile (Grabowski et al. 2014). Second, plants offer animal- and endotoxin-free platform to avoid the contamination of the final product (Magnusdottir et al. 2013). Third advantage from the use of plants as bioreactors is the production speed in case of transient expression system. The recombinant protein can be produced within several weeks after confirming the gene sequence (Shoji et al. 2011; Sainsbury 2020). This method has a great potential for emergency vaccines or biologics that was demonstrated for influenza (Ward et al. 2020), Ebola (The PREVAIL II Writing Group, for the Multi-National PREVAIL II Study Team 2016), and Covid-19 (Capell et al. 2020; Pillet et al. 2022) vaccines. Fourth, production of mucosal animal vaccines and therapeutics in plant cells and their administration as minimally processed plant material can reduce the manufacturing costs and in some cases enhance their effectiveness (Zimmermann et al. 2009; Virdi et al.

2013). Fifth, another benefit associated with the use of plants is a rapid scaling up in comparison to fermenters. Lastly, the production of recombinant proteins in plants, particularly cosmetic products, might improve the consumer acceptance of the final product (see also Chaps. 1 and 2).

Despite the promising advantages of molecular farming listed above, the implementation of plant-made proteins into the market was delayed (Fischer et al. 2014; Spiegel et al. 2018). The first commercialized proteins produced in plants were technical reagents and industrial enzymes such as avidin (Hood et al. 1997; Hood and Howard 2014), aprotinin (Pogue et al. 2010), lysozyme (Broz et al. 2013), trypsin (Woodard et al. 2003), and ß-glucuronidase (Witcher et al. 1998). To date, two plant-derived biopharmaceuticals for humans have been approved. The first one is Elelyso® produced in carrot suspension culture by Protalix Biotherapeutics (Karmiel, Israel) (Fox 2012). The second product is Covid-19 vaccine Covifenz®, which is manufactured by Medicago (Quebec, Canada) and has been approved by Health Canada on February 24, 2022 (see also Chap. 12). Table 14.1 shows the representative list of plant-produced proteins and their commercial status.

The limited number of molecular farming products on the market indicates several bottlenecks for commercial development, particularly for pharmaceutical proteins. The most widely recognized drawbacks include the relatively low expression levels, high costs of downstream processing, and regulatory hurdles (Schillberg and Finnern 2021; see also Chap. 13). In the EU, relevant molecular farming regulations refer to the legal standards applied to the production technology (transgenic plants, plant cell cultures, etc.) and to the regulation for specific classes of products (pharmaceutical and industrial proteins). An authorization can only be granted if no risk for environment, human, and animal health has been identified on the basis of risk assessment carried out by the independent risk assessment bodies. In the EU, the deliberate release of GMO (here: nonfood/non-feed plant production) in the environment is authorized involving the competent authorities of the member states and the European Food Safety Authority (EFSA). The European Medicines Agency (EMA) is responsible for the risk assessment of pharmaceutical products. Specific risks could be associated with molecular farming resulting in the specific challenges for the EU regulation. This chapter provides an overview of the current molecular farming landscape in terms of the risk assessment and regulation. Particular aspects discussed will include risk assessment, risk management, and authorization of molecular farming products in the EU.

14.2 Risk Analysis

Risk analysis follows a common regulatory procedure in the decision-making process for a proposed commercial release of recombinant proteins. It involves three interconnected components, risk assessment, risk management, and risk communication. Risk assessment identifies potential risks and evaluates the possible outcome associated with the specific activity on the basis of scientific data. Risk assessment consists of four steps including hazard identification, hazard characterization,

Table 14.1 Representative plant-produced proteins and their commercial status

Product	Application	Production host	Current status	Company	Reference
Trypsin Avidin	Technical reagents	Maize (seeds)	Commercialized	ProdiGene, USA	http://www.sigmaaldrich.com
Growth factors Cytokines	Research reagents	Barley (seeds)	Commercialized	ORF Genetics, Iceland	http://www.orfgenetics.com
Albumin Lactoferrin Lysozyme Transferrin	Research reagents	Rice (seeds)	Commercialized	Ventria Bioscience/ InVitria, USA	http://www.inVitria.com
Epithelial growth factor	Cosmetics	Barley (seeds)	Commercialized	Sif Cosmetics, Iceland	http://www.sifcosmetics.com
α-Amylase	Bioethanol production	Maize (seeds)	Commercialized	Syngenta	http://www.syngenta.com
Elelyso®	Treatment of Gaucher disease	Carrot (cells)	Commercialized	Protalix, Israel	https://protalix.com
PRX-102, alpha galactosidase	Treatment of Fabry disease	Carrot (cells)	Phase III completed	Protalix, Israel	https://protalix.com
Covifenz®	Covid-19 vaccine	*Nicotiana benthamiana* (transient)	Approved in Canada	Medicago/Mitsubishi Tanabe, Canada	https://medicago.com
QVLP vaccine	Influenza vaccine	*Nicotiana benthamiana* (transient)	Submitted for approval	Medicago/Mitsubishi Tanabe, Canada	https://medicago.com
Repleva AGAL	Treatment of Fabry disease	Moss	Phase I completed	Eleva, Germany	https://www.elevabiologics.com
ZMapp™	Vaccine against the Ebola virus	*Nicotiana benthamiana* (transient)	Randomized clinical trial completed	Mapp Biopharmaceutical/ Kentucky BioProcessing, USA	https://mappbio.com

(continued)

Table 14.1 (continued)

Product	Application	Production host	Current status	Company	Reference
Personalized idiotype, whole IgG	Treatment of non-Hodgkin's lymphoma	*Nicotiana benthamiana* (transient)	Phase I completed	Icon Genetics, Germany	https://www.icongenetics.com
Norovirus vaccine	Vaccine against norovirus	*Nicotiana benthamiana* (transient)	Phase I	Icon Genetics/Denka, Germany	https://www.icongenetics.com
HIV antibody	Microbicide	*Nicotiana tabacum* (transgenic)	Phase I completed	Pharma-Planta, Consortium, EU	Sparrow et al. (2007)
NMW01	Thaumatin	*Nicotiana tabacum* (transgenic)	GRAS registration in the USA	Nomad, Germany	https://www.nomad-bioscience.com
NMW02	Bacteriocin	*Nicotiana benthamiana* (transient)	GRAS registration in the USA	Nomad, Germany	https://www.nomad-bioscience.com

exposure assessment, and risk characterization. Risk management is based on the risk analysis and evaluates, selects, and executes options to manage the possible risks. Risk communication defines the interactive exchange of information and opinions during the risk analysis process among risk assessors, risk managers, consumers, feed and food businesses, academic community, and other interested parties. Risk communication also includes the elucidation of risk assessment findings and the basis of risk management decisions.

The differences between transgenic plants for feed and food use and molecular farming plants have been highlighted in several papers (Spök 2007; Spök et al. 2008; Sparrow et al. 2013). First, the accumulation of recombinant protein is considerably high by purpose in comparison to transgenic plants for feed and food application. Therefore, the environmental and human exposure could reach relevant effect levels. Second, genetically modified plants used for molecular farming can include several traits simultaneously (stacked events) including recombinant protein/proteins of interest and, e.g., resistance genes against pathogens and weeds, visual markers, and molecular confinement systems. Stacking of several events broadens the assessment of unintended effects on plants. Third, plant-made pharmaceuticals (PMPs) produce the pharmaceutically active compounds that have an intended biological effect on humans or animals. Thus, the properties of the recombinant protein might be of concern. Therefore, the specific features of the transgenic plants for nonfood and non-feed use impact their risk assessment.

14.2.1 GMO Risk Assessment

In the EU, EFSA GMO panel published a guiding opinion explaining the risk assessment of genetically modified (GM) plants for nonfood and non-feed purposes (EFSA 2009). Following the EU regulations, a principle for risk assessment of GM plants is a comparative assessment. It is based on datasets for (1) molecular characteristics of the GM plant; (2) agronomic, phenotypic, and compositional characteristics of the GM plant; and (3) toxicological and allergenic assessment of the foreign protein(s) and genetic modification. The environmental risk assessment requires a safety evaluation regarding potential changes in the weediness and invasiveness of GM plants, potential gene transfer to plants and microorganisms, interaction between the GM plant and target and nontarget organisms, and potential impacts on ecosystem functions (e.g., soil function through specific management measures) (Directive 2001/18/EC 2001; EFSA 2010). A case-by-case risk assessment is mandatory, because the features of the recombinant protein, protein expression, and exposure levels vary substantially (Shama and Peterson 2008a, b). While production in closed facilities such as greenhouses limits mandatory risk assessment efforts and simplifies authorization of transgenic products, approval for production in open field requires extended testing (field scale), datasets, and analyses of broader environmental impacts resulting in significant time and financial investment.

14.2.2 Ways to Minimize Gene Flow from Molecular Farming Plants

There are two main issues concerning open-field cultivation of PMP and PMI plants. The first one is the unintended spread of transgenic material into the environment and subsequent possible nontarget effects on the organisms and ecosystems. Three mechanisms are responsible for transgene escape: pollen dispersal, seed dissemination, and development of transgenic plants from volunteer vegetative organs. The transgene might spread from transgenic to non-transgenic population of the same plant species or to the related wild species (Liénard et al. 2007; Ryffel 2014). Another key challenge is the possible contamination of food and feed chains during harvesting, transport, processing, and waste disposal of transgenic plant material (Elbehri 2005). To prevent or minimize possible negative effects of PMI and PMP plants on human health and environment, a number of strategies have been proposed including choosing appropriate plant production host and physical and biological containment (Commandeur et al. 2003; Obembe et al. 2011; Rybicki 2010; Breyer et al. 2012).

Two types of factors should be considered when selecting the plant production host for molecular farming. The first group of factors includes economic considerations such as total biomass yield, length of production cycle, setup, scale-up, downstream, and containment costs. The second category takes into account the potential impact on the environment and food and feed chains (Sparrow and Twyman 2009; Breyer et al. 2012). Numerous plants that have been used for recombinant protein production can be divided into three classes, namely food, nonfood, and non-crop species (Twyman 2004; Sparrow et al. 2007). Food crops have the established cultivation and processing procedures that is favorable for using these plants for recombinant protein production. The food crops can be subdivided into three main groups: seed crops, vegetables, and food/leaf crops (Sparrow et al. 2007). Many seed crops including maize, rice, barley, pea, soybean, and oilseed rape are well suited for molecular farming. The accumulation of recombinant proteins in seeds offers a high protein storage capacity, weak protease activity, and long-term storage ability at ambient temperature (Boothe et al. 2010; Khan et al. 2020; see also Chaps. 5 and 6). The seed-specific expression can reduce the possible negative effects of the recombinant protein on nontarget organisms. The main concerns from the biosafety point of view are pollen-mediated gene flow and contamination of food and feed chains. The possible seed spillage and seed mixing during transport and handling should also be considered. The use of maize as a production crop has a number of benefits such as an available agricultural infrastructure, germplasm resources and genetic knowledge, and ease of transformation and scaling up. Maize has already been used for commercial production of trypsin, avidin, and β-glucuronidase (Tschofen et al. 2016). The main disadvantage of this crop is the potential crossing with non-transgenic maize (Luna et al. 2001). The second seed crop, which has been successfully used for molecular farming, is rice. This plant has a number of advantages including well-established transformation system and self-pollination feature. InVitria, a division of Ventria Bioscience Inc. (Junction City, USA), has produced rice seeds and commercialized several research reagents such as albumin, lactoferrin, lysozyme, and transferrin (see also chapter "The Use of Rice Seed as Bioreactor"). Barley is also self-pollinated and is used as a production host by ORF Genetics (Kopavogur, Iceland) to express growth factors and cytokines. Among the seed plants, oilseed rape has been considered as not suitable for field production of recombinant proteins due to the open pollination, compatibility with local weed species, and multiple-year seed dormancy (Sparrow et al. 2007). In the third group, vegetables such as potato and carrot can be used for recombinant protein production. One of the main advantages of these crops is the product stability in tubers or roots. Furthermore, vegetable crops have been considered as a suitable system for the production of oral vaccines. However, determination of dosage is a tedious task, and product accumulation can vary between individual tubers/roots resulting in batch-to-batch inconsistency (Sahoo et al. 2020; Naik 2022). Lettuce, spinach, and alfalfa represent the leaf crops. The possible consumption of uncooked, unprocessed, or partly processed material is a major benefit when pharmaceutical proteins are produced in these plants (see also chapter "Molecular Farming of Pharmaceutical Proteins in Different Crop Systems: A Way Forward"). Nevertheless, the recombinant proteins in leaves are in the aqueous environment and are subjected to rapid proteolytic degradation after harvesting. Therefore, plant tissue containing recombinant protein should be processed after collecting that requires corresponding logistic (Twyman 2004). From the biosafety perspective, the potential exposure of nontarget organisms can be a disadvantage. Harvesting the plant tissue before flowering can reduce the risk of pollen and seed escape. The next group of host plants for molecular farming is nonfood crops. Using nonfood crops helps to separate molecular farming products from the products used in the human and animal food chain. Tobacco is the most widely used plant from this group because of the well-established procedure for gene transfer and expression, high biomass yield, and available infrastructure for large-scale cultivation and processing. In terms of high level of toxic alkaloids, which are present in tobacco leaves, low alkaloid varieties could be exploited. Possible drawbacks of this expression host are the phenolic substances, which can interfere with the downstream processing (Rymerson et al. 2002, Tremblay et al. 2010; see also chapter "Tobacco Plants as a Versatile Host for the Expression of Glycoproteins").

The third category, non-crop plants, ranges from *N. benthamiana* to duckweed (*Lemna minor*), microalgae (*Chlamydomonas reinhardtii*), and moss (*Physcomitrella patens*) (Yao et al. 2015). The main advantages of these plants are their nonfood and non-feed status and possibility of contained cultivation. Currently, *N. benthamiana* is the most popular host plant for transient expression by agrodelivery. This plant is susceptible to infection with plant viruses and *Agrobacterium tumefaciens*, grows well under controlled indoor conditions, and can rapidly produce a large amount of biomass and a master seed bank (Goodin et al. 2008; Powell 2015; Bally et al. 2018). *N. benthamiana* is the core production host of many companies worldwide including Medicago (Quebec, Canada), Kentucky BioProcessing (Owensboro, USA), PlantForm (Ontario, Canada), Icon Genetics (Halle, Germany), iBio/Caliber

Biotherapeutics (Bryan, USA), CapeBio (Centurion, South Africa), Bioapp (Gyeongsangbuk-do, Korea), and Leaf Expression Systems (Norwich, UK) (LeBlanc et al. 2020).

In summary, the best production host from a biosafety perspective may not be the best from the commercial viewpoint. Therefore, the particular application will define the balance between biosafety and commercial considerations during the selection of the plant host (Clark and Maselko 2020).

In addition to the choice of the appropriate production host, physical isolation can be employed as a containment strategy. Physical containment involves the use of special equipment, facilities, and procedures to prevent the unintended transgene escape into the environment. The management of food and feed transgenic plants using isolation distances is also applicable to the plants used for recombinant protein production. The cross-fertilization levels decrease rapidly with increasing distance between the pollen donor (transgenic plant) and pollen recipient (non-transgenic plant) resulting in the reduced pollen-mediated transgene flow. In the EU, the member states can require appropriate buffer zones/isolation distances. The isolation distances depend on plant biology (self- or wind-pollinated) and meteorological/climate conditions. In addition to buffer zones, pollen barriers (non-transgenic trap plants) surrounding the GM crop can reduce the extend of cross-pollination between transgenic and non-transgenic plants (Murphy 2007). The second approach of physical containment is based on growing molecular farming plants under contained conditions. Several types of physical containment can be considered including underground facilities (mines), plastic tunnels, and greenhouses. Currently, plant-based *in vitro* and transient expression technologies can provide containment at all manufacturing stages, from the plant cultivation to the processing of the end product. *In vitro* technologies include cell suspensions, hairy root cultures, and cultivation of moss (*P. patens*), duckweed (*Lemna* sp.), and green microalga (*C. reinhardtii*) (Rosales-Mendoza et al. 2012; Decker and Reski 2020; Gutierrez-Valdes et al. 2020; Krasteva et al. 2021; Yang et al. 2021; see also chapters "Duckweed, an Efficient Green Bio-factory for the Production of Recombinant Proteins" and "Microalgae as a Bioreactor for Molecular Farming of Oral Edible Vaccines Against Infectious Diseases of Humans and Animals"). However, molecular farming manufacturing in contained facilities is connected with the increased production costs.

The biocontainment measures as well as the physical isolation can minimize the unwanted transgene escape and its potential negative consequences. The biological containment exploits the existing natural mechanisms to limit reproduction or introduces them by genetic engineering. The strategies relying on biological barriers preventing transgene flow include cleistogamy, apomixis, plastid transformation, and male and seed sterility (Daniell 2002; Breyer et al. 2009; Ryffel 2014). The naturally occurring reproduction without fertilization termed apomixis can be exploited as a genetic barrier for pollen-mediated transgene flow. About 400 plant species can produce seeds without fertilization (Spillane et al. 2004). In combination with the male sterility, this phenomenon might be an efficient biological containment tool. Nevertheless, it is limited by the necessity to investigate the involved genetic mechanisms, restriction to certain plant species, and a residual risk of seed dispersal (Ryffel 2014). Another option would be cleistogamy, self-pollination within a closed flower, found in barley, soybean, and rice (Hüsken et al. 2010). However, establishing cleistogamy is not practical in a number of important crops including maize, wheat, and cassava, and much more information is required to determine the genes that are responsible for this phenotype (Daniell 2002). The integration of transgene in the chloroplast genome is also a promising option for gene containment due to the maternal inheritance of plastids in many species. In addition to biosafety benefits, expression in chloroplasts offers high yield of recombinant proteins, lack of gene silencing and position effects, and polycistronic mRNA expression (Bock 2021). Plastid transformation technology as a confinement strategy has some limitations. The feasibility of plastid genetic engineering for biocontainment depends on the plant species. Maternal inheritance of plastids is not universal: more than one-third of the angiosperm species do not display a strict maternal inheritance. Furthermore, plastid transformation remains much more challenging than nuclear transformation and is not as widespread in plant research (Yu et al. 2020). Some studies have also shown the possibility of rare parental plastid transmission (Ruf et al. 2007; Svab and Maliga 2007) and the transgene transfer from the chloroplasts to nucleus (Stegemann and Bock 2006). In addition, proteins produced in chloroplasts are not glycosylated, limiting the potential product range (see also Chap. 9). Male sterility is another method to reduce pollen-mediated gene flow. There are a number of naturally occurring mechanisms of male sterility including cytoplasmic male sterility (CMS). CMS is a maternally inherited trait that fails to produce viable pollen (Chase 2007). This phenotype is encoded by the mitochondrial genome and can be suppressed by the nuclear restorer genes (Chen and Liu 2014). CMS is widely exploited in plant breeding to design hybrid seeds. However, the stability of the CMS phenotype depends on genotype and environmental conditions. In addition, male sterility can also be induced through introducing genes affecting pollen fertility. One of the best-studied systems of the engineered male sterility is the *barstar and barnase* system based on the ribonuclease (barnase) from *Bacillus amyloliquefaciens*, which inhibits pollen formation (Mishra and Kumari 2018). It has already been shown that this technology was useful to generate male-sterile maize, oilseed rape, tobacco, sugar beet, sunflower, potato, tomato, wheat, rice, cauliflower (Breyer et al. 2009), and Indian oilseed mustard (Bisht et al. 2007) plants. Basically, most of the biological containment strategies are far from the commercial application and display different levels of reliability. At their current stage of development, the biocontainment options will most probably not be able to achieve full protection of the environment from gene flow. To reach a high level of containment, a combination of approaches may be necessary. It is also obvious that the measures to manage risks associated with the plant-based production of recombinant proteins should be considered on a case-by-case basis.

14.3 Regulation of Molecular Farming

In the EU, the regulations relevant to molecular farming combine directives and regulations applied to genetically modified organisms (GMOs) as a production platform (process-triggered regulation) and to a specific class of products (product-specific regulation) (Spök et al. 2008). In the following section, these regulations are considered in detail.

14.3.1 *Process-Specific Regulation*

The regulation of GMOs worldwide is in most cases based on a process-specific (genetic engineering) legal regime, except in Canada, where a product-based regulation (based on the novelty of the trait in an organism) is in place. The EU regulatory framework for the release and placing on the market of GMOs is triggered by the method of production (by genetic engineering) and is one of the strictest in the world. It aims at protection of environment and human health and well-functioning EU internal market. The EU regulatory system relies on three main principles, namely risk assessment-based pre-market authorization, traceability, and labeling (Bruetschy 2019).

Molecular farming consolidates three main types of plant platforms, including cell suspension cultures/aquatic plants, transient expression, and transgenic plants that rely on the transfer of foreign genetic information into plant cells (Twyman et al. 2003; Paul and Ma 2011). The delivery method of recombinant DNA differs between the expression systems (Stöger et al. 2014; Shanmugaraj et al. 2020). Therefore, the same recombinant protein may fall under different regulations depending on the production method. The key directives of the EU, which govern the production process through GMO and are relevant to molecular farming, are shown in Table 14.2.

Directive 2009/41/EC covers the activities associated with the contained use of genetically modified microorganisms (GMMs), but several member states include

Table 14.2 Process-specific regulation in the EU

Directive/ guidelines	Scope of regulation	Covers	Authority
Directive 2009/41/EC	Contained use of GMMs	Use of GMMs in contained facilities (laboratory, greenhouse, etc.)	EU member states
Directive 2001/18/EC (Part B)	Deliberate release of GMOs into the environment	Experimental field trials	EU member states
Directive 2001/18/EC (Part C)	Deliberate release of GMOs into the environment	Commercial cultivation	EFSA, EU member states, EU Commission

handling of GM plants in contained use when adopting the Directive to their national legislation (EC 2021). It regulates several issues such as the initial laboratory development as well as use, transport, storage, and disposition of GMOs including bacteria, viruses, viroids, and animal and plant cells. Contained production reduces the risks of food and feed contamination. However, this type of manufacturing can lack some advantages of open-field cultivation of PMI and PMP plants (Spök 2007). According to the Directive 2009/41/EC, production in greenhouses and contained facilities is under the regulatory oversight of each EU member state. The contained use of GMOs must be notified to the relevant competent authorities of the member state. The applicant should perform an appropriate risk assessment and is responsible for organizing the contained level and minimizing any potential risks associated with the GMO.

To fully exploit the issue of commercial low-cost production of industrial and pharmaceutical proteins in plants, it may be desirable to grow molecular farming plants in the field. The deliberate release of GMOs including plants to the environment is regulated in the Directive 2001/18/EC. This Directive foresees two authorization levels: time- and area-limited field trials for research and development purposes (part B) and placing on the market of the GM plants including import, transport, processing, handling, storage, and cultivation (part C) (Spök and Karner 2008). Authorization under Part B does not allow the commercialization of pharmaceutical and industrial products.

The Directive 2001/18/EC (Part B) requires a prior authorization of GMO release into the environment. An application should be submitted to the competent national authority of the member state in which the field release will take place. It must include a technical dossier supplying information on the aim of the field release, host plant, introduced foreign gene, location and size of the field site, an environmental risk assessment, and management strategies. The competent authority of the member state must make a decision on whether or not to grant consent within 90 days of receiving the notification. Additionally, the national competent authority provides the EU Commission with a summary containing the most important information of the application (SNIF). To support the transparency of the proposed activities, the SNIF is published and can be accessed in the following URL: https://webgate.ec.europa.eu/fip/GMO_Registers/GMO_Part_B_Plants.php. In the EU, 21 experimental field trails relevant to molecular farming have been notified since 2013 (Table 14.3). Readers can find the data on the experimental field trials with the PMI and PMP plants in Europe for the period 1995–2012 in Sparrow et al. (2013).

Part C of the Directive 2001/18/EC covers the commercialization of GMOs. Placing on the market of GMOs refers to selling GMOs or products containing GMOs to third parties. Due to the free movement of goods within the EU, the decision about commercialization has to be granted at the EU level. This authorization is then obligatory for all member states. The application for cultivation or import of GMO must be submitted to a competent authority of one of the EU member states. Such an application comprises comprehensive documents about human and environmental risk assessment, monitoring plan, detection method, and labeling proposals.

Table 14.3 Experimental field trials of GM plants relevant to molecular farming in Europe (2013–2022) (source: https://webgate.ec.europa.eu/fip/GMO_Registers/GMO_Part_B_Plants.php)

Plant	Trait	Number of trials
Barley	Antimicrobial peptide	3
Barley	Thaumatin, human growth factors	1
Camelina	Modification of fatty acid content	5
Crambe	Modification of industrial oil content	1
Maize	Modification of starch content	1
Maize	Vitamins	3
Potato	Modification of starch content	4
Rice	Microbicide components	1
Tobacco	Thaumatin	3
Tobacco	Squalene	1

The competent authority may request additional information from the applicant and prepares an evaluation report about the full application dossier and a decision proposal. It also informs the other member states. If a member state disagrees with the proposed analyses, the EFSA is requested to provide an evaluation. Finally, the European Commission makes a decision about authorization based on EFSA opinion. Marked releases of GM crops intended for food or feed use are authorized according to the procedure laid out in the EU Regulation 1829/2003 with EFSA coordinating the evaluation of the application and risk assessment, consulting the member states' authorities and providing a decision proposal to the Commission that finally decides (Raybould and Poppy 2012). PMI and PMP plants have only been grown in field trials under Part B of Directive 2001/18/EC in the EU (Table 14.3). In contrast to insect- and herbicide-resistant transgenic plants for food and feed, the molecular farming plants are expected to be grown on small acreages. Moreover, cultivation, transport, and processing are carried out in one member state.

14.3.2 Product-Specific Regulation of Molecular Farming

Additionally to the process-specific GMO regulation, the plant-produced recombinant proteins need to adhere to the regulation covering the intended use of the product. Two main types of proteins can be produced in plants, namely, pharmaceutical and non-pharmaceutical proteins. The first category of the molecular farming products includes proteins for veterinary and human medicine. Placing of these products on the market is regulated by several EU directives und regulations, which are presented in Table 14.4.

Regulatory oversight for PMPs is covered by the Directive EC 2019/5. During the research and development stage, these products are overseen by the national authorities, whereas the European Commission grants the marketing authorization

Table 14.4 The EU regulation governing the intended use of molecular farming products

Directive/ guidelines	Scope of regulation	Covers	Authority
Clinical Trials Regulation (CTR)	Human clinical trials	Efficiency and safety evaluation of medicine product	EU member states at the research and development stage and EMA at marketing stage
Regulation (EC) 2019/5	Marketing of biotech products including pharmaceuticals	Authorization and supervision of pharmaceutical products for human and veterinary use	EMA (product evaluation) and EU Commission (market authorization)
Directive 2004/27/EC	Human pharmaceuticals	Authorization and supervision of human pharmaceuticals	EMA (product evaluation) and EU Commission (market authorization)
Directive 2004/28/EC	Veterinary pharmaceuticals	Authorization and supervision of veterinary pharmaceuticals	EMA (product evaluation) and EU Commission (market authorization)
Regulation (EU) 2017/745	Medical devices	Clinical investigation and sale of medical devices	EMA (product evaluation) and EU Commission (market authorization)
Regulation (EU) 2017/746	In vitro diagnostic devices	Clinical investigation and sale of in vitro diagnostic devices	EMA (product evaluation) and EU Commission (market authorization)
Regulation (EU) 1223/2009	Cosmetic products	Safety evaluation and sale of cosmetics	EU member states and EU Commission
Regulation 1907/2006 (REACH)	Chemical substances	Registration, evaluation, authorization, and restriction of chemicals	The European Chemicals Agency (ECHA)

following the scientific evaluation by the EMA. The efficiency, safety, and quality of medicinal products are assessed by reviewing the results reported in the marketing authorization dossier, which includes quality, preclinical, and clinical sections. The manufacturing of the medicine drugs must comply with the good manufacturing practice (GMP) standards. GMP provides an associated group of activities and norms, which are developed to guarantee that the product has a consistent high quality and meets the requirements for its intended use and clinical trials (Yusibov et al. 2011). Initially, the GMP guidance for PMP manufacturing was based on the GMP guidance for the well-established production platforms such as bacterial and mammalian cells. The plant suspension culture, moss, and algae cells are similar to animal cells. In this case, a number of concepts developed for the conventional expression systems are also applicable for plant system. These include master and working cell banks, batch-to-batch consistency, standard operating procedures, as well as downstream processing steps. Therefore, the first commercial successes in the pharmaceutical sector have been achieved using plant cell suspension cultures. For example, Dow AgroSciences (Corteva Inc) developed the CONCERT™ platform based on the tobacco BY-2 cells for the production of veterinary drugs. The first plant cell-derived vaccine against Newcastle disease virus produced using this expression system was approved by the USDA in 2006 (Fox 2006; see also chapter "Plant-Based Veterinary Vaccines"). The second example is the ProCellEx® platform utilizing carrot cells, which has been established by the Israeli company Protalix Biotherapeutics. In 2012, the taliglucerase alfa manufactured using the ProCellEx® platform was approved as a first plant-produced pharmaceutical for human use by the FDA (Fox 2012; Grabowski et al. 2014). Another important contribution is the moss expression system developed by the German company Eleva (Freiburg, Germany). The Repleva AGAL enzyme to treat Fabry disease produced in moss cells has completed the phase I clinical trial (Hennermann et al. 2019).

Updated EMA guidelines addressing the use of transgenic plants for molecular farming came into effect in 2009 (EMA 2009). In this document, a number of regulatory concepts developed for cell culture system have been revised. These modifications include particularly the replacement of master and working cell banks with master and working seed banks, taking into account the natural variation of recombinant protein accumulation between individual plants and plant organs and refinement of requirements for standard operating and downstream processing procedures (Fischer et al. 2012). In contained conditions, most of the factors, which affect plant development and recombinant protein yield including water supply, light, soil, and chemical treatments, can be controlled. Therefore, recombinant protein production in greenhouse can provide more sustainable batch-to-batch consistency during the manufacturing process. In this context, an EU-funded academic research consortium Pharma-Planta, which run from 2004 to 2011, demonstrated a proof of concept for the GMP-compliant manufacture of plant-produced HIV-neutralizing mAb 2G12 (Sparrow et al. 2007). The consortium identified the key regulatory issues for the GMP production and processing, and designed and performed a clinical trial. This clinical trial represented the first use of GMP-compliant transgenic plant-derived antibodies in humans. The plant-produced antibodies were well tolerated and safe in a healthy woman (Ma et al. 2015; see also Chap. 7). The consultation with the regulators, which was carried out in the frame of the Pharma-Planta project, established new guidelines for manufacturing of medicinal drugs in transgenic tobacco plants in contained facilities (Drake et al. 2017).

In case of open-field cultivation of the PMP transgenic plants, the GMP-compliant process validation is generally more challenging compared to the greenhouse cultivation. First, outdoor farming is more susceptible to climate, weather, and soil. Second, pest invasion, plant diseases, and application of pesticides are the sources of variability. Another challenge is an inconsistency in target protein expression, as well as the nature and amount of by-products between seasons, fields, and individual plants. Taking into account all these factors, the establishing of batch-to-batch consistency, one of the main standard requirements for producing biopharmaceutical drugs under GMP, is problematic for open-field manufacturing (Spök and Karner 2008). In addition, the low acceptance of GM plants in the EU resulted in de facto moratorium on their cultivation. Therefore, the EU companies working in the biotechnology sector focused on the contained plant-based expression systems.

Transient expression is one of the promising plant-based expression platforms (Gleba et al. 2005; Peyret and Lomonossoff 2015). Currently, transient expression is making rapid and impressive progress due to its high level of target expression, scalability, and time efficiency. It consists of two components: host plant and genetically modified bacteria (*A. tumefaciens*) and/or viruses. The genetic information encoding the protein of interest is introduced into plants by bacteria or viruses resulting in recombinant protein accumulation in the host plant. The manufacturing is performed in contained facilities, and the master and working cell banks should be defined for microbes carrying foreign genetic information (Spiegel et al. 2018). In contrast to the FDA/USDA (FDA/USDA 2002) guidelines accepting all plant-based production platforms, the EMA guidelines did not include the transient expression platform leading to slow development of commercial process based on this technique in the EU. In this sense, one of the main goals of the current EU-funded consortium Pharma-Factory (2017–2022) is the consultation with the regulatory bodies to establish the regulatory pathways for a range of plant production platforms, particularly transient expression (Schillberg and Finnern 2021; see also Chap. 15). This regulatory inconsistency has also a great impact on the development of large-scale facilities, which are necessary for commercial application of transient expression technology. For instance, several companies outside the EU operate such green factories, including Kentucky Bioprocessing (Owensboro, USA), Medicago (Quebec, Canada), Mapp Biopharmaceutical (San Diego, USA), iBio/Caliber Biotherapeutics (Bryan, USA), and Leaf Expression Systems (Norwich, UK). In contrast, only few pilot research facilities, namely Fraunhofer IME (Aachen, Germany) and Icon Genetics/Nomad (Halle, Germany), are available in the EU.

Placing on the market of the non-pharmaceutical products does not require clinical trials and thus generates faster returns. The approval of PMIs (technical and research reagents, industrial enzymes) is covered by the EU Regulation 1907/2006 on Registration, Evaluation, Authorisation and Restriction of Chemicals (REACH), which entered into force in 2007 (EU Regulation 1907/2006 2006). The European Chemicals Agency (ECHA) is the body responsible for the administration of REACH in the EU. The REACH is aimed at providing high protection level of human health and environment from the use of chemicals. The safety data requirements depend on the substance volume: the obligation to submit a registration dossier to the ECHA starts as soon as the production or import volume of a substance exceeds 1 tonne/year. In this dossier, the dangers of the substance and the measures to minimize the risks should be specified. Additional safety data are needed at higher tonnages (more than 10 tonnes/year). Authorization is granted by the EU Commission based on the opinion of the Committee for Risk Assessment (RAC) and the Committee for Socio-economic Analysis (SEAC) of the ECHA. The EU member states take part in the initial safety evaluation of substances (Cihák 2009). A number of plant-produced research reagents and enzymes are available on the EU market. For example, trypsin, avidin, and cellobiohydrolase I produced in transgenic maize are distributed by Merck (Merck, Darmstadt, Germany) (Table 14.1). These chemicals have not been manufactured in the EU though.

At present, the commercialized plant-produced non-pharmaceutical proteins outnumber the pharmaceutical products due to the lower regulatory burden. The approval of the therapeutic recombinant proteins requires the comprehensive and expensive preclinical studies and clinical trials as an essential component of product safety assessment adding additional costs and time.

14.4 Conclusions

Several bodies of evidences suggest that molecular farming has matured to commercial application. First, the technology now focuses on a relatively limited number of systems, which are relevant for the large-scale production. These expression systems include transient expression, transgenic/transplastomic plants, and single-cell or tissue culture systems (Sparrow et al. 2018). Second, the plant-derived recombinant proteins have become established in niche markets, which demonstrate the strength of the technology in comparison to the traditional animal- and bacterial cell-based expression platforms. These products are falling into several classes such as custom pharmaceutical molecules (Bendandi et al. 2010; McCormick 2011), rapid response vaccines (Qiu et al. 2014; Streatfield et al. 2015; LeBlanc et al. 2020; Tusé et al. 2020), bulk "commodity" antibody (Whaley et al. 2011; Stöger et al. 2014), and therapeutic medicine for rare diseases (Hintze et al. 2020). Third, transient expression is becoming an industrial technology (Kopertekh and Schiemann 2019). A number of companies operate large-scale facilities to produce pharmaceutical proteins using this method (Holtz et al. 2015; Huebbers and Buyel 2021). But most of these facilities have been built outside of the EU reflecting unfavorable regulatory framework for transient manufacturing of recombinant proteins in Europe. Fourth, techno-economic analysis of plant-based production for different proteins such as monoclonal antibody (Nandi et al. 2016), horseradish peroxidase (Walwyn et al. 2015), cellulases (Tusé et al. 2014), griffithsin (Alam et al. 2018), and antimicrobial proteins (McNulty et al. 2020) shows that such products can be commercialized successfully. Finally, the EU regulatory landscape has started changing to support translation of molecular farming products from research to the market. Particularly, the issues of risk assessment for nonfood and non-feed transgenic plants and industry standards of GMP have been addressed by the EU regulatory authorities. However, the interviews with the stakeholders involved into two current EU-funded projects Newcotiana and Pharma-Factory indicated that the EU regulatory environment and the perception of the public toward biotechnology are the main barriers to molecular farming commercialization (Menary et al. 2020a, b). The regulatory landscape and public perception of molecular farming affect business strategies and innovation worldwide and particularly in the EU. Within the last years, there has been a shift from the open-field cultivation of PMI and PMF plants to contained production. In terms of transient manufacturing in *N. benthamiana*, the

EU regulatory regime is still in development. On the other hand, the necessity of agricultural cultivation of PMI plants for the production of industrial enzymes and chemical reagents might renew the interest to revise a policy framework for such plants. In case of adoption of open-field PMI manufacturing, additional costs will arise from the need of controlling confinement measures and monitoring of food/feed contamination. Another challenge facing the regulatory framework for molecular farming in the EU is an involvement of multiple regulatory bodies, for example EFSA and EMA, in product authorization, making this process slow and more complex.

We believe that further improvement of confinement methods and a strong pipeline of plant-produced recombinant proteins will foster further development of an appropriate regulatory landscape and speed the translation of molecular farming products to the market.

References

Alam A, Jiang L, Kittleson GA, Steadman KD, Nandi S, Fuqua JL, Palmer KE, Tusé D, McDonald KA (2018) Technoeconomic modeling of plant-based griffithsin manufacturing. Front Bioeng Biotechnol 6:102. https://doi.org/10.3389/fbioe.2018.00102

Bally J, Jung H, Mortimer C, Naim F, Philips JG, Hellens R, Bombarely A, Goodin MM, Waterhouse PM (2018) The rise and rise of *Nicotiana benthamiana*: a plant for all reasons. Annu Rev Phytopathol 56:405–426

Bendandi M, Marillonnet S, Kandzia R, Thieme F, Nickstadt A, Herz S, Frode R, Inoges S, Lopez-Diaz de Cerio A, Soria E, Villanueva H, Vancanneyt G, McCormick A, Tuse D, Lenz J, Butler-Ransohoff JE, Klimyuk V, Gleba Y (2010) Rapid, high-yield production in plants of individualized idiotype vaccines for non-Hodgkin's lymphoma. Ann Oncol 21:2420–2427

Bisht NC, Jagannath A, Burma PK, Pradhan AK, Pental D (2007) Retransformation of a male sterile barnase line with the barstar gene as an efficient alternative method to identify male sterile-restorer combinations for heterosis breeding. Plant Cell Rep 26:727–733

Bock R (2021) Engineering chloroplasts for high-level constitutive or inducible transgene expression. Methods Mol Biol 2317:77–94

Boothe J, Nykiforuk C, Shen Y, Zaplachinski S, Szarka S, Kuhlman P, Murray E, Morck D, Moloney MM (2010) Seed-based expression systems for plant molecular farming. Plant Biotechnol J 8:588–606

Breyer D, Goossens M, Herman P, Sneyers M (2009) Biosafety considerations associated with molecular farming in genetically modified plants. J Med Plant Res 3:825–838

Breyer D, De Schrijver A, Goossens M, Pauwels K, Herman P (2012) Biosafety of molecular farming in genetically modified plants. In: Wang A, Ma S (eds) Molecular farming in plants: recent advances and future prospects. Springer, Dordrecht, pp 259–274

Broz A, Huang N, Unruh G (2013) Plant-based protein biomanufacturing. Genet Eng Biotechnol News 33:32–33

Bruetschy C (2019) The EU regulatory framework on genetically modified organisms (GMOs). Transgenic Res 28:169–174

Capell T, Twyman RM, Armario-Najera V, Ma JKC, Schillberg S, Christou P (2020) Potential applications of plant biotechnology against SARS-CoV-2. Trends Plant Sci 25:635–643

Chase CD (2007) Cytoplasmic male sterility: a window to the world of plant mitochondrial-nuclear interaction. Trends Genet 23:81–90

Chen L, Liu YG (2014) Male sterility and fertility restoration in crops. Annu Rev Plant Biol 65:579–606

Cihák R (2009) REACH—an overview. Interdiscip Toxicol 2:42–44

Clark M, Maselko M (2020) Transgene biocontainment strategies for molecular farming. Front Plant Sci 11:210. https://doi.org/10.3389/fpls.2020.00210

Commandeur U, Twyman RM, Fischer R (2003) The biosafety of molecular farming in plants. Ag Biotech Net 5:1–9

Daniell H (2002) Molecular strategies for gene containment in transgenic corps. Nat Biotechnol 20:581–586

Decker E, Reski R (2020) Mosses in biotechnology. Curr Opin Biotechnol 61:21–27

Drake PM, Szeto TH, Paul MJ, The AY, Ma JK (2017) Recombinant biologic products versus nutraceuticals from plants—a regulatory choice? Br J Clin Pharmacol 83:82–87

EC (2001) Directive 2001/18/EC of the European Parliament and of the Council on the deliberate release into the environment of genetically modified organisms. https://eur-lex.europa.eu/legal-content/EN/TXT/?uri=celex%3A32001L0018

EFSA Panel on Genetically Modified Organisms (GMO) (2010) Guidance on the environmental risk assessment of genetically modified plants. EFSA J 8:1879–1990

EFSA Scientific Panel on Genetically Modified Organisms (GMO) (2009) Scientific opinion on guidance for the risk assessment of genetically modified plants used for non-food or non-feed purposes on request of EFSA. EFSA J 1164:1–42

Elbehri A (2005) Biopharming and the food system: examining the potential benefits and risks. Ag Biol Forum 8:18–25

EMA (2009) Guideline on the quality of biological active substances produced by stable transgene expression in higher plants. www.ema.europa.eu/docs/en_GB/document_library/Scientific_guideline/2009/09/WC500003154.pdf

EU Regulation (EC) No 1907/2006 (2006) Registration, Evaluation, Authorisation and Restriction of Chemicals (REACH). https://osha.europa.eu/en/legislation/directives/regulation-ec-no-1907-2006-of-the-european-parliament-and-of-the-council

European Commission (2021) Commission Staff Document Accompanying the Document Report from the Commission Experience of Member States with Directive 2009/41/EC of the European Parliament and of the Council of 6 May 2009 on the contained use of genetically modified micro organisms for the period 2014–2018, Brussels, 31 May 2021, SWD(2021) 114 final. https://eur-lex.europa.eu/legal-content/EN/TXT/PDF/?uri=CELEX:52021SC0114&rid=7

FDA/USDA (2002) Draft guidance. Drugs, biologics, and medical devices derived from bioengineered plants for use in humans and animals. http://www.fda.gov/cber/gdlns/bioplant.pdf

Fischer R, Schillberg S, Hellwig S, Twyman RM, Drossard J (2012) GMP issues for recombinant plant derived pharmaceutical proteins. Biotechnol Adv 30:434–439

Fischer R, Schillberg S, Buyel JF, Twyman RM (2013) Commercial aspects of pharmaceutical protein production in plants. Curr Pharm Des 19:5471–5477

Fischer R, Buyel JF, Schillberg S, Twyman RM (2014) Molecular farming in plants: the long road to the market. In: Howard J, Hood E (eds) Commercial plant-produced recombinant protein products, Biotechnology in agriculture and forestry, vol 68. Springer, Berlin, pp 27–41

Fox JL (2006) Turning plants into protein factories. Nat Biotechnol 24:1191–1193

Fox JL (2012) First plant-made biologic approved. Nat Biotechnol 30:472

Gleba Y, Klimyuk V, Marillonnet S (2005) Magnifection-a new platform for expressing recombinant vaccines in plants. Vaccine 23:2042–2048

Goodin MM, Zaitlin D, Naidu RA, Lommel SA (2008) *Nicotiana benthamiana*: its history and future as a model for plant–pathogen interactions. Mol Plant-Microbe Interact 21:1015–1026

Grabowski GA, Golembo M, Shaaltiel Y (2014) Taliglucerase alfa: an enzyme replacement therapy using plant cell expression technology. Mol Genet Metab 112:1–8

Gutierrez-Valdes N, Häkkinen ST, Lemasson C, Guillet M, Oksman-Caldentey KM, Ritala A, Cardon F (2020) Hairy root cultures—a versatile tool with multiple applications. Front Plant Sci 11:33. https://doi.org/10.3389/fpls.2020.00033

Hennermann JB, Arash-Kaps L, Fekete G, Schaaf A, Busch A, Frischmuth T (2019) Pharmacokinetics, pharmacodynamics, and safety of moss-aGalactosidase A in patients with Fabry disease. J Inherit Metab Dis 42:527–533

Hiatt A, Cafferkey R, Bowdish K (1989) Production of antibodies in transgenic plants. Nature 342:76–78

Hintze S, Limmer S, Dabrowska-Schlepp P, Berg B, Krieghoff N, Busch A, Schaaf A, Meinke P, Schoser B (2020) Moss-derived human recombinant GAA provides an optimized enzyme uptake in differentiated human muscle cells of Pompe disease. Int J Mol Sci 21:2642. https://doi.org/10.3390/ijms21072642

Holtz BR, Berquist BR, Bennett LD, Kommineni VJ, Munigunti RK, White EL, Wilkerson DC, Wong KY, Ly LH, Marcel S (2015) Commercial-scale biotherapeutics manufacturing facility for plant-made pharmaceuticals. Plant Biotechnol J 13:1180–1190

Hood E, Howard J (2014) Commercial plant-produced recombinant avidin. In: Howard J, Hood E (eds) Commercial plant-produced recombinant protein products: case studies. Springer, Berlin, pp 15–25

Hood EE, Witcher DR, Maddock S, Meyer T, Baszczynski C, Baily M et al (1997) Commercial production of avidin from transgenic maize: characterization of transformant, production, processing, extraction and purification. Mol Breed 3:291–306

Huebbers JW, Buyel JF (2021) On the verge of the market—plant factories for the automated and standardized production of biopharmaceuticals. Biotechnol Adv 46:107681. https://doi.org/10.1016/j.biotechadv.2020.107681

Hüsken A, Prescher S, Schiemann J (2010) Evaluating biological containment strategies for pollen-mediated gene flow. Environ Biosaf Res 9:67–73

Khan MS, Joyia FA, Mustafa G (2020) Seeds as economical production platform for recombinant proteins. Protein Pept Lett 27:89–104

Komori R, Amano Y, Ogawa-Ohnishi M, Matsubayashi Y (2009) Identification of tyrosylprotein sulfotransferase in Arabidopsis. PNAS 106:15067–15072

Kopertekh L, Schiemann J (2019) Transient production of recombinant pharmaceutical proteins in plants: evolution and perspectives. Curr Med Chem 26:365–380

Krasteva G, Georgiev V, Pavlov A (2021) Recent applications of plant cell culture technology in cosmetics and foods. Eng Life Sci 21:68–76

LeBlanc Z, Waterhouse P, Bally J (2020) Plant-based vaccines: the way ahead? Viruses 13:5. https://doi.org/10.3390/v13010005

Liénard D, Sourrouille C, Gomord V, Faye L (2007) Pharming and transgenic plants. Biotechnol Annu Rev 13:115–147

Luna SV, Figueroa J, Baltazar MB, Gomez RL, Townsend R, Schoper JB (2001) Maize pollen longevity and distance isolation requirements for effective pollen control. Crop Sci 41:1551–1557

Ma JKC, Drake PMW, Christou P (2003) The production of recombinant pharmaceutical proteins in plants. Nat Rev Genet 4:794–805

Ma JK, Drossard J, Lewis D, Altmann F, Boyle J, Christou P, Cole T, Dale P, van Dolleweerd CJ, Isitt V, Katinger D, Lobedan M, Mertens H, Paul MJ, Rademacher T, Sack M, Hundleby PA, Stiegler G, Stoger E, Twyman RM, Vcelar B, Fischer R (2015) Regulatory approval and a first-in-human phase I clinical trial of a monoclonal antibody produced in transgenic tobacco plants. Plant Biotechnol J 13:1106–1120

Magnusdottir A, Vidarsson H, Björnsson JM, Örvar BL (2013) Barley grains for the production of endotoxin-free growth factors. Trends Biotechnol 31:572–580

Matsubayashi Y (2011) Post-translational modifications in secreted peptide hormones in plants. Plant Cell Physiol 52:5–13

McCormick AA (2011) Tobacco derived cancer vaccines for non-Hodgkin's lymphoma: perspectives and progress. Hum Vaccin 2011(7):305–312

McNulty MJ, Gleba Y, Tusé D, Hahn-Löbmann S, Giritch A, Nandi S, McDonald KA (2020) Techno-economic analysis of a plant-based platform for manufacturing antimicrobial proteins for food safety. Biotechnol Progress 36:e2896. https://doi.org/10.1002/btpr.2896

Menary J, Amato M, Sanchez AC, Hobbs M, Pacho A, Fuller SS (2020a) New hope for a "cursed" crop? Understanding stakeholder attitudes to plant molecular farming with modified tobacco in Europe. Front Plant Sci 11:791. https://doi.org/10.3389/fpls.2020.00791

Menary J, Hobbs M, Mesquita de Albuquerque S, Pacho A, Drake PMW, Prendiville A, Ma JK, Fuller SS (2020b) Shotguns vs lasers: identifying barriers and facilitators to scaling-up plant molecular farming for high-value health products. PLoS One 15:e0229952. https://doi.org/10.1371/journal.pone.0229952

Mishra S, Kumari V (2018) A review on male sterility-concepts and utilization in vegetable crops. Int J Curr Microbiol 7:3016–3034

Murphy DJ (2007) Improving containment strategies in biopharming. Plant Biotechnol J 5:555–569

Naik P (2022) Edible vaccines: current scenario and future prospects. In: Bhat R (ed) Future foods. Academic Press, London, pp 305–313

Nandi S, Kwong AT, Holtz BR, Erwin RL, Marcel S, McDonald KA (2016) Techno-economic analysis of a transient plant-based platform for monoclonal antibody production. MAbs 8:1456–1466

Obembe OO, Popoola JO, Leelavathi S, Reddy SV (2011) Advances in plant molecular farming. Biotechnol Adv 29:210–222

Paul M, Ma JKC (2011) Plant-made pharmaceuticals: leading products and production platforms. Biotechnol Appl Biochem 58:58–67

Paul MJ, The AY, Twyman RM, Ma JKC (2013) Target product selection—where can molecular pharming make the difference? Curr Pharm Des 19:5478–5485

Peyret H, Lomonossoff GP (2015) When plant virology met Agrobacterium: the rise of the deconstructed clones. Plant Biotechnol J 13:1121–1135

Pillet S, Arunachalam PS, Andreani G, Golden N, Fontenot J, Aye PP, Röltgen K, Lehmicke G, Gobeil P, Dubé C, Trépanier S, Charland N, D'Aoust MA, Russell-Lodrigue K, Monjure C, Blair RV, Boyd SD, Bohm RP, Rappaport J, Villinger F, Landry N, Pulendran B, Ward BJ (2022) Safety, immunogenicity, and protection provided by unadjuvanted and adjuvanted formulations of a recombinant plant-derived virus-like particle vaccine candidate for COVID-19 in nonhuman primates. Cell Mol Immunol 9:222–233

Pogue GP, Vojdani F, Palmer KE, Hiatt E, Hume S, Phelps J, Long L, Bohorova N, Kim D, Pauly M, Velasco J, Whaley K, Zeitlin L, Garger SJ, White E, Bai Y, Haydon H, Bratcher B (2010) Production of pharmaceutical-grade recombinant aprotinin and a monoclonal antibody product using plant-based transient expression systems. Plant Biotechnol J 8:638–654

Powell JD (2015) From pandemic preparedness to biofuel production: tobacco finds its biotechnology niche in North America. Agriculture 5:901–917

Qiu X, Wong G, Audet J, Bello A, Fernando L, Alimonti JB, Fausther-Bovendo H, Wei H, Aviles J, Hiatt E, Johnson A, Morton J, Swope K, Bohorov O, Bohorova N, Goodman C, Kim D, Pauly MH, Velasco J, Pettitt J, Olinger GG, Whaley K, Xu B, Strong JE, Zeitlin L, Kobinger GP (2014) Reversion of advanced Ebola virus disease in nonhuman primates with ZMapp. Nature 514:47–53

Raybould A, Poppy GM (2012) Commercializing genetically modified crops under EU regulations: objectives and barriers. GM Crops Food 3:9–20

Rosales-Mendoza S, Paz-Maldonado LMT, Soria-Guerra RE (2012) Chlamydomonas reinhardtii as a viable platform for the production of recombinant proteins: current status and perspectives. Plant Cell Rep 31:479–494

Ruf S, Karcher D, Bock R (2007) Determining the transgene containment level provided by chloroplast transformation. PNAS USA 104:6998–7002

Rybicki EP (2010) Plant-made vaccines for humans and animals. Plant Biotechnol J 8:620–637

Ryffel GU (2014) Transgene flow: facts, speculations and possible countermeasures. GM Crops Food 5:249–258

Rymerson RT, Menassa R, Brandle JE (2002) Tobacco, a platform for the production of recombinant proteins. In: Erickson L, Yu WJ, Brandle J, Rymerson R (eds) Molecular farming of plants and animals for human and veterinary medicine. Springer, Dordrecht, pp 1–31

Sabalza M, Christou P, Capell T (2014) Recombinant plant-derived pharmaceutical proteins: current technical and economic bottlenecks. Biotechnol Lett 36:2367–2379

Sahoo A, Mandal AK, Dwivedi K, Kumar V (2020) A cross talk between the immunization and edible vaccine: current challenges and future prospects. Life Sci 15(261):118343. https://doi.org/10.1016/j.lfs.2020.118343

Sainsbury F (2020) Innovation in plant-based transient protein expression for infectious disease prevention and preparedness. Curr Opin Biotechnol 61:110–115

Schillberg S, Finnern R (2021) Plant molecular farming for the production of valuable proteins—critical evaluation of achievements and future challenges. J Plant Physiol 258–259:153359. https://doi.org/10.1016/j.jplph.2020.153359

Schillberg S, Raven N, Spiegel H, Rasche S, Buntru M (2019) Critical analysis of the commercial potential of plants for the production of recombinant proteins. Front Plant Sci 10:720. https://doi.org/10.3389/fpls.2019.00720

Shama LM, Peterson RKD (2008a) Assessing risks of plant-based pharmaceuticals: I. Human dietary exposure. Hum Ecol Risk Assess 14:179–193

Shama LM, Peterson RKD (2008b) Assessing risks of plant-based pharmaceuticals: II. Non-target organism exposure. Hum Ecol Risk Assess 14:194–204

Shanmugaraj B, Bulaon CJI, Phoolcharoen W (2020) Plant molecular farming: a viable platform for recombinant biopharmaceutical production. Plants 9:842. https://doi.org/10.3390/plants9070842

Shoji Y, Chichester JA, Jones M, Manceva SD, Damon E, Mett V, Musiychuk K, Bi H, Farrance C, Shamloul M, Kushnir N, Sharma S, Yusibov V (2011) Plant-based rapid production of recombinant subunit hemagglutinin vaccines targeting H1N1 and H5N1 influenza. Hum Vaccin 7:41–50

Sparrow PAC, Twyman RM (2009) Biosafety and risk assessment of plant-made pharmaceuticals. Methods Mol Biol 483:341–353

Sparrow PA, Irwin JA, Dale PJ, Twyman RM, Ma JK (2007) Pharma-planta: road testing the developing regulatory guidelines for plant-made pharmaceuticals. Transgenic Res 16:147–161

Sparrow P, Broer I, Hood EE, Eversole K, Hartung F, Schiemann J (2013) Risk assessment and regulation of molecular farming—a comparison between Europe and US. Curr Pharm Des 19:5513–5530

Sparrow PA, Irwin JA, Hundleby PA, Sack M, Twyman RM (2018) Biosafety, risk assessment, and regulation of molecular farming. In: Kermode AR, Jiang L (eds) Molecular pharming: applications, challenges and emerging areas. John Wiley & Sons, Hoboken, pp 3–23

Spiegel H, Stöger E, Twyman RM, Buyel JF (2018) Current status and perspectives of the molecular farming landscape. In: Kermode AR, Jiang L (eds) Molecular pharming: applications, challenges and emerging area. John Wiley & Sons, Hoboken, pp 3–23

Spillane C, Curtis MD, Grossniklaus U (2004) Apomixis technology development-virgin births in farmers' fields? Nat Biotechnol 22:687–691

Spök A (2007) Molecular farming on the rise—GMO regulators still walking a tightrope. Trends Biotechnol 25:74–82

Spök A, Karner S (2008) Plant molecular farming, opportunities and challenges. JRC Technical Report: EUR 23383 EN. Office for Official Publications of the European Communities, Kopstal, Luxembourg

Spök A, Twyman RM, Fischer R, Ma JKC, Sparrow PA (2008) Evolution of a regulatory framework for plant-made pharmaceuticals. Trends Biotechnol 26:506–517

Stegemann S, Bock R (2006) Experimental reconstruction of functional gene transfer from the tobacco plastid genome to the nucleus. Plant Cell 18:2869–2878

Stöger E, Fischer R, Moloney M, Ma JKC (2014) Plant molecular pharming for the treatment of chronic and infectious diseases. Annu Rev Plant Biol 65:743–768

Strasser R (2016) Plant protein glycosylation. Glycobiology 26:926–939

Streatfield SJ, Kushnir N, Yusibov V (2015) Plant-produced candidate countermeasures against emerging and reemerging infections and bioterror agents. Plant Biotechnol J 13:1136–1159

Svab Z, Maliga P (2007) Exceptional transmission of plastids and mitochondria from the transplastomic pollen parent and its impact on transgene containment. PNAS 104:7003–7008

The PREVAIL II Writing Group, for the Multi-National PREVAIL II Study Team (2016) A randomized, controlled trial of ZMapp for Ebola Virus Infection. N Engl J Med 375:1448–1456. https://doi.org/10.1056/NEJMoa1604330

Tremblay R, Wang D, Jevnikar AM, Ma S (2010) Tobacco, a highly efficient green bioreactor for production of therapeutic proteins. Biotechnol Adv 28:214–221

Tschofen M, Knopp D, Hood E, Stöger E (2016) Plant molecular farming: much more than medicines. Annu Rev Anal Chem 9:271–294

Tusé D, Tu T, McDonald KA (2014) Manufacturing economics of plant-made biologics: case studies in therapeutic and industrial enzymes. Biomed Res Int 2014:1–16

Tusé D, Nandi S, McDonald KA, Buyel JF (2020) The emergency response capacity of plant-based biopharmaceutical manufacturing-what it is and what it could be. Front Plant Sci 11:594019. https://doi.org/10.3389/fpls.2020.594019

Twyman RM (2004) In: Fischer R, Schillberg S (eds) Host plants, systems and expression strategies for molecular farming. Wiley, Weinheim, pp 191–216. Molecular farming. Plant-made pharmaceutical and technical proteins.

Twyman RM, Stoger E, Schillberg S, Christou P, Fischer R (2003) Molecular farming in plants: Host systems and expression technology. Trends Biotechnol 21:570–578

Virdi V, Coddens A, De Buck S, Millet S, Goddeeris BM, Cox E, De Greve H, Depicker A (2013) Orally fed seeds producing designer IgAs protect weaned piglets against enterotoxigenic Escherichia coli infection. PNAS 110:11809–11814

Walwyn DR, Huddy SM, Rybicki EP (2015) Techno-economic analysis of horseradish peroxidase production using a transient expression system in *Nicotiana benthamiana*. Appl Biochem Biotechnol 175:841–854

Ward JB, Makarkov A, Séguin A, Pillet S, Trépanier S, Dhaliwall J, Libman MD, Vesikari T, Landry N (2020) Efficacy, immunogenicity, and safety of a plant-derived, quadrivalent, virus-like particle influenza vaccine in adults (18-64 years) and older adults (≥65 years): two multicentre, randomised phase 3 trials. Lancet 396:1491–1503

Whaley KJ, Hiatt A, Zeitlin L (2011) Emerging antibody products and Nicotiana manufacturing. Hum Vaccin 7:349–356

Witcher D, Hood E, Peterson D, Bailey M, Bond D, Kusnadi A, Evangelista R, Nikolov Z, Wooge C, Mehigh R et al (1998) Commercial production of β-glucuronidase (GUS): a model system for the production of proteins in plants. Mol Breeding 4:301–312

Woodard SL, Mayor JM, Bailey MR, Barker DK, Love RT, Lane JR, Delaney DE, McComas-Wagner JM, Mallubhotla HD, Hood EE, Dangott LJ, Tichy SE, Howard JA (2003) Maize (Zea mays)-derived bovine trypsin: characterization of the first large-scale, commercial protein product from transgenic plants. Biotechnol Appl Biochem 38:123–130

Yang GL, Feng D, Liu YT, Lv SM, Zheng MM, Tan AJ (2021) Research progress of a potential bioreactor: duckweed. Biomol Ther 11:93. https://doi.org/10.3390/biom11010093

Yao J, Weng Y, Dickey A, Wang KY (2015) Plants as factories for human pharmaceuticals: applications and challenges. Int J Mol Sci 16:28549–28565

Yu Y, Yu PC, Chang WJ, Yu K, Lin CS (2020) Plastid transformation: how does it work? Can it be applied to crops? What can it offer? Int J Mol Sci 21:4854. https://doi.org/10.3390/ijms21144854

Yusibov V, Streatfield SJ, Kushnir N (2011) Clinical development of plant-produced recombinant pharmaceuticals. Vaccines, antibodies and beyond. Hum Vaccin 7:313–321

Zimmermann J, Saalbach I, Jahn D, Giersberg M, Haehnel S, Wedel J, Macek J, Zoufal K, Glünder G, Falkenburg D, Kipriyanov SM (2009) Antibody expressing pea seeds as fodder for prevention of gastrointestinal parasitic infections in chickens. BMC Biotechnol 9:79. https://doi.org/10.1186/1472-6750-9-79

Chapter 15
Deep and Meaningful: An Iterative Approach to Developing an Authentic Narrative for Public Engagement for Plant Molecular Technologies in Human and Animal Health

Alison Prendiville, Rosie Hornbuckle, Silvia Grimaldi, Sara Mesquite De Albuquerque, and Sebastian Fuller

Abstract Public acceptance of plant molecular farming (PMF) for therapeutic and industrial proteins is a contentious public issue with stakeholder concerns focused on the social, environmental and regulatory challenges surrounding their development. For the public, cross-pollination with food crops, safety, potential side effects on human health and regulatory and policing issues are areas that require careful consideration when balanced with the benefits of the technology. Moreover, there is public concern over transparency and the role of business, motivation for profits and ownership and access to PMF technologies. With this background, Pharma-Factory, a 4-year EU-funded research project from the Horizon 2020-Biobased Innovation for sustainable goods and services call, is investigating new ways of producing pharmaceuticals. Concomitantly, the work is tasked with 'public engagement' to explore barriers to acceptance of PMF pharmaceuticals for human and animal health. In this chapter, we present the research undertaken to achieve the main objective of 'public engagement', the value of the process for stakeholders and science and technology partners. The approach incorporated a variety of perspectives and the development of many tools, including visuals, a glossary, an icon language, a narrative animation, posters and interactive exhibits, all relating to the perceived value and concerns of the technology under development, highlighting lived

A. Prendiville (✉) · R. Hornbuckle · S. Grimaldi
School of Design and Service Futures Lab, University of the Arts London, London, UK
e-mail: a.prendiville@lcc.arts.ac.uk

S. M. De Albuquerque
Applied Diagnostics Research and Evaluations Unit, Institute for Infection and Immunity, St George's University of London, London, UK

S. Fuller
Health Systems Collaborative, Nuffield Department of Medicine, University of Oxford, Oxford, UK

C. Kole et al. (eds.), *Tools & Techniques of Plant Molecular Farming*, Concepts and Strategies in Plant Sciences, https://doi.org/10.1007/978-981-99-4859-8_15

experiences of those who would be the recipients of the technologies. This extensive methodological approach not only builds scientific capacity and curiosity, through a process of participatory deliberations, but also offers a rich story around the technology which, we argue, provides the circumstances for a 'deep and meaningful' dialogue with the public and an authentic voice which has a legacy beyond the public engagement inside the funded project. Crucially, as the subsequent sections reveal, the public engagement story did not shy away from or 'hide' inconvenient questions or concerns, but rather highlighted them as opportunities for further discussion.

Keywords Co-design · Public engagement · Design methods

Abbreviations

BIMDG	British Inherited Metabolic Disease Group
BSE	Bovine spongiform encephalopathy
BSSA	British Sjögren's Syndrome Association
CNR	Consiglio Naztionale delle Ricerche—The Italian National Research Council
CSIC	The Spanish National Research Council
EU	European Union
FPCCSIDA	Fundacao Portuguesa a Comunidade contra a sida (Portuguese Foundation for the Community for AIDS)
GAT	Grupo De Ativistas em Tratamentos (Treatment Activist Group for HIV-Portugal)
GM	Genetically modified
GMO	Genetically modified organisms
HCP	Healthcare practitioners
IT	Information technologies
NPBT	New plant breeding techniques
PMF	Plant molecular farming
RRI	Responsible Research and Innovation
S&T	Science and technology
SGUL	St. George's University of London
UAL	University of the Arts London
UK	United Kingdom
UoR	University of Rouen
VTT	Valtion Teknillinen Tutkimuskeskus—Finnish Technical Research Centre
WP	Work package

15.1 Introduction

Plant molecular farming (PMF) offers new ways of researching and producing medicines, for vaccines, diagnostics and treatments in human and animal health. However, because of public opposition to genetically modified organisms in food production, there has been an inadvertent impact on other areas of bioengineering including PMF. Although there is less opposition towards these technologies for human health, there is still public concern that has crossed over from the GMO debate and impacted the public's acceptance of these medical products.

Against this backdrop, Pharma-Factory was a 4-year EU Horizon 2020 Innovation Action research project which took place between 2017 and 2022, which was tasked with developing medical, veterinary and diagnostic products for human and animal health using PMF technologies. Included in its programme of work was a diverse range of stakeholder engagement activities, to explore issues around social acceptance and resistances to these PMF products. To situate and frame the Pharma-Factory stakeholder engagement, this chapter offers in the first instance an overview of the rationale for the emergence and evolution of public engagement in scientific communication that developed in response to advances taking place in areas such as biotechnology. We then explain our methodological approach, in response to the different PMF technologies under development, and the use of co-design methods that were conceived as a dialogical and deliberative process that would embrace different ways of knowing amongst a diverse range of stakeholders such as patient groups, healthcare practitioners, clinicians, technology developers and the wider public. From the start, we designed the different steps of the research process to be iterative, to inform the follow-on engagement activities and to ultimately frame the content and activities for three public exhibitions.

Throughout the project, the focus of stakeholder engagement has been on building scientific capacity and criticality in relation to the Pharma-Factory products as a process of co-creating knowledge to better engage with issues around public acceptance. Thus, dialogue with a range of publics became central, not as an exercise to build consensus around PMF per se, but to meaningfully engage different groups, using co-design tools in ways that would be human centred and future facing and account for a diversity of citizen concerns and values. The co-design stakeholder engagement workshops enabled the emergence of concerns and values that were communicated through a series of interactive public exhibitions during 2021 and 2022. The findings from these various engagements suggest that the public are open to dialogue and, far from being resistant to PMF, are motivated to actively develop their knowledge so that they can better understand these technologies, not as sensationalist either-or options but as part of a more entangled range of personal and systemic issues that account for the complexities and trade-offs that need to be considered. Moreover, we found that this approach allowed an authentic narrative to evolve, which reaches beyond the assumptions of technology developers and enables meaningful dialogue with general audiences.

15.2 Background: Shifting Perspectives

From the 1970s through to the end of the twentieth century, the model for scientific communication with the public was constructed as a 'deficit' model (Wilsdon and Willis 2004; Stilgoe et al. 2014) in which 'the public' were imagined as ignorant, hostile to scientific developments and a hindrance and obstruction to scientific innovation. In this instance, scientists in white coats held authority and respect, with part of their professional responsibilities extending to improving the public's understanding and knowledge of science (Wilsdon and Willis 2004). In contrast, at the start of the millennium, with developments in information technologies, increase of access to information and scientific fall-out from BSE, there were calls for more scientific openness and better scientific governance (Irwin 2006). According to Wilsdon and Willis (2004), this change in perspective can also be attributed to the changing boundaries of science and business which, although never completely delineated, were becoming more entangled in business, societal and political relationships. For the authors, the consequence of this was a growing 'wariness towards scientists working in industry and government, and a suspicion of private ownership of scientific knowledge (Joly and Rip 2007)'. To address this, both Europe (European Commission 2001) and the UK (House of Lords Select Committee on Science and Society 2000) sought to overcome this legitimisation crisis by pushing for greater public engagement in scientific developments (Irwin 2006) in areas such as genetic engineering. However, Irwin notes how this rhetorical shift to a scientific governance based on public dialogue, transparency and democratic agency was similar to the previous deficit model of understanding, with this push for transparency and openness being an equivalent form of scientific authority, convincing a sceptical public of the need to trust decision-makers in their expertise, objectivity and impartiality. Thus, for Irwin, instead of the previous top-down approach to address public ignorance, the new form was still based on an information deficit model, only this time the deficit was a lack of trust. This shift from 'deficit to dialogue' for scientists, funders and policymakers is further critiqued by Stilgoe et al. (2014) as they question the validity of the approach in challenging and changing more fundamental structural institutional issues of science and its governance and see it as potentially reinforcing incumbent power structures (see also Chap. 13).

With much of the scientific research coming through taxpayer funding, Marris and Rose (2010) argue that the public should by necessity be involved in scientific decision-making relating to the technology and its governance. In other cases, they reiterate the need for public consultation as part of a process of building trust and additionally smoothing the way for new scientific innovations, ideally with the engagements taken early in the development to truly influence the direction of the work (Joly and Rip 2007). Yet, the authors stress that such approaches are not without tensions and opposition, and they should not be seen as exercises in agreement and acceptance. Instead, it is recommended that such encounters are framed as an opportunity to understand and recognise multiple viewpoints and to ultimately assist decision-making by accommodating different perspectives (Joly and Rip

2007). However, scientists have raised doubts about the capability of the public to contribute to scientific decision-making, especially if they are lacking scientific knowledge and expertise (Graur 2007). Equivalently, citizens and third-sector actors question and doubt their knowledge and skills to engage in participatory activities in science, technology and innovation (Dreyer et al. 2018).

For many scientists, the opposition by the public to GMO is attributed to a lack of appropriate knowledge, which also relates to doubts that scientists may have on the public's ability to be involved in scientific decision-making. For most of the time, people operate rationally on intuition, which occurs unconsciously, until we are confronted with abstract and complex scenarios that will then lead to irrationality (Blancke et al. 2015). For the authors, this irrationality occurs when ideas and concepts require effort and time to understand, with only education offering a solution to the entrenched biases. Furthermore, the authors argue that this irrationality comes to the fore when negative images such as those presented by anti-GMO campaigns work to undermine public trust, for example by linking socio-economic abuses to GM products whilst also playing to the notions of unnaturalness and contamination of the environment (Blancke et al. 2015). In contrast, Couée (2016) sees this argument as too simplistic and, instead of pitching scientists and citizens as rational and irrational against each other, suggests that an alternative approach is taken, that accounts for the complex interactions between biotechnologies, societies, industry, and capitalism. In many instances, what is in fact being exhibited is not irrationality but empirical scepticism and such complexity should be seen as a stimulating intellectual challenge (Couée 2016). MacPhetres et al. (2019) also make the connection between a lack of specific scientific domain knowledge of GM technology and attitudes to GM foods. However, acceptance can be improved by teaching people about the science behind GM foods, which leads to more knowledge and positive attitudes towards the foods and a greater willingness to consume the foods and to view them as less risky (MacPhetres et al. 2019). To achieve this, they argue that the information communicated must be presented as value neutral that avoids any ideological claims that GM is safe or good. By adopting this approach, participants are then encouraged to reflect about the information presented and to make their own decisions. Similarly, the authors point to the contribution of mechanistic knowledge of scientific processes so that participants build foundational information to support their decisions on whether a technology is safe or not. Overall, the authors recommend that communicating GM information in an engaging and accessible format to facilitate decisions is an effective way to build public engagement of these technologies, and by addressing gaps in knowledge, public opposition and acceptance may be critically informed.

15.3 Co-design for Empathetic and Flexible Stakeholder Engagement

An alternative perspective to public engagement for the purposes of scientific governance and scientific decision-making (Irwin 2006; Stilgoe et al. 2014) and one particularly salient to our work is offered by Selin et al. (2017). The authors note how public engagement for scientific governance, as a focus for scholarly research, is often criticised for producing questionable outcomes that are limited in their impact on scientific governance policies, through their various engagement processes (Stilgoe et al. 2014). Consequently, this has diminished other modes of public engagement that have weak or non-existent ties to governance models, such as science cafes, festivals, informal dialogues or online discussions. To counter this, the authors look instead to citizen capacity building by drawing on literature from other domains of scholarship such as public administration to offer additional perspectives on the value of engaging public. For Brodie et al. (2009), the rationale for participation in local and national governance is interconnected with the involvement offering a range of benefits from legitimacy and accountability of democratic institutions, empowering communities by coming together around a common cause or interest, which can assist in building social cohesion and as a tool for reforming public services that are more responsive and suited to people's needs.

Concurrently, much citizen engagement in these public administration contexts has looked to design research and especially co-design for its ability to deal with complex issues and to create active citizenship to address many of the systemic problems of late modernity (Buchanan 2001; Norman and Stappers 2015; Evans and Terrey 2016). Since the millennium, as an emerging and evolving practice, service design has been very much at the forefront of design's adoption within policy and local government (Bason 2014; Kimbell 2015) and with it the use of co-design as a methodological approach. Evans and Terrey (2016) see co-design as 'a methodology of research and professional reflection that supports inclusive problem solving and seeks solutions that will work for people'. Concomitantly by adopting these methods, the authors see opportunities for building trust with citizens and stakeholders, facilitating knowledge of policy and identifying delivery problems that public organisations do not possess. However, in formulating more relational government services, standardised approaches for tackling systemic and complex problems are less effective, as in-depth knowledge of personal circumstances is required (Muir and Parker 2014).

Co-design's role in Responsible Research and Innovation (RRI) is explored by Deserti et al. (2020) who highlight its designerly methodological characteristics, which are iterative, experimental, human centred and supported by prototyping. The application of these methods arose from the RRI framework (CNR, EU Commission 2015) that aimed to draw on a range of 'societal actors—researchers, citizens, policy makers, third sector organisations to collaborate during the research and innovation phases in order to better align both the process and its outcomes with the values, needs and expectations of society and to engage citizens and end users in the co-creation of the solutions they wish for and need (Rizzo et al 2020)'. By using such methods, the authors note how the practice takes policymaking beyond its more utilitarian approach of problem-solving with experts, to a space which enables a better alignment between the technology and the situated nature of the context in which a policy will be implemented. In such experimental environments, stakeholders may be engaged in new and far-future-making with the technology, mapping their own current experiences and relationships, including those with the technology, whilst exploring concerns over the social, safety and risk dimensions.

Specifically relating to the biosciences and PMF, Hornbuckle et al. (2020) present design as a distinctly different methodology from the biosciences and social sciences, offering something fundamental that science communication has been missing. For the authors, design attributes are seen as being 'flexible, problem-orientated and empathic, with co-design providing the tools to build a bridge between the highly specific but conceptually abstract science with its codified language, and the values of specific stakeholders or wider audiences (Hornbuckle et al. 2020)'. Similarly Michael (2012) sees designerly public engagement as being thoughtful within a context of complexity, in contrast to other forms of public engagement that aim to channel public opinion into existing institutions to influence policymaking. Consequently, material objects used in public engagement from a design perspective are thus 'meant to evoke in their audiences less a need for clarity, than a desire for, and exploration of complexity' (541). Describing design's role further within these co-design processes, Hornbuckle (2022) expands and reflects on the contribution of design as changing proximities between different actors. Recognising the complexity of a multi-stakeholder system where pronounced distances and differences exist between expert knowledge and lived experiences of different stakeholders such as patients, healthcare practitioners and the wider public; the author recommends a range of co-design methods to facilitate knowledge flow as a set of translatory practices to build closer proximity.

For the purposes of our research, Pharma-Factory focused on stakeholder engagement as informed by service design (Hornbuckle et al. 2020) rather than public engagement, to account for the mix of the audiences that we would be interacting with. The overall structure of the different research phases was also designed to be highly iterative with each phase informing subsequent activities and culminating in the final public exhibitions. This research frame offered an experimental and flexible dialogical approach involving prototyping possible futures through individual experiences. It also created a relational and human-centred perspective that accounted for the lived experiences for the people with health conditions that would be affected by the PMF products under development. In the following section, we present the co-design methodological steps undertaken to build scientific capacity and curiosity for stakeholders, to build confidence and criticality when engaging in issues of acceptance on PMF.

15.4 Methodological Approach

Increasingly, emphasis is being placed on 'public engagement' by funders of science and technology (S&T) innovation projects. However, as described in the Background section of this chapter, public engagement can take many different forms, varying in the approach, the resource invested, the time taken and the involvement of different viewpoints. In the case of Pharma-Factory, social scientists and design researchers proposed that, rather than relying on assumptions about the potential value of the technology to stakeholders, the project could adopt an iterative methodological approach, which aims to build dialogue between technology developers and stakeholders through a series of workshops and ultimately build scientific capacity around PMF. This would have a multiplier effect: building trust with stakeholders (such as patients and HCPs), better understanding the value and therefore being able to communicate more effectively with the public, developing an authentic narrative and providing technology developers with insights about the value of their products to inform their future work. This iterative methodology is diagrammatised in Fig. 15.1.

The iterative process adopted in Pharma-Factory involved three phases, which were executed using a mixed-methods approach. Site visits and literature reviews were conducted in phase 1 in preparation for the stakeholder engagement workshops in phase 2. Design researchers from the University of the Arts London (UAL) applied their expertise in co-design tools and methods, whilst social science researchers from St. George's University of London (SGUL) undertook a series of partner and patient interviews typical of a social science methodology. The combined approach of the two disciplines was an important feature of the approach; the research team felt that this would strengthen the methodology as well as build trust in the research (for example that a university hospital was involved in research involving patients). Each discipline's role is coded within the diagram in Fig. 15.1. In phases 1 and 2 of the research process, the qualitative social science research and the co-design workshops contextualised the S&T researchers' understanding of public perceptions and the opinions of sections of the public about PMF processes and products. In addition to the data collection phases, there are important 'in-between' moments of analysis, synthesis, sensemaking, translation and tool development in preparation for the next set of engagements, including the final phase involving interactive public exhibitions and partner feedback. This typifies the reflexive and responsive approach of design research and some modes of the social sciences.

Table 15.1 outlines full range of activities undertaken during the three research phases to meet the objective of 'public engagement'. The details of the research methods are described in the subsequent subsections. Due to the COVID-19 pandemic, some of the planned activities had to be cancelled, postponed or moved online. The greatest loss to the process was the public engagements planned in phase 2, which were intended as 'pop-up' events, but due to COVID-19 social restrictions, it was not permitted by national and local authorities at that time.

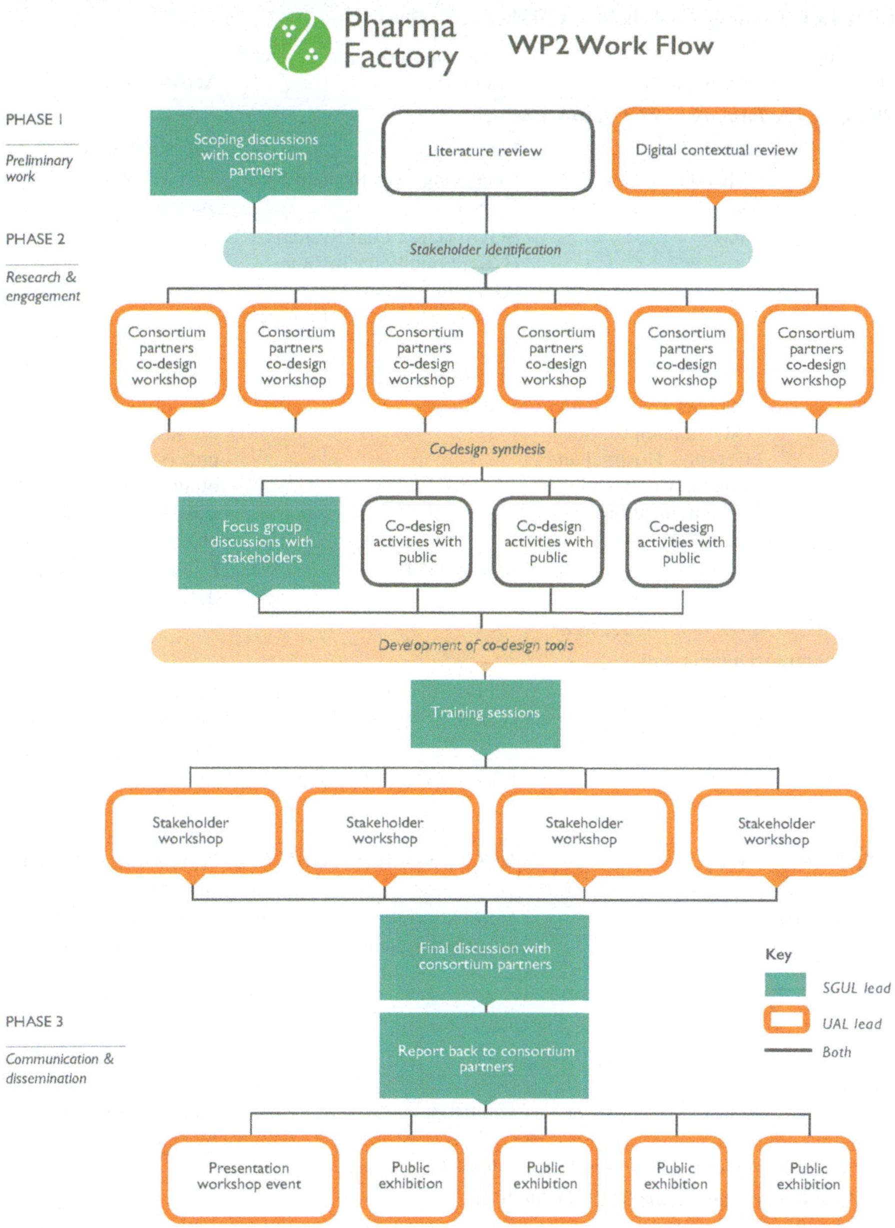

Fig. 15.1 WP2 workflow diagram

Table 15.1 Co-design stakeholder workshops completed

Activity date/s	Participants	Organisation (project partners/ external)	Activity type
Phase 1: Preliminary work 11/2018–02/2020			
	n/a	n/a	Literature review
	All WPs	Representatives from all S&T WPs	S&T scoping interviews
	n/a	SGUL (UK); Leaf Systems (UK); Fraunhofer Institute (Germany); TransAlgae (Israel); Samabriva (France) (all project partners)	Site visits
	CEO	Leaf Systems	Discussion
	Members of the public (SGUL campus, St. George's Hospital and London College of Communication campus)	SGUL	Five-minute person-on-the street interviews to understand knowledge of and perception of PMF process and products
		SGUL; UAL	In-person exhibition at design festival
Phase 2: Research and engagement			
S&T partner engagement			
January 2019	WP3	VTT, University of Rouen (UoR), CSIC, SGUL	Ecosystems, values, persona pitches and prototypes
	WP4	Fraunhofer, CSIC and SGUL	Ecosystems, personas, service storyboard and innovation canvas
March 2019	WP5	UCL, TransAlgae, SGUL	Ecosystem, values and innovation canvas
January 2019	WP6	UoR, AlbaJuna, Leaf Systems, Fraunhofer, Samabriva, SGUL	Ecosystem, values, persona pitches, prototypes, regulatory timeline, innovation canvas and storyboarding
March 2019	WP7	SGUL, Leaf Systems, Diamante	Ecosystems, personas, service storyboard and innovation canvas
Co-design with public [not possible due to lockdown measures]			
Stakeholder engagement: 02/2020–06/2021			
26/02/2020	IMD Pharmacists $N = 8$	British Inherited Metabolic Disease Group (UK—external)	In-person workshop
29/06/2020–03/07/2020	Rheumatoid arthritis patients $N = 13$	British Rheumatoid Arthritis Society (UK—external)	Online (asynchronous) workshop

(continued)

Table 15.1 (continued)

Activity date/s	Participants	Organisation (project partners/ external)	Activity type
14/09/2020–18/09/2020	Sjögren's syndrome patients $N = 8$	British Sjogren's Syndrome Association (UK—external)	Online (asynchronous) workshop
16/12/2020–20/12/2020	HIV patients $N = 7$ $N = 4$	ABRAÇO; Fundacao Portuguesa a Comunidade contra a sida (FPCCSIDA); Associação Positivo; Liga Portuguesa Contra a SIDA; Grupo De Ativistas em Tratamentos (GAT); SERES (Portugal—external)	Online (asynchronous) and in-person workshop
17/05/2021–14/06/2021	Conscious consumers $N = 5$ (UK) (survey: $N = 39$ UK, $N = 13$ ESP, $N = 11$ FRA)	Personal networks (UK, Spain, France, Italy—external)	Online (asynchronous) workshop
Phase 3: Communication and dissemination			
Public exhibitions			
07/10/21–09/10/21	School children, university students	Rouen Fete de la Sciences at Rouen University (French)	In-person exhibition stand at science fair
25/10/21–30/10/21	Families	Norwich Science Festival, Norwich, UK, in English	In-person exhibition stand at science fair
6/07/22–7/07/22	Patients, university students, healthcare workers	SGUL	In-person exhibition in university hospital
S&T partner feedback			
17/06/22	Project partners $n = 5$	**Samabriva,** CSIC UofR, SGUL	Online focus group
14/06/22	Project partners $n = 6$	**TransAlgae,** UCL	Online focus group
14/06/22	Project partners $n = 4$	**AlbaJuna,** UoR VTT, Leaf Systems	Online focus group
21/06/22	Project partners $n = 4$	**Diamante,** SGUL, Leaf Systems	Online focus group
			Online focus group

15.4.1 Phase 1: Preliminary Work

Design researchers undertook a period of primary and secondary research in the form of a literature review, field visits, telephone interviews and co-development of a project glossary with scientific partners. The purpose of this phase was to develop the literacy of the design researchers in the technology, to better understand and make sense of the technologies, their potential benefits and the challenges to implementation. Field visits to the technology sites allowed design researchers to gather experiential knowledge, which helped them to understand the processes better, recording these in various media such as photographs, field notes and sketches, methods typical of an anthropological approach. In turn, this enabled researchers to translate the complex scientific information into appropriate language and develop a narrative so that stakeholders can quickly understand and 'access' the concepts presented by the technologies. For example, concepts such as 'transient expression', 'diagnostic kit' or 'genetic modification' are relatively simple for a scientist to rationalise, whereas non-scientists may have little or no knowledge of these concepts, or they may have a viewpoint formed from media sources, which often presents a bias and does not convey the 'facts' or context to the audience. Therefore, the challenge for design researchers was to try to reveal and communicate 'truths' about these concepts, based on scientific information, so that participants in the research can understand the context quickly and effectively within the workshop setting and can respond with a more accurate reflection on the value of the technology to them. This phase overlapped the beginning of phase 2 as needs arose during the process; in this way, the approach is reflexive, responding to themes emerging through the research.

15.4.2 Phase 2: Research and Engagement

Based on their experience as service designers and design facilitators, the researchers developed a general narrative for the workshops taking participants on a journey from their own experiences through several carefully designed steps. This is depicted in Fig. 15.2.

15.4.2.1 Science and Technology Partner Engagement

Four consortium co-design sessions were held at SGUL in January 2019 for Work Packages (WP) 3, 4 and 6 with two further workshops held at a review meeting in Valencia for WP 5 and 7. For each activity, the WP consortium members collaborated with their team members to address each of the co-design tasks. The purpose of the exercise was to take the scientists through a series of deliberative and dialogical processes relating to their PMF technologies and to consider diverse stakeholders and wider systemic issues. The workshops also included creative exercises for the consortium members to prototype ideas on how these PMF platforms could have greater visibility within human and animal healthcare environments.

15.4.2.2 Partner Co-design Tools

The co-design tools were designed to actively engage the consortium members in situating their WP technological developments within broader more systemic frames and to consider and respond to the diversity of opinions that exist. In addition, for WP2 members, these workshops were part of a sensemaking process to

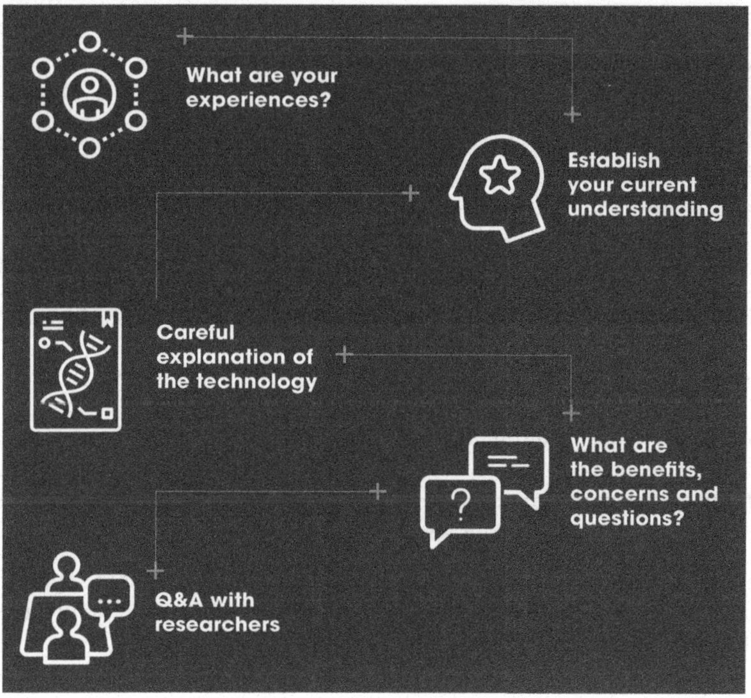

Fig. 15.2 Co-design narrative used in stakeholder workshops

better understand how the scientists conceived the PMF technologies in terms of near and far futures, and the benefits and values to different stakeholders, and how the scientists could attend to the challenges of communicating the value of their technology to different audiences. The following list summarises the key activities included in each of the workshops:

- **Ecosystem mapping near and far futures**: to situate the technology within more systemic frames of stakeholders, production methods, regulatory implications and distribution channels.
- **Regulatory mapping**: to map out the regulatory pathway for each WP product, including the time frames, and challenges for each of the PMF technologies.
- **Stakeholder value tools for near and far futures**: to reflect on how three stakeholders—a WP partner, an end user and an influencer—would be affected by the PMF technology. Questions explored: What is their need for PMF? What is the value to them?
- **Persona pitches**: relational and empathetic tools were designed to represent diverse groups of actors who will benefit and be opposed to the technology. The consortium members were encouraged to think about the value of their technology to a persona profile and the values and beliefs that the person may have, which may influence their acceptance or rejection.

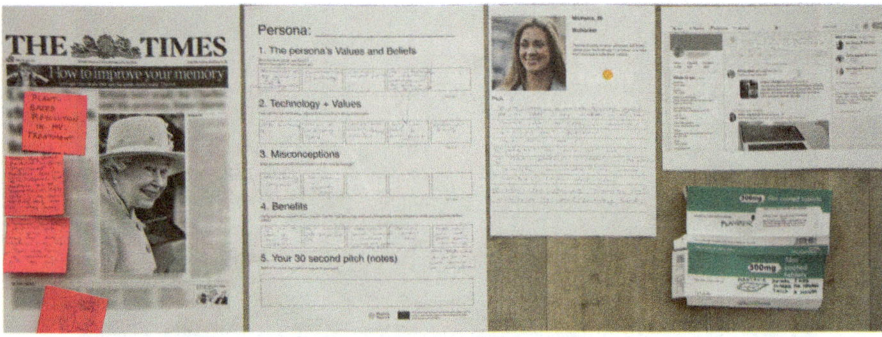

Fig. 15.3 An example of a workshop output by one of the consortium teams

- **Prototyping**: consortium members were asked to conceptualise ways in which their PMF technology could become more visible within a particular context such as healthcare and to generate newspaper headlines that would communicate the values and benefits of their specific technology.

The synthesised data from these workshop activities was used to support the co-design activities with the stakeholders and frame the exhibition around the benefits and values of PMF to diverse audiences (Fig. 15.3).

15.4.2.3 Stakeholder Recruitment

Stakeholders were identified for each of the four technologies under development. A wide range of the stakeholder communities identified were approached for engagement, including healthcare professionals (nurses and doctors), industry (pharmaceutical companies), health service organisations (NICE in the UK), regulators and regulatory consultants, fish farmers and veterinary professionals, and supermarkets. However, there were difficulties in engaging stakeholders for whom the value of the technology is less clear, and for large organisations where there was no previous or existing contact, or where this type of engagement might be seen as low priority compared with more pressing concerns.

Patients have the most to gain from new technologies and are convened in the form of patient organisations, which is more complicit to recruitment in research. Vouchers were provided to participants as an incentive.

15.4.2.4 Stakeholder Co-design Tools

In addition to the S&T partner engagement co-design tools, a set of relational tools were developed for the palette of existing tools, to translate the co-design narrative into a series of activities for participants to respond to step-by-step. These were then adapted to the needs of the group of participants in each workshop:

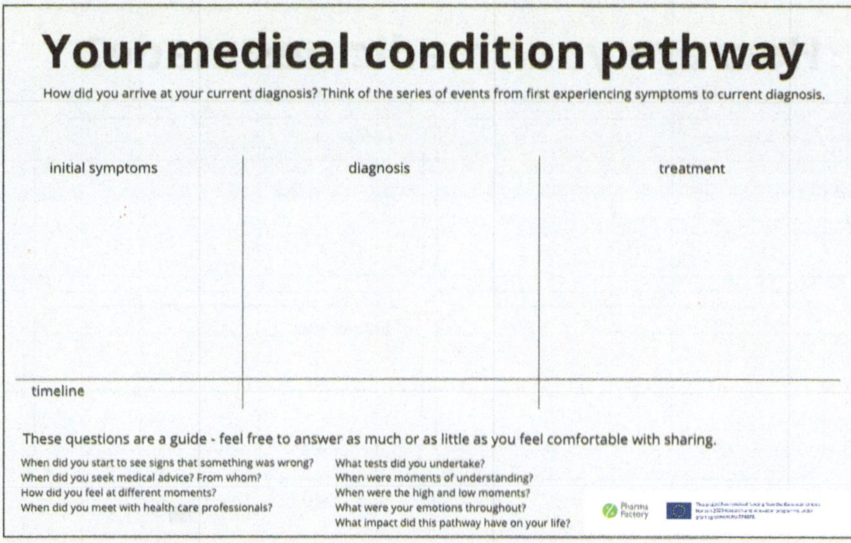

Fig. 15.4 Example of co-design tools for 'experience gathering'

- **'Experience gathering' tools**: these are designed to encourage participants to reflect on their own experiences, including pathways of diagnosis and treatment, diaries, educational histories and ecosystem maps (Fig. 15.4).
- **'Establishing understanding' tools**: these were designed to understand the entry point of the participants to the subject, often focusing on their knowledge of how different therapies are produced, or on particular terms central to the technology, such as 'genetic modification' (Fig. 15.5).
- **'Explanation of the technology' tools**: based on the glossary and research conducted in the previous phase, participants were presented with a series of 'cards' which outlined key concepts in simple language and icons, such as 'how are medical proteins produced?', which positions PMF with the spectrum of genetically engineered organisms used to express proteins (Fig. 15.6).
- **'Perspective gathering' tools**: worksheets with simple matrices for gathering participants' views on the benefits, concerns and questions raised by the information they had been presented with (Fig. 15.7). This included a specific GM proximities tool, which was used for all the workshops as a penultimate task, which is discussed later.

15.4.2.5 Stakeholder Engagement

We planned workshops lasting 3-h to be conducted in a convenient location for participants; however, due to lockdown restrictions, only two of the workshops could take place in-person. For the remote workshops, Facebook was chosen as a

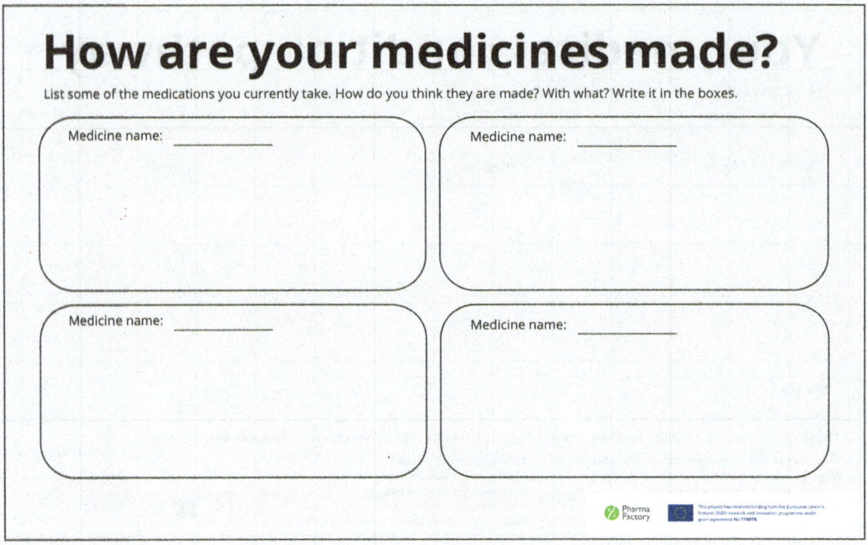

Fig. 15.5 Example of co-design tools for 'establishing understanding'

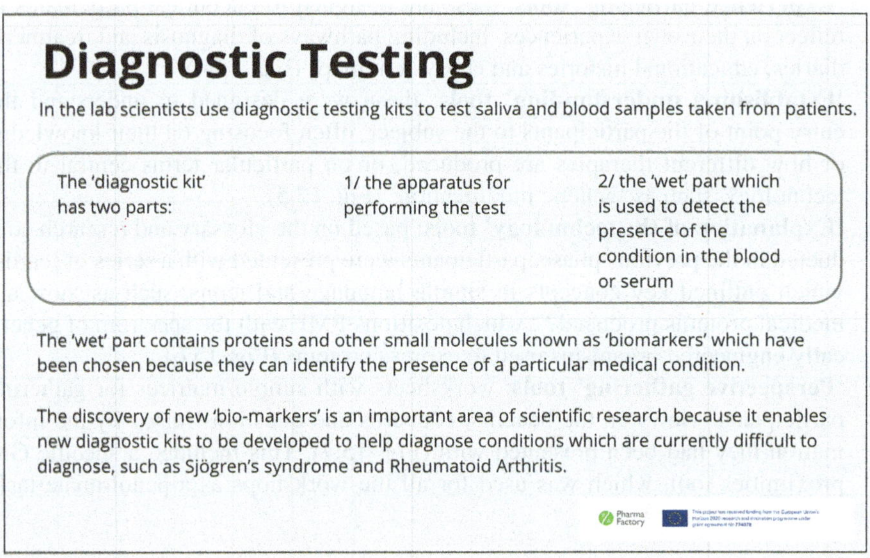

Fig. 15.6 Example of co-design tools for 'explaining the technology'

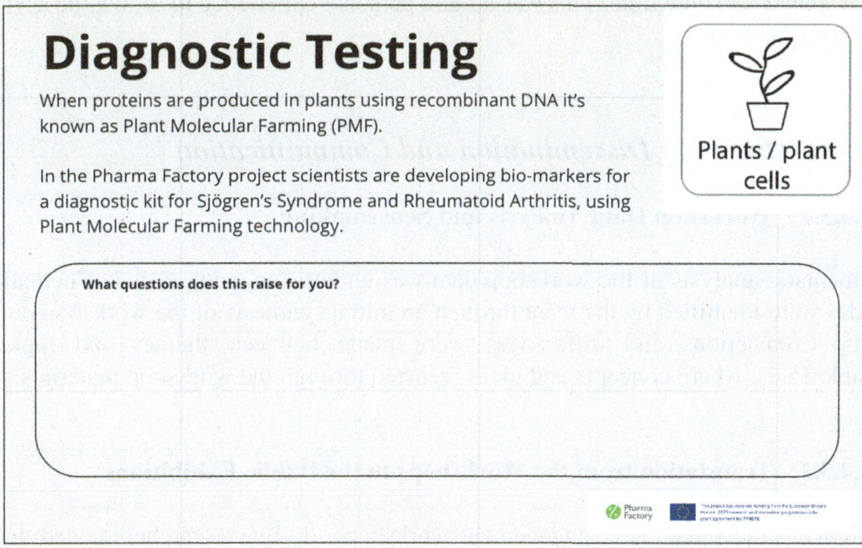

Fig. 15.7 Example of co-design tools for 'gathering perspectives'

'base camp' for participants because our first group were already familiar with this platform and operated through Facebook via a private discussion group. Generally, we found that 'snowball' recruitment through Facebook was slow; we first needed to gain the trust of a 'gatekeeper' in the form of a community group who owned an active Facebook group. We invited participants to closed Facebook groups where they could be informed of updates and workshop steps, and to post introductory videos and links to our online collaborative space in the MIRO platform (https://miro.com/), which is free for educators to use. Facebook was useful as a meeting and discussion space when hosting the online workshops, and participants were supportive of each other and actively posted questions.

The online workshops were 'asynchronous', meaning that participants were set daily 30-min tasks over a period of 3 or 4 days but could complete them independently without the need for direct facilitation. Participant feedback was largely positive, citing that they found it less intimidating and more convenient to work at their own pace. The online workshops were more successful when participants were confident IT users, but less so for some users with limited experience. It also proved less satisfactory for Sjögren's patients who experience 'dry eyes'. We had taken measures to mitigate these issues such as preparing participants with trial tasks and checking if their health would make online participation difficult, but there were still a couple of participants who found the tasks challenging.

Participants in the online workshops were given the opportunity to attend a live Q&A with researchers. This was an important moment for them to get quick feedback to their questions when the other outputs from the research would be unavailable for some time. This was a more personal experience to follow the asynchronous activities.

15.4.3 *Phase 3: Dissemination and Communication*

15.4.3.1 Workshop Data Analysis and Sensemaking

A thematic analysis of the workshop data was undertaken using NVivo. Thematic codes were identified by the team through an initial synthesis of the workshop outputs. Connections and differences were made between themes and topics (Table 15.2), where concepts and ideas recurred through the workshop materials.

15.4.3.2 Translation from the Workshops to the Public Exhibitions

Following the stakeholder engagement workshops, design researchers coded and analysed the worksheets, drawing out the key themes that could then be translated into meaningful content for the public exhibitions (Table 15.3). The following research questions were central to this phase of sensemaking and translation:

- What is the perceived value of these technologies for the stakeholders?
- What are the main barriers to acceptance identified by the stakeholders?
- What are the stakeholders' main questions and concerns, and how might we best frame the exhibition content to address these?

The main challenge in sensemaking and translating the research for a general audience was to represent the rich data collected around stakeholder experiences and

Table 15.2 Details of the coding themes and topics for data analysis

Coding themes	Coping topics
Agency	Conventional medicines
Choice	COVID-19
Cost, efficacy	Diagnostic investigation
Human experience (anxiety, certainty, proximity, side effects, stigma and discrimination, symptoms and uncertainty)	Features
Knowledge (ethics, not known, trust)	Genetic modification
Learning through the workshop	How medicines work?
Safety	Manufacturing
Scale	PMF
Security of supply	

Table 15.3 The key topics drawn from the co-design workshops that were then used to inform the content of the exhibition posters

Co-design workshop key issues	Poster content
Patient lived experience	Conveying the patient journey
Time and uncertainty	The benefits of how PMF may alter diagnostic, treatment and choices in human healthcare
Gaps in understanding the technologies and medicines more generally	Types of PMF techniques as a key thread throughout the posters
Safety	Conveying the containment and secure facilities of the growing and processing of PMF and regulatory pathway
Food security and health	New ways to administer vaccines in human and animal health
Build knowledge that counters the sensationalist approach in the media	Communicate the human, production and pharmaceutical opportunities
Ethics and ownership in pharma (Exhibition 3 only)	Engaging with the debate over who should invest in PMF and other ethical considerations

perspectives, whilst also telling the story of the project and the science. The research team was also mindful that the exhibition content had to be communicated in a language that a lay audience could understand and relate to, and delivered using methods that would engage and stimulate discussion. Therefore, the approach was to take visitors on a journey through a series of themed posters, whilst also providing interactive activities as a 'way-in' to the more detailed information for a diverse range of visitors. This also enabled a dialogical approach to the engagement, with a key aim of the exhibition being to enable conversations between the exhibition stand 'hosts' and the visitors. The posters needed to be easy to access out of sequence. For example, if a scientist on the stand was explaining the use of PMF to develop a diagnostic kit, they should be able to refer to the poster about benefits for patients with Sjögren's syndrome and Rheumatoid Arthritis.

Personas, a method typical of service design and co-design, were developed from the stakeholder engagement and added a relatable 'human dimension' to help the public to understand how each technology may transform an individual's life. An 'icon language' was commissioned to create and articulate a visual language that could be understood and interacted with at different points in the exhibitions. This was used throughout the exhibition to aid cognition and familiarity and help visitors to make connections between the different elements. An animation was also commissioned and shown on the exhibition stand to broadly explain in an imaginative way the two different plant molecular techniques used—transient expression and transgenic—used in Pharma-Factory; a French and English version was produced. Here, the longevity of the communication outcomes was also taken into consideration, as the animation could be used in many different contexts after the funding period. These are available for download from the Pharma-Factory website https://pharmafactory.org/.

Following the exhibitions, all individuals who hosted the stand and engaged with members of the public who were visiting were invited to provide feedback, which is discussed in the following sections.

15.5 Science and Technology Partner Feedback

This involved presenting the findings from the stakeholder engagement to the S&T partners so that they could reflect on the implications for their future work during an online focus group that was recorded and transcribed. This was also an opportunity to receive feedback on the value of the process undertaken. This data will now be considered, along with the feedback from the exhibition hosts.

15.5.1 Results and Discussion: The Value of a Deep and Meaningful Narrative

The multifaceted approach taken prior to reaching the final objective of 'public engagement' included engaging a diverse range of actors in in-depth activities to explore PMF from their perspectives as scientists, patients, healthcare experts or members of the public. As part of this process, the research explored new ways of representing PMF through the development of participatory tools, visuals, a glossary, an icon language, an animation, posters and interactive elements. These methods not only built scientific capacity around the technical mechanisms of PMF, but also enabled participants to critically reflect on their current knowledge and its sources. Furthermore, the co-design methods offered a 'deep and meaningful' dialogue with the public which did not avoid or 'hide' difficult or inconvenient questions or concerns, but instead created deliberations around societal challenges, personal stories, and opportunities for further discussion.

The following section presents the data collected about the value of this methodological approach to four stakeholder groups: patients, healthcare professionals, general audiences (publics) and scientists (technology developers). This data is based on the small sample of participants in our research and is therefore not generalisable as such, but provides insights into the value of this approach so that we can confidently propose 'public engagement' recommendations for future S&T innovation projects and applied science more generally, which incorporates relational tools for meaningful engagement through an iterative process of translation and sense-making as described in the methodology and illustrated as a core framework in Fig. 15.8.

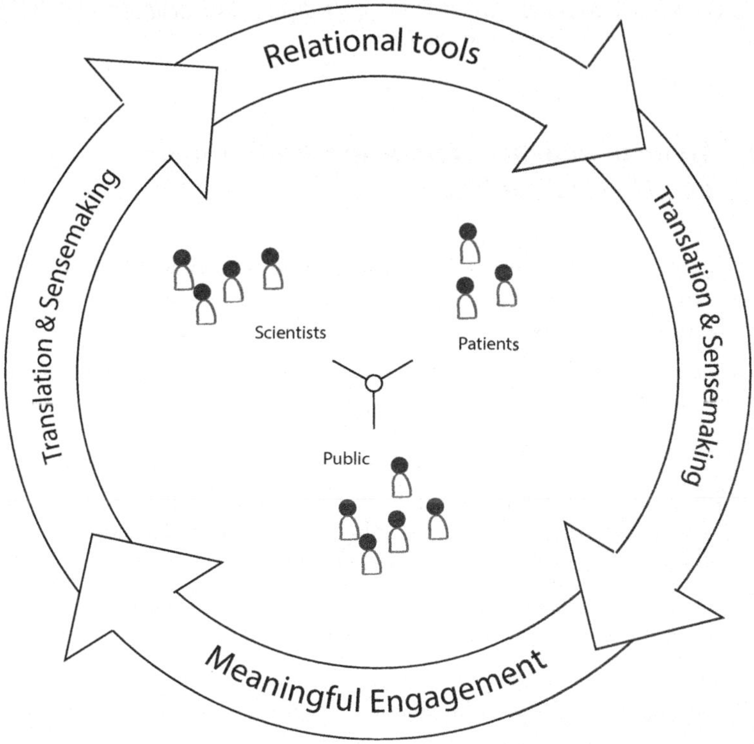

Fig. 15.8 A dialogical and iterative framework for deep and meaningful public engagement, based on co-design principles and practices

15.5.2 Value to Scientists at the Start of the Project

The Pharma-Factory scientists were unfamiliar with co-design methods, and this is reflected in their responses to the process used at the early stage of the research, but it is important to note the less sceptical tone by the end of the work when they were well versed in this dialogical and deliberative approach:

> *I found the activities very interesting. This helped us identify points we will probably need to take into account in the future. I'd like to know how this information is analysed and which conclusions can be taken from that.*

> *Great work, and a relaxed way to contextualise the work.*

> *Interesting exercises, regulatory issues identified as key. Need to define next steps. A meeting with different stakeholders seems to be necessary.*

> *Timeline activities helped with awareness of key stakeholders.*

Innovation and storyboard discussions were very helpful tools to understand the needs of the company and the existing barriers to PMF.

15.5.3 Value to Patients, Healthcare Professionals and General Audiences

Many of the participants reported that the workshops had enabled them to reflect on their own experiences and reveal aspects of their healthcare journey, which had not come to the fore previously.

I learned more about myself than about the project per se but the scale and scope of GM plants as biomarkers is just beyond comprehension and it's a thrill to see out of the box thinking applied to our autoimmune issues of diagnosis and hopefully treatment in the longer term. (Patient)

Furthermore, the workshops raised awareness of the scientific research and gave a voice to the patients in assessing the value of something that could impact their lives, which they apparently had not experienced before:

It was great to know that researchers actually went out and talked to ordinary citizens to understand their concerns. My trust in science strengthened further.

One finding from the research is that the co-design stakeholder engagement both revealed and enabled multiple dimensions of learning and agency: about the self, and an awareness of the wider healthcare ecosystem.

Most participants had very little prior knowledge of how medicines are made, even amongst the pharmacists, and revealed that they put their trust almost entirely with the regulatory and healthcare service organisations. The information provided through the workshop in most cases enabled people to understand the role of genetic engineering and rationalise the role of PMF. In some cases, this was not enough for people to understand that using plants did not necessarily remove the need for animal testing during the process of development and regulatory approval. However, participants reported that the workshops had raised questions about the way that medicines are produced, which may impact their future actions:

Really interesting workshop: learnt a lot and raised awareness of PMF! Made me reflect on current transparency of how drugs are made and what info should be given to patients. Good to be made aware of what future drug manufacturing could look like. (Pharmacist)

Participants reported that the workshops had challenged their previous views of GM:

The most value to me is having my misconceptions about GM plants usage in medicine challenged and taken forward (Patient)

I had to reassess my views on GM and ask myself why I had the opinions I had and how they were formed. It was a good personal exercise. (Conscious consumer)

From this research, it is apparent that stakeholders need reassurance about the steps being taken to avoid negative impacts of PMF on the environment and on human health. There is also a wider topic relating to trust in the profit-driven companies that use the technologies, which may be harder to address. Importantly, providing participants with more accessible information about the context in which GM technologies are used, and why, enabled them to rationalise the safety and value for themselves and challenge the views that mainly form through sensationalist media reports.

Therefore, a clear value to these participants was access to scientific knowledge and rationale, which allowed them the agency to do their own research and to improve their understanding about medicines and how they are made, and about themselves and their medical condition. In the case of the stakeholder workshops, participants also exhibited confidence during the expert Q&A, allowing them to ask pertinent questions on a topic they knew nothing about previously. During the partner feedback sessions, scientists were surprised that people wanted to talk to them about medicine production:

> *I think what is unexpected to me is like actually people want to communicate with scientists. So because like when we come up with new ideas, in most cases it got rejected by the public because people are more concerned about risk and environment impact. But actually they want to talk with scientists*

In the case of the exhibitions, the authentic narrative allowed members of the public to engage in dialogue with experts when visiting the exhibition stands:

> *A group of 3 oncology nurses, all originally from different countries in Africa, but working now in the NHS ... immediately understood the problem of accessibility that we are trying to solve and the potential advantages for people in developing countries. They all became very excited about the work, stayed at the stand for at least 20 minutes, and took away copies of every postcard to share with their friends. [Scientist host]*

The design tools, narratives and visualisations—developed following extensive research of the terminology and the technologies using design research methods in phase 1—enabled the scientists to have meaningful conversations with diverse members of the public, thus increasing this access to dialogue with experts:

> *The different stamps were quite useful in illustrating the different parts of the PMF process and the products that we can make. Many of the children had heard of antibodies, cells and DNA so the stamps gave a simplistic demonstration of what these things might look like. This in turn led to conversation about what their functions were. [Scientist host]*

> *The posters tended to be observed and read by a few of the parents and older children that had a genuine interest in the subject or wanted to find out more. By asking them 'Does it all make sense?' normally stimulated more in-depth questions. [Scientist host]*

15.5.4 Value to Scientists and Technology Innovation Developers, Project Partners at the End of the Projects

Scientist hosts on the exhibition stands reported some interesting conversations with visitors, noting that the interactions led to new potential collaborations in one case:

> *I spoke to a consultant ophthalmologist who noted access to anti-angiogenic antibodies for AMD could be improved through the plant system and suggested to prime collaboration with colleagues at UCL/Moorfields (Scientist host)*

The role of the discussion cards was surprising in some ways, as they became the vehicle by which the hosts exchanged knowledge with visitors, handing them out to allow people to have a quick reference for the meaning behind PMF. The PMF explanatory cards were reprinted to allow for more to be distributed. The interactive poster also allowed the hosts to bring visitors into an engagement around ethical questions, which would have been near impossible otherwise:

> *I think the poster with different perspectives + red/yellow/green stickers was a hit. Was simple but very effective and I think people like being able to see how their opinion compares with others. (Scientist host)*

> *I'm surprised how easy it was to get people to spend time reading the participant statements and giving their opinion. There was a lot of reading to do so I was surprised people were willing to take the time. (Scientist host)*

There was also some scepticism and difficult questions raised by exhibition visitors, which prompted reflection from the scientific hosts:

> *The conversation that stood out for me was a woman asking me how we know our technology won't be used for unethical purposes … I thought it was an interesting question. What could one do that's unethical with our platform? (Scientist host)*

Importantly, this feedback suggests that the scientists' assumptions about public perceptions, attitudes and willingness to engage were challenged through their experiences of the exhibition, allowing them to have conversations about challenging topics that they would not have had otherwise.

In the S&T partner feedback sessions, there was a general acknowledgement that dialogue can be an effective strategy towards achieving a reasonable and 'sensible' response from stakeholders and general audiences. This was one of the elements that surprised some partners, that stakeholders had been able to rationalise the use of GM within the context of PMF, in the development of therapies, diagnostics and vaccine production:

> *I actually thought that most of what they said was quite sensible and quite thoughtful (S&T partner)*

There was an acknowledgement that peoples' concerns may go beyond the motivations of the scientific research and therefore understanding these perspectives can help to guide communication: for example, concerns over safety for people and the environment, which appeared to be based on having little understanding of the mechanisms and infrastructure in place. In the exhibitions, these concerns were addressed with images of contained facilities with clear definitions of terminology, and a regulatory timeline which showed the complexity of stakeholder interactions rather than oversimplifying the process.

S&T partners developed an understanding of the power of visual and narrative methods of communication for quickly and simply bringing people into closer proximity with the technology, giving access and agency. Expanding on the reported impact of including images of the secure PMF facilities, one senior scientist discussed the possibility of more interactive and hands-on experiences for members of the public.

> *The stakeholders or the patients visualize, and you show them those pictures of what it means 'the contained facility'. And then what I understood is that, in a way, produced more trust on them. I'm not sure if it's because of the picture that it looks more 'under control' or is simply because they really can imagine it and they really can see it and this is this other way of connecting with the reality that is not only about reasons, but also about images, about feelings. And I think we are under-exploiting those who want to advocate technology. (S&T partner)*

> *I know we are scientists […] we know the pathway of rationalizing and explaining but in addition to that I think there's a group of the population, for which, [researcher] was explaining, this is probably not the main route pathway, but feeling, touching and proximity bring it closer. So, whether we like it or not this is nowadays one of the main pathways to get people acquainted or accepting. (S&T partner)*

The research presented led to a general acknowledgement that alternative methods to 'science communication' are worth investing in to move the technology forward. Partners felt that the service design and co-design approach had been a valuable exercise, which provided an authenticity to the public engagement phase of the project (exhibitions). For example, one partner aligned the benefits of the technology to the environmental concerns raised by stakeholders:

> *one of the biggest advantages of using the oral vaccines that we are, if we are talking about the challenge of sustainability right now: we are going to replace all the needles, the gloves, the expert men that needs to be educated in order to vaccine each fish individually. So also regarding the environmental impact we are bringing a unique advantage. (S&T partner)*

> *The GM problem was always badly managed by the scientists and well managed by the opposers. I mean, it was basically a one-sided argument that the scientists never got involved and so it was just that 'this is bad. Stop doing it'. (S&T partner)*

S&T partners developed an understanding that exploring future narratives and bringing people into that discussion can build confidence and trust: being the ones to ask the difficult questions and guide the process of answering them. This appeared to give the partners an energy and enthusiasm to explore previously difficult topics with general audiences, which they might have found it difficult to navigate previously. In this way, the resources created for the 'public engagement' provided the S&T partners with tools with which to engage in meaningful dialogue with stakeholders and general audiences.

> *I think what you've done has been really interesting and I think it's valuable for bench scientists to realize what the public think about what you're doing. And you know there's*

always this perception from people working with plants that there are big anti-GM lobbies still out there (S&T partner)

the temptation as a scientist is always to present the science, the technology as the main story. But actually, the main story when talking to the public is you know 'this is a problem, a worldwide problem that that needs to be addressed'. And you know 'this is the regulations that we need to meet in order to address it' and then the technology you know should follow. And as [researcher] was saying, also the opportunity to talk directly to scientists to realize that actually we are, you know, rational, normal, responsible individuals and that we do consider all the risks and the benefits of what we're trying to do. (S&T partner)

As expressed in the previous quote, one finding that surprised some of the S&T partners was that people want to talk to scientists. However, the research suggests that this was most accessible once they have been taken through the co-design process, which allows them to acquire sufficient understanding to engage in meaningful and equitable exchanges.

[the general public] get so much information and [they] don't know even if what [they're] hearing is right or wrong and so definitely to talk to the scientists would help (S&T partner)

Finally, partners appreciated that building an authentic narrative around value as a foundation for dialogue with general audiences takes time, process and investment (as demonstrated by the service design and co-design research process). However, they also recognised that this process is worthwhile and vital for new technologies such as PMF. In particular, partners said that they felt that this is something that should be done more in future projects:

It has great integrity and people have confidence in the results because of that. I think we can certainly use that as you say. We did do that through the public exhibitions and I think that's really helped us have meaningful discussions at those exhibitions rather than people feel like they're taking the party line from a group of people who are invested in the development of a particular technology. They saw the impact from those other groups and that was fantastically valuable to the project and continues to be. (S&T partner)

When a more immersive exhibition was suggested, one senior scientist commented:

Next project, we need to budget for that. (S&T partner)

this public engagement it's absolutely necessary but it's sort of a never ending story. I think we should be able to find somehow tools to educate the new generations earlier. (S&T partner)

In summary, the S&T partners recognised some limitations in engaging stakeholders in dialogue around novel technologies without prior contextual awareness. However, partners recognised that the research had revealed some interesting questions, and in some cases had challenged their preconceptions of the relationship between stakeholders and scientists. The partners also reported that the way that the research had built an authentic foundation around stakeholder values had been hugely helpful for the public engagement and gave scientists the confidence and tools to engage in meaningful dialogue with people who have no prior knowledge of PMF.

15.6 Conclusions

As design researchers and social scientists, we instinctively understand the importance of meaningful engagement when building trust in unfamiliar, or even feared, technologies and products. Through experience, we understand that meaningful engagement cannot be built on superficial and ill-considered assumptions about the value to stakeholders. Indeed, we have found that a deeper level of understanding, achieved through a well-resourced, dialogical and iterative co-design research process, can achieve an authentic narrative based on the lived experiences of those people who will ultimately be impacted by the technologies. Through this chapter, we have presented research processes to demonstrate the value and impacts of participating in the process, for scientists and stakeholders such as patients, healthcare professionals, the wider public and S&T partners.

Therefore, our main finding is that design research and social science can provide the research methods and co-design tools to build proximity between developers of S&T innovations, such as PMF technologies, and wider audiences so that they can engage in meaningful dialogue. An iterative process allows for a deep and meaningful narrative to be developed that tells the story of a novel technology from multiple perspectives and allows the voices and concerns of diverse stakeholders to be heard. This approach builds an authenticity, which reaches beyond the assumptions of technology developers and enables meaningful dialogue with general audiences.

The approach has the potential to fundamentally shift how the sciences engage with specific and general audiences, mitigating the damage that has previously been caused by avoidance, and the consequential impact of sensationalised media. However, deep and meaningful iterative approaches to stakeholder and public engagement must be prioritised in novel technology development projects. There is no short-cut to public engagement of this type; if deep and meaningful engagement research around the value of technologies to stakeholders such as patients and healthcare professionals is to be achieved, project coordinators need to resource and invest in this approach and bring in researchers with appropriate expertise in these methods.

One shortcoming of the research was that we were unable to engage with more sceptical members of the public who may not immediately appreciate the direct value of the technology on human or animal health. This was due to lockdown restrictions, which prevented the planned 'pop-up' events. This could be a next step for PMF public engagement as S&T partners in the Pharma-Factory project also expressed an interest in engaging more widely to build dialogue with general audiences, raise awareness and literacy in PMF and discover more about their concerns and questions. Recognising the value of sensory and experiential public engagements, S&T partners were also keen to trial more experiential engagements in future projects, utilising interaction design expertise to further explore PMF safety with the public. The PMF community also hopes to make the icon language developed during the project a 'standard' visual language in the field, led by Professor Julian Ma.

Acknowledgements We would like to thank the European Commission for funding this important research and emphasising the importance of public engagement in their S&T innovation calls. Thank you to the project coordinator for stepping into the unknown and trusting us, and to the Pharma-Factory consortium partners who have been open-minded and generous with their time and attention. Huge gratitude goes to the participants in the research and to the following organisations for supporting the research through their networks: British Inherited Metabolic Disease Group (BIMDG), British Rheumatoid Arthritis Society (BRAS), British Sjögren's Syndrome Association (BSSA), ABRAÇO, Fundacao Portuguesa a Comunidade contra a sida (FPCCSIDA), Associação Positivo, Liga Portuguesa Contra a SIDA, Grupo De Ativistas em Tratamento (GAT), and SERES Portugal.

Bibliography

Bason C (2014) Design for policy. In: Cooper R (ed) Design for social responsibility series. Gower Publishing Limited

Blancke S, Van Breusegem F, De Jaeger G, Braeckman J, Van Montagu M (2015) Fatal attraction: the intuitive appeal of GMO opposition. Trends Plant Sci 20(7):414–418. https://doi.org/10.1016/j.tplants.2015.03.011

Brodie E, Cowling E, Nissen N, Paine AE, Warburton D (2009) Understanding participation: a literature review. Pathways through participation, Dec 2009. Institute for Volunteering Research, involve, NCVO, Lottery funded, p 50. http://www.sp.gov.tr/upload/Sayfa/47/files/Pathways-literature-review-final-version.pdf

Buchanan R (2001) Design research and the new learning. Design Issues 17(4):3. https://doi.org/10.1162/07479360152681056. Accessed 24 Oct 22

CNR (2015) Science, innovation and society: achieving responsible research and innovation SIS–RRI, p 6. https://ec.europa.eu/research/participants/documents/downloadPublic?documentIds=080166e5a222b4b1&appId=PPGMS

Couée I (2016) Hidden attraction: empirical rationality in GMO opposition letter. Trends Plant Sci 21(2):91. https://doi.org/10.1016/j.tplants.2015.12.002

Deserti A, Rizzo F, Smallman M (2020) Experimenting with co-design in STI policy making. Policy Des Pract 3:135–149. https://doi.org/10.1080/25741292.2020.1764692

Dreyer M, Kosow H, Dratsdrummer F (2018) Engaging society for responsible research and innovation: lowering barriers—innovating policies and practices. The PROSO consortium, Engaging Society for RRI. EU Grant Agreement 665947

European Commission (2001) European Governance: a white paper. Commission of the European Communities, Brussels

Evans M, Terrey N (2016) Co-design with citizens and stakeholders. In: Stoker G, Evans M (eds) Evidence-based policy making in the social sciences: methods that matter. Policy Press, Bristol, pp 243–262. https://doi.org/10.1332/policypress/9781447329367.003.0014

Graur D (2007) Public control could be a nightmare for researchers. Correspondence. Nature 450:1156. https://www.nature.com/articles/4501156b

Hornbuckle R (2022) Project proximities: a meta review of how design addresses distance in complex collaborations. In: DRS2022: Bilbao. https://doi.org/10.21606/drs.2022.677

Hornbuckle R, Grimaldi S, Prendiville A (2020) Beyond science communication: a service design approach to building mutual stakeholder understanding in the development of novel biotechnologies. In: Proceedings of the 6th international conference on Design4Health, Amsterdam, 1–3 Jul 2020

House of Lords Select Committee on Science and Society (2000). https://publications.parliament.uk/pa/ld199900/ldselect/ldsctech/38/3801.htm. Accessed 2 Aug 22

Irwin A (2006) The politics of talk: coming to terms with the "New" Scientific Governance. Social Stud Sci 36(2):299–320. https://doi.org/10.1177/0306312706053350

Joly PB, Rip A (2007) A timely harvest. Nature 450(7167):174. https://doi.org/10.1038/450174a

Kimbell L (2015) Applying design approaches to policy making: discovering policy lab. University of Brighton-Centre for Research and Development Faculty of Arts, University of Brighton, pp 1–43. https://researchingdesignforpolicy.files.wordpress.com/2015/10/kimbell_policylab_report.pdf

MacPhetres J, Rutjens BT, Weinstend N (2019) Modifying attitudes about modified foods: increased knowledge leads to more positive attitudes. J Environ Psychol 64:21–29. https://centaur.reading.ac.uk/93588/1/2019_McPhetres_etal_GM_food.pdf. Accessed 20 Sept 22

Marris C, Rose N (2010) Open engagement: exploring public participation in the biosciences. PLoS Biol 8(11). https://doi.org/10.1371/journal.pbio.1000549

Michael M (2012) "What are we busy doing?": engaging the idiot. Sci Technol Human Values 37(5):528–554. https://doi.org/10.1177/0162243911428624

Muir R, Parker I (2014) Many to many: how the relational state will transform public services. Institute of Public Policy Research. https://www.ippr.org/files/images/media/files/publication/2014/02/Many-to-many_Feb2014_11865.pdf

Norman DA, Stappers PJ (2015) DesignX: complex sociotechnical systems. She Ji 1(2):83–106. https://doi.org/10.1016/j.sheji.2016.01.002

Selin C, Rawlings KC, de Ridder-Vignone K, Sadowski J, Altamirano Allende C, Gano G, Davies SR, Guston DH (2017) Experiments in engagement: designing public engagement with science and technology for capacity building. Public Underst Sci 26(6):634–649. https://doi.org/10.1177/0963662515620970

Stilgoe J, Lock SJ, Wilsdon J (2014) Why should we promote public engagement with science? Public Underst Sci 23(1):4–15. https://doi.org/10.1177/0963662513518154

Wilsdon J, Willis R (2004) See-through science. DEMOS, Technology and Science, Policymaking, Environment, London

Correction to: Plant-Based Antibody Manufacturing

Gregory P. Pogue, Kelsi Swope, Joseph Rininger, Lauren Schoukroun-Barnes, Josh Morton, Steve Hume, Krystal Hamorsky, Josh Fuqua, Joshua M. Royal, Michael H. Pauly, Max Brennan, Larry Zeitlin, Kevin Whaley, Sean Stevens, and Barry Bratcher

Correction to:
Chapter 7 in: C. Kole et al. (eds.), *Tools & Techniques of Plant Molecular Farming*, Concepts and Strategies in Plant Sciences, https://doi.org/10.1007/978-981-99-4859-8_7

The following corrections have been made after original publication:

1. The original version of this chapter was inadvertently published with incorrect authorship. An author's name has been removed after publication. The authorship has been updated with this correction. The correct authorship now reads as follows:

Gregory P. Pogue, Kelsi Swope, Joseph Rininger, Lauren Schoukroun-Barnes, Josh Morton, Steve Hume, Krystal Hamorsky, Josh Fuqua, Joshua M. Royal, Michael H. Pauly, Max Brennan, Larry Zeitlin, Kevin Whaley, Sean Stevens, and Barry Bratcher

2. The authors have included an addendum to the original publication of this chapter.

The updated version of this chapter can be found at https://doi.org/10.1007/978-981-99-4859-8_7

C. Kole et al. (eds.), *Tools & Techniques of Plant Molecular Farming*, Concepts and Strategies in Plant Sciences, https://doi.org/10.1007/978-981-99-4859-8_16

SPRINGER NATURE

GPSR Compliance

The European Union's (EU) General Product Safety Regulation (GPSR) is a set of rules that requires consumer products to be safe and our obligations to ensure this.

If you have any concerns about our products, you can contact us on ProductSafety@springernature.com

In case Publisher is established outside the EU, the EU authorized representative is:

Springer Nature Customer Service Center GmbH
Europaplatz 3
69115 Heidelberg, Germany

The manufacturer's authorised representative in the EU is Springer Nature Customer Service Centre GmbH, Europaplatz 3, 69115 Heidelberg, Germany. If you have any concerns regarding our products, please contact ProductSafety@springernature.com

Printed and bound by CPI Group (UK) Ltd, Croydon, CR0 4YY
30/09/2026
02225321-0019